T0189493

Springer Series in Optical Sciences

Volume 234

Springer Series in Optical Sciences is led by Editor-in-Chief William T. Rhodes, Florida Atlantic University, USA, and provides an expanding selection of research monographs in all major areas of optics:

- lasers and quantum optics
- ultrafast phenomena
- optical spectroscopy techniques
- optoelectronics
- information optics
- applied laser technology
- industrial applications and
- other topics of contemporary interest.

With this broad coverage of topics the series is useful to research scientists and engineers who need up-to-date reference books.

More information about this series at http://www.springer.com/series/624

Thomas Kürner • Daniel M. Mittleman
Tadao Nagatsuma
Editors

THz Communications

Paving the Way Towards Wireless Tbps

Editors
Thomas Kürner
Institute for Communications Technology
Technische Universität Braunschweig
Braunschweig, Germany

Daniel M. Mittleman
Brown University
Providence, RI, USA

Tadao Nagatsuma
Graduate School of Engineering Science
Osaka University
Toyonaka, Japan

ISSN 0342-4111 ISSN 1556-1534 (electronic)
Springer Series in Optical Sciences
ISBN 978-3-030-73740-5 ISBN 978-3-030-73738-2 (eBook)
https://doi.org/10.1007/978-3-030-73738-2

This Springer imprint is published by the registered company Springer Nature Switzerland AG
The registered company address is: Gewerbestrasse 11, 6330 Cham, Switzerland

Contents

Chapter 1
Introduction to THz Communications

Thomas Kürner, Daniel M. Mittleman, and Tadao Nagatsuma

Abstract Terahertz (THz) systems are seen as an enabling technology for terabits per second (Tbps) wireless communications. This chapter provides a brief overview on the need for THz communications and the history of its development. Some typical applications and their requirements are described, and the structure of the book is introduced.

1.1 Need for THz Communications

The idea of using millimetre and submillimetre waves for the purpose of wireless communications has been a topic of speculation since at least the 1960s. Several authors identified numerous advantages, including copious bandwidth, enhanced security, and diminished scattering compared to optical signals, more than 50 years ago. These ideas inspired early research efforts, for example, into the properties of the atmosphere at frequencies above 100 GHz (e.g. see [1]). Yet, it is only more recently that technologies have progressed to the point where such wireless systems can be considered to be feasible on the relatively near-term horizon. The combination of this technological evolution and the rapid growth of wireless traffic, which has continued to accelerate over the last two decades, has inspired many researchers to recognize that a move to higher frequencies (above the currently occupied bands which mostly lie below 6 GHz) can provide an effective solution to alleviate the overcrowding of the spectrum and can enable new applications which rely on ultra-broadband wireless links enabling data rates of 100 gigabits per second

T. Kürner (✉)
Institute for Communications Technology, Technische Universität Braunschweig, Braunschweig, Germany
e-mail: t.kuerner@tu-bs.de

D. M. Mittleman
Brown University, Providence, RI, USA

T. Nagatsuma
Osaka University, Toyonaka, Japan

T. Kürner et al. (eds.), *THz Communications*, Springer Series in Optical Sciences 234, https://doi.org/10.1007/978-3-030-73738-2_1

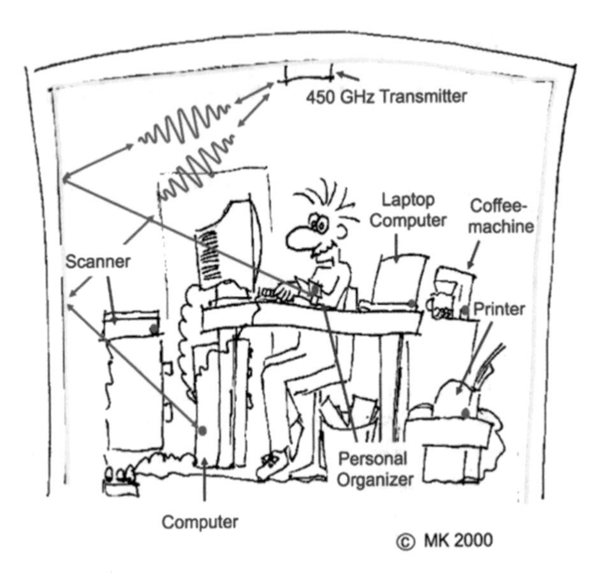

Fig. 1.1 Vision of a point-to-multipoint short-range wireless link [4]. (© 2000 Martin Koch, reproduced with permission)

(Gbps) and beyond, even reaching data rates of terabits per second (Tbps). The research trend devoted to the study of wireless communications above 100 GHz began about 20 years ago; the first demonstration of sending data on a THz carrier dates from the early 2000s [2, 3].

Figure 1.1 indicates an early diagram drawn by one of the pioneering researchers in the field, dating from 2000, which illustrates a vision of a point-to-multipoint short-range wireless link at 450 GHz [4]. At the time this picture was drawn, few would have envisioned the smartphones of today or the Internet of Things, a future in which wireless devices are innumerable and wireless connectivity is ubiquitous. Yet, in the intervening two decades, wireless technologies have proven to be an enabling platform for countless new applications, driving us towards an increasingly interconnected society. The accompanying growth in demand has driven the development of each new generation of wireless systems. Today, we are witnessing the roll-out of the fifth generation (5G), which will for the first time include a standard for short-range links in the millimetre-wave range [5].

Of course, throughout the history of wireless technology, new capabilities have always inspired new (and often unanticipated) uses, generating even greater demand. This has been true since the days of Marconi, and it seems likely to remain true as 5G evolves from a novelty to the global workhorse wireless standard. As a result, there appears to be little doubt that subsequent generations will be needed beyond 5G and, eventually, that these will include standards and operations at even higher frequencies, above 100 GHz. We do not presume to know when this transition will occur – many are already assuming that it will be a component of the sixth generation (6G) [6] enabling links with data rates of up to 1 Tbps, but it may be too early to make that prediction. Nevertheless, the research community has embraced the challenge. Activity in this field has grown very rapidly within the last few years. This growth has been inspired by both key technological advances and burgeoning wireless demand, as noted above, but also perhaps by the ongoing 5G roll-out, which offers some validation for the idea of using millimetre waves in a mobile networking context. Many new capabilities and ideas have been described, and a host of exciting demonstration experiments has been performed. The purpose of this book is to provide a 'snapshot' of the global state of the art in wireless communications research, at frequencies above 100 GHz.

To start the discussion, it is useful to consider the rationale for the frequency range of interest. In this text, we focus exclusively on frequencies above 100 GHz. The low-frequency limit of our range is to some extent arbitrary; however, there are some rationales for the choice. First and most obvious, it clearly distinguishes this work from considerations of 5G systems, which will contain no frequency bands above about 60–80 GHz. In subsequent generations of wireless systems, if higher-frequency bands are needed, it will be difficult to find any unallocated spectrum that lies at frequencies below 100 GHz, which is not already incorporated into the 5G standards. Therefore, this frequency is a reasonable choice as the dividing line between the millimetre-wave bands that are already contemplated for the next generation of wireless and those lying at higher frequencies that will be used in future generations.

A second consideration is based more on the physics of propagation. As one moves to higher frequencies, wireless transmission will necessarily become more directional, as high-gain antennas will be needed to overcome the rapidly growing (in proportion to frequency squared) free-space path loss. Although the millimetre-wave bands of 5G are already considered to be very directional (in comparison with legacy bands below 6 GHz), they are still fairly wide-angle fans of typically several tens of degrees [7]. In contrast, at frequencies above 100 GHz, much narrower almost pencil-like beams with angular widths of only a few degrees are typically expected [7–9]. This narrow beam cone will necessitate many changes to the design and operation of wireless systems, including not only the physical layer but also the control plane and the multiple access (MAC) layer, for example. Things are really very different at these higher frequencies.

Finally, at frequencies above 100 GHz, the attenuation of signals by atmospheric water vapour becomes a significant issue, whereas it can often be ignored at lower frequencies. The absorption spectrum of the atmosphere is dominated by a number of relatively narrow (but in some cases quite strong) absorption lines (see Fig. 1.2), arising from discrete rotational-vibrational transitions in water molecules in the gas phase (in the figure, the lone peak below 100 GHz is due to O_2; all the remaining peaks are due to water vapour). These lines ride on top of a 'continuum' background which increases with increasing frequency. This background attenuation is due to effects such as absorption by water dimers and trimers [11]. Typically, one would imagine that broadband wireless communication systems will be designed to operate in the 'windows' between the strong but narrow loss peaks, where the continuum is the dominant atmospheric effect (at least, in clear weather). As frequency increases, these windows become increasingly sparse and narrow. This fact dictates the upper frequency limit for most practical purposes. Above ~1 THz

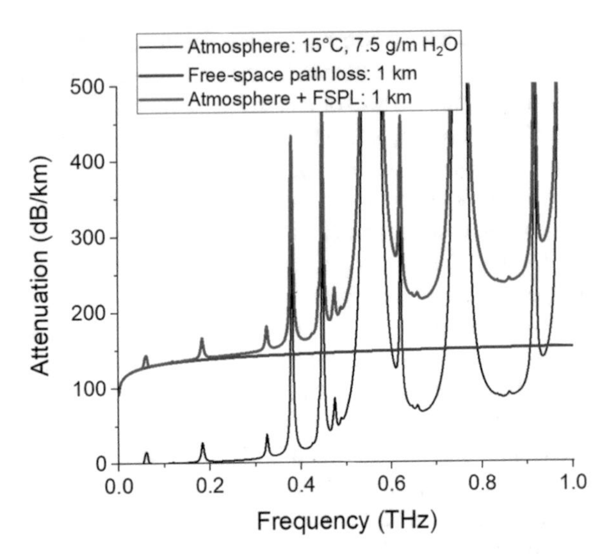

Fig. 1.2 Atmospheric attenuation for a specific water vapour density (black) along with the free-space path loss $(4\pi df/c)^2$, for a range of 1 km (blue), assuming frequency-independent antenna gain. The red curve shows the sum of the two effects. Water vapour attenuation computed using the ITU-R-P676 model [10]

(1000 GHz), it may be very difficult to find a window with low enough loss over a broad enough bandwidth for practical communications applications. Thus, the THz wireless research community has mostly focused on the frequency range 100–1000 GHz, with more interest concentrated on the more readily accessible low-frequency side of that range. In this text, we refer to this as the 'terahertz range' and to the 'terahertz links' that operate with carrier frequencies in this range.

1.2 History of THz Communications

A number of review articles have been published over the years, providing valuable summaries of this rapidly evolving research landscape. One of the earlier examples appeared in 2007 [12]. Here, the authors considered the possibility of using an engineered environment, such as a specially designed multi-layer dielectric stack as a wall covering [13] which would enable low-loss reflections of a terahertz beam from the wall. This work recognized one of the key system design considerations for THz networks – the range of possible ray paths connecting a transmitter to a receiver is much smaller than what is encountered in conventional wireless systems operating at lower frequencies. The so-called rich multipath scattering environment which underlies many of the signal processing operations that are fundamental to the operation of 4G and earlier networks is replaced by a sparse environment at higher frequencies, with only one or perhaps a few possible paths. This distinction arises because of several factors: the aforementioned atmospheric attenuation, but also the relatively larger absorption (dielectric loss) of many solid materials [14]. As a result, the idea of a 'non-line-of-sight' path which relies on scattering of radiation over a wide range of angles is replaced by that of a 'specular non-line-of-sight' path which relies on a small number of specular reflections (e.g. one or two), which can add significant loss due to absorption [8]. These paths, although lossy, may still play a critical role as networks dynamically adapt to steer beams around a transient blockage of the direct line-of-sight link [15], as envisioned in Fig. 1.1. However, this creates a set of demanding requirements on the design of antenna systems, which need to provide high-gain steerable or switchable antennas, as well as on the signal processing algorithms. Today, ideas of using reflection from walls have been further developed by introducing the concept of intelligent reflecting surfaces (IRS) to improve coverage of THz communication systems by adjusting the discrete phase shifts of the IRS elements and hence smartly reconfigure the propagation of electromagnetic waves [16]. The corresponding antenna aspects will be covered in Chaps. 17, 18, and 19.

Other review articles have provided additional valuable information and perspectives. For example, ref. [17] includes an early comprehensive discussion of one of the key advantages of THz wireless links: enhanced security and resilience against eavesdropping and jamming. This aspect of THz systems remains one of the important motivations for the field and is recently a topic of increasing discussion [18]. This review also touched on crucial issues such as the influence of precipitation and the effects of atmospheric scintillation, many of which have subsequently been the focus of more intense study. That article also attempted to summarize some of the key demonstration experiments that had been performed up to that time. In this book, we are to some extent inspired by this effort, and so we have included an updated set of discussions of recent demonstration experiments, which is by now a much longer list (see Chaps. 32, 33, 34, 35, 36, 37, 38, 39, 40, 41, 42, 43, 44, 45, 46, 47, 48, 49, 50 and 51). Meanwhile, a contemporary review article [19] provided a discussion of numerous complementary issues, such as the use of ray tracing for evaluation of indoor link scenarios, as an alternative to statistical approaches that are more commonly used at lower frequencies. This review also touched on a number of physical layer issues, such as the prevalence of planar antennas, and their importance in integration for, for example, MIMO arrays. This work also noted the foundational demonstration of a 120-GHz-band link at the 2008 Beijing Olympic Games, which paved the way for much future development [20]. Finally, the article concluded with a brief mention of the regulatory questions that must inevitably be faced when considering any new radio technology, though the frequency allocation from 116 GHz to 134 GHz was officially made in Japan in January 2014. The path towards standardization and regulation is described in another overview paper [21], focusing on the necessary steps to making THz communication a reality. In recent years, significant progress has been made in this area. In 2017, IEEE 802 Std. 15.3d-2017 – the first wireless communication standard at carrier frequencies around 300 GHz – has been published [22], and in 2019, the World Radiocommunication Conference (WRC 2019) has identified 137 GHz of spectrum in the frequency range 275–450 GHz, which can be used for communications [23]. We also recognize the importance of this issue, which remains a looming concern for the field. Consequently, this book provides an updated view of the standardization and regulatory landscape in Chaps. 52, 53 and 54.

One of the enduring research themes in this field involves the development of new or improved sources and detectors. Indeed, one could argue that the recent explosion of activity in THz wireless research has been enabled by recent impressive developments in source and detector subsystems. Prior to the emergence of practical sources and detectors, it was much harder to justify a system vision such as shown in Fig. 1.1. Of course, this is still an extremely active area of research. There are numerous distinct strategies for generating (and detecting) radiation in the THz frequency range, each with its own set of advantages and disadvantages. A number of review articles have focused on specific subsets. For example, photonics-based approaches, reviewed in [24], have been the basis for many of the demonstrations with the highest data rates. Meanwhile, approaches based on the scaling of semiconductor electronics have also advanced rapidly and are now competitive especially in the lower-frequency range, below 400 GHz [25]. The ultimate choice of technology platform will depend on the details of the application and will be influenced by many factors [26]. Discussions of the state of the art for these and other technology aspects are prominently featured in this book in Chaps. 20, 21, 22, 23, 24, 25, 26 and 27.

Whereas the first two decades of research on THz communications have primarily focused on RF aspects, namely, channel characterization and the development of components for the RF front-ends, more recent research is emerging on baseband, multiple access, and networking aspects, as potential applications have to be demonstrated. Demanding requirements on baseband processing, multiple access, and networking are coming not only from the use of ultrahigh carrier frequencies but also from the ultrahigh data rates in the Tbps regime. This will require new concepts in analogue-to-digital conversion [27] and bring the focus to implementation efficiency, implementation cost, and power consumption, which are of key relevance for the development of forward error correction schemes [28]. Initial access and beam tracking as key elements in THz communications will require new MAC concepts [29–30]. All these aspects are covered in Chaps. 28, 29, 30 and 31.

Since around the year 2010, several research groups have built demonstrators to show that it is possible to fabricate transceiver systems at carrier frequencies of 200 GHz and above, enabling the transmission of tens of Gbps with link distances of several meters up to almost 1 km. These transceiver systems use either electronic [31, 32] or photonic [33] approaches to generate THz signals. In the year 2013 for the first time, with a single-input and single-output wireless communication system operating at 237.5 GHz, researchers demonstrated a data rate of 100 Gbps over 20 m [34]. This breakthrough result was achieved by combining terahertz photonics and electronics. In 2016, electronic beam steering at 300 GHz has been demonstrated in the German TERAPAN project [35, 36]. Recently, demonstrators have been built targeting proof of concepts for backhaul and fronthaul links at 300 GHz [37, 38]. While the aforementioned demonstrators mainly rely on expensive III–V semiconductor technology, significant progress has also been made in Si-based demonstrators [39, 40]. A first 300-GHz CMOS demonstrator has been built in 2017 [41, 42], showing a real-time transmission of video over a range of 10 m. Si-based and CMOS transceivers are a prerequisite to pave the way for mass-market applications. Numerous experiments and research projects targeting at demonstrators are described in Chaps. 32, 33, 34, 35, 36, 37, 38, 39, 40, 41, 42, 43, 44, 45, 46, 47, 48, 49, 50 and 51 in this book.

1.3 Applications and Requirements

The demonstrators mentioned above have triggered activities to exploit the THz spectrum for numerous applications. The propagation conditions at these frequencies and the consequent need for high-gain antennas have some important implications for the potential applications, which can be split into two categories:

1. *Fixed-point-to-point applications*, where the positions of the antennas at both ends of the links are a priori known. This avoids the need for beam steering both during link establishment and while the link is active. Such applications are, for example [43, 44]:

   (a) Intra-device communication, i.e. a communication link within a device like a computer, camera, or video projector including inter-chip communication
   (b) Close proximity point-to-point applications like kiosk downloading or file exchange between two electronic products such as smartphones, digital cameras, camcorders, computers, TVs, game products, and printers
   (c) Wireless links in data centres, where complementary THz links enable a faster reconfiguration of the data centre.
   (d) Wireless backhaul and fronthaul links in cellular networks, being a connection between a base station and a more centralized network element (backhauling) or a connection between a base band unit and a remote radio head (fronthauling)

   All these applications have been the target of IEEE 802.15.3d-2017 [22, 45], which does not include any procedures for beam steering. Also due to either the short link distance or the high-gain antennas used (or both), the requirements on multiple access and interference mitigation can be relaxed. At the time when the project to start the development of IEEE 802.15.3d-2017 was initiated, the technology was considered mature enough to develop a standard.
2. *Applications with at least one end of the link being mobile*. This requires algorithms for device discovery during link establishment [29, 30], beam forming [46], and beam tracking [47]. Examples for such applications are:

   (a) WLAN (Wireless Local Area Network)-type applications [21], where users in conference or lecture rooms or hotspots in public areas can enjoy data rates up to 10s of Gbps. Further examples include in-flight [48] or in-train entertainment [49] or the use of augmented or virtual reality in indoor environments.
   (b) Backhauling for the aggregated data rate covering users in the moving vehicles [49].

(c) Vehicle-to-X communication allowing ultrahigh data rate exchange between cars or between cars [50, 51] and an infrastructure.

At the time of writing this book, the technology is not yet considered mature enough to start developing a standard for the second category of applications. This suggests the need for research to achieve solutions enabling THz communications for this latter category of applications. To achieve the needed breakthrough results in antenna technology, signal processing in the context of antennas and protocols will also be required. Solutions in semiconductor technology allowing cost-efficient production of chipsets are mandatory to introduce THz-based mass-market products. The growing momentum in all aspects of the research discussed in this book may inspire effective solutions to these numerous vexing challenges.

In almost all environments mentioned in the list of applications from both categories above, research on propagation and channel characterization is already underway, as described in Chaps. 10, 11, 12, 13, 14, 15 and 16 of this book. Although stand-alone THz communications have not yet reached either their entry into the market or the end of the basic research stage, new visions are already appearing on the horizon: The integration of communications with sensing, imaging, and localization will open the door for even more applications in the future, undoubtedly including many that are currently unforeseeable [52].

1.4 Structure of the Book

We recognize that a book on this burgeoning topic cannot be assembled without making choices about which topics to include. There is simply too much material to be contained in a single volume. Therefore, we have attempted to select a subset which represents some of the most significant and compelling examples. We hope that readers find this overview of the field to be a useful guide to the state of the art, as of the end of 2020.

Beyond this introductory chapter, this book is split into Parts I–XI containing 53 additional chapters, as illustrated in Fig. 1.3.

The first three parts are dedicated to 'propagation and channel modelling', and it distinguishes the basics of 'channel measurement techniques', the description of 'basic propagation phenomena', and the application of these parts in 'modelling and measurements in complex environments'. Part IV is dedicated to 'antenna concepts and realization' and completes the area on antennas and propagation.

The following four parts, V to VIII, are dedicated to 'transceiver technologies' and split into 'silicon-based electronic', 'III–V-based electronics', 'photonics', and 'vacuum electronic devices'.

Four chapters on various aspects of 'baseband processing and networking interface' are the subject of Part IX of the book.

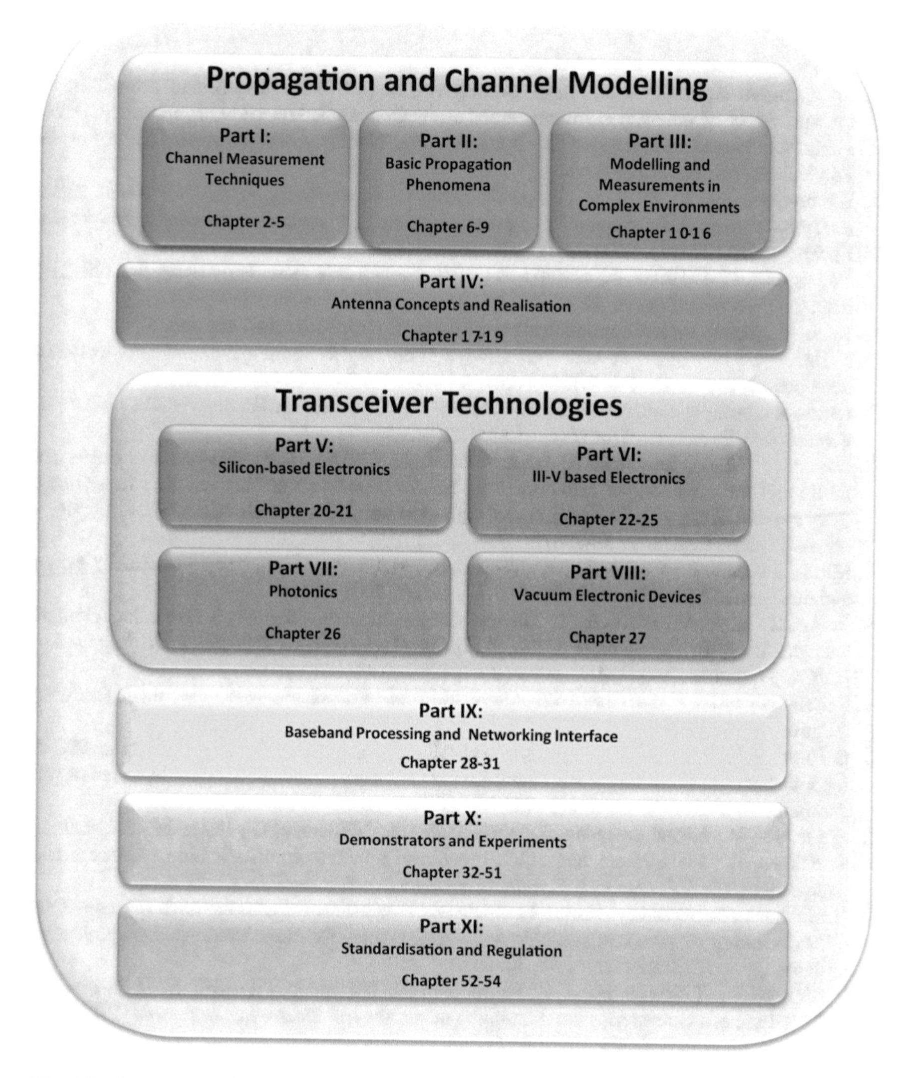

Fig. 1.3 Structure of the book

A large number of demonstrators and experiments are briefly described in Part X. Although the list of demonstrators and experiments contained in these 20 chapters is not exhaustive, Part X of the book provides a good overview of achievements so far in the field of THz communications.

Part XI completes the book by describing the status and outlook in the area of standardization and regulation.

References

1. Sheppard, A. P., Breeden, K. H., & McSweeney, A. (1970). High resolution submillimeter measurements of atmospheric water vapor absorption. In J. Fox (Ed.), *Proceedings of the symposium on Submillimeter waves* (pp. 445–454). Brooklyn, NY: Polytechnic Press of the Polytechnic Institute of Brooklyn.
2. Kleine-Ostmann, T., Pierz, K., Hein, G., Dawson, P., & Koch, M. (2004). Audio signal transmission over THz communication channel using semiconductor modulator. *Electronics Letters, 40*(2), 124–126. https://doi.org/10.1049/el:20040106.
3. Nagatsuma, T., & Hirata, A. (2004). 10-Gbit/s wireless link technology using the 120-GHz band. *NTT Technical Review, 2*(11), 57–61.
4. Koch, M. (2000). Private communication 2000.
5. NGMN 5G White Paper. https://www.ngmn.org/fileadmin/ngmn/content/downloads/Technical/2015/NGMN_5G_White_Paper_V1_0.pdf
6. Lattva-aho, M., & Lappänen, K.. *Key drivers and research challenges for 6G ubiquitous wireless intelligence*. http://urn.fi/urn:isbn:9789526223544
7. Nassar, A. M. T., Sulyman, A. I., & Alsanie, A. (2015). Radio capacity estimation for millimeter wave 5G cellular networks using narrow beamwidth antennas at the base stations. *International Journal of Antennas and Propagation, 2015*, Article ID 878614, 6 pages. https://doi.org/10.1155/2015/878614.
8. Ma, J., Shrestha, R., Moeller, L., & Mittleman, D. M. (2018). Channel performance for indoor and outdoor terahertz wireless links. *APL Photonics, 3*, 051601.
9. Priebe, S., Jacob, M., & Kürner, T. The impact of antenna directivities on THz indoor channel characteristics. In *Proceedings of 6th European Conference on Antennas and Propagation (EuCAP)* (pp. 483–487) Praque, March 2012. [9] EuCAP 2012 papers.
10. Recommendation ITU-R P.676-12: Attenuation by atmospheric gases and related effects; August 2019.
11. O'Hara, J. F., & Grischkowsky, D. R. (2018). Comment on the veracity of the ITU-R recommendation for atmospheric attenuation at terahertz frequencies. *IEEE Transactions on Terahertz Science and Technology, 8*(3), 372–375.
12. Piesiewicz, R., Kleine-Ostmann, T., Krumbholz, N., Mittleman, D., Koch, M., Schöbel, J., & Kürner, T. (2007). Short-range ultra broadband terahertz communications: Concept and perspectives. *IEEE Antennas & Propagation Magazine, 49*, 24–39.
13. Krumbholz, N., Gerlach, K., Rutz, F., Koch, M., Piesiewicz, R., Kürner, T., & Mittleman, D. (2006). Omnidirectional terahertz mirrors: A key element for future terahertz communication system. *Applied Physics Letters, 88*, 202905.
14. Piesiewicz, R., Kleine-Ostmann, T., Krumbholz, N., Mittleman, D., Koch, M., & Kürner, T. (2005). Terahertz characterisation of building materials. *IEE Electronics Letters, 41*(18), 1002–1003.
15. Petrov, V., Komarov, M., Moltchanov, D., Jornet, J. M., & Koucheryavy, Y. (2017). Interference and SINR in millimeter wave and terahertz communication systems with blocking and directional antennas. *IEEE Transactions on Wireless Communications, 16*(3), 1791–1808.
16. Xinying, M. A., Chen, Z., Chen, W., Chi, Y., Li, Z., Han, C., & Wen, Q. (2020). Intelligent reflecting surface enhanced indoor terahertz communication systems. *Nano Communication Networks, 24*, 100284, ISSN 1878-7789. https://doi.org/10.1016/j.nancom.2020.100284.
17. Federici, J., & Möller, L. (2010). Review of terahertz and subterahertz wireless communications Journal of. *Applied Physics, 107*, 111101. https://doi.org/10.1063/1.3386413.
18. Ma, J., Shrestha, R., Adelberg, J., Yeh, C.-Y., Hossain, Z., Knightly, E., Jornet, J. M., & Mittleman, D. M. (2018). Security and eavesdropping in terahertz wireless links. *Nature, 563*, 89–93.
19. Kleine-Ostmann, T., & Nagatsuma, T. (2011). A review on terahertz communications research. *Journal of Infrared Milli Terahz Waves, 32*, 143–171.

20. Hirata, A., Takahashi, H., Kukutsu, N., Kado, Y., Ikegawa, H., Nishikawa, H., Nakayama, T., & Inada, T. (2009). Transmission trial of television broadcast materials using 120-GHz-band wireless link. *NTT Technical Review, 7*(March).
21. Kürner, T., & Priebe, S. (2014). Towards THz communications – Status in research, standardization and regulation. *Journal of Infrared, Millimeter, and Terahertz Waves, 35*, 53–62.
22. IEEE Standard for High Data Rate Wireless Multi-Media Networks–Amendment 2: 100 Gb/s Wireless Switched Point-to-Point Physical Layer,“ in IEEE Std 802.15.3d-2017 (Amendment to IEEE Std 802.15.3-2016 as amended by IEEE Std 802.15.3e-2017) , pp.1–55, Oct. 18, 2017.
23. Kürner, T., & Hirata, A. (2020). On the Impact of the Results of WRC 2019 on THz Communications. In *2020 Third International Workshop on Mobile Terahertz Systems (IWMTS)* (pp. 1–3). Essen, Germany. https://doi.org/10.1109/IWMTS49292.2020.9166206.
24. Nagatsuma, T., Ducournau, G., & Renaud, C. (2016). Advances in terahertz communications accelerated by photonics. *Nature Photon, 10*, 371–379. https://doi.org/10.1038/nphoton.2016.65.
25. Fujishima, M., & Amakawa, S. (2019). Design of Terahertz CMOS integrated circuits for high-speed wireless communication; IET digital. *Library*.
26. Sengupta, K., Nagatsuma, T., & Mittleman, D. M. (2018). Terahertz integrated electronic and hybrid electronic–photonic systems. *Nature Electronics, 1*, 622–635. https://doi.org/10.1038/s41928-018-0173-2.
27. Meier, J., Misra, A., Preußler, S., & Schneider, T. (2019). Orthogonal full-field optical sampling. *IEEE Photonics Journal, 11*(2), 1–9. Art no. 7800609.
28. H2020 EPIC Project Technical Report. (March 2018). B5G Wireless Tb/s FEC KPI Requirement and Technology Gap Analysis: https://epic-h2020.eu/downloads/EPIC-D1.2-B5G-Wireless-Tbs-FEC-KPI-Requirement-and-Technology-Gap-Analysis-PU-M07.pdf
29. Xia, Q., & Jornet, J. M. (2019). Expedited neighbor discovery in directional terahertz communication networks enhanced by antenna side-lobe information. *IEEE Transactions on Vehicular Technology, 68*(8), 7804–7814. https://doi.org/10.1109/TVT.2019.2924820.
30. Peng, B., Guan, K., Rey, S., & Kürner, T. (2019). Power-angular spectra correlation based two step angle of arrival estimation for future indoor terahertz communications. *IEEE Transactions on Antennas and Propagation, 67*(11), 7097–7105. https://doi.org/10.1109/TAP.2019.2927892.
31. Kallfass, I., Antes, J., Schneider, T., Kurz, F., Lopez-Diaz, D., Diebold, S., Massler, H., Leuther, A., & Tessmann, A. (2011). All active MMIC based wireless communication at 220GHz. *IEEE Transactions Terahertz Science and Technology, 1*(2), 477–487.
32. Kallfass, I., Boes, F., Messinger, T., Antes, J., Inam, A., Lewark, U., Tessmann, A., & Henneberger, R. (2015). 64Gbit/s transmission over 850m fixed wireless link at 240 GHz carrier frequency. *Journal of Infrared Millimeter and Terahertz Waves, 36*(2), 221–233.
33. Nagatsuma, T., Horiguchi, S., Minamikata, Y., Yoshimizu, Y., & Hisatake, S. (2013). Terahertz wireless communications based on photonics technologies. *Optics Express, 21*(20), 23736–23747.
34. Koenig, S., Lopez-Diaz, D., Antes, J., et al. (2013). Wireless sub-THz communication system with high data rate. *Nature Photon, 7*, 977–981. https://doi.org/10.1038/nphoton.2013.275.
35. Kallfass, I., Dan, I., Rey, S., Harati, P., Antes, J., Tessmann, A., Wagner, S., Kuri, M., Weber, R., Massler, H., Leuther, A., Merkle, T., & Kürner, T. (2015). Towards MMIC-Based 300 GHz Indoor Wireless Communication Systems. *Transactions Institute of Electronics, Information and Communication Engineers, E98-C*(12), 1081–1090.
36. Merkle, T., Tessmann, A., Kuri, M., Wagner, S., Leuther, A., Rey, S., Zink, M., Stulz, H.-P., Riessle, M., Kallfass, I., & Kürner, T. (2017). Testbed for phased array communications from 275 to 325 GHz. In *2017 IEEE Compound Semiconductor Integrated Circuit Symposium (CSICS), Miami, FL.* (electronic publication, 4 pages).
37. Dan, I., Ducournau, G., Hisatake, S., Szriftgiser, P., Braun, R., & Kallfass, I. (Jan. 2020). A terahertz wireless communication link using a superheterodyne approach. *IEEE Transactions on Terahertz Science and Technology, 10*(1), 32–43. https://doi.org/10.1109/TTHZ.2019.2953647.

38. Castro, C., Elschner, R., Merkle, T., Schubert, C., & Freund, R. (2020). Experimental demonstrations of high-capacity THz-wireless transmission systems for beyond 5G. *IEEE Communications Magazine, 58*(11), 41–47. https://doi.org/10.1109/MCOM.001.2000306.vehicularpaper.
39. Rodríguez-Vázquez, P., Grzyb, J., Heinemann, B., & Pfeiffer, U. R. (2020). A QPSK 110-Gb/s polarization-diversity MIMO wireless link with a 220–255 GHz Tunable LO in a SiGe HBT technology. *IEEE Transactions on Microwave Theory and Techniques*, 3834–3851.
40. Rodríguez-Vázquez, P., Grzyb, J., Heinemann, B., & Pfeiffer, U. R. (2019). A16-QAM 100-Gb/s 1-M wireless link with an EVM of 17% at 230 GHz in an SiGe technology. *IEEE Microwave and Wireless Components Letters*, 297–299.
41. Takano, K., Katayama, K., Hara, S., Dong, R., Mizuno, K., Takahashi, K., Kasamatsu, A., Yoshida, T., Amakawa, S., & Fujishima, M. (Jan. 2018). 300-GHz CMOS transmitter module with built-in waveguide transition on a multilayered glass epoxy PCB. *IEEE Radio Wireless Symposium*, 154–156.
42. Hara, S., Takano, K., Katayama, K., Dong, R., Mizuno, K., Takahashi, K., Watanabe, I., Sekine, N., Kasamatsu, A., Yoshida, T., Amakawa, S., & Fujishima, M. (2018). 300-GHz CMOS receiver module with WR-3.4 waveguide interface. In *Proceedings on 2018 European Microwave Conference, Madrid* (pp. 396–399).
43. Kürner, T. et al. (2015). Applications Requirement Document (ARD). DCN: 15-14-0304-16-003d, IEEE 802.15 TG3d. https://mentor.ieee.org/802.15/documents
44. Peng, B., Guan, K., Kuter, A., Rey, S., Patzold, M., & Kuerner, T. (2020). Channel modeling and system concepts for future terahertz communications: Getting ready for advances beyond 5G. *IEEE Vehicular Technology Magazine, 15*(2), 136–143. https://doi.org/10.1109/MVT.2020.2977014.
45. Petrov, V., Kürner, T., & Hosako, I. (2020). IEEE 802.15.3d: First standardization efforts for sub-terahertz band communications toward 6G. *IEEE Communications Magazine, 58*(11), 28–33. https://doi.org/10.1109/MCOM.001.2000273.
46. Rey, S., Merkle, T., Tessmann, A., & Kürner, T. (2016). A phased array antenna with horn elements for 300 GHz communications. In *Proceedings of 2016 International Symposium on Antennas and Propagation (ISAP)* , , electronic publication (2 pages), Ginowan, Okinawa, Japan.
47. Peng, B., Jiao, Q., & Kürner, T. (2016). Angle of arrival estimation in dynamic indoor THz channels with Bayesian filter and reinforcement learning. In *Proceedings of 24th European signal processing conference (EUSIPCO 2016)* (pp. 1975–1979). Budapest, Ungarn.
48. Eckhardt, J. M., Doeker, T., & Kürner, T. (2020). Indoor-to-outdoor path loss measurements in an aircraft for terahertz communications. In *2020 IEEE 91st vehicular technology conference (VTC2020-spring)* (pp. 1–5). Antwerp, Belgium. https://doi.org/10.1109/VTC2020-Spring48590.2020.9128849.
49. Guan, K., Li, G., Kürner, T., Molisch, A. F., Peng, B., He, R., Bing, H., Kim, J., & Zhong, Z. (2017). On millimeter wave and THz mobile radio channel for smart rail mobility. *IEEE Transactions on Vehicular Technology, 66*(7), 5658–5674.
50. Eckhardt, J. M., Petrov, V., Moltchanov, D., Koucheryavy, Y., & Kurner, T. (June 2021). Channel measurements and modeling for Low-Terahertz band vehicular communications *IEEE Journal on Selected Areas in Communications, 39*(6), 1590–1603. https://doi.org/10.1109/JSAC.2021.3071843.
51. Petrov, V., et al. (2019). On unified vehicular communications and radar sensing in millimeter-wave and low terahertz bands. *IEEE Wireless Communications, 26*(3), 146–153.
52. Sarieddeen, H., Saeed, N., Al-Naffouri, T. Y., & Alouini, M. (2020). Next generation terahertz communications: A rendezvous of sensing, imaging, and localization. *IEEE Communications Magazine, 58*(5), 69–75. https://doi.org/10.1109/MCOM.001.1900698.

Part I
Propagation and Channel Modelling 1: Channel Measurement Techniques

Chapter 2
Terahertz Time-Domain Spectroscopy

Peter Uhd Jepsen, Tobias Olaf Buchmann, Binbin Zhou, Edmund John Railton Kelleher, and Martin Koch

Abstract Basic principles of transmission THz spectroscopy are reviewed, with a focus on characterization of atmospheric transmission and dispersion. Important literature on the use of the THz frequency range for wireless communication studies is highlighted.

2.1 Introduction

Since its introduction in 1989 [1], terahertz time-domain spectroscopy (THz-TDS) has been a standard tool in research laboratories for the experimental determination of optical properties of materials in the THz frequency range. In this technique, a beam of femtosecond laser pulses generates and detects the temporal shape of ultrashort bursts of broadband electromagnetic radiation. Due to their frequency content covering the THz range and short temporal duration, these radiation bursts are commonly known as THz pulses. Figure 2.1 shows a schematic overview of a typical experimental realization of a THz-TDS system. The femtosecond laser beam is split into two parts with a beam splitter. One part is used for generation of the THz pulses, and the other part is used for sampling of the temporal shape of the THz pulse waveform. The key point of THz-TDS is that the electric field of the THz pulse is measured as a function of time. Hence, the amplitude and phase of each of its frequency components are known. This is the fundamental difference from many other spectroscopic techniques, where only the intensity if the detected light is recorded and thus phase information lost.

The sample is placed in the THz beam path, either in an intermediate focal plane (not shown in Fig. 2.1) or in the case of an extended sample such as gas

---

P. U. Jepsen (✉) · T. O. Buchmann · B. Zhou · E. J. R. Kelleher
Department of Photonics Engineering, Technical University of Denmark, Kongens Lyngby, Denmark
e-mail: puje@fotonik.dtu.dk

M. Koch
Phillips-Universität Marburg, Marburg, Germany

T. Kürner et al. (eds.), *THz Communications*, Springer Series in Optical Sciences 234, https://doi.org/10.1007/978-3-030-73738-2_2

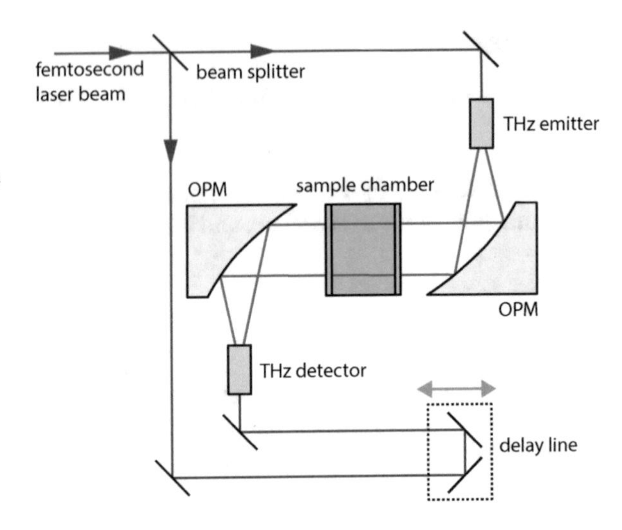

Fig. 2.1 Schematic of a standard THz-TDS experimental setup to determine the index of refraction and the absorption coefficient of materials in the THz spectral range

volume, in a collimated section of the beam as shown in Fig. 2.1. See [2] for a more detailed description of the experimental technique, including an overview of different sources and detector types used in THz-TDS.

2.2 Atmospheric Measurements with Terahertz Pulses

A thorough understanding of the propagation of THz radiation through the atmosphere is naturally of high importance to the development of THz wireless communication systems. The main concern here is the water vapour content of the atmosphere, as the water molecule exhibits a rich absorption spectrum due to rotational transitions across the 0.5–12 THz region. Figure 2.2 shows a severe example of pulse distortion due to water vapour absorption and dispersion after propagating through 64-cm laboratory atmosphere at 30% relative humidity. The excitation of the many rotational transitions in the water molecules lead to a long-lived re-emission (free induction decay [3]) following the main pulse (Fig. 2.2a) and corresponding sharp absorption lines across the spectrum (Fig. 2.2b).

THz wireless communications is expected to exploit the transparency windows in the lowest parts of this spectrum, specifically below 1 THz. The very first demonstration of THz-TDS [1] used the laboratory atmosphere as a test case, and in the following decades, the optical properties of the atmosphere and other representative gases in this region have been studied to great detail with THz time-domain techniques, for instance, by the Grischkowsky team and the Nuss team [4–9]. Using either custom-built long-pass cells or even transmission between buildings, THz-TDS measurements have also been used directly to investigate propagation over rather long distances in the atmosphere [10–12].

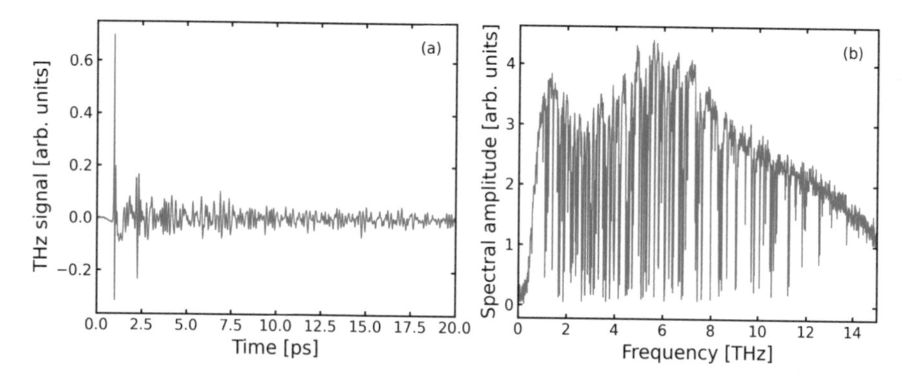

Fig. 2.2 (**a**) Ultrafast THz pulse distortion after 64-cm propagation in a laboratory atmosphere (30% relative humidity). All oscillations after the main pulse at t = 1 ps are due to water vapour. (**b**) Spectral content of the signal. Strong water vapour absorption lines fill the spectrum from 0.5 THz to above 12 THz

2.3 THz-TDS Methodology

At high field strengths, additional nonlinear interactions may take place, but such interactions can be safely ignored when using standard, commercial THz spectroscopy tools or custom-built THz spectrometers based on regular femtosecond laser oscillators. Hence, here we will consider only linear propagation of the THz field through the sample. For further simplification, we will only consider the THz pulse to propagate as a plane wave, i.e. ignore effects due to tight focusing of the THz beam that may complicate the analysis [13, 14].

The ultrashort THz pulse has a frequency content that can be determined by the discrete Fourier transform $\tilde{E}(\omega)$ of the time trace of the pulse. The medium in which the THz pulse propagates will have a (frequency-dependent) complex-valued refractive index $\tilde{n} = n + i\kappa$, where $\kappa = \alpha c/2\omega$ is the extinction coefficient related to the power absorption coefficient α. If we select one frequency $\omega = 2\pi\nu$ within the bandwidth of the THz pulse, propagation over a distance d in the medium will modify the plane wave according to $\tilde{E}(\omega, d) = \tilde{E}(\omega, 0)\exp\left(i\tilde{n}\omega d/c\right)$. A complete THz-TDS measurement consists in general of at least one measurement of the THz pulse after propagation through the medium of interest and at least one measurement of the same THz pulse after propagation through the same distance in a reference medium. If we consider the atmosphere as the sample, then a reference environment can be dry air, a pure nitrogen atmosphere, or even vacuum. In any case, the measurement relies on the knowledge of the refractive index of the reference medium. For simplicity, we will here assume that the reference refractive index is $n_{ref} = 1$. In this case, the experimentally measured transmission ratio of the THz field component after propagation through the sample and the reference will be $\tilde{T}(\omega) = \left|T(\omega)\right|\exp\left(i\Delta\phi(\omega)\right)$. This experimental value is compared to the theoretical prediction $\exp(-\alpha(\omega)d/2)\exp(i(n(\omega) - 1)\omega d/c)$ so that the index of refraction (Eq. (2.1)) and the absorption coefficient (Eq. (2.2)) can be found:

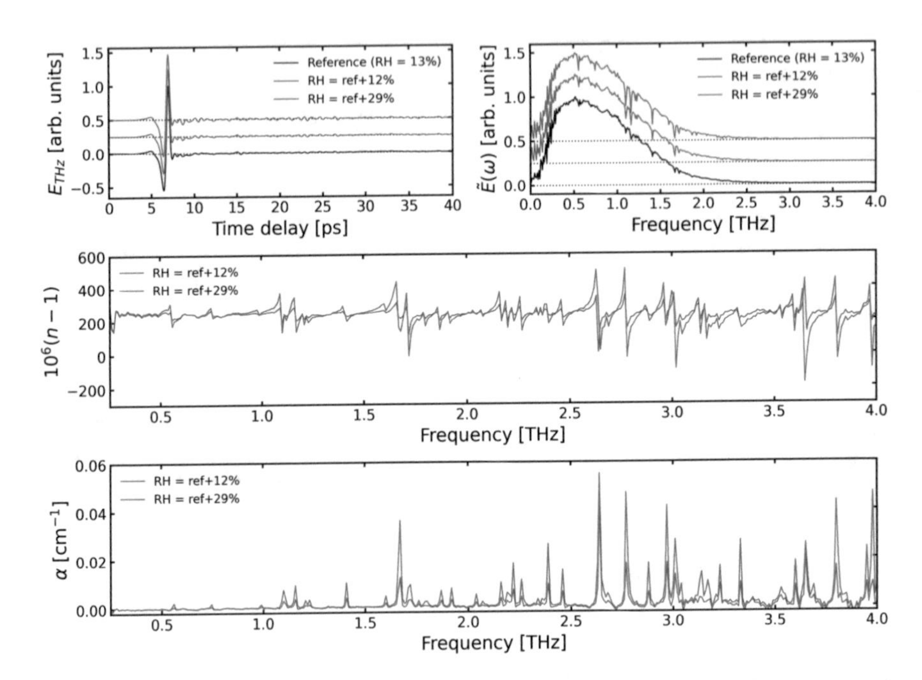

Fig. 2.3 (Top) Time domain (left) and spectral amplitude (right) of THz pulses propagated through a relatively dry (RH = 13%, black) atmosphere and in RH = 25% and 42% (blue and red, respectively). The middle and lower panel shows the extracted index of refraction and the absorption coefficient at the two sets of relative humidity in the 0.25–4 THz range

$$n(\omega) = \frac{\Delta\phi(\omega)\, c}{\omega d} - 1 \tag{2.1}$$

$$\alpha(\omega) = -\frac{2}{d} \ln \mid T(\omega) \mid \tag{2.2}$$

The above equations are excellent approximations for spectroscopy of gases, where the refractive index is close to unity and the absorption coefficient is rather low. In the more general case, the transmission coefficient for the THz field in and out of the sample region is not the same as for the reference region, and the transmission coefficients must explicitly be taken into account [2].

Figure 2.3 shows an example of a THz-TDS measurement of the index of refraction and the absorption coefficient of the atmosphere at two different levels of relative humidity for the 105-mm-long beam path of the THz spectrometer. The top row of the figure shows the time- and frequency-domain representation of the raw data, and the middle and lower panels show the index of refraction and absorption coefficient, extracted by using Eqs. (2.1) and (2.2).

2.4 Modelling of THz Pulse Propagation in the Atmosphere

The complex-valued refractive index of the atmosphere can be modelled precisely with the aid of data from the HITRAN database [15]. HITRAN reports the temperature- and pressure-dependent strength and linewidth of absorption lines of the atmospheric constituents, where mainly H_2O is relevant for the description of THz pulse propagation. The HITRAN database reports line strengths S_{ij} (absorption coefficient per column density of molecules), linewidths $\tilde{\gamma}_H$, and frequencies $\tilde{\nu}_{ij}$ (reported in wavenumber units [cm^{-1}]), leading to a spectral absorption profile $\alpha\left(\tilde{\nu}\right) = \rho S_{ij} f\left(\tilde{\nu}\right)$, where ρ is the density of the molecular species and $f\left(\tilde{\nu}\right) = \left(\tilde{\gamma}_H/\pi\right)/\left(\tilde{\gamma}_H^2 + \left(\tilde{\nu} - \tilde{\nu}_{ij}\right)^2\right)$ is a Lorentzian line profile. Summation over all absorption lines then gives the transmission spectrum of the atmosphere in a given frequency band.

The HITRAN data thus gives no direct information about the dispersion (refractive index profile) of the atmosphere. For Lorentzian line profiles, the standard expression for the permittivity is

$$\varepsilon\left(\omega\right) = 1 + \frac{A}{\omega_0^2 - \omega^2 + i\omega\gamma_L}. \tag{2.3}$$

However, under the assumption of weak absorption and dispersion (as is the case of the atmosphere compared to, for example, condensed matter), the complex-valued permittivity (Eq. (2.3) needed for modelling of THz pulse propagation is related to the HITRAN parameters as [16]:

$$\gamma_L = 4\pi c\tilde{\gamma}_H, \tag{2.4}$$

$$A = 4c^2\rho S_{ij}. \tag{2.5}$$

In this manner, HITRAN data can be used to construct a permittivity model for the atmosphere at a given temperature, density, and pressure. This model permittivity is the square of the complex refractive index of the atmosphere. A given THz pulse can thus be numerically propagated through a model atmosphere by multiplying its spectrum with the frequency-dependent transfer function $\exp\left(i\tilde{n}\left(\omega\right)\omega d/c\right)$ followed by an inverse Fourier transformation back to the time domain. This principle, including a careful evaluation of the lineshape profile, has been carried out by the Grischkowsky group [3, 6].

2.5 THz-TDS in Areas Relevant to Wireless Communications

Terahertz time-domain spectroscopy is an important technique not only to characterize the transmission of THz waves through the atmosphere or THz communication channels, respectively, but also to characterize the reflective properties of building materials. These properties are important as it is expected that ideal line-of-sight (LOS) link between the emitter and receiver of THz communication systems might be temporarily obstructed, for example, by people moving through the room. In this case, the THz system will have to rely on non-line-of-sight (NLOS) channels which involve one or more reflections off walls or furniture as a backup. This concept has been studied numerically by Kürner and coworkers more than a decade ago in 2007 and 2008 [17, 18] and very recently in 2018 and 2019 experimentally by the Mittleman group [19, 20]. Hence, a comprehensive knowledge of the reflection properties of typical indoor building materials is required to allow for accurate predictions about the link performance [17]. As terahertz time-domain spectroscopy is a broadband technique, it is ideally suited for material characterization regarding their complex refractive index in the lower THz range. Early work on building materials considered 'smooth' showed that the surface reflections could be described by Fresnel's formulas [21, 22]. Subsequently, rough surfaces were subject to further studies. It was noticed that scattering in the specular direction has a severe, frequency-dependent impact on the received power [23–25]. In 2011, Jansen et al. used a fibre-coupled terahertz time-domain spectroscopy system to perform angle- and frequency-dependent measurements on rough surfaces to study diffuse scattering [26]. They compared their experimental results to an extended Kirchhoff scattering model as proposed by Beckmann [27]. Accompanying coverage simulations based on a ray tracing algorithm clearly showed that the diffuse scattered power has a strong impact on the non-line-of-sight propagation paths.

In addition, more complex geometries which can be encountered in an indoor environment were studied. In particular, it was demonstrated that multiple reflections from painted walls or double-pane windows can lead to strong alterations of the transmission and reflection behaviour [28].

Terahertz time-domain spectroscopy was also used in 2004 to demonstrate the first THz communication link at several hundred GHz. Here, Kleine-Ostmann et al. used an electrically driven room temperature semiconductor THz modulator which was based on the depletion of a two-dimensional electron gas [29] to modulate the transmission through a terahertz time-domain spectrometer [30]. In the data transmission experiment, audio signals up to 25 kHz were imprinted onto a 75 MHz train of broadband THz pulses. The communication channel had a length of 48 cm spanning from the modulator to the receiver antenna. The signal strength was sufficient to transmit pieces of music with a quality comparable of that of a phone call. Now, 16 years later, after we have seen a remarkable fast progress in the field of short-range THz communication systems with >600 GBit/s capacity [31, 32], this first demonstration appears antediluvian. Yet, it was never thought to be a serious approach towards a THz communication system since the transmittable data rate is limited by the RC time of the modulator. Finally, it is worth mentioning that THz-TDS is an important tool for characterization of semiconductor materials used in contemporary and future high-speed electronics [33, 34]. The switching time of modern transistors approach the picosecond regime, and thus knowledge of material properties in the THz range becomes more and more relevant.

References

1. van Exter, M., Fattinger, C., & Grischkowsky, D. (1989). Terahertz time-domain spectroscopy of water vapor. *Optics Letters, 14*, 1128–1130.
2. Jepsen, P. U., Cooke, D. G., & Koch, M. (2011). Terahertz spectroscopy and imaging – Modern techniques and applications. *Laser & Photonics Reviews, 5*, 124–166.
3. Harde, H., Keiding, S., & Grischkowsky, D. (1991). THz commensurate echoes: Periodic rephasing of molecular transitions in free-induction decay. *Physical Review Letters, 66*, 1834–1837.
4. Cheville, R. A., & Grischkowsky, D. (1995). Far-infrared terahertz time-domain spectroscopy of flames. *Optics Letters, 20*, 1646–1648.
5. Flanders, B. N., Cheville, R. A., Grischkowsky, D., & Scherer, N. F. (1996). Pulsed terahertz transmission spectroscopy of liquid CHCl3, CCl4, and their mixtures. *The Journal of Physical Chemistry, 100*, 11824–11835.
6. Harde, H., Cheville, R. A., & Grischkowsky, D. (1997). Terahertz studies of collision-broadened rotational lines. *The Journal of Physical Chemistry. A, 101*, 3646–3660.
7. Cheville, R. A., & Grischkowsky, D. (1999). Far-infrared foreign and self-broadened rotational linewidths of high-temperature water vapor. *Journal of the Optical Society of America B: Optical Physics, 16*, 317–322.
8. Jacobsen, R. H., Mittleman, D. M., & Nuss, M. C. (1996). Chemical recognition of gases and gas mixtures with terahertz waves. *Optics Letters, 21*, 2011–2013.
9. Mittleman, D. M., Jacobsen, R. H., Neelamani, R., Baraniuk, R. G., & Nuss, M. C. (1998). Gas sensing using terahertz time-domain spectroscopy. *Applied Physics B: Lasers and Optics, 67*, 379–390.
10. Moon, E., Jeon, T., & Grischkowsky, D. R. (2015). Long-path THz-TDS atmospheric measurements between buildings. *IEEE Transactions on Terahertz Science and Technology, 5*, 742–750.
11. Kim, G.-R., Moon, K., Park, K. H., O'Hara, J. F., Grischkowsky, D., & Jeon, T.-I. (2019). Remote N_2O gas sensing by enhanced 910-m propagation of THz pulses. *Optics Express, 27*, 27514–27522.
12. Kim, G. R., Lee, H. B., & Jeon, T. I. (2020). Terahertz time-domain spectroscopy of low-concentration N_2O using long-range multipass gas cell. *IEEE Transactions on Terahertz Science and Technology, 10*, 524–530.
13. Ruffin, A. B., Rudd, J. V., Whitaker, J. F., Feng, S., & Winful, H. G. (1999). Direct observation of the Gouy phase shift with single-cycle terahertz pulses. *Physical Review Letters, 83*, 3410–3413.
14. Kužel, P., Němec, H., Kadlec, F., & Kadlec, Č. (2010). Gouy shift correction for highly accurate refractive index retrieval in time-domain terahertz spectroscopy. *Optics Express, 18*, 15338–15348.
15. Rothman, L. S., Jacquemart, D., Barbe, A., Benner, D. C., Birk, M., Brown, L. R., Carleer, M. R., Chackerian, C., Chance, K., Coudert, L. H., Dana, V., Devi, V. M., Flaud, J. M., Gamache, R. R., Goldman, A., Hartmann, J. M., Jucks, K. W., Maki, A. G., Mandin, J. Y., Massie, S. T., Orphal, J., Perrin, A., Rinsland, C. P., Smith, M. A. H., Tennyson, J., Tolchenov, R. N., Toth, R. A., Auwera, J. V., Varanasi, P., & Wagner, G. (2005). The HITRAN 2004 molecular spectroscopic database. *Journal of Quantitative Spectroscopy and Radiative Transfer, 96*, 139–204.

16. Mayerhöfer, T. G., & Popp, J. (2019). Quantitative evaluation of infrared absorbance spectra – Lorentz profile versus Lorentz oscillator. *ChemPhysChem, 20*, 31–36.
17. Piesiewicz, R., Kleine-Ostmann, T., Krumbholz, N., Mittleman, D., Koch, M., Schoebel, J., & Kürner, T. (2007). Short-range ultra-broadband terahertz communications: Concepts and perspectives. *IEEE Antennas and Propagation Magazine, 49*, 24–39.
18. Piesiewicz, R., Jacob, M., Koch, M., Schoebel, J., & Kürner, T. (2008). Performance analysis of future multigigabit wireless communication systems at THz frequencies with highly directive antennas in realistic indoor environments. *IEEE Journal of Selected Topics in Quantum Electronics, 14*, 421–430.
19. Ma, J., Shrestha, R., Moeller, L., & Mittleman, D. M. (2018). Invited article: Channel performance for indoor and outdoor terahertz wireless links. *APL Photonics, 3*, 051601.
20. Ma, J., Shrestha, R., Zhang, W., Moeller, L., & Mittleman, D. M. (2019). Terahertz wireless links using diffuse scattering from rough surfaces. *IEEE Transactions on Terahertz Science and Technology, 9*, 463–470.
21. Piesiewicz, R., Kleine-Ostmann, T., Krumbholz, N., Mittleman, D., Koch, M., & Kürner, T. (2005). Terahertz characterisation of building materials. *Electronics Letters, 41*, 1002–1004.
22. Piesiewicz, R., Jansen, C., Wietzke, S., Mittleman, D., Koch, M., & Kürner, T. (2007). Properties of building and plastic materials in the THz range. *International Journal of Infrared and Millimeter Waves, 28*, 363–371.
23. Piesiewicz, R., Jansen, C., Mittleman, D., Kleine-Ostmann, T., Koch, M., & Kürner, T. (2007). Scattering analysis for the modeling of THz communication systems. *IEEE Transactions on Antennas and Propagation, 55*, 3002–3009.
24. Dikmelik, Y., Spicer, J. B., Fitch, M. J., & Osiander, R. (2006). Effects of surface roughness on reflection spectra obtained by terahertz time-domain spectroscopy. *Optics Letters, 31*, 3653–3655.
25. Ortolani, M., Lee, J. S., Schade, U., & Hübers, H.-W. (2008). Surface roughness effects on the terahertz reflectance of pure explosive materials. *Applied Physics Letters, 93*, 081906.
26. Jansen, C., Priebe, S., Moller, C., Jacob, M., Dierke, H., Koch, M., & Kürner, T. (2011). Diffuse scattering from rough surfaces in THz communication channels. *IEEE Transactions on Terahertz Science and Technology, 1*, 462–472.
27. Beckmann, P., & Spizzichino, A. (1987). *The scattering of electromagnetic waves from rough surfaces*. Artech House.
28. Jansen, C., Piesiewicz, R., Mittleman, D., Kürner, T., & Koch, M. (2008). The impact of reflections from stratified building materials on the wave propagation in future indoor terahertz communication systems. *IEEE Transactions on Antennas and Propagation, 56*, 1413–1419.
29. Kleine-Ostmann, T., Dawson, P., Pierz, K., Hein, G., & Koch, M. (2004). Room-temperature operation of an electrically driven terahertz modulator. *Applied Physics Letters, 84*, 3555–3557.
30. Kleine-Ostmann, T., Pierz, K., Hein, G., Dawson, P., & Koch, M. (2004). Audio signal transmission over THz communication channel using semiconductor modulator. *Electronics Letters, 40*, 124–125.
31. Yu, X., Jia, S., Hu, H., Galili, M., Morioka, T., Jepsen, P. U., & Oxenløwe, L. K. (2016). 160 Gbit/s photonics wireless transmission in the 300-500 GHz band. *APL Photonics, 1*, 081301.
32. Jia, S., Zhang, L., Wang, S., Li, W., Qiao, M., Lu, Z., Idrees, N. M., Pang, X., Hu, H., Zhang, X., Oxenløwe, L. K., & Yu, X. (2020). 2 × 300 Gbit/s line rate PS-64QAM-OFDM THz photonic-wireless transmission. *Journal of Lightwave Technology, 38*, 4715–4721.
33. Buron, J. D., Petersen, D. H., Bøggild, P., Cooke, D. G., Hilke, M., Sun, J., Whiteway, E., Nielsen, P. F., Hansen, O., Yurgens, A., & Jepsen, P. U. (2012). Graphene conductance uniformity mapping. *Nano Letters, 12*, 5074–5081.
34. Bøggild, P., Mackenzie, D. M. A., Whelan, P. R., Petersen, D. H., Buron, J. D., Zurutuza, A., Gallop, J., Hao, L., & Jepsen, P. U. (2017). Mapping the electrical properties of large-area graphene. *2D Materials, 4*, 042003.

Chapter 3
Measurements with Modulated Signals

Daniel M. Mittleman

Abstract Historically, the majority of measurements to characterize devices, components, channels, or systems in the terahertz range have been made using either narrowband continuous sources or ultrabroadband pulse trains. In many cases, new information is revealed when the characterization is performed using a modulated data stream. This chapter discussed a few illustrative examples.

3.1 THz Communication Measurements

As described in Chap. 2, many of the early measurements in the THz range have been performed using time-domain spectroscopy techniques. These include studies to characterize components and devices such as filters and modulators, as well as measurements of channel properties such as atmospheric attenuation and effects of precipitation. The TDS technique offers many advantages and has become a widespread laboratory tool, especially in the years since the first commercial TDS instruments emerged in 2000.

However, in the context of wireless systems intended for high-speed data transmission, it is important to recognize the shortcomings of time-domain spectroscopy as a tool for device and system characterization. The technique inherently produces a train of phase-locked ultrashort (and therefore ultrabroadband) pulses, at a rate determined by the repetition rate of the mode-locked laser used to generate and detect them. This rate is typically in the range of 100 MHz [3], although higher rates are possible [4]. By the nature of a pulse train, it is impossible to modulate a TDS source at a rate exceeding this repetition rate. Moreover, the spectral resolution of TDS measurements is usually determined by the length of a scanning mechanical delay and is therefore typically a few GHz, ultimately limited by the laser repetition rate, since the regular pulse train corresponds to a frequency comb with a comb spacing equal to this repetition rate. Therefore, it is generally not feasible to generate

D. M. Mittleman (✉)
Brown University, Providence, RI, USA
e-mail: daniel_mittleman@brown.edu

T. Kürner et al. (eds.), *THz Communications*, Springer Series in Optical Sciences 234, https://doi.org/10.1007/978-3-030-73738-2_3

a carrier with a GHz-scale (or faster) modulation rate using TDS techniques. Even if one did so, the power remaining after the required spectral filtering would be extremely low, since TDS systems typically produce very low power per Hz of bandwidth. For a typical commercial TDS system generating ~10 μW of average power (integrated over the entire spectral range of the generated radiation), one can estimate that this corresponds to about −130 dBm/Hz at the spectral peak of the signal. As a result of these considerations, it is clear that TDS systems cannot readily be used to characterize components or systems with modulated signals at high data rates.

It is an interesting question to consider whether a broadband low spectral resolution characterization of a component or system is sufficient to understand how this component or system will perform when it is used with signals carrying modulated data at a high bit rate. In some cases, this broadband characterization is adequate. For example, a measurement which provides access to the complex frequency-dependent dielectric function of a material is sufficient to completely characterize the propagation of any complicated waveform in the medium, as long as the spectrum of the signal falls within the spectral range of the measurement and as long as its peak intensity is low enough to avoid inducing nonlinear effects. One good illustration is the precise characterization of atmospheric absorption and dispersion over the entire THz range, as reported in [5] (and references therein). On the other hand, there can be situations where the behavior of modulated signals can produce unanticipated results. Here, we discuss one example of each of these situations in which specific device configurations were tested.

3.2 3D Printed THz Dielectric Waveguides

A first example involves the characterization of waveguides, which have been designed for operation near 200 GHz. Dielectric waveguides represent one important platform for both signal transport and signal processing. With relatively low absorption losses and considerable design flexibility, dielectric structures offer great promise as both active and passive devices. One particular advantage is the possibility of fabrication by 3D printing, which offers a low-cost and highly versatile method for production of nearly arbitrary waveguiding structures [6]. For example, Weidenbach et al. [7] recently demonstrated a suite of 3D printed waveguide designs with low loss in the THz range, including a fixed power splitter and a variable waveguide splitter based on evanescent-wave coupling [8] between two parallel dielectric waveguides.

One example of these evanescent-wave coupling structures was recently investigated using both a continuous-wave signal and a modulated signal with a pseudorandom bit stream at 1 Gbps [1]. The waveguides are printed out of polystyrene (PS), a material which is reasonably transparent at THz frequencies with a nearly frequency-independent refractive index of 1.56 and which can be printed using a Ultimaker Original 3D printer [9]. The printed waveguide has a square cross

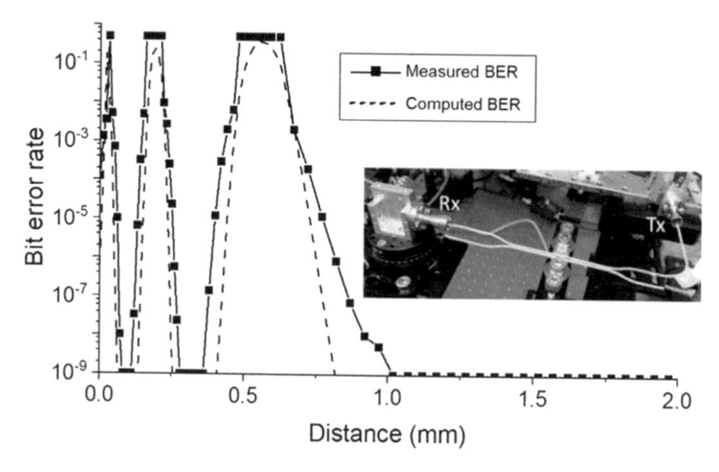

Fig. 3.1 Bit error rate as a function of separation between two evanescently coupled waveguides. The data points show measured BER when a modulated signal is injected into the waveguide input. In these measurements, the BER detector saturates at a rate of 10^{-9}. The dashed curve show the result obtained by measuring the power transmission using unmodulated data and then computing the BER. The transmitter and receiver antennas are labeled in the inset as Tx and Rx. (Adapted from [1])

section of size 0.8 mm by 0.8 mm, which assures single-mode operation in the frequency range of interest. As shown in the inset to Fig. 3.1, it consists of a 90° turn followed by a coupling section in which two identical waveguides run parallel to each other with a mechanically variable spacing between them. The 90° bend prevents direct free-space coupling between the transmitter and receiver, and its bending loss is negligible for a curvature radius $r = 20$ mm [7]. The coupling region has a nominal length of 60 mm, although the effective coupling length is somewhat larger due to nonzero coupling even in the curved regions.

For the data transmission experiments, the source is a frequency multiplier chain (Virginia Diodes), which up-converts a modulated baseband signal to 196 GHz producing about 60 mW of power at the output of the horn antenna. This carrier wave is modulated at 1 Gbps via on-off keying (OOK) modulation. The signal is detected (amplitude only) using a zero-biased Schottky diode intensity detector. This signal is amplified to drive a bit error rate (BER) tester (Anritsu MP1764A) for real-time signal analysis.

Figure 3.1 summarizes the comparison between the power measurements and bit error rate measurements. The solid lines and data points indicate the measured bit error rate as a function of the separation between the two waveguides in the coupling region. As the separation increases, the signal oscillates between the two waveguides, as expected [7]. For large separations ($d > 1$ mm), all the power remains in the primary waveguide, and the BER saturates at 10^{-9}. Meanwhile, the dashed line shows the predicted bit error rate, computed using measurements of the power vs. separation, according to the conventional result for an incoherent detector and OOK modulation, as given in [10]. The oscillatory power results in a prediction of an oscillatory BER (dashed), in excellent agreement with the measured BER (squares). This is a good illustration of a situation where the unmodulated signal can be used to accurately predict the performance with a modulated signal.

3.3 THz Demultiplexing

A recent report of demultiplexing of THz signals provides an example of the alternative situation, where the results obtained using modulated data are not obvious from measurements using unmodulated sources. The example in this case is a proposed frequency-division demultiplexer based on a leaky-wave parallel-plate waveguide [11]. Signals injected into this waveguide, propagating in the TE_1 mode, emerge through a slot aperture at an angle which depends on their frequency. This effect can be exploited for multiplexing and demultiplexing, by assigning specific frequencies to individual users in a local area network according to their angular location relative to the access point.

For radiation emitted from a leaky waveguide, the relation between frequency and emission angle is determined by a phase-matching constraint on the parallel component of the wave vector. A simple analysis gives the requirement:

$$\beta = k_0 \cos \phi \tag{3.1}$$

where β is the wave vector of the guided TE_1 mode, k_0 is the free-space wave vector, and ϕ is the emission angle relative to the waveguide's propagation axis. However, because of the effects of diffraction, the radiation that emerges from the aperture has a finite angular divergence, even if it consists of only a single-frequency component. This angular width depends on geometrical parameters such as the length and width of the slot aperture.

Under realistic conditions, the angular width of a single-frequency beam emerging from the aperture can be large compared to the angular separation between a signal at frequency f and its modulation sidebands at $f \pm \delta f$. This situation is illustrated in Fig. 3.2a, which shows the radiation patterns computed for single tones at a carrier frequency of 300 GHz, plus two sidebands at 290 GHz and 310 GHz (assuming a modulation rate of 10 Gbps, for illustration purposes). In a case like this, one might estimate the minimum necessary detector aperture based on the angular width of the carrier frequency's beam. One would then predict that lower modulation rates (<10 Gbps), for which the side bands are more closely spaced in angle, should exhibit no unusual changes as the detector's angular position is varied. In particular, the bit error rate should have the same dependence on detector angle for any lower modulation rate. Yet, as shown in Fig. 3.2b, the measured BER vs. detector angle shows a clear trend, even for much lower data rates [2]. This results from a small asymmetry in the detection sensitivity across the detector's aperture, which causes the side bands on one side of the carrier to be detected slightly more efficiently than on the other side. This asymmetry, although small enough to escape

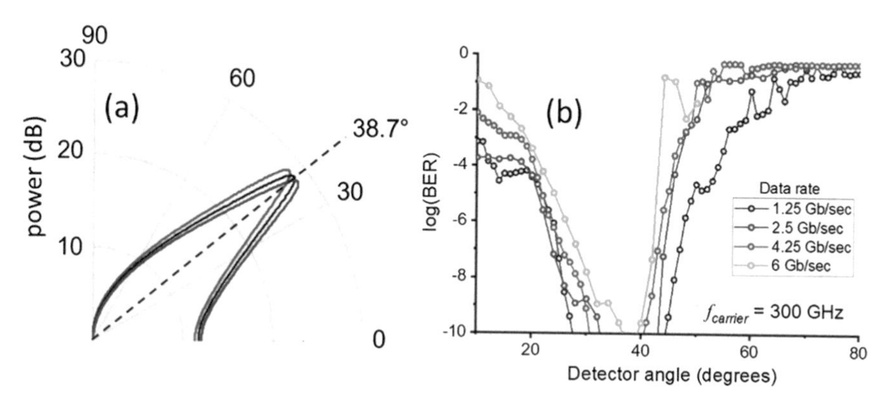

Fig. 3.2 (**a**) Radiation patterns computed for three different frequencies emerging from a slot aperture in a leaky-wave parallel-plate waveguide: 290 GHz (blue), 300 GHz (black), and 310 GHz (red). The dashed line indicates the angle predicted by Eq. (3.1) for 300 GHz. (**b**) Bit error rates for a modulated signal emerging from a leaky-wave waveguide, as a function of the angular location of the detector. The bit error rate's dependence on detector position is different for different data modulation rates. (Adapted from [2])

notice in conventional power calibration measurements, has a significant effect on the bit error rates for different modulation rates. One might expect similar effects to be very common in THz wireless systems, where diffraction (and therefore a strong coupling between frequency and propagation direction) will be ubiquitous. The result illustrated in Fig. 3.2b emphasizes the need for characterization of THz components and systems using a variety of complementary techniques.

References

1. Ma, J., Weidenbach, M., Guo, R., Koch, M., & Mittleman, D. M. (2017). Communication with THz waves: Switching data between two waveguides. *Journal of Infrared Millimetre THz Waves, 38*, 1316–1320.
2. Ma, J., Karl, N. J., Bretin, S., Ducournau, G., & Mittleman, D. M. (2017). Frequency-division multiplexer and demultiplexer for terahertz wireless links. *Nature Communications, 8*, 729.
3. Mittleman, D. M. (2003). *Sensing with terahertz radiation*. Berlin: Springer.
4. Bartels, A., Thoma, A., Janke, C., Dekorsy, T., Dreyhaupt, A., Winnerl, S., & Helm, M. (2005). High-resolution THz spectrometer with kHz scan rates. *Optics Express, 14*, 430–437.
5. O'Hara, J. F., & Grischkowsky, D. R. (2018). Comment on the veracity of the ITU-R recommendation for atmospheric attenuation at terahertz frequencies. *IEEE Transactions on THz Science and Technology, 8*, 372–375.
6. Liu, J., Mendis, R., & Mittleman, D. M. (2013). A Maxwell's fish eye lens for the terahertz region. *Applied Physics Letters, 103*, 031104.
7. Weidenbach, M., Jahn, D., Rehn, A., Busch, S. F., Beltrán-Mejía, F., Balzer, J. C., & Koch, M. (2016). 3D printed dielectric rectangular waveguides, splitters and couplers for 120 GHz. *Optics Express, 24*, 28968–28976.
8. Reichel, K., Sakoda, N., Mendis, R., & Mittleman, D. M. (2013). Evanescent wave coupling in terahertz waveguide arrays. *Optics Express, 21*, 17249–17255.

9. Busch, S. F., Weidenbach, M., Fey, M., Schäfer, F., Probst, T., & Koch, M. (2014). Optical properties of 3D printable plastics in the THz regime and their ap-plication for 3D printed THz optics. *Journal of Infrared Millimeter THz Waves, 35*, 993–997.
10. Tang, Q., Gupta, S. K. S., & Schwiebert, L. (2005). BER performance analysis of an on-off keying based minimum energy coding for energy constrained wireless sensor application. In *IEEE International Conference on Communications (ICC)* (pp. 2734–2738). Seoul, South Korea.
11. Karl, N. J., McKinney, R. W., Monnai, Y., Mendis, R., & Mittleman, D. M. (2015). Frequency division multiplexing in the terahertz range using a leaky wave antenna. *Nature Photonics, 9*, 717–720.

Chapter 4
Vector Network Analyzer (VNA)

Thomas Kleine-Ostmann

Abstract This chapter discusses the fundamentals of vector network analysis including the definition of scattering parameters, the hardware architecture of vector network analyzers and frequency extenders, and methods for system error correction. The metrological approach of establishing traceability to the international system of units (SI) is described as needed to measure with known measurement uncertainty. The chapter ends with a brief view on using vector network analysis for quasi-static channel measurements.

Vector network analysis is a well-established technique to determine the radio frequency transmission and reflection properties of circuits, devices, setups, or linear time-invariant networks in general. A vector network analyzer (VNA) determines the frequency-dependent and complex-valued *scattering parameters* as defined for an n-port network as measurands. These scattering parameters are used to derive component properties (e.g., filter transmission, amplification, mixer attenuation), antenna properties (gain, antenna factor), or transmission channel properties (channel impulse response after transformation of measurement results into the time domain).

4.1 Scattering Parameters

For a two-port network, the scattering parameters S_{11}, S_{21}, S_{12}, and S_{22} are defined as complex ratios between forward (a_1, a_2) and backward (b_1, b_2) traveling waves at the ports 1 and 2 as shown in Fig. 4.1 [1, 2]:

T. Kleine-Ostmann (✉)
Physikalisch-Technische Bundesanstalt, Braunschweig, Germany
e-mail: thomas.kleine-ostmann@ptb.de

T. Kürner et al. (eds.), *THz Communications*, Springer Series in Optical Sciences 234, https://doi.org/10.1007/978-3-030-73738-2_4

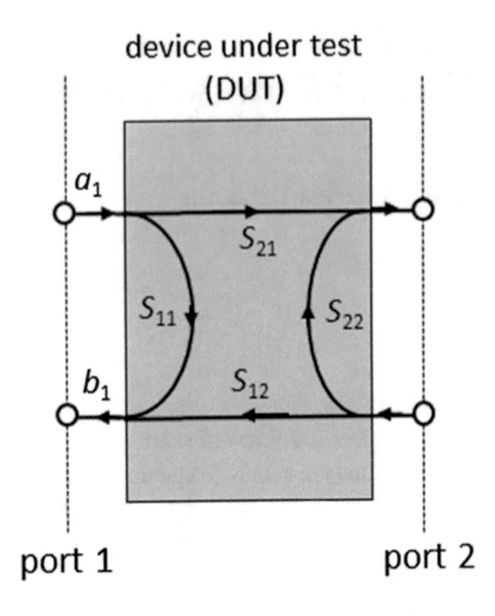

Fig. 4.1 Two-port network with wave quantities

$$S_{11} = \left.\frac{b_1}{a_1}\right|_{a_2=0} \qquad \textit{reflection} \text{ at port 1 with match at port 2}$$

$$S_{21} = \left.\frac{b_2}{a_1}\right|_{a_2=0} \qquad \textit{forward transmission} \text{ with match at port 2}$$

$$S_{12} = \left.\frac{b_1}{a_2}\right|_{a_1=0} \qquad \textit{backward transmission} \text{ with match at port 1}$$

$$S_{22} = \left.\frac{b_2}{a_2}\right|_{a_1=0} \qquad \textit{reflection} \text{ at port 2 with match at port 1}$$

with $a_1 = \frac{1}{\sqrt{Z}}V_{1f} = \sqrt{Z}I_{1f}$, $a_2 = \frac{1}{\sqrt{Z}}V_{2f} = \sqrt{Z}I_{2f}$,

$b_1 = \frac{1}{\sqrt{Z}}V_{1b} = \sqrt{Z}I_{1b}$, $b_2 = \frac{1}{\sqrt{Z}}V_{2b} = \sqrt{Z}I_{2b}$.

Here, Z is the *characteristic impedance* of the transmission lines (usually 50 Ohm), and V_{1f}, V_{2f}, I_{1f}, and I_{2f} are the complex phasors of the forward travelling voltage and current waves, whereas V_{1b}, V_{2b}, I_{1b}, and I_{2b} denote the backward travelling waves. The scattering parameters S_{ij} form the scattering matrix S which maps the vector a of forward propagating waves into the vector b of backward travelling waves:

$$b = \begin{pmatrix} b_1 \\ b_2 \end{pmatrix} = S \cdot a = \begin{bmatrix} S_{11} & S_{12} \\ S_{21} & S_{22} \end{bmatrix} \cdot \begin{pmatrix} a_1 \\ a_2 \end{pmatrix}$$

The scattering parameters are complex relative quantities, which mean that they are dimensionless. The complex number without unit can be represented in amplitude and phase or as real and imaginary part, e.g., in the Touchstone® format. The amplitude is very often given on a logarithmic scale: $|S_{ij}|_{dB} = 20\ dB\ \lg |S_{ij}|$.

For passive networks not containing anisotropic media, the scattering matrix S is reciprocal: $S_{ij} = S_{ji}$ for $i \neq j$. It can easily be calculated from the impedance matrix Z or admittance matrix G commonly used to describe networks [2]. The concept of the scattering matrix can be extended to n-port networks by replacing the 2×2 scattering matrix by $n \times n$ matrix.

4.2 Network Analyzer Architecture

The main components of a vector network analyzer (VNA) are the *generator with source switch* which generates the stimulus signal, the *test set* which separates forward and backward travelling waves, and the *measurement and reference channel receivers* which are necessary to precisely measure the scattering parameters as relative quantities. Figure 4.2 shows the architecture of a simple two-port network analyzer [3]. The RF generator can be switched to ports 1 and 2, consecutively. The signal is split in the test set. Half of it hits the device under test (DUT), and half of it is measured in the reference receiver of the active port. In addition, the measurement receiver at the active port measures the reflected signal returned from the device under test, and the measurement receiver at the inactive port measures the signal transmitted through the DUT. The scattering parameters are derived from the ratio of the respective receiver readings.

The upper frequency limit of a VNA is usually defined by the coaxial connector system used, e.g., 40 GHz (2.92 mm), 50 GHz (2.4 mm), or 67 GHz (1.85 mm). To extend the available frequency range to the waveguide bands, a set of frequency converters specific for the respective band will be used. Usually it is operated with a four-port VNA which has access to reference and measurement channel at two of the ports as shown in Fig. 4.3.

4.3 System Error Correction and Traceability

The couplers as main components of the test sets shown in Fig. 4.2 represent a three-port network with scattering parameters S_{21}, S_{31}, and S_{32} themselves. Their nonideal properties are described by the *reflection tracking* $R = S_{32} \bullet S_{21}$ and the

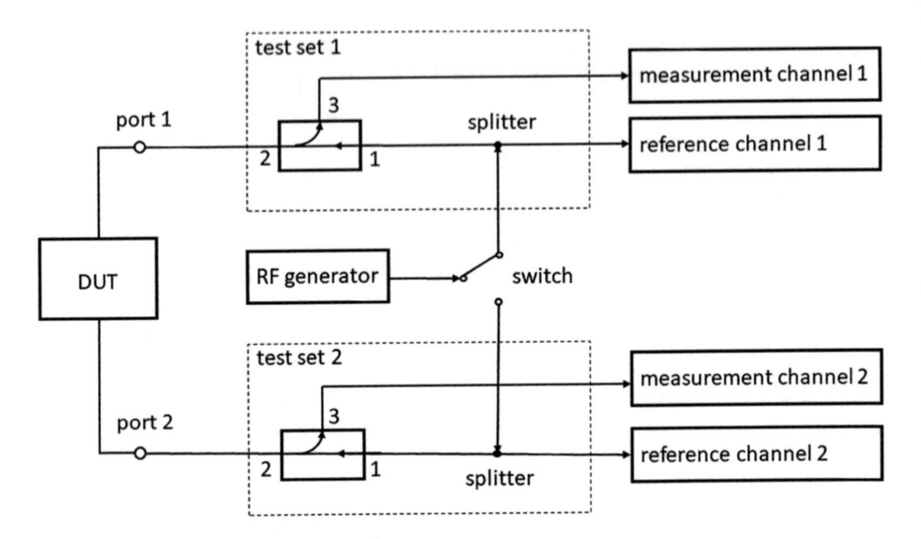

Fig. 4.2 Two-port network analyzer architecture

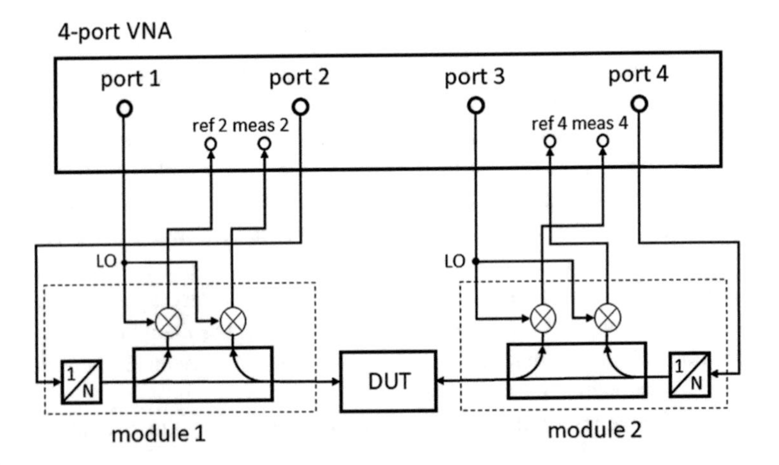

Fig. 4.3 Frequency extension into the waveguide bands

reflectivity $D = {S_{31}}/{R}$. In addition to this, the *source match* S of the measurement port deteriorates the measurement. In case of a one-port measurement of a DUT with reflection factor Γ_{DUT}, the measurement yields:

$$M = R\left(D + \frac{\Gamma_{\mathrm{DUT}}}{1 - S \cdot \Gamma_{\mathrm{DUT}}}\right).$$

The systematic errors caused by R, D, and S can be corrected by the procedure of *systematic error correction*, often denoted as calibration of the VNA. Here,

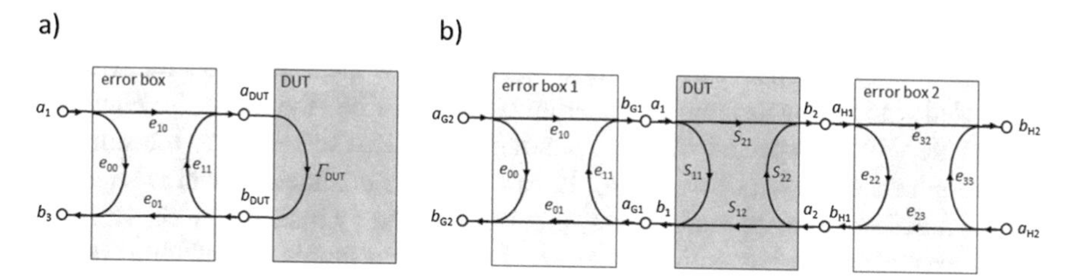

Fig. 4.4 Error box models for (**a**) one-port and (**b**) two-port system error correction [3] without cross talk and not distinguishing between forward and backward model

after waiting for thermal stabilization of the VNA, *calibration standards* with known properties are connected to the VNA and measured to calculate the *error box parameters* as shown in Fig. 4.4 as needed for a numerical correction of the measurement results.

In case of a one-port measurement, the task is to determine the four error box parameters according to Fig. 4.4a, of which only three are independent (three-term model). This can be done by consecutive measurements of an open, short, and match (OSM method). If two-port measurements are intended, eight error box parameters have to be determined according to Fig. 4.4b, of which only seven are independent. Here, a number of methods exist involving additional calibration standards such as through connection (T), unknown through connection (U), attenuation standard (A), line standard (L), reflect standard (R), and symmetrical network standard (N): TOM, TRM, TRL, TNA, and UOSM (see [3] for details). Very often, the VNA architecture differs from that shown in Fig. 4.2 so that several ports share a reference receiver. In that case, the switch needs to be modeled so that two seven-term models are required that comprise twelve independent terms. Often, this twelve-term model is reduced to 10 terms by neglecting crosstalk terms. The most common method for system error correction in this case is the TOSM method.

After system error correction, the systematic errors are significantly reduced when measuring with the VNA. However, a residual error remains. In addition to the residual systematic error, other systematical and statistical errors contribute to the measurement uncertainty when measuring with a VNA, as, for example, thermal drift, noise floor and trace noise, repeatability of connectors, cable movement, nonlinearities, and cross talk.

To be able to rely on measurements, an unbroken *traceability* chain of measurements has to be established that connects the individual VNA measurement to the representation of SI units kept at the National Metrology Institutes (such as NIST in the USA, NPL in the UK, PTB in Germany, and many others) with highest precision. The calculation of the resulting measurement uncertainty for the VNA measurement is a rigorous analysis according to the rules internationally defined by the *Guide to the Expression of Uncertainty in Measurement* [4] published by broadly based international organizations working in the field of metrology. It requires a complete modeling of the measurement process and determination of the input quantities with attributed measurement uncertainties that are propagated through the model. The result is the best estimate of the measurand together with an attributed standard measurement uncertainty that specifies the range, in which the true value will be expected with an uncertainty of 68% (according to the standard deviation of a Gaussian distribution). In case a higher confidence level is required, an expanded measurement uncertainty can be specified by multiplying the standard measurement uncertainty with an extension factor (for a simple analysis considering the magnitude, only $k = 2$ leads to a confidence interval of 95% in case of a sufficiently high degree of freedom).

The modeling of the measurement process required for VNA traceability and specification of the overall measurement uncertainty is described in the Euramet Calibration Guide cg-12 [5]. It includes description of the traceability chain, determination of all relevant measurement uncertainty contributions, and best measurement practice and practical advice. While traceability is well established in coaxial line systems with standard measurement uncertainties reaching down to 0.0015 dB (calibration and measurement capability (CMC) of PTB as published by the International Bureau of Weights and Measures (BIPM)), traceability in the upper waveguide bands is an ongoing research topic [6, 7]. Here, calibration standards for system error correction are lacking mechanical precision and are more difficult to model correctly which leads to measurement uncertainties in the range of a dB.

4.4 Quasi-static Channel Characterization

Vector network analyzers in conjunction with horn antennas can be used for channel and propagation measurements if the channel is varying slowly, only. The channel properties are determined in a frequency sweep (duration T_{sweep}) measuring $S_{21}(f)$, which is proportional to the transfer function $H(f)$. From this, the (band limited) impulse response $h(t)$ or the *power delay profile* (PDP) ~ $h^2(t)$ can be determined using an inverse Fourier transform. In order not to violate the Nyquist sampling theorem, the maximum frequency of change has to be below $1/(2T_{\text{sweep}})$. If angle of departure (AoD) under which the transmitter (Tx) emits and angle of arrival (AoA) under which the receiver (Rx) receives the incoming waves are not fixed, the sweep time is not restricted to a single-frequency sweep but comprises the whole angular scans of transmitter and receiver antenna. This leads to the determination of an impulse response $h(t, \varphi_{\text{AoD}}, \varphi_{\text{AoA}})$ depending on the angles φ_{AoD} and φ_{AoA} in the propagation plane. Depending on the orientation of the horn antenna polarization with regard to reflecting surfaces, transversal electric (TE) or transversal magnetic (TM) measurements can be performed.

Figure 4.5 shows a setup that has been used for quasi-static channel characterization [8]. The setup has been used to study diffraction [9] and basic propagation properties for intra-device communication [10]. A four-port VNA with frequency extension modules and 20 dB standard gain horn (SGH) antennas for the waveguide bands WR 15 (50 GHz–75 GHz), WR 10 (75 GHz–110 GHz), WR 7 (110 GHz–

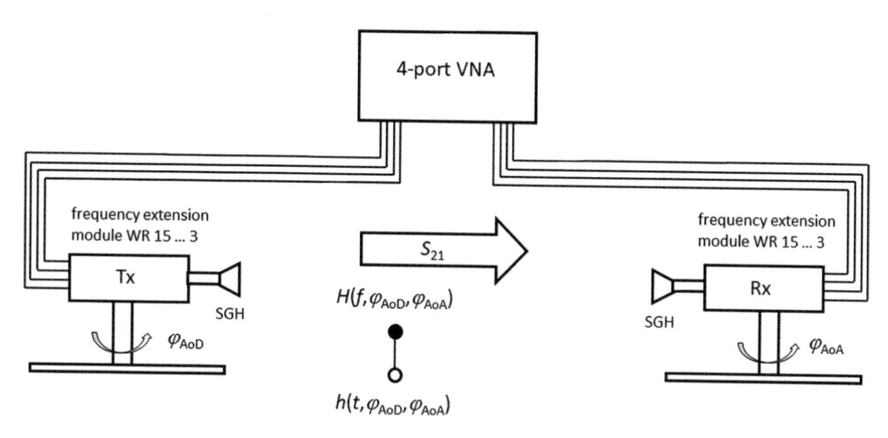

Fig. 4.5 Quasi-static channel measurement setup

170 GHz), WR 5 (140 GHz–220 GHz), and WR 3 (220 GHz–325 GHz) have been used for measurements. The output power of the converters is decreasing with frequency from 4 dBm in the WR 15 band to −20 dBm in the WR 3 band. The length of the cables connecting transmitter and receiver to the VNA is restricted to several meters. With typical resolution bandwidths between 100 Hz in the WR 15 band and 10 kHz in the WR 3 band, measurements with high dynamic range larger than 80 dB are possible in all five frequency bands. Depending on the desired maximum excess delay τ_{max} and delay resolution $\Delta\tau$, a sufficient number of frequency points N are needed. With $N = 6401$ measurement points, a frequency spacing of $\Delta f = (f_{max}\text{-}f_{min})$ / $(N\text{-}1)$ =105 GHz/6400 $\approx$ 16.4 MHz can be achieved in the WR 3 band leading to a delay resolution of $\Delta\tau = 1$ / $(f_{max}\text{-}f_{min}) \approx 9.5$ ps and a maximum excess delay of $\tau_{max} = N \cdot \Delta\tau \approx 61$ ns. While the delay resolution of $\Delta\tau \approx 9.5$ ps corresponds to a spatial resolution of $\Delta d = c \cdot \Delta\tau \approx 2.85$ mm, the maximum excess delay $\tau_{max} \approx 61$ ns corresponds to a maximum detectable path length of $d_{max} = c \cdot \tau_{max} \approx 18.3$ m (with speed of light $c \approx 2.998 \cdot 10^8$ m/s). Such a measurement takes several seconds.

The angular scan is realized using rotational stages with stepper motors. With typical angular scanning resolutions of several degrees, complete circular scans of Tx and Rx can take hours to days.

References

1. Thumm, M., Wiesbeck, W., & Kern, S. (1998). *Hochfrequenzmesstechnik – Verfahren und Messsysteme* (2nd ed.). Vieweg + Teubner.
2. Pozar, D. M. (2005). *Microwave engineering* (3rd ed.). Wiley.
3. Hiebel, M. (2007). *Fundamentals of vector network analysis*. Rohde & Schwarz.
4. Evaluation of measurement data – Guide to the expression of uncertainty in measurement, Joint Committee for Guides in Metrology (JCGM) 100:2008, GUM 1995 with minor corrections.

5. Guidelines on the Evaluation of Vector Network Analysers (VNA), EURAMET Calibration Guide No. 12, European Association of National Metrology Institutes (Euramet), Version 3.0, 03/2018.
6. Schrader, T., et al. (2011). Verification of scattering parameter measurements in waveguides up to 325 GHz including highly-reflective devices. *Advances in Radio Science, 9*, 9–17.
7. Williams, D. F. (2011). 500 GHz −750 GHz rectangular-waveguide vector-network-Analyzer calibrations. *IEEE Transactions on THz Science and Technology, 1*, 364–377.
8. Salhi, M., Kleine-Ostmann, T., & Schrader, T. (2015). Antenna characterization and channel measurements in the mm wave and sub-mm wave region. *Microwave Journal, 58*, 124–134.
9. Jacob, M., Priebe, S., Dickhoff, R., Kleine-Ostmann, T., Schrader, T., & Kürner, T. (2012). Diffraction in mm and sub-mm wave indoor propagation channels. *IEEE Transactions on Microwave Theory and Techniques, 60*, 833–844.
10. Kürner, T., Fricke, A., Rey, S., Le Bars, P., Mounir, A., & Kleine-Ostmann, T. (2015). Measurements and modeling of basic propagation characteristics for intra-device communications at 60 GHz and 300 GHz. *Journal of Infrared Millimeter Terahz Waves, 36*, 144–158.

Chapter 5
THz Broadband Channel Sounders

Robert Müller and Diego Dupleich

Abstract This chapter is the perfect introduction to get an overview of THz channel sounder technologies. Additionally, all relevant state of the art and references for the field of THz channel sounding are summarized. The aim of the THz sounder chapter is to create a basic understanding of measurement setups and challenges for the measurement of the electromagnetic wave propagation in the THz range. All necessary principles, from generating the transmit signal over different mixing principles to the THz band and the data acquisition, are compact summarized.

The use of millimeter frequency bands in 5G allows the implementation of instantaneous bandwidths (BWs) with a couple of GHz [1–5] and opens the possibility to reach the goal of 100 Gbit/s and higher data rates in mobile networks. Additionally, in the field of Wi-Fi systems (IEEE802.11/802.15), different activities from mmWave to THz and in the light spectrum are being carried out to enable broadband communications with several Gbit/s. Current wireless systems can reach several hundreds of Gbit/s using BWs larger than 2 GHz in the 5G mmWave and in the 60 GHz ISM band. The advantages of THz systems in comparison to 5G mmWave and 60 GHz ISM frequency bands are a great increase of the available BW. However, channel sounders (CSs) and channel models can only be developed with a clear idea on the use cases for future THz communication systems. These range from networks for data centers [6], M2M communications in industry environments [7], inter-device communication for smart devices [8], Wi-Fi applications [9], THz backhauling for "6G" [10], body networks [11], and so on. Currently, the focus is set on the extension of the existing wireless LAN systems providing higher data rates. However, secondary applications such as localization and RADAR can also be greatly improved in terms of resolution, thanks to the enormous available BW in the THz frequency range. To analyze the potential and to develop suitable technologies in these frequency bands, the equipment to measure the propagation

R. Müller (✉) · D. Dupleich
TU Ilmenau, Ilmenau, Germany
e-mail: mueller.robert@tu-ilmenau.de

T. Kürner et al. (eds.), *THz Communications*, Springer Series in Optical Sciences 234, https://doi.org/10.1007/978-3-030-73738-2_5

of electromagnetic waves needs to provide a BW of more than 20 GHz [12–14]. Another important property of channel sounders is the ability to capture dynamic propagation effects with the target BW of THz systems.

5.1 THz CS Overview

The main challenge on the design of a CS for THz measurements is the target BW, which exceeds 20 GHz. The second question is the coverage range. THz communications are foreseen to be applied in different scenarios and applications which require distances from few centimeters up to kilometers. The third aspect is to provide a stable reference clock distribution when the TX and RX units are separated over several hundred meters in large scenarios, such as backhauling or small cell access point. Additionally, a high measurement rate is necessary to analyze the Doppler effects in THz communication. As an example, a walking person generates, in average, a Doppler shift of around 400 Hz at 30 GHz (mmWave in 5G) and of 4000 Hz at 300 GHz.

The analysis of Doppler shift requires, at least, a sampling rate following the Nyquist criteria with the highest Doppler frequency which is present in the measurement. In combination with the large BWs and the required amplitude dynamic range (in the optimum of 60 dB which is around 12-bit resolution of the ADC [15]), the data value per second at 20 GHz BW is around 223.5 GB/s per receiver channel. Depending on the selected measurement method and the period length of the transmit signal, the measurement speed and the maximum Doppler resolution of the CS system are determined.

From current knowledge of the authors, there is no measuring system available that currently meets all the requirements for THz channel sounding in view of coverage of large distances, measurement BW, measurement speed, spatial resolution in azimuth/elevation, and polarization resolution.

Nevertheless, there are different approaches for the development of CSs for THz applications, which will be discussed in the following subchapters. To get an overview of common THz channel sounder systems, we focus on THz applications for small- and medium-size scenarios like offices, data center, entrance halls, etc. In these environments, many channel measurements with different CSs and VNAs are done [6, 9, 16–19].

Most of the analyses were carried out using VNAs with frequency expansions in the THz range as described in Chap. 4. Further analysis was performed using different sounder architectures, e.g., FMCW baseband units [20] and correlation sounders, or using broadband arbitrary waveform generator (AWG) in combination with broadband DSO/DPO (digital storage oscilloscope/digital phosphor oscilloscope) [21]. Furthermore, the use of broadband ADC and DAC with several GHz of BW for the baseband units from different suppliers such as National Instruments [22], Teledyne SP Devices [23], Xilinx [24], and other companies is a possible solution for a THz CS. With flexible solutions like the already mentioned DAC and

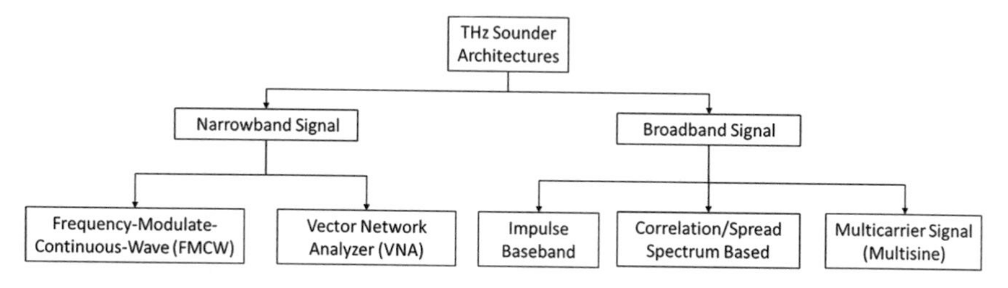

Fig. 5.1 Overview of different basic principles for THz channel sounding

ADC or AWG or DPO/DSO, any of these previous operating modes can be enabled. An overview of different channel sounder architectures is summarized in Fig. 5.1.

5.2 Broadband Baseband Units for THz Channel Sounding

In comparison to narrowband architectures (stepping of sinusoid) which are used for VNA and FMCW [25] systems, broadband architectures offer the capability to measure the full instantaneous BW of a THz system and enable the measurements of dynamic propagation effects of the THz channels. The majority of broadband THz sounder systems utilize a broadband and periodic pseudorandom binary sequence (PRBS). The advantage of PRBS is the simple generation of the transmit signals with flip-flops or faster digital outputs. For the realization of PRBS-based sounders, different approaches are possible, and the most important types of PRCS systems are explained in the following subsections.

Advantages of averaging gain, correlation gain, etc. of using different periodic sequences are described in [26–28]. Alternatively, a multicarrier signal [29, 30] can be used which requires additional broadband DACs instead of fast binary flip-flops or digital outputs at the transmitter side.

5.3 Pseudo-Noise (PN)-based correlation: Sliding Correlator

There are two commonly used architectures for PN-sequence correlation-based sounders. The first is an analog version of a correlation-based sounder. The "sliding correlator" principle [27] generates the same PN (PN-G) with slightly different clock frequencies at the transmitter and receiver (Fig. 5.2). The received signal is mixed with a replica of the transmitted signal with a frequency offset that generates the real-time channel impulse responses at the ADC. The sliding correlator generates a time shift in the channel measurement when the received PN-sequence slides past the PN-sequence which is generated at the receiver in the correlator. To correct the time shift of the sliding correlator, multiple samplings of

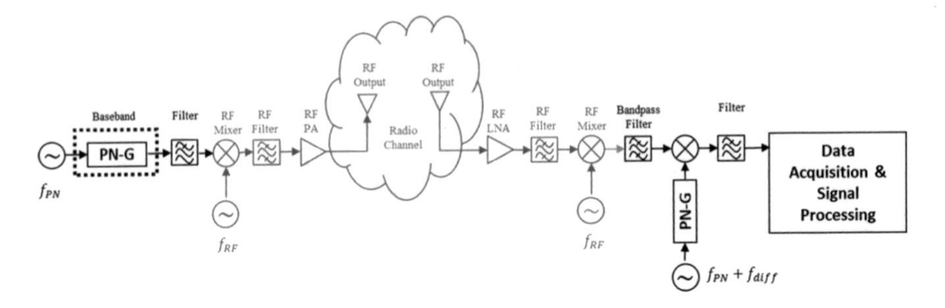

Fig. 5.2 Principle architecture of a sliding correlator THz channel sounder

the PN signal are required (measurement of several periods) or a parallel correlation (with different clock shifts at every correlator) at the receiver hardware is necessary.

5.4 Real-Time Sampling Broadband Baseband Units for THz CS

Another baseband realization for broadband sounders is based on the digitization of the full measurement bandwidth of the received signal with a high sampling rate in real time [26, 31] based on Nyquist sampling. Depending on the used signal (multisine, PRBS), further processing takes place in digital domain. In most of the THz channel sounder based on the PRBS principle (Fig. 5.3 red part), the correlation is performed after the digitalization. At this point, different approaches for a PRBS sounder can be followed: from a fast online correlation calculated directly in the FPGA to an offline correlation after the measurement campaign. The correlation process can be performed by multiplying the Fourier transformation of the received signal with the conjugated response of the transmitted signal. In addition, filters to suppress out-of-band mixing products of the sounder hardware or to limit the BW can be applied in this stage. Finally, the CIR is obtained by applying the inverse FFT to the result.

5.5 PN-Sequence Correlation-Based Subsampling System

A special compact realization of the direct PN-sequence correlation-based sounder is the subsampling system which is used in [28, 32] (Fig. 5.3 green part). This unit uses the periodic PN signal and a track and hold (T&H) to reach several GHz BWs for THz measurements. The subsampling operation also reduces the problem of fast data storage and enables a compact building factor of the baseband units. The main disadvantages of the subsampling CS system are the longer measurement

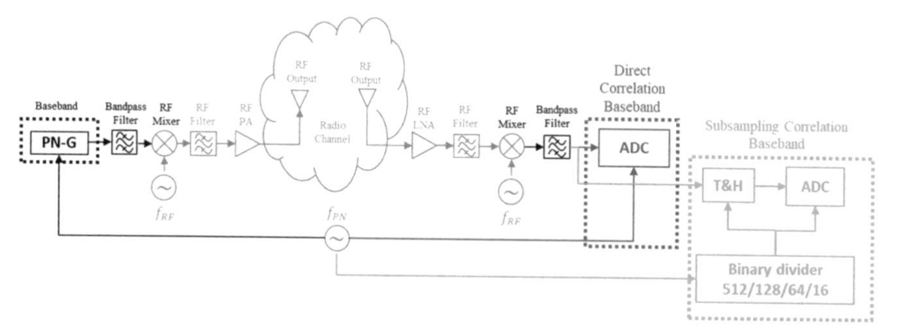

Fig. 5.3 Principle architectures of a direct correlation channel sounder in red and in green the subsampling version PN-sequence correlation-based at receiver for THz channel sounders

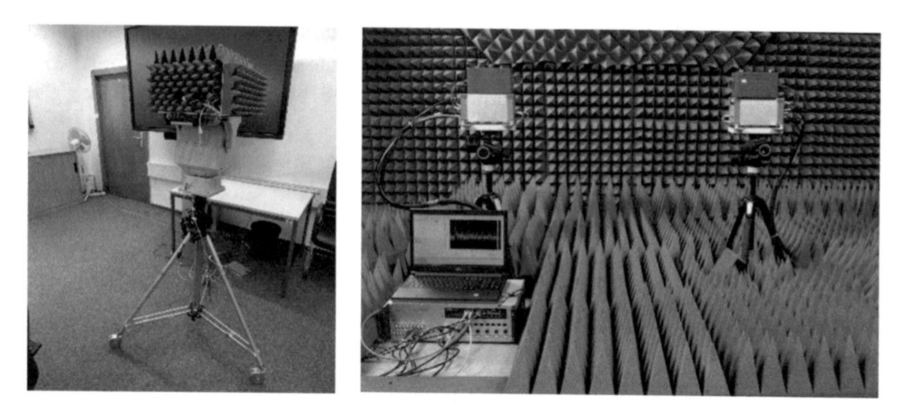

Fig. 5.4 Left, the 200 GHz multiband sounder (180–220 GHz/71–78 GHz/3.5–10.5 GHz) of the Technische Universität Ilmenau, right, the 300 GHz channel sounder (5.2–13.2 GHz/60.32–68.32 GHz/300.2–308.2 GHz) of Technische Universität Braunschweig

time and the requirement of a highly stable and coherent clock feed in the GHz range. On the other hand, a stable high-frequency clock can also be an advantage for THz CSs, since it generates a low-phase noise RF signal when it is used for the frequency conversion. Compared to other measurement methods that use reference clocks in the MHz range, the phase noise influences can be smaller which improves the quality of the measurement results. Another disadvantage, depending on the subsampling factor, is the ability to record real-time events. Depending on the subsampling factor, the measurement of Doppler frequencies is still possible up to a few hundred of MHz.

5.6 Techniques for Expanding Bandwidth of THz Channel Sounder

A further increase of the measurement BW for broadband channel sounder systems can be achieved by different technologies. One is the I/Q sampling. This can be implemented in a parallel or sequential structure. The I/Q sampling can take place at IF frequency, mostly away by a couple of GHz from the baseband signal, or in the baseband. The extension of the BW using the I/Q mixing process requires a calibration of the deviation from the ideal 90° phase shift between the I and Q stages. The differences between the ADCs must also be considered during calibration with the parallel I/Q method.

The parallel sampling of larger BWs can also take place with several ADCs in the baseband, which is called time-interleaved (TI) sampling. The main challenge of interleaving broadband ADCs is the untreated mismatch between the units that generate spurs in the output spectrum. This is based on the fact that during the manufacturing procedure of semiconductors, there is an unavoidable slight random variation on every mid- to high-resolution ADC. This produces a mismatch in the gain, and the offset can be considered as DC effects. The effects of timing (jitter in the ADC) and BW (each ADC has a deviation from the target BW) mismatch become more severe with larger BWs. To eliminate this influence of the ADCs, the standard procedures "foreground" or "background" are often used to calibrate time-interleaved broadband ADCs.

BW extensions of the transmission signal can be easily done by using a double sideband up-conversion method. Depending on the receiver architecture, the mixing clock at the receiver and transmitter can differ. Alternatively, the multi-gigabit transceivers (MGTs) of the FPGA can be used to generate PRBS signals with BWs of more than 50 GHz. For the generation of broadband multisine signals, TI-DACs can be used.

5.7 Up- and Down-Conversion Principles for THz Sounders

In the next step, the baseband signal needs to be mixed up to the THz band at the transmitter and mixed down at the receiver. Depending on the position of the baseband signal, different mixing processes can be used. Historically, the heterodyne principle is the most known in the RF techniques. The heterodyne conversion uses an IF conversion step to reach the target radio frequency (RF). Depending on the structure, heterodyne converters can also include several IF stages for the mixing procedure. The heterodyne conversion principle offers a high sensitivity, frequency, stability, and selectivity at the receiver. Another advantage is the tunable RF position of the signal by changing the local oscillator frequency.

In contrast to the heterodyne conversion, the homodyne conversion (known as direct conversion or zero IF) places the RF signal to zero. In comparison to the heterodyne principle, the conversion to the IF band is omitted here. The homodyne conversion is a simple design at a first glance but has a disturbing effect, which demands an extreme performance on the LO and the mixer. Instabilities or imperfections in the LO chain and any DC offsets or imperfections in the mixer corrupting the zero-IF output result in poor system performance. The zero-IF principle has issues of achieving sufficient dynamic range, which is an important factor for THz channel sounder for the received signals. Currently, we can find the homodyne conversion in many devices like cellphones, software-defined radio, and so on. The low-IF principle converts the RF signal down to a frequency close to zero. The distance between zero and the baseband signal depends on the BW of the useful signal and the target suppression of undesired mixing products and is typically of a few hundred MHz. The low-IF conversion combines the advantage of the heterodyne and homodyne architectures and avoids the DC offset and 1/f noise problems of the homodyne conversion.

Mostly, a mixed architecture of heterodyne and low-IF conversion is used in the design of measurement and communication systems in the THz range with BWs greater than 10 GHz. However, the requirements for a channel sounder are different from those for a communication system in terms of dynamics and BW. The measuring system should always have at least the maximum BW of the target communication system. To measure all propagation effects of the THz channel, the measurement sensitivity should be at least in the dynamic range of the target modulation. For QAM 1024 signals, the minimum required instantaneous dynamic range should be greater than 30 dB [33] within a committed information rate (CIR). In the THz frequency range and under the use of BWs larger than 20 GHz, different noise processes (shot noise, phase noise, thermal noise) become increasingly noticeable. In order to minimize the influence of these noise processes, the maximum possible transmission power that is allowed and technically available should be used. Furthermore, the implementation of high-gain LNAs and AGCs can further improve the dynamic measurement capabilities of THz CS systems. An example of a mixed frequency conversion setup of a THz channel sounder is displayed in Fig. 5.5.

5.8 Correction of RF-Hardware Influences of THz Channel Sounder

After the discussion of the hardware design of channel sounders, it is necessary to correct all remaining hardware influences which cannot be compensated by the RF design. The calibration of CSs is crucial to avoid the influence of the hardware, such as the antenna characteristics, attenuators, AGC, noise level, etc., on the interpretation of the measured data. Nonlinear distortions in the frequency response can be easily removed using a back-to-back [34] or in situ calibration [35]. During the back-to-back calibration, the frequency response is measured between

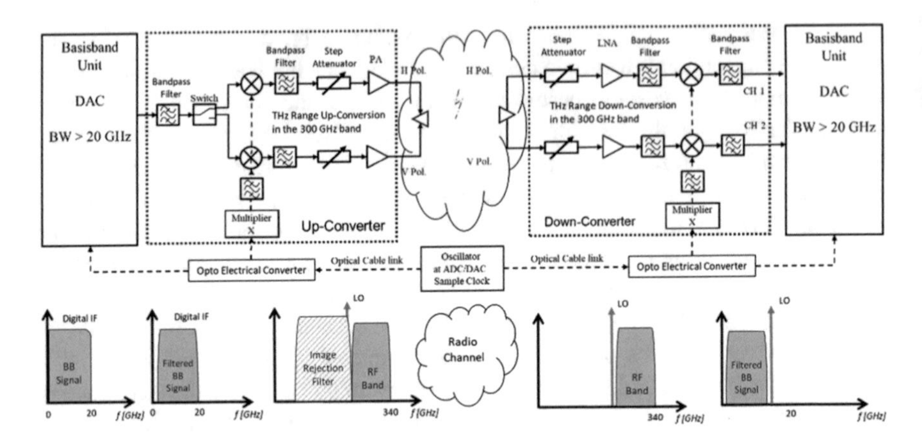

Fig. 5.5 A possible ideal setup for THz channel sounder which tries to minimize nonlinear broadband effects at the RF components

the transmitter and receiver, using calibrated attenuators which cover the dynamic range of the channel sounder system. For the in situ calibration, the frequency response of the system is measured over the free space. Ideally, this reference measurement has to be performed in an ideal environment as an anechoic chamber or in an open field with as less reflections as possible. The correction of the measurement data takes place afterward in post-processing to reduce or eliminate internal reflection of the system and other linear distortions. In addition, a large number of nonlinear components that are used in frequency-converting systems introduce nonlinear distortions. The calibration of such a distortion in broadband RF-systems requires knowledge of the whole system and a model which handles the nonlinear distortions in dependency of BW and input power at the receiver. Some approaches for the correction of nonlinear distortions can be found in [36, 37].

5.9 Measurement Setup and Procedure at THz

Up to this point, only the RF measurement system has been described. But in comparison to a standard VNA, a channel sounder needs to characterize the spatial parameters of the wave propagation. The spatial parameters consist of the angle of departure (AoD) at TX and RX, the angle of arrival (AoA) at TX and RX, the transmitted and received polarizations, the frequency, and the time variation of the channel. A complete description of the polarization consists of two orthogonal components at TX and RX. For the measurements, a positioner can be used to sample the spatial domain [9, 17, 38]. Polarization can be measured using dual-polarized antennas or linear polarized antennas which need to be rotated mechanically by 90°in the E-field plane.

Alternatively, the directional information can also be recorded with antenna arrays and under using high-resolution parameter estimator (HRPE) algorithms such as RiMAX [39], SAGE, or MUSIC [40] in combination. In the lower GHz bands, it is possible to resolve the directional information with a resolution well below the half-power beamwidth of the opening angle of the antenna elements. These channel sounding techniques using different 3D and 2D antenna arrays [39] are structured for sampling of the spatial information. An important design point of these channel sounding measurement antennas is the element spacing of approximately $\lambda\backslash 2$ to resolve the phase difference at the highest measured frequency. However, in the THz bands, only planar arrays with an antenna element spacing of approximately $\lambda\backslash 2$ of the highest measured frequency can be used with HRPE. Currently, the physical design, manufacturing, and feed of the RF signal to antenna elements represent a challenge for the design of 3D channel sounding antennas at THz. More detailed information about the boundary conditions and necessary calibration steps to apply HRPE can be found in the abovementioned publications.

5.10 Conclusion

The research field of broadband THz CSs is challenging. Issues such as real-time sampling, enormous instantaneous BWs, dynamic range (measurable path loss), and direction- and polarization-resolved acquisition of the THz channels still require some technological developments. Nevertheless, the THz channel is an important medium for future communication systems, not only for inter-device communication over a short distance or backhaul for 6G but also for secondary applications such as RADAR, sensing of vital data, and so on. Due to the enormous frequency resources which are available in the THz range, the full potential of this technology cannot yet be foreseen. Channel measurements reflecting the propagation characteristics at THz are an indispensable requirement to provide a platform for the development of future and applications.

Finally, even the large manufacturers of communication technologies must be convinced of the advantages and possibilities of THz. Here, the comparison of several frequency bands in the same environment can help to show the potential of THz technologies. Furthermore, only with an accepted channel model, communication systems can be standardized at 3GPP, ITU, and IEEE and be successful in the mass market.

References

1. T. S. Rappaport, G. R. MacCartney, M. K. Samimi, and S. Sun. Wideband millimeter-wave propagation measurements and channel models for future wireless communication system design, 2015.

2. Solomitckii, D., Semkin, V., Karttunen, A., Petrov, V., Nguyen, S. L. H., Nikopour, H., Haneda, K., Andreev, S., Talwar, S., & Koucheryavy, Y. (2020). Characterizing radio wave propagation in urban street canyon with vehicular blockage at 28Â GHz. *IEEE Transactions on Vehicular Technology, 69*(2), 1227–1236.
3. Müller, R., Häfner, S., Dupleich, D., Thomä, R. S., Steinböck, G., Luo, J., Schulz, E., Lu, X., & Wang, G. (2016). Simultaneous multi-band channel sounding at mm-wave frequencies. In *2016 10th European Conference on antennas and propagation (EuCAP)* (pp. 1–5, April 2016).
4. Salous, S., Degli Esposti, V., Fuschini, F., Thomae, R. S., Mueller, R., Dupleich, D., Haneda, K., Molina Garcia-Pardo, J., Pascual Garcia, J., Gaillot, D. P., Hur, S., & Nekovee, M. (2016). Millimeter-wave propagation: Characterization and modeling toward fifth-generation systems. [wireless corner]. *IEEE Antennas and Propagation Magazine, 58*(6), 115–127.
5. Papazian, P. B., Remley, K. A., Gentile, C., & Golmie, N. (2015). Radio channel sounders for modeling mobile communications at 28 GHz, 60 GHz and 83 GHz. In *Global Symposium on Millimeter-Waves (GSMM)* (pp. 1–3, May 2015).
6. Cheng, C., & Zajic, A. (2020). Characterization of propagation phenomena relevant for 300 GHz wireless data center links. *IEEE Transactions on Antennas and Propagation, 68*(2), 1074–1087.
7. Chowdhury, M. Z., Shahjalal, M., Ahmed, S., & Jang, Y. M. (2020). 6g wireless communication systems: Applications, requirements, technologies, challenges, and research directions. *IEEE Open Journal of the Communications Society, 1*, 957–975.
8. Sen, P., Pados, D. A., Batalama, S. N., Einarsson, E., Bird, J. P., & Jornet, J. M. (2020). The teranova platform: An integrated testbed for ultra-broadband wireless communications at true terahertz frequencies. *Computer Networks, 179*, 107370.
9. Dupleich, D., Müller, R., Skoblikov, S., Landmann, M., Galdo, G. D., & Thomä, R. (2020). Characterization of the propagation channel in conference room scenario at 190 GHz. In *2020 14th European Conference on antennas and propagation (EuCAP)* (pp. 1–5, March 2020).
10. Claudio Paoloni, Angeliki Alexiou, Oliver Bouchet, Alan Davy, Vladimir Ermolov, Thomas Kürner, Bruce Napier, Onur Sahin, and European Conference on Networks and Communications (EuCNC), Valencia, 18. - 21. Juni 2019. Ict beyond 5g cluster: Seven h2020 for future 5g. 2019.
11. Yang, K., Abbasi, Q. H., Qaraqe, K., Alomainy, A., & Hao, Y. (2014). Body-centric nanonetworks: Em channel characterisation in water at the terahertz band. In *2014 Asia-Pacific microwave Conference* (pp. 531–533).
12. You, X., Wang, C.-X., Huang, J., Gao, X., Zhang, Z., Wang, M., Huang, Y., Zhang, C., Jiang, Y., Wang, J., Zhu, M., Sheng, B., Wang, D., Pan, Z., Zhu, P., Yang, Y., Liu, Z., Zhang, P., Tao, X., Li, S., Chen, Z., Ma, X., Chih-Lin, I., Han, S., Li, K., Pan, C., Zheng, Z., Hanzo, L., Shen, X. (S.), Guo, Y. J., Ding, Z., Haas, H., Tong, W., Zhu, P., Yang, G., Wang, J., Larsson, E. G., Ngo, H. Q., Hong, W., Wang, H., Hou, D., Chen, J., Chen, Z., Hao, Z., Li, G. Y., Tafazolli, R., Gao, Y., Poor, H. V., Fettweis, G. P., & Liang, Y.-C. (2020). Towards 6g wireless communication networks: Vision, enabling technologies, and new paradigm shifts. *Science China Information Sciences, 64*(1), 110301.
13. Corre, Y., Gougeon, G., Dore, J.-B., Bicais, S., Miscopein, B., Faussurier, E., Saad, M., Palicot, J., & Bader, F. (2019). *Sub-THz spectrum as enabler for 6g wireless communications up to 1 tbit/s*. 03.
14. Giordani, M., Polese, M., Mezzavilla, M., Rangan, S., & Zorzi, M. (2020). Toward 6g networks: Use cases and technologies. *IEEE Communications Magazine, 58*(3), 55–61.
15. Manolakis, D. G., & Ingle, V. K. (2011). *Applied digital signal processing: Theory and practice*. Cambridge University Press.
16. Kleine-Ostmann, T., Jastrow, C., Priebe, S., Jacob, M., Kürner, T., & Schrader, T. (2012). Measurement of channel and propagation properties at 300 GHz. In *2012 Conference on precision electromagnetic measurements* (pp. 258–259) July 2012.
17. Priebe, S., Jastrow, C., Jacob, M., Kleine-Ostmann, T., Schrader, T., & Kürner, T. (2011). Channel and propagation measurements at 300 GHz. *IEEE Transactions on Antennas and Propagation, 59*(5), 1688–1698.

18. Zantah, Y., Sheikh, F., Abbas, A. A., Alissa, M., & Kaiser, T. (2019). Channel measurements in lecture room environment at 300 GHz. In *2019 Second International Workshop on Mobile Terahertz Systems (IWMTS)* (pp. 1–5) July 2019.
19. Guan, K., Peng, B., He, D., Yan, D., Ai, B., Zhong, Z., & Kürner, T. (2019). Channel sounding and ray tracing for train-to-train communications at the THz band. In *2019 13th European Conference on antennas and propagation (EuCAP)* (pp. 1–5) March 2019.
20. Park, J., Lee, J., Kim, K., & Kim, M. (2020). 28-GHz high-speed train measurements and propagation characteristics analysis. In *2020 14th European Conference on antennas and propagation (EuCAP)* (pp. 1–5) March 2020.
21. Schmieder, M., Keusgen, W., Peter, M., Wittig, S., Merkle, T., Wagner, S., Kuri, M., & Eichler, T. (2020). THz channel sounding: Design and validation of a high performance channel sounder at 300 GHz. In *2020 IEEE wireless Communications and networking Conference workshops (WCNCW)* (pp. 1–6) April 2020.
22. https://www.ni.com/en-us/innovations/white-papers/19/creating-sub-terahertz-testbed-with-ni-mmwave-transceiver-system.html.
23. https://www.spdevices.com/.
24. https://www.xilinx.com/products/silicon-devices/soc/rfsoc.html.
25. Salous, S., Lee, J., Kim, M. D., Sasaki, M., Yamada, W., Raimundo, X., & Cheema, A. A. (2020). Radio propagation measurements and modeling for standardization of the site general path loss model in international telecommunications union recommendations for 5g wireless networks. *Radio Science, 55*(1), e2019RS006924. e2019RS006924 2019RS006924.
26. Papazian, P. B., Gentile, C., Remley, K. A., Senic, J., & Golmie, N. (2016). A radio channel sounder for mobile millimeter-wave communications: System implementation and measurement assessment. *IEEE Transactions on Microwave Theory and Techniques, 64*(9), 2924–2932.
27. W. G. Newhall, T. S. Rappaport, and D. Sweeney. A spread spectrum.sliding correlator system for propagation measurements. 1996.
28. Müller, R., Herrmann, R., Dupleich, D. A., Schneider, C., & Thomä, R. S. (2014). Ultrawideband multichannel sounding for mm-wave. In *The 8th European Conference on Antennas and Propagation (EuCAP 2014)* (pp. 817–821) April 2014.
29. Carvalho, N. B., Remley, K. A., Schreurs, D., & Gard, K. G. (2008). Multisine signals for wireless system test and design [application notes]. *IEEE Microwave Magazine, 9*(3), 122–138.
30. Remley, K. A. (2003). Multisine excitation for acpr measurements. In *IEEE MTT-S International Microwave Symposium Digest, 2003* (Vol. 3, pp. 2141–2144).
31. MacCartney, G. R., & Rappaport, T. S. (2017). A flexible millimeter-wave channel sounder with absolute timing. *IEEE Journal on Selected Areas in Communications, 35*(6), 1402–1418.
32. Rey, S., Eckhardt, J. M., Peng, B., Guan, K., & Kürner, T. (2017). Channel sounding techniques for applications in THz communications: A first correlation based channel sounder for ultra-wideband dynamic channel measurements at 300 GHz. In *2017 9th International Congress on Ultra Modern Telecommunications and Control Systems and Workshops (ICUMT)* (pp. 449–453).
33. Yang, L., Lin, Z., & Zhongpei, Z. (2009). High order qam signals recognition based on layered modulation. In *2009 International Conference on Communications, Circuits and Systems* (pp. 73–76).
34. Landmann, M., Kaske, M., & Thomä, R. S. (2012). Impact of incomplete and inaccurate data models on high resolution parameter estimation in multidimensional channel sounding. *IEEE Transactions on Antennas and Propagation, 60*(2), 557–573.
35. Zajic, A. (2013). *Mobile-to-Mobile wireless channels*. Artech House.
36. Ntouni, G. D., Boulogeorgos, A. A., Karas, D. S., Tsiftsis, T. A., Foukalas, F., Kapinas, V. M., & Karagiannidis, G. K. (2014). Inter-band carrier aggregation in heterogeneous networks: Design and assessment. In *2014 11th International Symposium on Wireless Communications Systems (ISWCS)* (pp. 842–847).

37. Liu, Y., Li, C., Quan, X., Roblin, P., Rawat, M., Naraharisetti, N., Tang, Y., & Kang, K. (2019). Multiband linearization technique for broadband signal with multiple closely spaced bands. *IEEE Transactions on Microwave Theory and Techniques, 67*(3), 1115–1129.
38. Kleine-Ostmann, T., Jacob, M., Priebe, S., Dickhoff, R., Schrader, T., & Kürner, T. (2012). Diffraction measurements at 60 GHz and 300 GHz for modeling of future THz communication systems. In *2012 37th International Conference on Infrared, Millimeter, and Terahertz Waves* (pp. 1–2).
39. Reiner Thomä, Markus Landmann, and A. Richter. Rimax-a maximum likelihood framework for parameter estimation in multidimensional channel sounding. 2004.
40. Yin, X., Ouyang, L., & Wang, H. (2016). Performance comparison of sage and music for channel estimation in direction-scan measurements. *IEEE Access, 4*, 1–1.

Part II
Propagation and Channel Modelling 2: Basic Propagation Phenomena

Chapter 6
Free Space Loss and Atmospheric Effects

Tae-In Jeon, Janne Lehtomäki, Joonas Kokkoniemi, Harri Juttula, Anssi Mäkynen, and Markku Juntti

Abstract Free space path loss for THz communication is discussed. Molecular absorption mainly caused by water vapor is a major effect in terahertz propagation in the atmosphere. Both experimental and theoretical results are also discussed using the ITU models and the am model. A simplified model for molecular absorption that enables analytical study of THz communication is discussed without needing numerical calculations using huge number of coefficients. The time delay (shift) of the THz pulse is due to an increase in water vapor in the atmosphere, that is, an increase in refractive index. The time delay was measured and compared with theoretical calculation. Also, measured and calculated power attenuation losses for snowy and rainy weather conditions are presented.

6.1 Experimental Measurements

6.1.1 Power Attenuation in the Atmosphere (Free Space Loss)

When THz wave propagates in the atmosphere, the water vapor content of the atmosphere is the main barrier to THz propagation. Recently, using THz time-domain spectroscopy (TDS) system, there have been studies of broadband coherent THz pulse propagation in 186- and 910-m open paths under atmospheric weather conditions [1–3]. With a 30.5-cm-diameter telescope mirror, the THz beam can be collimated over a few 100 m distance in the atmosphere.

Figure 6.1a shows THz pulses propagated through 159-m outdoor free space with two different water vapor densities (WVD), which can be calculated using the

T.-I. Jeon
Korea Maritime and Ocean University, Yeongdo-gu, South Korea
e-mail: jeon@kmou.ac.kr

J. Lehtomäki · J. Kokkoniemi · H. Juttula · A. Mäkynen · M. Juntti (✉)
University of Oulu, Oulu, Finland
e-mail: markku.juntti@oulu.fi

T. Kürner et al. (eds.), *THz Communications*, Springer Series in Optical Sciences 234, https://doi.org/10.1007/978-3-030-73738-2_6

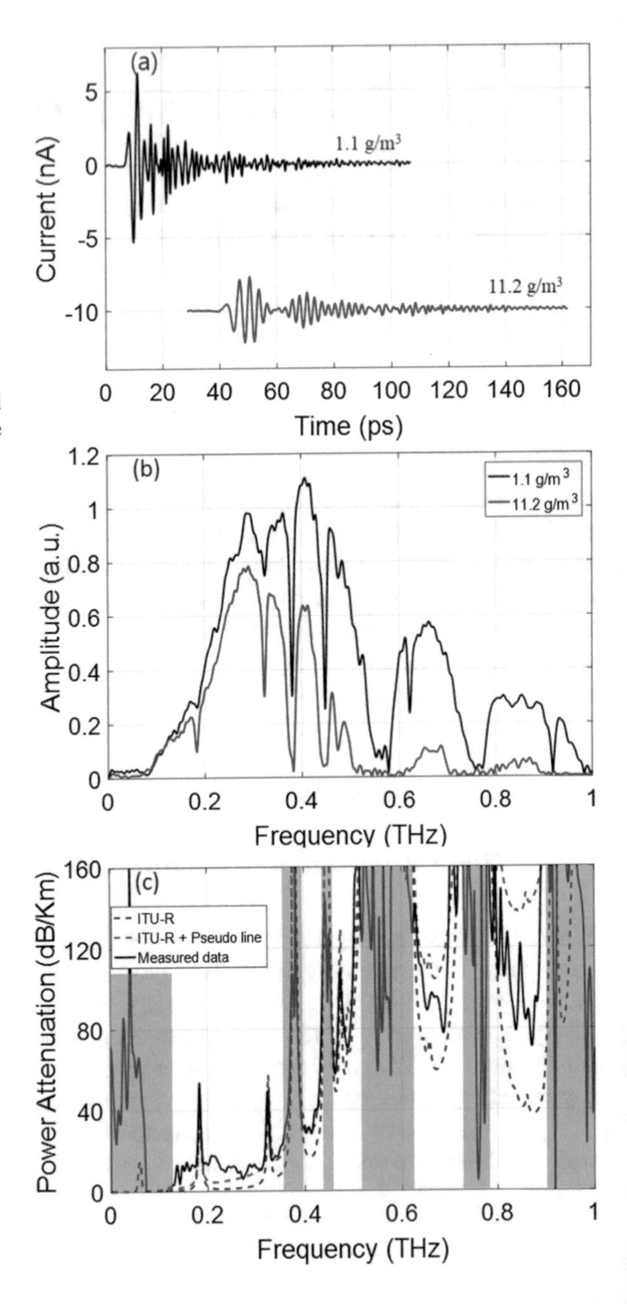

Fig. 6.1 (**a**) Measured THz pulses through 27-m indoor and 158-m outdoor free spaces for WVD values of 1.1 g/m^3 (RH and temperature of 12.5% and 9.6 °C, respectively) and 11.2 g/m^3 (RH and temperature of 73.9% and 17.6 °C, respectively). (**b**) Corresponding amplitude spectra for the measured THz pulses. (**c**) Measured power attenuation in dB/km. Dashed red and blue lines indicate the power attenuation calculated by ITU-R with and without compensation for continuum absorption at 1.78 THz [1]

relative humidity and temperature. THz pulse delay due to WVD with respect to the first maximum of the pulse is 37.4 ps. The pulses with WVD of 1.1 g/m^3 oscillate at much higher frequencies than those of pulses with WVD of 11.2 g/m^3, because THz pulses with the lower WVD have larger bandwidth. Because it has a larger power attenuation, the amplitude of the THz pulse with a WVD of 11.2 g/m^3 is smaller than that of the pulse with a WVD of 1.1 g/m^3. The corresponding amplitude spectra for the THz pulses with 1 THz spectrum bandwidth are shown in Fig. 6.1b. The amplitude spectrum of high WVD (E_H) is further reduced compared to that of low WVD (E_L). The power attenuation is given by $10\log(E_H/E_L)/0.159$ km. The measured power attenuation in dB/km is shown in Fig. 6.1c. The gray regions cover the no-signal areas caused by strong water vapor absorption lines. The strong water vapor attenuation values in the low-frequency regime due to the resonant rotational lines are 0.183 and 0.325 THz, which are much stronger than the continuum absorption and proportional to the square of the frequency. The dashed red and blue lines indicate the power attenuation calculated by ITU-R with and without compensation for continuum absorption at 1.78 THz (ITU-R P.676-12: Attenuation by atmospheric gases and related effects) [4]. The measured water resonant rotation lines are in a good agreement with the calculated power attenuations by both ITU-R calculations. The power attenuation determined by ITU-R + pseudo-line at 1.78 THz calculation is slightly higher than that determined by ITU-R calculation in the low-frequency range. The measured continuum absorption is about 5 dB/km higher than the calculated continuum absorption determined by ITU-R + pseudo-line at 1.78 THz around 0.2 THz, which is a 10% increase of the input amplitude spectrum for the WVD 1.1 g/m^3 reference pulse. Because outdoor weather cannot be controlled, the 2 THz pulses were measured more than a month apart. However, the measured power attenuation exists between ITU-R + pseudo-line at 1.78 THz calculation line and ITU-R calculation line in the high-frequency range. Because the measurements were made in low WVD in the atmosphere, the measured power attenuation is less than that found in the ITU-R + pseudo-line at 1.78 THz calculation. Since the power attenuation from 0.13 to 0.33 THz has very low loss, this frequency bandwidth will in the future be good for THz applications such as communications, sensing, and imaging.

6.1.2 *Time Delay in the Atmosphere*

When the concentration of WVD in the atmosphere is high, the refractivity of the atmosphere increases. Therefore, the time delay (shift) of the THz pulse appears as shown in Fig. 6.1a. The theoretical total time shift can be calculated from the refractivity of the atmosphere, as shown in the formula based on the following Essen and Froome equation [2, 3, 5]:

$$(n-1)10^6 = \frac{103.49}{T}p_1 + \frac{177.4}{T}p_2 + \frac{86.26}{T}\left(1+\frac{5748}{T}\right)p_3 \tag{6.1}$$

in which atmospheric pressure (p) $= p_1 + p_2 + p_3$ where p_1 is the partial pressure of dry air, p_2 is the partial pressure of CO_2, and p_3 is the partial pressure of water vapor. Although CO_2 has a high refractive index compared to other gases in the air,

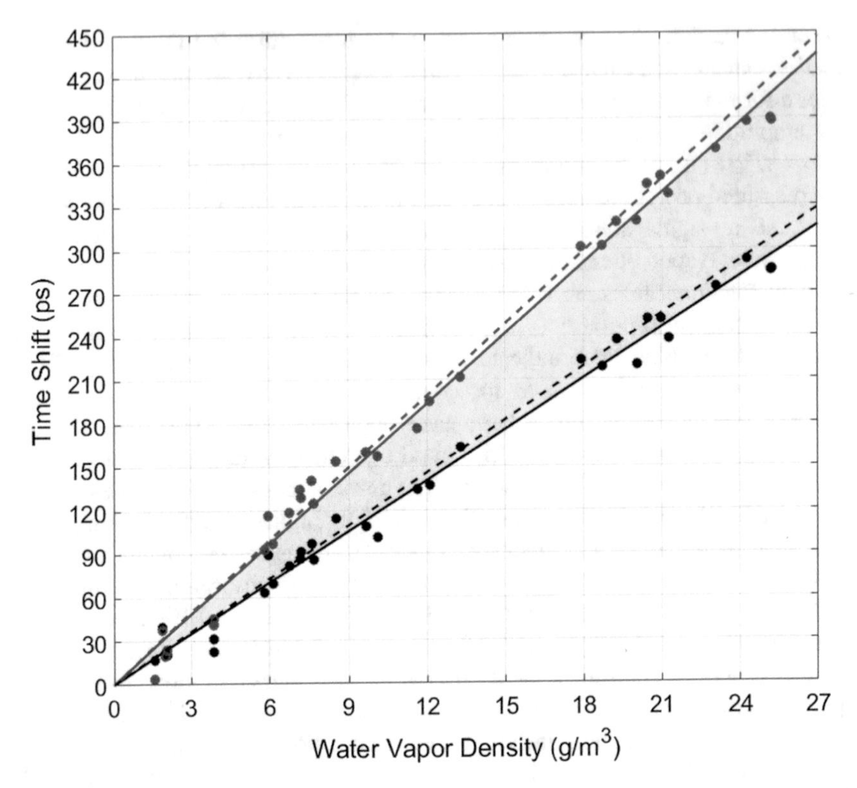

Fig. 6.2 Comparison of the measured and calculated time shifts as a function of WVD. Black dots indicate the measured time shifts caused by outdoor water vapor and dry air. Red dots indicate the measured time shifts caused by outdoor water vapor only. Solid and dashed lines indicate the fitting lines for measured data and theoretically calculated data, respectively. The yellow area of the figure indicates the time shift caused by the dry air [3]

we ignore the refractive index of CO_2 because of the low concentration (~0.03%) in the outdoor air. Therefore, the partial pressure of dry air is the atmospheric pressure minus the partial pressure of water vapor ($p_1 = p - p_3$). The partial pressure of dry air depends on the partial pressure of water vapor in the atmosphere.

Figure 6.2 provides a comparison of the measured and calculated time shifts as a function of WVD. We measured the time shift by measuring the THz pulse propagation at distances of 186 and 910 m [1, 3]. The black dots indicate the measured time shift derived by the phase difference for the lowest WVD and the highest WVDs at 0.25 THz; the solid line is a fitting line for the measured data. The measured time shift is caused by outdoor water vapor and dry air. If the time shift takes into account only the water vapor in the atmosphere, the time shift caused by dry air between water molecules must be eliminated. The red dots and solid line indicate the measured time shift due to water vapor only and a fitting line, respectively. The dashed lines indicate the theoretical calculation performed using

Eq. (6.1). Measurement and theoretical calculation for water vapor agreed very well, with values of 16.1 and 16.5 ps/(g/m^3), respectively. The yellow area of the figure indicates the time shift caused by dry air in the atmosphere.

6.2 Theoretical Model

6.2.1 *Free Space Path Loss*

Formally, free space path loss assumes the dielectric constant and the permeability of the propagating medium are those of the vacuum [6]. Received power ratio (ratio between received power P_r and transmitted power P_t) including free space path loss (FSPL) and the effect of the generic transmit and receive antennas is

$$\frac{P_r}{P_t} = \frac{A_r A_t f^2}{d^2 c^2} \tag{6.2}$$

where f is frequency, d is the distance, and A_r and A_t are effective antenna apertures. FSPL assumes isotropic antennas for which the aperture decreases as a function of frequency. For isotropic antenna $A_{\text{isotropic}} = c^2 \big/ 4\pi f^2$, where c is the speed of light. By using $A_{\text{isotropic}}$ in (6.2), we get

$$\frac{P_r}{P_t} = \frac{c^2}{4\pi f^2} \frac{c^2}{4\pi f^2} \frac{f^2}{d^2 c^2} = \left(\frac{c}{4\pi f d}\right)^2.$$

The inverse of this is the FSPL. Let us decompose it into two parts:

$$FSPL = \left(\frac{4\pi f d}{c}\right)^2 = \left(4\pi d^2\right)\left(\frac{4\pi f^2}{c^2}\right) \tag{6.3}$$

The spreading loss in the first term is independent of the frequency. The loss due to increasing frequency comes from the effective aperture of the receiving antenna (not from the channel itself). In the decibel scale, the FSPL is

$$\begin{aligned}
\text{FSPL[dB]} &= 10\log_{10}\left(\left(\frac{4\pi d f}{c}\right)^2\right) \\
&= 20\log_{10}(d) + 20\log_{10}(f) + 20\log_{10}\left(\frac{4\pi}{c}\right) \\
&= 20\log_{10}(d) + 20\log_{10}(f) - 147.56,
\end{aligned}$$

where distance d is in meters and frequency f is in Hz. So far, we see that FSPL increases as a function of frequency. For example, for 10-m distance, the FSPL at 2.4 GHz is 60 dB, and at 300 GHz, it is already 102 dB. However, there are some special cases where this behavior is not true. In fact, when taking into account FSPL + antenna gains, the loss can decrease as a function of frequency. The reason for this is that for fixed effective antenna apertures A_r and A_t, the receiver power ratio (6.2) increases as a function of frequency due to the f^2 term. For example, for parabolic antennas (and conical horn antennas)

$$A = e_a \frac{\pi d_{\text{dia}}^2}{4}$$

where d_{dia} is the diameter of the reflector (or the diameter of a conical horn aperture) and e_a is aperture efficiency. This is not a function of frequency (except possible variation in e_a).

After taking into account the gain of the antennas, the path loss becomes

$$\text{FSPL[dB]} - \text{G}_\text{t} - \text{G}_\text{r},$$

where G_t and G_r are the transmit and receive antenna gains in dB, respectively. It should be noted that some authors include antenna gains in FSPL, so the above expression would be called FSPL, leading to much reduced FSPL for high antenna gains. The gain for parabolic antenna can be calculated using

$$G = \frac{4\pi A_{\text{eff}}}{\lambda^2} = e_a \left(\frac{\pi d_{\text{dia}} f}{c} \right)^2,$$

where A_{eff} is the effective antenna aperture and λ is the wavelength. For example, parabolic antennas with 20 cm diameter and 80% efficiency on both sides at 365 GHz lead to >110 dB gain compared with FSPL as defined in (6.3). Parabolic antennas are suitable, for example, for fixed point-to-point outdoor links.

Regarding distance, path loss increases by a factor of 6 dB if link distance is doubled (square-law behavior). This is totally a different behavior than the exponential losses which would be doubled in the dB domain.

The first Fresnel zone should be mostly free from obstructions to avoid issues with radio reception. Luckily, the Fresnel zone reduces as a function of frequency. For example, for a 100-m link at 2.4 GHz radius of the first Fresnel zone is 1.77 m, but at 300 GHz, it is only 0.16 m. This means that at higher frequencies, we need to worry less about the Fresnel zone being empty.

6.2.2 Molecular Absorption

Molecular absorption is caused by the energy of the electromagnetic (EM) wave exciting the molecules to higher energy states. At THz frequencies, water vapor is the main cause of the absorption losses.

The Beer–Lambert law tells that transmittance (the ratio between received and transmitted powers) is

$$\tau\left(v,r\right)=e^{-\sum_j k_a^j(v)r}=e^{-k_a^{\mathrm{TOT}}(v)r}$$

where v is the wavenumber in cm (wavenumber in cm is f/c, where f is frequency in Hz and c is the speed of light in cm/s), $k_a^j(v)$ is the absorption coefficient of jth gas at wavenumber v, and r is the distance from transmitter to receiver in cm. The absorption coefficients depend on pressure, temperature, and molecular composition of the channel $k_a^j\left(v\right)=N_j\sigma_j\left(v\right)$, where N_j is the number density and $\sigma_j(v)$ is the absorption cross section for jth gas. The absorption cross section describes the effective area for absorption of an element, and the number density tells the number of elements. Line intensity describes the strength of the spectral lines, and spectral line shape tells us the width and shape of the spectral lines. By using the line intensity and line shape, we get absorption cross section with $\sigma_j\left(v\right)=S^j\left(T\right)F^j\left(p,v,T\right)$, where S is line intensity and F is spectral shape. The Lorentz line shape is

$$F_L^i\left(v-v_c^i\right)=\frac{1}{\pi}\frac{a_L^i}{\left(v-v_c^i\right)^2+\left(\alpha_L^i\right)^2}$$

where α_L^i is line width. The exact calculations and necessary scaling of the line intensity and the line shape can be done following [7, 8].

Van Vleck-Weisskopf (VVW) line shape includes correction for negative resonances. Van Vleck-Huber (VVH) is modification of VVW. Both VVW and unmodified VVH are applicable at millimeter wave and lower frequencies [7]. As discussed in [7], a line wing cutoff at 750 GHz can be introduced leading to VVH_750 line shape.

Molecular absorption can make channel frequency selective similar to multipath channel without actual multipath [9]. Molecular absorption does not affect very much short-distance links, but due to its exponential nature, it starts to affect, for example, outdoor backhaul links. For easy calculations over the whole THz band, ready-made programs/calculators can be used. For example, am: https://www.cfa.harvard.edu/~spaine/am/ (open source, please note that installing in Linux can be easier than in Windows).

example.amc: configuration file for am

```
f 0 GHz 10000 GHz 50 MHz # frequency grid for computation
layer
P 1013.25 mbar # pressure
T 20 C # temperature in Celsius
h 100 m # link distance
column dry_air vmr
column h2o RH 50% # relative humidity
lineshape VVH_750 strict_self_broadening h2o_lines
```

We can process this configuration file by running in command line: am example.amc > am.out. The resulting file am.out can be processed in MATLAB (such as by adding FSPL). The first column in the file is frequency and the second column is transmittance. Figure 6.3 shows the calculated molecular loss (without FSPL) and compares results with the MATLAB's implementation (function gaspl) of the ITU P.676-10 model [10]. The FSPL is also shown separately. Please note that ITU model is only specified up to 1000 GHz. With am, loss can be calculated for the whole THz band (100 GHz–10 THz). We can see that molecular absorption has the trend to increase with frequency but that even at high frequencies there are areas with less loss (called "windows") where communication could potentially occur.

Molecular absorption loss depends of course on the relative humidity but also on temperature. Loss is very different at the same relative humidity but at different temperatures such as 0° and 40 °C.

It is important to note that molecular absorption loss is additive in the dB scale. For example, loss at 1 km could be obtained from loss at 100 m by multiplying the loss (in dB) with 10. Therefore, the molecular loss in Fig. 6.3 could be called dB per 100 m. However, free space path loss cannot be calculated this way, because it cannot be specified as dB per 100 m. Instead, when going from 100 m to 1 km, the FSPL will increase by $20\log_{10}(10) = 20\,\text{dB}$. Sooner or later, the exponential losses such as molecular absorption loss will dominate. For example, at 100 m at 325 GHz, the molecular absorption with 50% relative humidity is (by ITU model) 4.4 dB, and free space path loss is 122.7 dB. We can calculate from these that the FSPL and molecular absorption loss will be equivalent at distance of 3484 m. Of course, the effect of the molecular absorption loss is significant already when it is less than FSPL. For example, we will reach 10 dB molecular absorption loss at distance $100 \times 10/4.4 = 227\,\text{m}$.

To get impulse response, we need also phase response in addition to amplitude response. Commonly assumed linear phase response leads to noncausal impulse response. Method for causal phase response has been presented in [11].

Researchers/engineers typically would like to have a closed form expression for the channel model for analytical studies. The existing models either use huge numbers of tabulated coefficients or are too complex to be used in practice. A

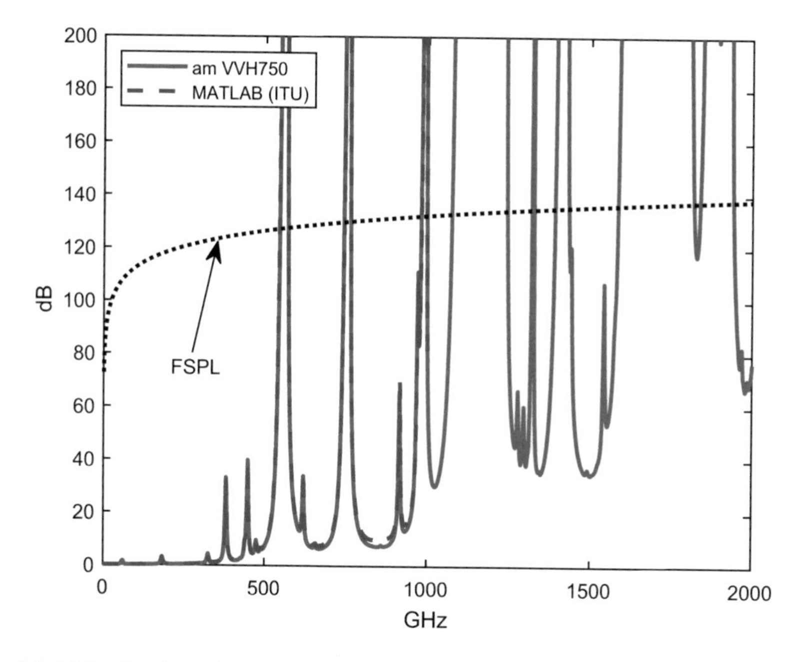

Fig. 6.3 Molecular absorption loss, 50% relative humidity, 100-m link distance

simplified channel model enables analytical study of THz communication without needing to resort to numerical calculations using huge number of coefficients.

A polynomial absorption loss model has been obtained by searching the strongest absorption lines and extracting their parameters. Since the absorption on the frequencies above 100 GHz is mainly caused by the water vapor, the volume mixing ratio of water vapor is left floating. The Beer–Lambert model becomes [12, 13]

$$\mathrm{PL}_{\mathrm{abs}}(f,\mu)=e^{d\left(\sum_{i} y_i(f,\mu)+g(f,\mu)\right)},\tag{6.4}$$

where f is the desired frequency, y_i is an absorption coefficient for the ith absorption line, $g(f,\mu)$ is a polynomial to fit the expression to the actual response, and μ is the volume mixing ratio of water vapor. Listing of the polynomials y_i and g for model for 100–450 GHz frequency band can be found in [13]. There are six polynomials y_i corresponding to lines at frequencies at 119, 183, 325, 380, 439, and 448 GHz. For example,

$$y_3(f,\mu)=\frac{E(\mu)}{F(\mu)+\left(\frac{f}{100c}-p_3\right)^2},\tag{6.5}$$

where $E(\mu)$ and $F(\mu)$ are second-order polynomials in terms of μ given in [13] and p_3 is constant given in [13]. Since the correction polynomial g is being used to handle wing absorption, the model can be reduced by using only the polynomials y_i in/close to range of interest. For example, to cover frequency range 275–325 GHz, we can use $y_3(f, \mu) + g(f, \mu)$ in (6.4) instead of including all six polynomials. By including polynomials $y_3(f, \mu)$ and $y_4(f, \mu)$, we can already cover 200–390 GHz frequency range.

There is also ITU simplified model, but that model is only specified to be valid up to 350 GHz, and it uses more polynomials and different line shape than the model in [13].

6.2.3 Rain and Fog Loss

It is possible to reduce the effect of molecular absorption by operating between the absorption peaks. However, for outdoor fixed point-to-point links, rain and fog attenuations cannot be avoided, since they are present at all frequencies. Attenuation due to raindrops increases strongly with increasing frequency up to 100 GHz, after which the attenuation becomes a nearly constant function of frequency. ITU-R has proposed models for rain and fog losses up to 1 THz. ITU-R model represents an "average," but for worst case analysis, we may need additional margin, since the attenuation's dependency on drop size distribution increases as a function of frequency [14]. MATLAB has implemented the ITU-R P.838-3 raindrop attenuation and ITU-R P.840-6 fog and cloud droplet attenuation models in functions "rainpl" and "fogpl," respectively, as part of the phased array system toolbox. These attenuation models can be used to evaluate average losses, but if one wants to take into account local atmospheric conditions, attenuation can be calculated with the Mie scattering theory [15].

Physically water drops in rain and fog attenuate terahertz waves by absorbing and scattering the incident wave. Both of these interactions depend on the size of the drops and their amount in a unit volume of air. The impact of absorption and scattering by water drops leads to exponential decay, and it can be written with Beer–Lambert law as a function of distance z as:

$$I(z) = I_0 e^{-\mu_{\text{ext}} z}, \tag{6.6}$$

where μ_{ext} is the extinction coefficient of the water drops suspended in air. The extinction coefficient includes both absorption and scattering, and it can be solved with appropriate scattering theory of electromagnetic waves. Since typical diameters of fog droplets ranges from 1 to 10 μm and for raindrops from 0.1 to 10 mm, wavelength of terahertz waves can be in the same order of magnitude as the drops, and the Mie scattering should be used. In case of drops that are much smaller than incident wave, scattering can be treated with Rayleigh scattering. However, it should be noted that the Mie scattering is more general case and it provides same results as Rayleigh scattering in case of small drops but the opposite is not true for large drops. Therefore, while mathematically more complex than Rayleigh scattering, the Mie theory is a safer option of the two.

The Mie scattering provides an analytical solution for scattering and absorption characteristics of a perfectly spherical particle with known complex refractive index. The raindrops are nearly spherical up to 2 mm diameter, after which they become slightly flattened spheroids as their size increases. If we assume the drops to be spherical, the Mie scattering theory gives a solution for extinction, scattering, and absorption cross sections for water drops:

$$\sigma_{\text{ext}} = \frac{2\pi}{k^2}\sum_{n=1}^{\infty}(2n+1)\Re(a_n+b_n) \tag{6.7}$$

$$\sigma_{\text{sca}} = \frac{2\pi}{k^2}\sum_{n=1}^{\infty}(2n+1)(|a_n|^2+|b_n|^2) \tag{6.8}$$

$$\sigma_{abs} = \sigma_{\text{ext}} - \sigma_{\text{sca}} \tag{6.9}$$

Here, k is the wavenumber and a_n and b_n are scattering coefficients related to the scattered electric field of electromagnetic wave. Coefficients a_n and b_n are functions of complex refractive index and dimensionless size parameter, and they require solving the Riccati-Bessel functions. Detailed formulations of a_n and b_n are outside the scope of this chapter, but luckily there are numerous implementations of the Mie theory in many common programming languages that can be used to solve the cross sections.

The extinction cross section (6.7) can be used to solve the extinction coefficient in (6.6) by multiplying the cross section by the number of water drops in a unit volume of air. Since rain and fog are actually composed of different numbers of drops with different sizes, we must integrate over all possible drop sizes. The extinction coefficient becomes now

$$\mu_{\text{ext}} = \int_0^{\infty} N(D)\sigma_{\text{ext}}(D)dD. \tag{6.10}$$

Here $N(D)$ is the drop size distribution, which represents the number of drops with diameter $D + dD$ in a unit volume of air.

The drop size distributions can be either measured or taken from the literature. Many drop size distributions for rain are typically expressed as functions of rain rate R in mm/h. Two common distributions for raindrops are the Weibull distribution [16] and the Marshall–Palmer distribution [17]. The Weibull distribution can be written as:

$$N(D) = N_0\frac{c}{b}\left(\frac{D}{b}\right)^{c-1} e^{-(D/b)^c}, \tag{6.11}$$

where $N_0 = 1000\,\text{m}^{-3}$, $b = 0.26R^{0.44}$ mm, and $c = 0.95R^0.14$. The Marshall–Palmer distribution can be written as:

$$N(D) = N_0 e^{\Lambda D}, \tag{6.12}$$

where $N_0 = 8000\ \text{m}^{-3}\text{mm}^{-1}$ and $\Lambda = 4.1R^{-0.21}\ \text{mm}^{-1}$. Generally, the Marshall–Palmer distribution is valid only for drops larger than 2 mm, and it greatly overestimates the number of smaller drops. In GHz and THz regime, the scattering efficiency of small drops may be significant, and hence, the Marshall–Palmer distribution is not recommended, while it may be a valid approximation in radio wave applications in MHz frequencies.

In Fig. 6.4, the rain attenuation is shown for three different rain models together with the attenuation of the air. ITU-R model is calculated with MATLAB's rainpl function. The attenuations labeled Weibull and Marshall–Palmer are calculated by combining Eq. (6.6)–(6.10) with corresponding drop size distributions (6.11) and (6.12). For comparison, specific attenuation of air with 60% relative humidity at 15 °C temperature is included in Fig. 6.4.

Figure 6.5 shows the measured power attenuations for snowy and rainy weather conditions. Reference THz pulses were measured for cloudy weather. Because the differences of the WVDs between the reference and sample pulses are very small for both snowy and rainy weather, power attenuations are also small compared to

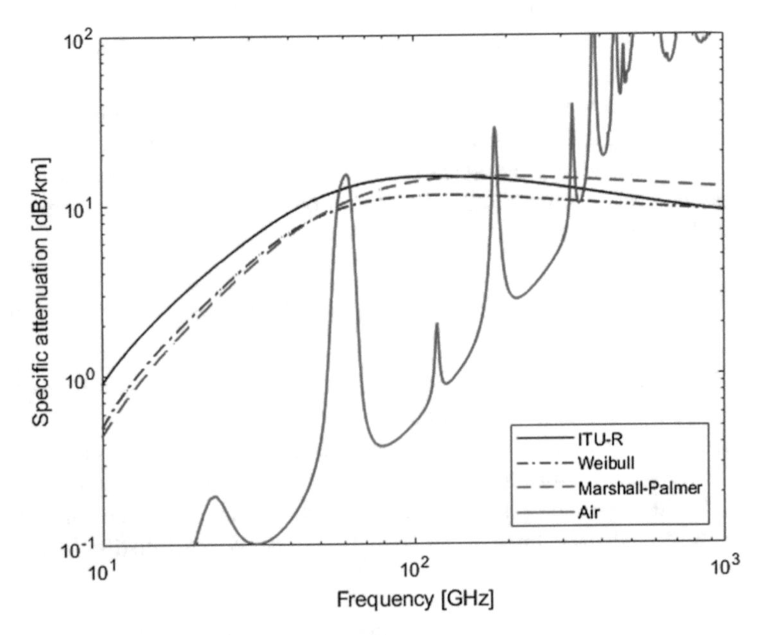

Fig. 6.4 Specific attenuation of three different rain models with 20 mm/h rainrate. For comparison, attenuation spectrum of air at 15 °C with 7.5 g/m^3 of water vapor density is shown

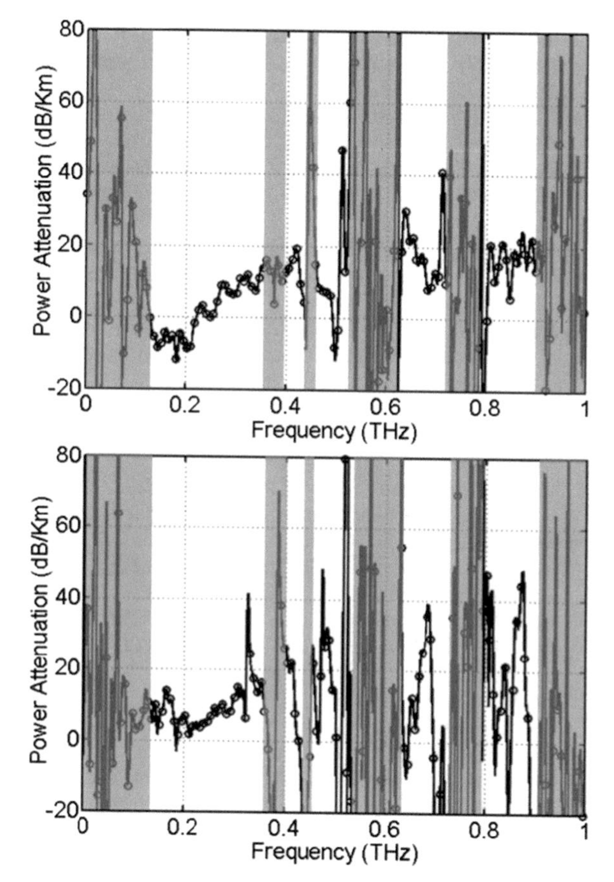

Fig. 6.5 The WVD differences for snow and rain are 0.1 and 0.6, respectively. The amount of snowfall and rainfall during the measurement were approximately 2 cm/h (which corresponds to 2 mm/h rainfall) and 3.5 mm/h, respectively [1]

the data shown in Fig. 6.1c. Therefore, power attenuations in Fig. 6.5 are caused by snow and rain in the THz beam path. The measured power attenuations for snowy and rainy weather were around 20 dB/km. If power attenuations caused by snow and rain are added to the power attenuation caused by WVD in the atmosphere, the total power attenuation in the low-frequency range is still small. These measurements demonstrate the potential of line-of-sight THz communications, sensing, and imaging through snow and rain.

References

1. Moon, E. B., Jeon, T.-I., & Grischkowsky, D. (2015). Long-path THz-TDS atmospheric measurements between buildings. *IEEE Transactions on Terahertz Science and Technology, 5*(5), 742–750.
2. Kim, G. R., Jeon, T. I., & Grischkowsky, D. (2017). 910-m propagation of THz ps pulses through the atmosphere. *Optics Express, 25*(21), 25422–25434.

3. Kim, G. R., Moon, K., Park, K. H., O'Hara, J. F., Grischkowsky, D., & Jeon, T. I. (2019). Remote N_2O gas sensing by enhanced 910-m propagation of THz pulses. *Optics Express, 27*(20), 27514–27522.
4. Radiocommunication Sector of International Telecommunication Union. (2019). Recommendation ITU-R P.676-12: Attenuation by atmospheric gases and related effect.
5. Essen, L., & Froome, K. D. (1951). The refractive indices and dielectric constants of air and its principle constituents at 24,000 Mc/s. *Proceedings of the Physical Society. Section B, 64*(10), 862–875.
6. McClaning, K. (2012). *Wireless receiver design for digital communications*. IET.
7. Paine, S. (2017). The am atmospheric model. Technical Report 152, Smithsonian Astrophysical Observatory.
8. Jornet, J., & Akyildiz, I. (2011). Channel modeling and capacity analysis for electromagnetic nanonetworks in the terahertz band. *IEEE Transactions on Wireless Communications, 10*(10), 3211–3221.
9. Wu, Z., Ebisawa, H., Umebayashi, K., Lehtomäki J., & Zorba, N. (2021). Time domain propagation characteristics with causal channel model for terahertz band, in *IEEE ICC 2021 Workshops*.
10. Radiocommunication Sector of International Telecommunication Union. (2013). Recommendation ITU-R P.676-10: Attenuation by atmospheric gases.
11. Tsujimura, K., Umebayashi, K., Kokkoniemi, J., Lehtomäki, J., & Suzuki, Y. (2018). A causal channel model for the Terahertz band. *IEEE Transactions on Terahertz Science and Technology, 8*(1), 52–62.
12. Kokkoniemi, J., Lehtomäki, J., & Juntti, M. (2019). Simple molecular absorption loss model for 200–450 gigahertz frequency band. In *2019 European Conference on Networks and Communications (EuCNC)* (pp. 219–223).
13. Kokkoniemi, J., Lehtomäki, J., & Juntti, M. (2021). A line-of-sight channel model for the 100–450 gigahertz frequency band. *EURASIP Journal on Wireless Communications and Networking 2021*, 88.
14. Juttula, H., Kokkoniemi, J., Lehtomäki, J., Mäkynen, A., & Juntti, M. (2019). Rain induced co-channel interference at 60 GHz and 300 GHz frequencies. In *2019 IEEE International Conference on Communications Workshops (ICC Workshops)*, Shanghai, China (pp. 1–5).
15. Bohren, C. F., & Huffman, D. R. (1983). *Absorption and scattering of light by small particles*. Wiley.
16. Sekine, M., & Lind, G. (1982). Rain attenuation of centimeter, millimeter and submillimeter radio waves. In: *12th European Microwave Conference*, Helsinki, Finland (pp. 584–589).
17. Marshall, J. S., & Palmer, W. M. K. (1948). The distribution of raindrops with size. *Journal of Meteorology, 5*(4), 165–166.

Chapter 7
Reflection, Scattering, and Transmission (Including Material Parameters)

Jianjun Ma, Rui Zhang, and Daniel Mittleman

Abstract Propagation modeling is fundamental in terahertz (THz) wireless communications due to the insurmountable computational burden of electromagnetic simulation methods when applied to THz propagation problems in the presence of obstacles, as well as the high cost of measurement campaigns. This has promoted the development of several propagation and channel models for various scenarios, which require the knowledge of the reflection, transmission, and scattering properties of the main obstacles present in the propagation environment. In this chapter, we will present a review on the characteristics of such fundamental interaction mechanisms that take place when obstacles are present, including the electromagnetic characteristics of materials at THz frequencies.

7.1 Introduction

Most of the research on THz channel modeling focuses on the line-of-sight (LOS) communications, which are obviously important due to very high path loss. That can best be countered by having an LOS path available. However, it has been proved, e.g., by us in [1, 2], that non-line-of-sight (NLOS) communications are possible at least in some cases, due to tolerable path losses via reflections and scattering.

Reflection, scattering, and transmission are fundamental and important propagation phenomena which should be characterized thoroughly for establishing possible THz wireless links. Related channel models are still a necessity due to the insurmountable numerical complexity of realistic problems, the high cost of measurement campaigns, and the formidable computational burden of electromagnetic simulation methods in the presence of dielectric or conductive multi-scale obstacles.

J. Ma (✉) · R. Zhang
School of Information and Electronics, Beijing Institute of Technology, Beijing, China
e-mail: jianjun_ma@bit.edu.cn

D. Mittleman
School of Engineering, Brown University, Providence, RI, USA

T. Kürner et al. (eds.), *THz Communications*, Springer Series in Optical Sciences 234, https://doi.org/10.1007/978-3-030-73738-2_7

This necessity has promoted the development of many propagation and channel models for various scenarios, including human blockage models, path loss models, intro-body models, and others [3]. All of these models require the knowledge of the reflection, scattering, and transmission properties of the main obstacles present in the propagation environment.

The purpose of this chapter is to review the characteristics of such fundamental interaction mechanisms that take place when obstacles are present. The electromagnetic characteristics (e.g., the complex permittivity) of these obstacle materials at THz frequencies are summarized also for the channel modeling with more or less careful derivations from Maxwell's equations, which are usually used as a starting point in the models for reflection, scattering, and transmission due to obstacles. In this chapter, we will not show how the equations are derived, but derivation can be found in the references.

7.2 Material Characterization

In order to obtain accurate channel models and predictions, material parameters, i.e., refractive index and absorption coefficient, are required. The complex refractive index n is derived from the complex dielectric permittivity ε of Maxwell's equations as $\varepsilon = n^2$. The real and imaginary parts of n are related to refractive index and absorption coefficient, respectively. Hence, the exact knowledge of these parameters in THz frequency range allows for reliable channel modeling in future THz wireless communication systems.

There have been several methods designed in efforts for material characterization, such as Fabry-Perót resonance [4, 5], Mach-Zehnder spectrometer [6], waveguide reflectometer [7], oversize cavity resonator [8], rotation of a parallel slab specimen with input and output devices [9], vector network analyzer (VNA) [10], and THz time-domain spectroscopy (THz-TDS) technique [11]. Among all of this methods, the THz-TDS and VNA (discussed in Chaps. 2 and 4, respectively) are the most commonly used for material studies in the millimeter and THz range, while others have their own particular applications such as other wavelength ranges, odd specimen sizes, and different physical properties such as liquids and gases [12].

Although many studies addressed material characterization at THz frequencies [13–15], a few studies have dealt with the characterization of common materials found in nature or man-made building, objects, and biological tissues. For example, a list of building materials, such as wood, brick, plastic, and glass, with measured parameters is presented in [11]. More measurements for common indoor materials (wallpaper, plaster, sanded birch board, concrete, and so on) and biological tissues (can be regarded as different layer with different dielectric properties) are presented in [10, 16–23].

7.3 Specular and Nonspecular Reflection

Additional to the characterization of their basic electromagnetic properties, obstacles can be characterized in terms of reflection and scattering characteristics. When specular reflection dominates, which is the case for very smooth surfaces (when the surface roughness deviations are much smaller than the wavelength) (Fig. 7.1(a)), classical Fresnel equations should be sufficient in most cases to describe reflectivity properties in the case of TE- or TM-polarized waves, given the frequency-dependent dielectric properties [24, 25].

However, when surface roughness deviations are comparable to the wavelength (as can often occur for wavelengths in the millimeter and submillimeter range), the scattering losses in the specular direction should be considered as shown in Fig. 7.1(b). The Fresnel reflection coefficients should be modified by introducing a Rayleigh roughness factor [1, 20] as

$$r_i' = \rho_R \cdot r_i, \tag{7.1}$$

with i for TE or TM mode. The Rayleigh roughness factor $\rho_R = e^{-g/2}$ with $g = (4\pi \cdot \sigma_h \cdot \cos\theta_i/\lambda)^2$. Here, λ is the incident wavelength in free space, and σ_h is the root mean square value of the roughness (deviation from perfectly flat). A rougher surface has a larger σ_h and therefore exhibits a larger value of g.

When surface roughness is much more pronounced, nonspecular reflection, also called diffuse scattering (Fig. 7.1(c)), should also be considered. This is a much more challenging calculation, which can be approached by employing proper small perturbation method (SPM) [26], radar cross-section (RCS) models [27], and Kirchhoff approximation [28, 29] or using integral equation model (IEM) [2]. Such methods work well when surface roughness can be described as Gaussian with a correlation length much greater than the wavelength. This approximation is valid for many indoor construction materials [30]. In case of surfaces with sharp edges or abrupt material discontinuities, heuristic solutions can be used, such as the effective roughness model, already applied with success at lower-frequency bands [31]. Here, we briefly review each of these approaches.

The SPM was first introduced for sound waves on sinusoidal corrugated surfaces [32] and then extended to conducting and dielectric surfaces [33–35]. It formulates the scattering as a partial differential equation boundary value problem, and the

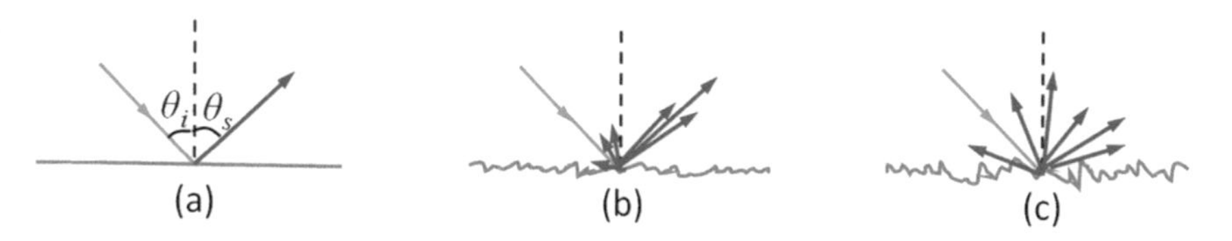

Fig. 7.1 Relative contributions of specular and diffuse scattering components for different surface roughness conditions: (**a**) specular, (**b**) slightly specular, (**c**) very rough

basic idea is to find a solution that matches the surface boundary conditions, where the tangential component of the field must be continuous across the boundary. The surface fields are expanded in a perturbation series with respect to surface height [26] as

$$\mathbf{E} = \mathbf{E}_0 + \mathbf{E}_1 + \mathbf{E}_2 + \cdots . \tag{7.2}$$

$\mathbf{E}_0$ is the surface field of a smooth and flat surface. In this approach, small effective surface currents on a mean surface are assumed to replace the role of small-scale roughness. So this method is only suitable for a surface with height standard deviation much less than the incident wavelength and an average surface slope comparable to or less than the surface standard deviation multiplied by the wave number. The small-scale roughness is expanded in a Fourier series, and the contribution to the field is therefore analyzed in terms of different wavelength components.

It has been argued that the SPM does account for multiple scattering up to the order of the perturbative expansion. This means that some multiple scattering effects can be obtained in the higher-order solutions [36]. Theoretical calculations have been compared to experimental measurements of aluminum plates with different surface roughness [37], where correlation length needed to be slightly enlarged for the relative lower frequency to get the best fit.

The Kirchhoff approach is commonly used to describe the scattering from rough surfaces. According to the geometry of the scattering problem in [2], the scattering field can be written in terms of the tangential surface fields in the medium above the separating surface [26] as

$$E_{qp}^{s} = K \int \left\{ \hat{\mathbf{q}} \cdot \left[\hat{\mathbf{k}}_s \times \left(\hat{\mathbf{q}} \times \mathbf{E}_p \right) + \eta \left(\hat{\mathbf{n}} \times \mathbf{H}_p \right) \right] \right\} \; e^{j(\mathbf{k}_s \cdot \mathbf{r})} ds, \tag{7.3}$$

with $K = -jke^{-jkr}/4\pi r$ and $\mathbf{k}_s$ as the wave number for scattered wave. So the Kirchhoff approach is also sometimes referred to as the tangent plane approximation.

However, this equation cannot in general be solved analytically, and therefore approximations have to be introduced as Kirchhoff approximation. For surface with large σ_h, the stationary-phase approximation has been used, and for surfaces with small σ_h, a scalar approximation should be considered first [26].

The radar cross-section model is a measure of the power density scattered in the direction of the receiver relative to the power density of the radio wave illuminating the scattering object [38]. It can be thought of as a combination of contributions from small-scale and large-scale roughness, and the surface scattering can be calculated by dividing the surface into small-scale and large-scale patches [39]. Thus,

$$\text{RCS } \sigma \;\; = \sigma_{rough} + \left| \chi_s^2 \right| \sigma_{smooth}, \tag{7.4}$$

with σ_{rough} and σ_{smooth} as small-scale and large-scale RCS σ, respectively [25]. The weighting factor χ_s is the rough surface height characteristic function. This factor is given as $\chi_s = \exp(-k_0{}^2 \cdot \sigma_h{}^2 \cos^2\theta_i)$ and approaches to 1 as frequency decreases, which implies that σ_{smooth} dominates the RCS σ. However, as frequency reaches to THz range, χ_s becomes negligible for rough surfaces, and the impact of σ_{rough} becomes much more significant.

The value of σ_{smooth} depends on the large-scale scattering object, where only the width of the scattering object is considered. It is found to be in an inverse relationship with the wavelength of the incident wave [38]. The σ_{rough} can be obtained by calculating the weighted average cross section of the individual, randomly orientated small patches. Then it should be modulated by the slope of small-scale patches, as described in [40].

The RCS model traces its origin to radar theory where it was originally designed to detect large, metallic objects such as aircrafts and ships in the far field [41]. It usually assumes that the scattering object is a perfect electrical conductor, which may not be applicable for many practical wireless network scenarios.

The integral equation method is a relatively new method for calculating scattering of electromagnetic waves from rough surfaces. It has been used extensively in the microwave region in recent years and proved to offer good predictions for a wide range of surface profiles [42]. The method can be viewed as an extension of the Kirchhoff method and the SPM since it has been shown to reproduce results of these two methods in appropriate limits.

The scattered field from the rough surface can be expressed as a combination of the Kirchhoff and the complementary term [43] as

$$E_{qp}^{s} = E_{qp}^{k} + E_{qp}^{c}, \tag{7.5}$$

which correlates to the Kirchhoff field coefficient and the complementary field coefficient. Expressions for both are reported in [44], and the diffuse scattering coefficient can be found in [43].

The IEM employs four correlation functions and multiple roughness scales to characterize the surface height profile, which represent different approaches on how to characterize the statistical correlation between two locations separated by a small distance. It can therefore provide greater flexibility in matching the model based on measured surface height profiles. A theoretical calculation is shown in Fig. 7.2 by employing an x-exponential correlation function. It's close to the measured data.

7.4 Transmission and Volume Effects

Sometimes, the theoretical models, which only consider surface scattering, could not fit the measured data even when the correlation length was expanded to the millimeter or THz range [42]. This can happen if the dielectric material is a mixture with different types and quantities of particles with sizes comparable to

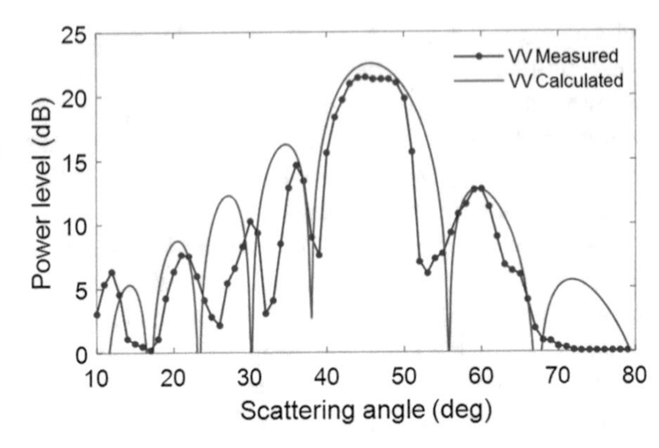

Fig. 7.2 Comparison between theoretical predictions for a metallic surface sample and measured data. (Adapted from reference [2])

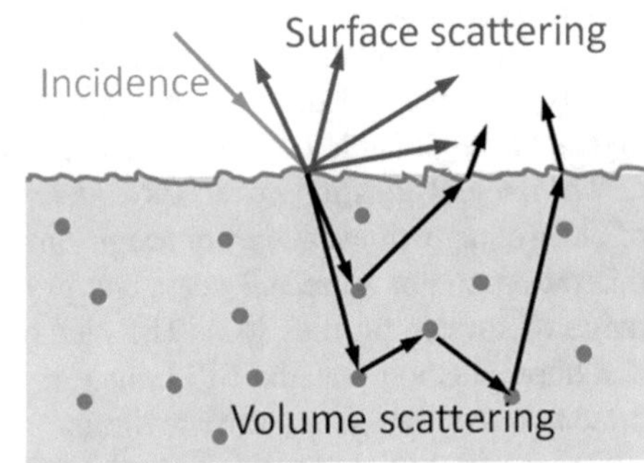

Fig. 7.3 Surface and volume scattering contributions by a dielectric sample

the wavelength. That implies that the effects of volumetric inclusions must be taken into account and volume scattering should be considered as shown in Fig. 7.3. It is caused by discrete particles present in a homogeneous background medium, which is different from surface scattering.

This phenomenon can be of particular relevance in the case of an intra-body scenario [22, 45]. Since the body is composed of cells, organelles, proteins, and molecules with various geometries and arrangements, the electromagnetic wave is scattered by the microscopic nonuniformities of the particles present in living tissue. So volume scattering inside medium should play the main role.

Two methods that can be used to describe volume scattering effects are an exact analytical theory and a vector radiative transfer (VRT) theory. The analytical approach, starting with Maxwell's equations, can explain all multiple scattering, diffraction, and interference effects, but it is mathematically complicated and computationally demanding [46]. In contrast, the VRT approach deals with the transport of energy through a medium containing particles and assumes that there is no correlation between the fields scattered by different particles. This assumption allows for the calculation of the incoherent scattering contributions, rather than the addition of their electric field.

Most media are not composed of spherical particles, which means that scattering by the particles is, in general, wave polarization-dependent. Also, the scattering medium is bounded by a lower surface and an upper surface. Reflection by and transmission through surface boundaries are also polarization-dependent. So the volumetric scattering contribution can be described by the VRT equation with both incident and radiation intensities fully described by Stokes parameters [47] as

$$\frac{d\mathbf{I}\left(\mathbf{R},\hat{\mathbf{s}}\right)}{ds}=-\boldsymbol{\kappa}_e\left(\hat{\mathbf{s}}\right)\cdot\mathbf{I}\left(\mathbf{R},\hat{\mathbf{s}}\right)-\kappa_{a,b}\mathbf{I}\left(\mathbf{R},\hat{\mathbf{s}}\right)+\iint_{4\pi}\boldsymbol{\Psi}\left(\hat{\mathbf{s}},\hat{\mathbf{s}}'\right)\cdot\mathbf{I}\left(\mathbf{R},\hat{\mathbf{s}}'\right)d\Omega', \tag{7.6}$$

where $\boldsymbol{\kappa}_e(\hat{\mathbf{s}})$ is the extinction matrix that defines the attenuation of the Stokes vector-specific intensity $\mathbf{I}(\mathbf{R},\ \hat{\mathbf{s}})$ with $\hat{\mathbf{s}}$ as the propagation direction and $\mathbf{R}$ as the distance vector of the location of the differential volume. $\boldsymbol{\Psi}$ $(\hat{\mathbf{s}},\ \hat{\mathbf{s}})$ is the phase matrix, and $\kappa_{a,b}$ is the background absorption coefficient.

Except for some special cases, the VRT equation does not have an analytical solution. The interactive approach is usually employed to cast the equation into an integral form and then solve it interactively to obtain the zero-, first-, and second-order solutions [26, 43]. In principle, an accurate solution can be obtained by iterating many times.

We presented a review of models for reflection, scattering, and transmission properties in the presence of obstacles in the propagation path with the characterization of related materials being summarized. Classical Fresnel equations are enough for specular reflection by a smooth surface, while a Rayleigh roughness factor should be introduced when the surface roughness deviations are comparable to the wavelength. If the deviations increase further and diffuse scattering dominates, the small perturbation method, the Kirchhoff approximation, the radar cross-section model, and the integral equation method are much used with the last one having a larger domain of validity. When the effects of volumetric inclusions in a dielectric layer cannot be not be neglected, the vector radiative transfer theory should be employed together with the surface scattering methods for theoretical predictions.

Acknowledgments We appreciate the support by Beijing Institute of Technology Research Fund Program for Young Scholars. DMM also acknowledges partial support from the US National Science Foundation and from the Air Force Research Laboratory.

References

1. Ma, J., Shrestha, R., Moeller, L., & Mittleman, D. M. (2018). Invited article: Channel performance for indoor and outdoor terahertz wireless links. *APL Photonics, 3*, 12.
2. Ma, J., Shrestha, R., Zhang, W., Moeller, L., & Mittleman, D. M. (2019). Terahertz wireless links using diffuse scattering from rough surfaces. *IEEE Transactions on Terahertz Science and Technology, 9*, 460.
3. Elayan, H., Amin, O., Shihada, B., Shubair, R. M., & Alouini, M. (2020). Terahertz band: The last piece of RF Spectrum puzzle for communication systems. *IEEE Open Journal of the Communications Society, 1*, 1–32.

4. Degli-Esposti, V., Zoli, M., Vitucci, E. M., Fuschini, F., Barbiroli, M., & Chen, J. (2017). A method for the electromagnetic characterization of construction materials based on Fabry-Pérot resonance. *IEEE Access, 5*, 24938–24943.
5. Cook, R. J., & Jones, R. G. (1976). A correction to open-resonator permittivity and loss measurements. *Electronics Letters, 12*, 1–2.
6. Birch, J. R. (1980). Free space video detection of harmonic content of 100 GHz IMPATI oscillator. *Electronics Letters, 16*, 799–800.
7. VanLoon, R., & Finsy, R. (1974). Measurement of complex permittivity of liquids at frequencies from 60 to 150 GHz. *Review of Scientific Instruments, 45*, 523–525.
8. Stumper, U. (1973). A TE01n cavity resonator method to determine the complex permittivity of low loss liquids at millimeter wavelengths. *Review of Scientific Instruments, 44*, 165–169.
9. Shimabukuro, F. I., Lazar, S., Chernick, M. R., & Dyson, H. B. (1984). A quasi-optical method for measuring the complex dielectric constant of materials. *IEEE Transactions on Microwave Theory and Techniques, 32*, 659–665.
10. Piesiewicz, R., Jacob, M., & Kürner, T. *Overview of challenges in channel and propagation characterization beyond 100 GHz for wireless communication systems*.
11. Piesiewicz, R., Jansen, C., Wietzke, S., Mittleman, D., Koch, M., & Kürner, T. (2007). Properties of building and plastic materials in the THz range. *International Journal of Infrared and Millimeter Waves, 28*, 363–371.
12. Kolbe, W. F., & Leskovar, B. (1982). Sensitivity and response time improvements in a millimeter-wave spectrometer. *Review of Scientific Instruments, 53*, 769–775.
13. Birch, J. R., & Clarke, R. N. (1982). Dielectric and optical measurements from 30 to 1000 GHz. *The Radio and Electronic Engineer, 52*, 565–584.
14. Afsar, M. N., & Button, K. J. (1985). Millimeter-wave dielectric measurement of materials. *Proceedings of the IEEE, 73*, 131–153.
15. Lamb, J. W. (1996). Miscellaneous data on materials for millimetre and submillimetre optics. *International Journal of Infrared and Millimeter Waves, 17*, 1997–2034.
16. Hiromoto, N., Fukasawa, R., & Hosako, I. (2006). *Measurement of optical properties of construction materials in the terahertz region*. Presented at the 2006 joint 31st international conference on infrared millimeter waves and 14th international conference on Teraherz electronics, 2006.
17. Piesiewicz, R., Jansen, C., Koch, M., & Kuerner, T. (2008). *Measurements and modeling of multiple reflections effect in building materials for indoor communication systems at THz frequencies*. Presented at the microwave conference (GeMIC), German, 2008.
18. Kokkoniemi, J., Lehtomäki, J., & Juntti, M. (2018). *Reflection coefficients for common indoor materials in the terahertz band*. Presented at the Proceedings of the 5th ACM International Conference on Nanoscale Computing and Communication, 2018.
19. Piesiewicz, R., Kleine-Ostmann, T., Krumbholz, N., Mittleman, D., Koch, M., & Kürner, T. (2005). Terahertz characterisation of building materials. *Electronics Letters, 41*, 1002.
20. Jansen, C., Piesiewicz, R., Mittleman, D., Kurner, T., & Koch, M. (2008). The impact of reflections from stratified building materials on the wave propagation in future indoor terahertz communication systems. *IEEE Transactions on Antennas and Propagation, 56*, 1413–1419.
21. Simonis, G. J. (1982). Index to the literature dealing with the near-millimeter wave properties of materials. *International Journal of Infrared and Millimeter Waves, 3*, 439–469.
22. Zhang, R., Yang, K., Yang, B., AbuAli, N. A., Hayajneh, M., Philpott, M., et al. (2019). Dielectric and double Debye parameters of artificial Normal skin and melanoma. *Journal of Infrared, Millimeter and Terahertz Waves, 40*, 657–672.
23. Vilagosh, Z., Lajevardipour, A., & Wood, A. W. (2018). *Modelling terahertz radiation absorption and reflection with computational phantoms of skin and associated appendages*. Presented at the Nanophotonics Australasia Society of Photo-Optical Instrumentation Engineers (SPIE) conference series, 2018.
24. Saunders, S. R. (1999). *Antennas and propagation for wireless communication system*. New York: Wiley.

25. Rappaport, T. S., Robert, H. W., Daniels, R. C., & Murdock, J. N. (2014). *Millimeter wave wireless communications*. Prentice Hall.
26. Ulaby, F. T., Moore, R. K., & Fung, A. K. (1982). *Microwave remote sensing: Active and passive*. Artech House: Norwood.
27. Ju, S., Shah, S. H. A., Javed, M. A., Li, J., Palteru, G., Robin, J., et al. (2019). *Scattering mechanisms and modeling for terahertz wireless communications*. Presented at the IEEE international communications conference (ICC), Shanghai, China, 2019.
28. Beckmann, P., & Spizzichino, A. (1987). *The scattering of electromagnetic waves from rough surfaces*. Norwood: Artech House.
29. Jansen, C., Priebe, S., Moller, C., Jacob, M., Dierke, H., Koch, M., et al. (2011). Diffuse scattering from rough surfaces in THz communication channels. *IEEE Transactions on Terahertz Science and Technology, 1*, 462–472.
30. Piesiewicz, R., Jansen, C., Mittleman, D., Kleine-Ostmann, T., Koch, M., & Kurner, T. (2007). Scattering analysis for the modeling of THz communication systems. *IEEE Transactions on Antennas and Propagation, 55*, 3002–3009.
31. Degli-Esposti, V., Fuschini, F., Vitucci, E. M., & Falciasecca, G. (2007). Measurement and modelling of scattering from buildings. *IEEE Transactions on Antennas and Propagation, 55*, 143–153.
32. Rayleigh, L. (1896). *The theory of sound*. London: Macmillan.
33. Rice, S. O. (1951). Reflection of electromagnetic waves from slightly rough surfaces. *Communications on Pure and Applied Mathematics, 4*, 351–378.
34. Rice, S. O. (1963). *Reflection of EM from slightly rough surfaces*. New York: Interscience.
35. Peake, W. H. (1959). Theory of radar return from terrain IRE. *National Convention Record 7, Pt 1*, 27–41.
36. Zhurbenko, V. (2011). *Electromagnetic waves*. Rijeka: InTech.
37. DiGiovanni, D. A., Gatesman, A. J., Goyette, T. M., & Giles, R. H. (2014). *Surface and volumetric backscattering between 100 GHz and 1.6 THz*. Presented at the Proc. SPIE, Passive and Active Millimeter-Wave Imaging XVII, 2014.
38. Balanis, C. (2012). *Advanced Eng*. Electromagnetics: Wiley.
39. Johnson, J. T., Shin, R. T., Kong, J. A., Tsang, L., & Pak, K. (1998). A numerical study of the composite surface model for ocean backscattering. *IEEE Transactions on Geoscience and Remote Sensing, 36*, 72–83.
40. Rees, J. V. (1987). Measurements of the wide-band radio channel characteristics for rural, residential, and suburban areas. *IEEE Transactions on Vehicular Technology, 36*, 2–6.
41. Skolnik, M. I. (1980). *Introduction to Radar systems*. New York: McGraw Hill Book Co..
42. Wei, J. C., Chen, H., Qin, X., & Cui, T. J. (2017). Surface and volumetric scattering by rough dielectric boundary at terahertz frequencies. *IEEE Transactions on Antennas and Propagation, 65*, 3154–3161.
43. Fung, A. K. (1994). *Microwave scattering and emission models and their applications*. Norwood: Artech House.
44. Brogioni, M., Macelloni, G., Paloscia, S., Pampaloni, P., Pettinato, S., Pierdicca, N., et al. (2010). Sensitivity of Bistatic scattering to soil moisture and surface roughness of bare soils. *International Journal of Remote Sensing, 31*, 4227–4255.
45. Gabriel, S., Lau, R., & Gabrie, C. (1996). The dielectric properties of biological tissues: III. Parametric models for the dielectric Spectrum of tissues. *Physics in Medicine and Biology, 41*, 2271–2293.
46. Ulaby, F. T., Haddock, T. F., Austin, R. T., & Kuga, Y. (1991). Millimeter-wave radar scattering from snow: 2. Comparison of theory with experimental observations. *Radio Science, 26*, 343–351.
47. Jin, Y.-Q. (1993). *Electromagnetic scattering modeling for quantitative remote sensing*. Singapore: World Scientific.

Chapter 8
Diffraction and Blockage

Thomas Kürner

Abstract In case the line-of-sight between a transmitter and a receiver is blocked, diffraction and alternative propagation paths exploiting scattering and reflections become relevant effects to be considered. In this chapter, first experimental and theoretical descriptions of diffraction phenomena at 300 GHz are described. This description is followed by models taking into account the effect of human blockage considering both the diffraction effect and the statics of the spatial distribution and movement of humans. Finally mitigation techniques to enable wireless communication in obstructed line-of-sight scenarios are briefly described.

8.1 Introduction

With the wavelength being on the order of 1 mm in THz communications, the influence of even small objects will become relevant. This becomes especially important, if the line-of-sight (LOS) path of the wireless link is blocked by an object, which brings up the question concerning the extent to which a communication link can be established via diffraction. Early publications mention diffraction at THz frequencies already around the millennium [1, 2]. First investigations in the context of wireless communications describe the effect of diffraction at 300 GHz both experimentally and theoretically [3–5]. From these investigations it becomes very clear that the observed path loss will exceed the limits to set-up a reliable link. Among the expected operational environments for THz communications (see Chap. 1), many scenarios exist, where blocking of the LOS path by humans or other large static objects like building walls occurs frequently. In order to model these scenarios properly, human blockage models for THz communications are required, which have been proposed already in literature [5, 6]. Apart from modelling the

T. Kürner (✉)
Institute for Communications Technology, Technische Universität Braunschweig, Braunschweig, Germany
e-mail: t.kuerner@tu-bs.de

T. Kürner et al. (eds.), *THz Communications*, Springer Series in Optical Sciences 234, https://doi.org/10.1007/978-3-030-73738-2_8

blocking itself, mitigation techniques enabling communication via reflection and scattering in obstructed line-of-sight (OLOS) have been proposed [7–9].

The remaining part of the chapter addresses the above-mentioned aspects in Sects. 8.2 (Diffraction), 8.3 (Human Blockage) and 8.4 (Communication in obstructed Line-of Sight).

8.2 Diffraction

Diffraction phenomena have been widely investigated at frequencies below 6 GHz. First diffraction measurements at 60 GHz have been reported in the framework of developing first 60 GHz wireless Local-area-network (WPAN) systems [10] and it turned out, that simple knife-edge models can be used to model the effect of people moving through rays [11, 12]. First systematic diffraction measurements at 60 GHz and 300 GHz along with the comparison with theoretical models have been reported in [3, 4] and will be briefly described in the following.

8.2.1 Diffraction Models

Knife-Edge Diffraction (KED) KED relies on perfect conductivity and provides an approximation formula for the edge diffraction at a semi-infinite half-plane. The diffracted electric field with reference to the electric field received from an isotropic radiator in the absence of the obstructing edge can be calculated using the Fresnel integral [13]. Although KED does not take into account either the effect of polarisation or the exact shape of a real diffracting object, reasonable accuracy is reported both below 6 GHz [14] and at 60 GHz [11].

Uniform Geometrical Theory of Diffraction (UTD) UTD provides heuristic solutions for diffraction on various canonical objects. In [3], a dielectric wedge as described in [15] and a conducting cylinder as described [16, 17] have been applied. In both cases, the diffracted field is calculated w.r.t. the incident field at the edge. UTD takes into account the effect of polarisation.

8.2.2 Diffraction Measurements

Diffraction measurements have been performed in [3, 4] at 60 GHz and 300 GHz at both horizontal and vertical polarisation using a Rohde & Schwarz ZVA50 vector network analyzer (VNA) in combination with external transmitting and receiving test heads using standard gain horn antennas with similar antenna patterns in both frequency bands. The measurement set-up is shown in Fig. 8.1a. This set-up allows

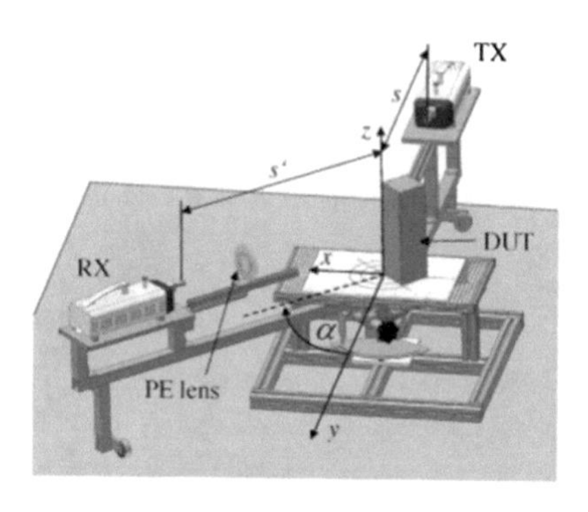

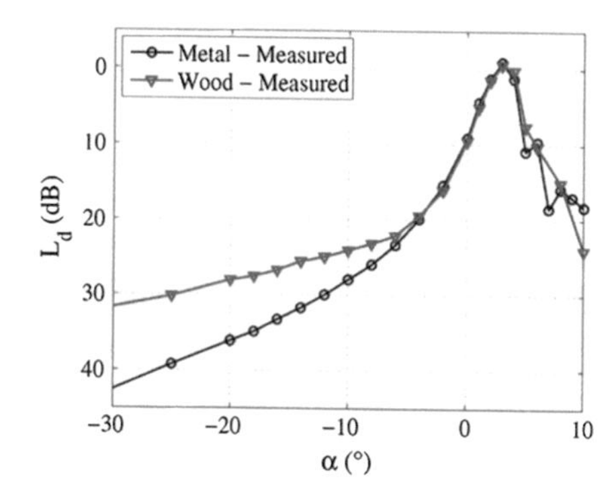

Fig. 8.1 (**a**) left: Measurement Set-up; (**b**) right: Comparison of measured diffraction loss of the wooden and metal wedge [3] (© 2012 IEEE, reproduced with permission)

both an angular-dependent measurement by changing angle α as well as a translation in directions x and y.

The goals of these measurements have been to (i) validate the theoretical diffraction models and (ii) compare the effects at 60 GHz and 300 GHz. Three types of measurements have been performed:

Angular-Dependent Diffraction: In this set-up, the transmitter (Tx) is kept fixed pointing always to the direction of the rotation axis. The receiver is rotated, with positive angles corresponding to the lit region and negative angles corresponding to the shadow region. In this measurement set-up, an additional polyethylene lens (PE) at the receiver has been used. Figure 8.1b shows an exemplary result, where the diffraction loss has been measured at a wooden and metal wedge at 300 GHz and vertical polarisation. At the light shadow and the lit region, the diffraction loss is almost independent of the material, whereas in the deep shadow region ($\alpha = -30°$), the diffraction loss at the metal wedge as more than 10 dB higher compared to the wooden one.

Translation Stage Measurements: In this set-up, both Tx and Rx are kept fixed with $\alpha = 0°$. The measurement object is positioned on a translation stage and moved in x- and y-direction. In these measurements only the standard gain horns have been used. Figure 8.2 shows both the configuration used to measure the diffraction loss for $y = 0$ and an exemplary measured and predicted diffraction loss for the UTD and KED models for horizontal polarisation. For both UTD and KED – although the diffracting object is not edge-shaped - results are in good agreement with the measurements. Similar findings are reported in [3] for other polarisations and also for metallic cylinders.

In a slightly modified 2D translator diffraction measurement set-up [4], the object is moved both in x- and y-direction. Figure 8.3 shows the measured and predicted UTD diffraction loss in horizontal polarisation using this set-up. The increasing noise in the deep shadowing area due to the limited dynamic range and the slight asymmetry of the interference pattern can be observed. The latter

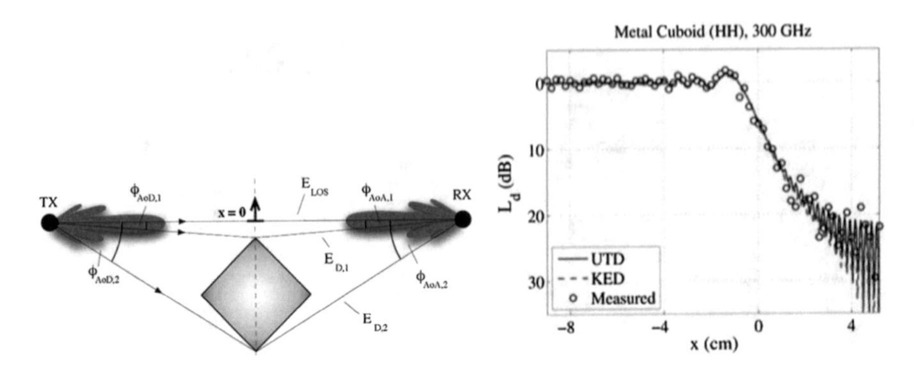

Fig. 8.2 (**a**) left: Translation stage measurement set-up; (**b**) right: Comparison of diffraction measurements and modelling of a metallic cuboid at 300 GHz [3] (© 2012 IEEE, reproduced with permission)

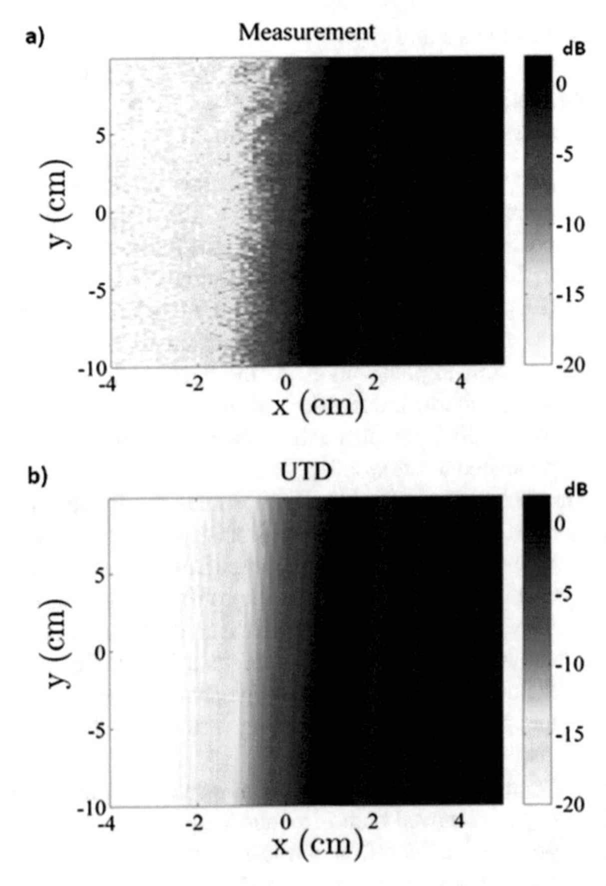

Fig. 8.3 Diffraction loss at 300 GHz (in dB) (**a**) measured and (**b**) simulated with the UTD at a metallic cuboid (8x8x8 cm3) in horizontal polarisation for different displacements orthogonal (x direction) and parallel (y direction) to the transmission link [4] (© 2012 IEEE, reproduced with permission)

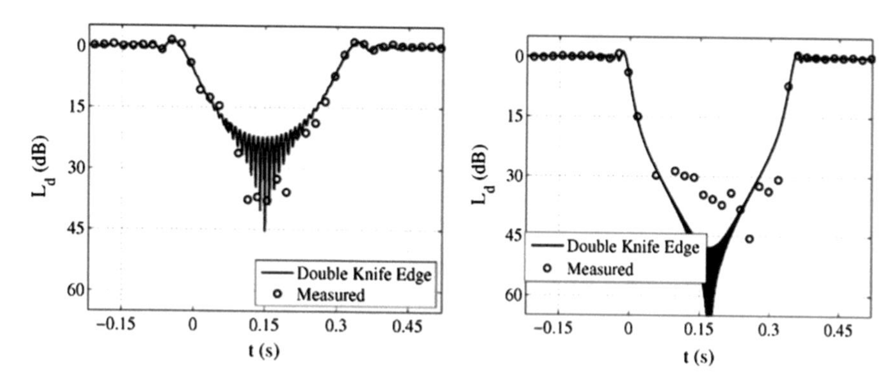

Fig. 8.4 Comparison of measured human-induced shadowing events and prediction by a double knife-edge model at 60 GHz (left) and 300 GHz (right) [3] (© 2012 IEEE, reproduced with permission)

is due to non-ideal antenna alignment, which has been taken into account for the simulations as well.

Diffraction by moving persons: In this set-up, shadowing caused by human bodies has been measured. Similar to the second set-up depicted in Fig. 8.2, a person crosses the link between TX and RX, while the time-dependent losses have been measured. The distance between Tx and Rx was chosen to 2.7 m at 60 GHz and 1 m at 300 GHz. The test heads have been mounted at a height at 1.10 m. For the prediction, the Double Knife-Edge Model from [14] has been applied, where a person is modelled by two knife edges at the front and the back of the body. The results are depicted in Fig. 8.4 showing a good agreement at 60 GHz for both the lit and the shadow regions, whereas at 300 GHz, the measured diffraction loss is lower than the predicted one in the deep shadow area, which is due to the limited dynamic range at 300 GHz. Similar findings are reported in [5], where a correlation-based channel sounder has been used in a similar set-up.

8.3 Modelling of Human Blockage

As shown in the previous section, the high diffraction loss at 300 GHz does not allow one to maintain a wireless link, in the case that a human blocks the line-of-sight and no other paths are available. In order to investigate the impact of human blockage on the performance of a 300 GHz wireless communication system, the determination of the probability of LoS and the duration of the blockage, respectively, is required. Various methods for this task have been proposed and will be briefly described in the following.

Ray Tracing/Ray Launching Based Models In this deterministic approach, a realistic scenario consisting of 3D environmental and a movement model of persons

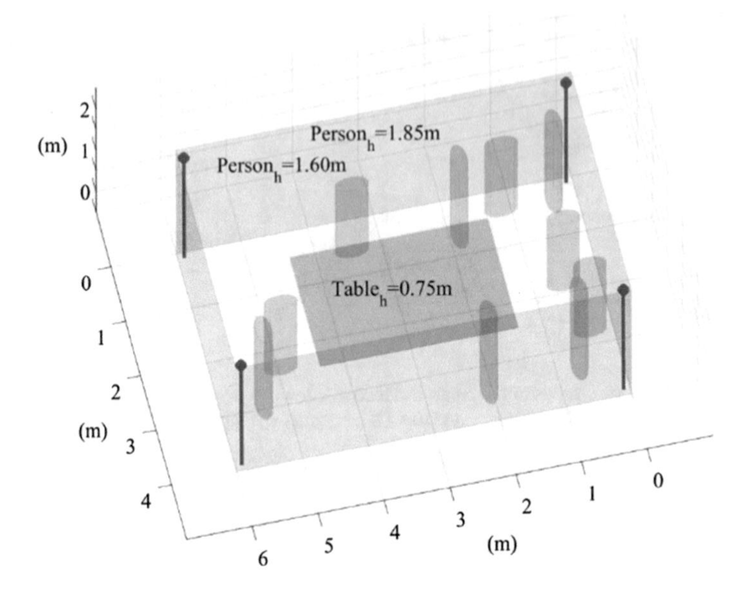

Fig. 8.5 Meeting room scenario with up to four antennas mounted in the corners of the room [5] (© 2018 IEEE, reproduced with permission)

moving around in this environment are required. A ray tracing or ray launching algorithm is applied to determine the LoS probability and the probability for the duration of blockage. An example for such an investigation has been performed in [5] for the meeting room scenario using a distributed antenna system (DAS) depicted in Fig. 8.5. In this scenario, humans are modelled by elliptic cylinders with two different heights of 1.85 m and 1.60 m. Tx antennas are mounted on up to four corners of the room at a height of 2,40 m and the Rx is moved in the whole 2D area at a height of 1 m.

Figure 8.6 shows the LOS coverage probability for three antenna configurations, 10 moving persons and a simulation of 1 h with a temporal resolution of 1 s. This means 3600 different time instants have been evaluated for each location. The spatial resolution is 20 cm $\times$ 20 cm. Here, the LoS coverage probability is defined as the ratio of time where the LoS path between the Rx and at least one of the Tx antennas is available. The cumulative distribution function (CDF) of the LoS duration is shown in Fig. 8.6 for the same simulation scenario with 5 and 10 moving persons, respectively, showing a clear dependency of the LoS probability on the number of moving persons and the number of antennas in the DAS.

Empirical Models Based on Ray Tracing or Ray Launching Instead of performing time-consuming ray tracing or ray launching, a further simplification consists of deriving analytical formulas from the results of ray-based simulation. Such concepts have been already successfully applied at 60 GHz to derive channel models used for standardisation [11, 12]. In [5], such a method has been applied at 300 GHz to derive

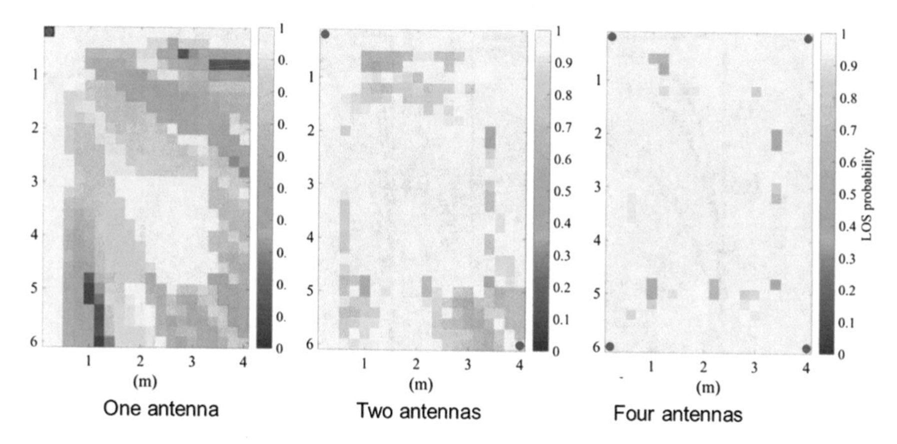

Fig. 8.6 Spatial distribution of LOS coverage probabilities for three different antenna settings (based on results presented in [5]), red dots indicated the positions of the Tx antennas

CDFs for the LoS duration, the blockage duration and the number of blockages per hour (Fig. 8.7).

Analytical LoS Probability Models In [6], an analytical model has been derived for the determination of the LoS probability between Tx and Rx in room crowded with persons. This method allows to adjust:

- The room dimensions
- The number of persons in the room
- The height of the Tx and Rx antennas
- The number of antennas (one antenna or two diagonally placed antennas)

enabling an easy assessment of the dependency of LoS on these parameter changes.

8.4 Communication in Obstructed Line-of Sight

Although a proper communication link at 300 GHz cannot be maintained through diffraction, communication may be still possible using a single reflection or scattering process (see Fig. 8.8). Such a scenario is called obstructed line-of-sight (OLoS). Such scenarios do not only occur in case of shadowing by a person, but also by building corners, which block the LoS path. The use of OLoS is possible even by typical building materials. This was already investigated by some early publications on the properties of building materials at 300 GHz [18–20]. Ma et al. have experimentally demonstrated, for the first time, that data links via NLOS reflection in a non-specular direction can be established at frequencies above 100 GHz, with low bit error rates [9].

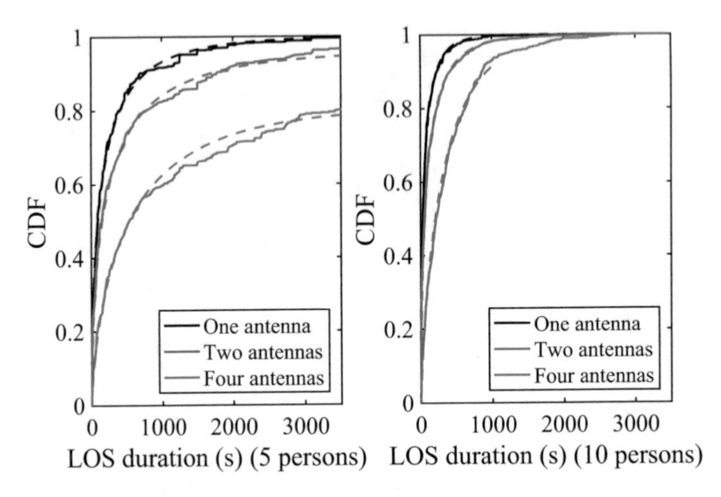

Fig. 8.7 Cumulative Distribution Function of LoS durations with solid lines simulation results and dashed lines regressions [5] (© 2018 IEEE, reproduced with permission)

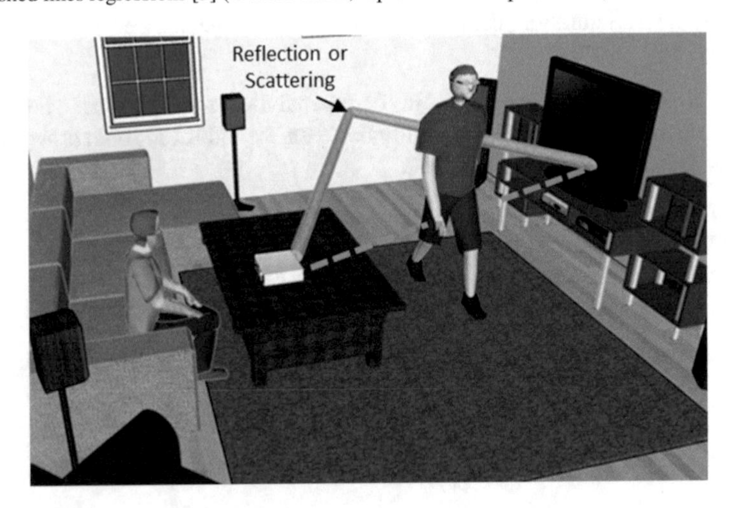

Fig. 8.8 Communication Scenario for obstructed line-of-sight

Since the scattering loss might be high for some building materials or certain directions, various groups have investigated the possibility of deploying artificial reflectors:

– Turchinovich et al. [8] have proposed the use of flexible all-plastic mirrors building on a well-known concept of dielectric mirrors in the far infrared. In this approach, alternating layers of different polymer materials with a typical thickness of several tens of micrometers are used to build the plastic mirror. The mirrors are easy and cheap to produce. Furthermore, they have an almost constant low reflection coefficient over a large range of incidence angles. In a simulative study [20], it was shown how coverage can be significantly improved by using all-plastic mirrors.

– Barros et al. [8] have introduced another concept using dielectric mirrors, where dielectric mirrors are coupled to a THz patch antenna in order to reflect signals from neighbouring base stations. The mirrors can be adapted mechanically by motors in order to adjust their preferred reflection angle.
– The concept of intelligent reflecting surface (IRS) makes use of a large number of reflecting elements with adjustable phase-shifts to enhance the reflection of the THz communication beam in certain directions [21].

References

1. Nahata, A., & Heinz, T. F. (1996, September). Reshaping of freely propagating terahertz pulses by diffraction. *IEEE Journal of Selected Topics in Quantum Electronics, 2*(3), 701–708. https://doi.org/10.1109/2944.571770.
2. Pearce, J., & Mittleman, D. (2002, November). Defining the Fresnel zone for broadband radiation. *Physical Review. E, Statistical, Nonlinear, and Soft Matter Physics, 66*(5 Pt 2), 056602. https://doi.org/10.1103/PhysRevE.66.056602. Epub 2002 Nov 15. PMID: 12513618.
3. Jacob, M., Priebe, S., Dickhoff, R., Kleine-Ostmann, T., & Kürner, T. (2012). Diffraction in mm and sub-mm wave indoor propagation channels. *IEEE Transactions on Microwave Theory and Techniques, 60*(3), 833–844.
4. Kleine-Ostmann, T., Jacob, M., Priebe, S., Dickhoff, R., Schrader, T., & Kürner, T. (2012). Diffraction measurements at 60 GHz and 300 GHz for modeling of future THz communication systems. In *2012 37th international conference on infrared, millimeter, and terahertz waves, Wollongong, NSW* (pp. 1–2). https://doi.org/10.1109/IRMMW-THz.2012.6380411.
5. Peng, B., Rey, S., Rose, D. M., Hahn, S., & Kürner, T. (2018, October). Statistical characteristics study of human blockage effect in future indoor millimeter and sub-millimeter wave wireless communications. In *Proceedings of IEEE 87th Vehicular Technology Conference (VTC Spring), electronic publication (5 pages), Porto, Portugal.*
6. Bilgin, B. A., Ramezani, H., & Akan, O. B. (2019). Human blockage model for indoor terahertz band communication. In *2019 IEEE international conference on communications workshops (ICC workshops), Shanghai, China* (pp. 1–6). https://doi.org/10.1109/ICCW.2019.8757158.
7. Barros, M. T., Mullins, R., & Balasubramaniam, S. (2017, July). Integrated terahertz communication with reflectors for 5G small-cell networks. *IEEE Transactions on Vehicular Technology, 66*(7), 5647–5657. https://doi.org/10.1109/TVT.2016.2639326.
8. Turchinovich, D., Kammoun, A., Knobloch, P., Dobberitn, T., & Koch, M. (2002). Flexible all-plastic mirrors for the THz range. *Applied Physics A: Materials Science & Processing, 74*, 291–293.
9. Ma, J., Shrestha, R., Zhang, W., Moeller, L., & Mittleman, D. M. (2019). Terahertz wireless links using diffuse scattering from rough surfaces. *IEEE Transactions on Terahertz Science and Technology, 9*, 463–470.
10. Maltsev, A., Maslermikov, R., Sevastyanov, A., Lomayev, A., Khoryaev, A., Davydov, A., & Ssorin, V. (2010, April). Characteristics of indoor millimeter-wave channel at 60 GHz in application to perspective WLAN system. In *Proceedings of 4th European Conference on Antennas and Propagation (EuCAP), Barcelona, Spain* (pp. 1–5).
11. Jacob, M., Priebe, S., Maltsev, A., Lomayev, A., Erceg, V., & Kürner, T. (2011, April). A ray tracing-based stochastic human blockage model for the IEEE 802.11ad 60 GHz channel model. In *Proceedings of European Conference on Antennas and Propagation (EuCAP), Rome, Italy* (pp. 3084–3088).

12. Jacob, M. (2014). The 60 GHz indoor radio channel – Overcoming the challenges of human blockage. Aachen 2014. Shaker. [online] https://doi.org/10.24355/dbbs.084-201404281135-0
13. Vaughan, R., & Andersen, J. (2003). *Channels, propagation and antennas for Mobile communications*. New York: IET.
14. Kunisch, J., & Pamp, J. (2008, September). Ultra-wideband vertical double knife-edge model for obstruction of a ray by a person. In *Proceedings of IEEE international conference on ultra-wide band (ICUWB)* (Vol. 2, pp. 17–20).
15. Kouyoumjian, R., & Pathak, P. (1974). A uniform geometrical theory of diffraction for an edge in a perfectly conducting surface. *Proceedings of the IEEE, 62*(11), 1448–1461.
16. McNamara, D., Pistorius, C., & Malherbe, J. *Introduction to the uniform geometrical theory of diffraction*. Norwood: Artech House.
17. Pathak, P., Burnside, W., & Marhefka, R. (1980, May). A uniform GTD analysis of the diffraction of electromagnetic waves by a smooth convex surface. *IEEE Transactions on Antennas and Propagation, AP-28*(5), 631–642.
18. Piesiewicz, R., Jansen, C., Wietzke, S., Mittleman, D., Koch, M., & Kürner, T. (2007, May). Properties of building and plastic materials in the THz range. *International Journal of Infrared, Millimeter, and Terahertz Waves, 28*(5), 363–371.
19. Piesiewicz, R., Jansen, C., Mittleman, D., Kleine-Ostmann, T., Koch, M., & Kürner, T. (2007, November). Scattering analysis for the modeling of THz communication systems. *IEEE Transactions on Antennas and Propagation, 55*(11), 3002–3009.
20. Piesiewicz, R., Jacob, M., Koch, M., Schoebel, J., & Kürner, T. (2008, March-April). Performance analysis of future multigigabit wireless communication systems at THz frequencies with highly directive antennas in realistic indoor environments. *IEEE Journal of Selected Topics in Quantum Electronics, 14*(2), 421–430. https://doi.org/10.1109/JSTQE.2007.910984.
21. Chen, W., Ma, X., Li, Z., & Kuang, N. (2019). Sum-rate maximization for intelligent reflecting surface based terahertz communication systems. In *2019 IEEE/CIC international conference on communications workshops in China (ICCC workshops), Changchun, China* (pp. 153–157). https://doi.org/10.1109/ICCChinaW.2019.8849960.

Chapter 9
Noise and Interference

Josep Miquel Jornet, Zahed Hossain, and Vitaly Petrov

Abstract The performance of a communication system does depend not only on the strength of the signal at the receiver but on its relative strength when compared to any other non-desired signal, such as noise and interference. Given the limited power of THz transceivers and the high propagation losses of the THz channel, understanding and either minimizing or leveraging the noise and interference at the receiver are of key importance. In this chapter, the main noise and interference sources at THz frequencies are described, with a special emphasis on the molecular absorption noise and the multi-user interference in different application scenarios at the nano- and macro-scales.

9.1 Molecular Absorption Noise

Beyond the conventional noise sources intrinsic to the transceiver device technology, such as the thermal noise associated with electronic receivers or the photon-counting or shot noise associated with photonic receivers, the THz channel introduces a new type of noise, namely, the molecular absorption noise. Reciprocal to the process by which the absorption of THz radiation leads to internal vibrations in different types of molecules (e.g., water vapor), vibrating molecules result in the emission of

J. M. Jornet (✉)
Ultrabroadband Nanonetworking Laboratory, Institute for the Wireless Internet of Things, Department of Electrical Engineering, Northeastern University, Boston, MA, USA
e-mail: jmjornet@northeastern.edu

Z. Hossain
Present Address:
Intel Corporation, Hillsboro, OR, USA

The State University of New York, Buffalo, NY, USA
e-mail: zahed.hossain@intel.com

V. Petrov
Unit of Electrical Engineering, Tampere University, Tampere, Finland
e-mail: vitaly.petrov@tuni.fi

T. Kürner et al. (eds.), *THz Communications*, Springer Series in Optical Sciences 234, https://doi.org/10.1007/978-3-030-73738-2_9

THz radiation [1]. This property is known as the emissivity of the channel and for a homogeneous gaseous medium is given by:

$$\varepsilon\left(f, r\right) = 1 - \tau\left(f, \tau\right), \tag{9.1}$$

where f and r refer to frequency and distance, respectively, and τ is the transmittance of the medium, defined as

$$\tau\left(f, r\right) = e^{-\sum_{i,g} k^{i,g}(f) r}, \text{ with} \tag{9.2}$$

$$k^{i,g}\left(f\right) = \frac{p}{p_0}\frac{T_{STP}}{T} Q^{i,g} \sigma^{i,g}\left(f\right), \tag{9.3}$$

where p is the system pressure, p_0 refers to the reference pressure (1 atm), T_{STP} is the standard pressure temperature (273.15 K), T is the system temperature, and $Q^{i,g}$ and $\sigma^{i,g}$ stand for the molecular volumetric density and absorption cross section of isotopologue i of gas g, respectively.

There are two types of molecular absorption noise, namely, background noise and self-induced noise [2, 3].

9.1.1 Background Noise

The background molecular absorption noise or sky noise is caused by the temperature of the absorbing medium (e.g., the atmosphere), which makes it behave as a black body radiator. This noise is independent of the transmitted signal and is always present as long as the medium temperature is above 0 K. The background noise temperature is given by

$$T_{back}\left(f\right) = \lim_{r \to \infty} T \varepsilon\left(f, r\right), \tag{9.4}$$

where ε is the emissivity given by (9.1) and the limit takes into account that this noise is contributed by the whole medium. The corresponding noise power density (PSD) in Watts/Hz can be written as

$$S_{N_{back}}\left(f\right) = k_B T_{back}\left(f\right) A_{rx}\left(f\right), \tag{9.5}$$

where k_B is the Boltzmann constant and A_{rx} is the frequency-dependent effective area or aperture of the receiver's antenna. Finally, the background noise power can be obtained by integrating the noise frequency response over the receiver's bandwidth B:

$$N_{back} = \int_B S_{N_{back}}\left(f\right) df. \tag{9.6}$$

9.1.2 Self-Induced Noise

The induced molecular absorption noise is created by ongoing transmissions between one or more transmitters and one or more receivers. For simplicity, let us consider the simplest scenario with one transmitter and one receiver. The self-induced noise is correlated to the actual signal being transmitted: if no signal is being transmitted, this noise is zero. The self-induced noise power density is given by

$$S_{N_{induced}}\left(f,r\right)=S_X\left(f\right)D_{tx}\left(f\right)\varepsilon\left(f,r\right)\frac{1}{4\pi r^2}A_{rx}\left(f\right), \tag{9.7}$$

where S_X is the PSD of the transmitted signal and D_{tx} is the frequency-dependent directivity of the transmitter. As before, the corresponding noise power is obtained by integrating the PSD over the receiver's bandwidth.

It is relevant to note that in a gaseous medium, the power lost due to molecular absorption is in fact converted into molecular absorption noise, i.e., the total power remains constant. While this is true, because the re-emission of the power by vibrating molecules is random and out of phase, the resulting radiation needs to be treated as noise. Nevertheless, because it is correlated to the transmitted signal, its presence is an indicator of an ongoing transmission and can be leveraged in a communication system [4].

Beyond gaseous media, the use of THz communications in intra-body applications requires the study of noise in liquid and solid media. The main difference between the two scenarios arises from the fact that molecules in a liquid or a solid cannot freely vibrate, but their motion is constrained. Excited absorbing molecules attempt to vibrate, but such vibrations lead to friction and, ultimately, heat [5]. To properly characterize the photothermal effects, including noise, in intra-body scenarios, an accurate multi-physics analysis combining the diffusive heat flow equation and Maxwell's equations needs to be conducted[6]. Such analysis needs to take into account the absorption, scattering and thermal conductivity of the tissues and body fluids involved (e.g., skin, fat, blood).

Finally, whether inside or outside of the body, from the stochastic perspective, we can characterize the molecular absorption noise as being:

1. *Additive.* The noise linearly contributes to the received signal (i.e., it is not multiplicative, as the noise found in optical systems).
2. *Not independent.* At least a fraction of it (i.e., the self-induced noise) is triggered by the actual transmitted signal.
3. *Gaussian.* Due to the very large number of molecules present in a standard medium, under the central limit theorem, it can be considered to be Gaussian.
4. *Not white.* Because of the frequency-selective nature of absorption and, thus, emissivity, the noise is colored or pink.

9.2 Multi-User Interference

Simultaneous transmissions by different nodes can result in multi-user interference. Such interference can be modeled in different ways. At the physical layer, multi-user interference can be accounted as an additional noise term affecting the signal-to-interference-plus-noise-ratio (SINR) and, thus, the bit error rate (BER). At the link layer, multi-user interference leads to frame or packet collisions, which might trigger different retransmission strategies. Therefore, understanding and accurately modeling multi-user interference is a necessary step toward enabling practical THz communication systems.

Compared to the many existing works on interference modeling in wireless networks, the peculiarities of the THz channel and the capabilities of THz devices require the development of new models. Next, we describe the state of the art in interference modeling for two key scenarios, namely, ultra-broadband short-range and directional long-range THz communication systems.

9.2.1 Ultra-Broadband Short-Range Terahertz Communications

In short-range THz communication application, such as wireless nanosensor networks [7] or wireless networks on chip [8], the THz-band channel provides very large bandwidths, exceeding several contiguous THz. As a way to maximize the utilization of such bandwidth, the transmission of 100-femtosecond-long pulses by following an on-off keying modulation spread in time has been proposed [2]. The PSD of such pulses has its main frequency components under 4 THz. By utilizing an on-off keying modulation, the transmitter can reduce its energy consumption by remaining silent during the transmission of "0s." By spreading the transmission of pulses in time, concurrent transmissions can be multiplexed in time.

The transmission of ultrashort pulses minimizes the probability of collisions due to the very small time that the channel is occupied by each user. However, given that many of the envisioned applications of THz communications involve very large node densities, multi-user interference is unavoidable. Given the shape of such pulses, the transmitted signals have large fluctuations between positive and negative amplitude values. The existing interference studies for traditional narrowband carrier-based wireless communication systems [9] are centered on modeling the received signal power instead of the received signal amplitude and hence ignore the fact that the interference can be constructive or destructive. This can cause unrealistically large values of interference due to the high amplitude of the pulses.

To properly account for multi-user interference in ultra-broadband pulse-based communication systems, the focus should be on modeling the interference amplitude and its fluctuations through time. In this direction, in [10], a stochastic model of multi-user interference in pulse-based THz-band communications is proposed and experimentally validated. This model is developed by considering the fact that the interference power at the receiver is not a combination of the received powers from the individual nodes, rather the power of the combination of the signal amplitudes.

In particular, the interference i_u generated by one interfering node u at the receiver is given by

$$i_u(t) = \sqrt{e_{p,u}}\, p\left(t - \tau_{t,u}\right), \tag{9.8}$$

where $e_{p,u}$ stands for the energy of the received pulse, p is the shape of the transmitted pulse with unitary energy, and $\tau_{t,u}$ refers to the total delay. Considering that nodes are randomly distributed in space and transmit in an uncoordinated manner, interference can be modeled as a random process. More specifically, the single-node interference i_u is a product of two independent random variables, namely, the pulse amplitude and the pulse shape. The probability density function (PDF) f_{E_s} of the pulse amplitude e_s, which depends on the transmission distance and, thus, network topology, is given by

$$f_{E_s}(e_s) = \begin{cases} \frac{4\xi^{2/\eta}|e_s|^{-\frac{\eta+4}{\eta}}}{(a^2-b^2)\eta} & \text{for } \sqrt{\xi a^{-\eta}} < e_s < \sqrt{\xi b^{-\eta}} \\ 0 & \text{otherwise,} \end{cases} \tag{9.9}$$

where $\xi r^{-\eta}$ is the approximated energy loss function as a function of distance r and a and b are the maximum and minimum distances at which interfering nodes are located from the receiver. The PDF f_P of the pulse shape p, which depends on the waveform as well as on the timing of the transmission, is given by

$$f_P(p) = \begin{cases} \frac{1}{\pi \upsilon \sqrt{1-\frac{p^2}{\upsilon^2}}} & \text{for } -\upsilon < p < \upsilon \\ 0 & \text{otherwise,} \end{cases} \tag{9.10}$$

where $\upsilon = \sqrt{\frac{2}{T_p}}$ and T_p is the pulse duration. Now, by redefining the interference as $Y = E_S P$ and recalling the definition of PDF of a function of two PDFs, the PDF f_Y of the interference y can be obtained as [10]

$$f_Y(y) = \begin{cases} \int_{\frac{-y}{r}}^{\sqrt{\xi b^{-\eta/2}}} f_{E_s}(e_s)\, f_P\left(\frac{y}{e_s}\right) \frac{1}{|e_s|} de_s & -\sqrt{\xi} r b^{-\eta/2} < y < -\sqrt{\xi} r a^{-\eta/2} \\ \int_{\sqrt{\xi a^{-\eta/2}}}^{\sqrt{\xi b^{-\eta/2}}} f_{E_s}(e_s)\, f_P\left(\frac{y}{e_s}\right) \frac{1}{|e_s|} de_s & -\sqrt{\xi} r a^{-\eta/2} < y < \sqrt{\xi} r a^{-\eta/2} \\ \int_{\frac{y}{r}}^{\sqrt{\xi b^{-\eta/2}}} f_{E_s}(e_s)\, f_P\left(\frac{y}{e_s}\right) \frac{1}{|e_s|} de_s & \sqrt{\xi} r a^{-\eta/2} < y < \sqrt{\xi} r b^{-\eta/2} \\ 0 & \text{otherwise.} \end{cases} \tag{9.11}$$

Finally, since the signals from individual nodes add up at the receiver and are independent, the PDF of the total interference can be determined by the convolution

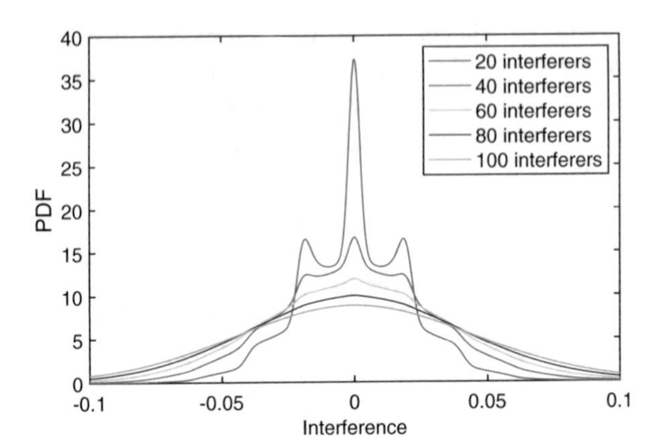

Fig. 9.1 The probability density function of the multi-user interference as the number of interferers increases

of the individual PDFs [10]. As shown in Fig. 9.1, for large number of interferers, the multi-user interference appears to follow a Gaussian distribution (as first assumed in [2]).

9.2.2 *Directional Long-Range Terahertz Communications*

Given the limited power of THz transceivers, highly directional antennas are needed at the transmitter and the receiver to overcome the severe propagation losses and increase the communication beyond a few meters. A common belief is that the high directivity of antennas eventually leads to a noise-limited regime of communication systems [11]. However, razor-sharp-beam interference-free communications are not on the immediate horizon. The reasons range from the complexity of high directivity beamforming antennas to the synchronization challenges that they introduce, in addition to an increasing network densification and the pervasive adoption of device-to-device communications.

Another effect to take into account when modeling multi-user interference is line-of-sight (LoS) blockage. This phenomenon has been addressed in a number of papers in the context of microwave communication systems [12], where buildings block the path between the transmitter and the receiver. Terahertz systems are expected to operate over much shorter distances than microwave cellular systems, and thus, buildings are not expectedly a major problem in outdoor deployments. However, at these frequencies, users themselves may block the LoS path between the transmitter and the receiver, as almost any object whose volume is larger than several wavelengths (millimeters in the bands of interest) is effectively an obstacle. Therefore, the process of LoS possible blockage by users also needs to be taken into account.

With these observations in mind, in [13], an analytical model of multi-user interference and SINR for THz communication systems is developed by using the tools of stochastic geometry. The model explicitly captures the following three effects inherent for these frequencies: (1) path loss component caused by molecular absorption, (2) directivity of the transmitter and the receiver, and (3) blockage of the link by obstacles. Two radiation pattern models of directional antennas are considered, namely, the cone model representing an ideal directional antenna and the cone plus sphere model capturing specifics of a nonideal directional antenna with side lobes. For the latter model, in a field of interfering nodes with density λ_I and blocking radius r_B, the mean multi-user interference I can be written as

$$E\left[I\right] = \frac{A_1\alpha^2\lambda_I}{2\pi}\Theta(R, r_B, \lambda_I, K) + \frac{A_2[2\pi - \alpha^2]\lambda_I}{2\pi}\Theta(R, r_B, \lambda_I, K \tag{9.12}$$

with

$$A_i = P_{Tx}G_{Tx_i}G_{Rx}\left(\frac{c}{4\pi f}\right)^2 \text{ and} \tag{9.13}$$

$$\Theta = e^{-\lambda_I r_B^2} Ei(-R[K + \lambda_I r_B]) - Ei(-r_B^2[K + r_B\lambda_I]), \tag{9.14}$$

where G_{Tx_1} refers to the gain of the main lobe of non-blocked interfering nodes, G_{Tx_2} refers to the gain of the side lobes of the same, α is the antenna main lobe beam width, and $Ei(\cdot)$ is the exponential integral function.

In Fig. 9.2a, the mean interference is shown as a function of the antenna main lobe beam width for different transmitter and receiver configurations. It can be seen that when utilizing directional antennas simultaneously at the transmitter and the receiver, the interference drastically increases when decreasing the beam width. While a narrow beam is more likely to be blocked, when it is not, the

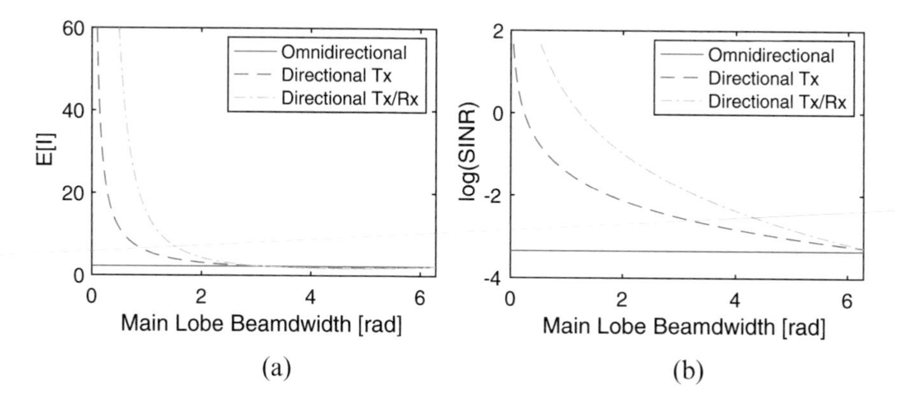

Fig. 9.2 Interference analysis in long-range THz communications in the presence of blockage for different transmitter and receiver antenna configurations. (**a**) Mean interference $E\left[I\right]$. (**b**) Signal to noise and interference ratio

interference can be very high. However, as can be seen in Fig. 9.2b, the SINR increases for narrower beams, which is a result of the intended transmitter signal strength increasing much more than the interference.

9.3 Conclusion

Noise and multi-user interference play a key role in the performance of any wireless communication system. For THz communications, the role of molecular absorption noise and the ultra-broadband and ultra-directional nature of the channel need to be taken into account both when studying noise and interference but, more importantly, when developing tailored physical layer and networking solutions. In addition, as higher power THz transceivers and more directional smart antenna systems are developed, additional scenarios become relevant. For example, enabling the coexistence of passive satellite-based THz sensors with on-the-ground high-power THz communication links (e.g., in wireless backhaul applications) will require accurate noise and multi-user interference analyses and, consequently, innovative communication system designs, to prevent any disruption in the operation of Earth and space exploration research. The work described in this chapter provides the foundations to do so.

Acknowledgment This work was supported by the US National Science Foundation Grant CNS-2011411.

References

1. Jornet, J. M., & Akyildiz, I. F. (2011). Channel modeling and capacity analysis of electromagnetic wireless nanonetworks in the terahertz band. *IEEE Transactions on Wireless Communications, 10*(10), 3211–3221.
2. Jornet, J. M., & Akyildiz, I. F. (2014). Femtosecond-long pulse-based modulation for terahertz band communication in nanonetworks. *IEEE Transactions on Communications, 62*(5), 1742–1754.
3. Kokkoniemi, J., Lehtomäki, J., & Juntti, M. (2016). A discussion on molecular absorption noise in the terahertz band. *Nano Communication Networks, 8*, 35–45.
4. Hoseini, S. A., Ding, M., & Hassan, M. (2017). Massive MIMO performance comparison of beamforming and multiplexing in the terahertz band. In *2017 IEEE Globecom Workshops (GC Wkshps)* (pp. 1–6). IEEE.
5. Elayan, H., Johari, P., Shubair, R. M., Jornet, J. M. (2017). Photothermal modeling and analysis of intra-body terahertz nanoscale communication. *IEEE Transactions on NanoBioscience, 16*(8), 755–763.
6. Elayan, H., Stefanini, C., Shubair, R. M., & Jornet, J. M. (2018). End-to-end noise model for intra-body terahertz nanoscale communication. *IEEE Transactions on Nanobioscience, 17*(4), 464–473.
7. Akyildiz, I. F., & Jornet, J. M. (2010). Electromagnetic wireless nanosensor networks. *Nano Communication Networks (Elsevier) Journal, 1*(1), 3–19.

8. Abadal, S., Alarcón, E., Cabellos-Aparicio, A., Lemme, M. C., & Nemirovsky, M. (2013). Graphene-enabled wireless communication for massive multicore architectures. *IEEE Communications Magazine, 51*(11), 137–143 (2013)
9. Cardieri, P. (2010). Modeling interference in wireless ad hoc networks. *IEEE Communications Surveys & Tutorials, 12*(4), 551–572 (2010)
10. Hossain, Z., Mollica, C. N., Federici, J. F., & Jornet, J. M. (2019). Stochastic interference modeling and experimental validation for pulse-based terahertz communication. *IEEE Transactions on Wireless Communications, 18*(8), 4103–4115.
11. Andrews, J. G., Buzzi, S., Choi, W., Hanly, S. V., Lozano, A., Soong, A. C., & Zhang, J. C. (2014). What will 5G be? *IEEE Journal on Selected Areas in Communications, 32*(6), 1065–1082.
12. Bai, T., Vaze, R., & Heath, R. W. (2014). Analysis of blockage effects on urban cellular networks. *IEEE Transactions on Wireless Communications, 13*(9), 5070–5083.
13. Petrov, V., Komarov, M., Moltchanov, D., Jornet, J. M., & Koucheryavy, Y. (2017). Interference and SINR in millimeter wave and terahertz communication systems with blocking and directional antennas. *IEEE Transactions on Wireless Communications, 16*(3), 1791–1808.

Part III
Propagation and Channel Modelling 3: Modelling and Measurements in Complex Environments

Chapter 10
Indoor Environments

Vittorio Degli-Esposti

Abstract Indoor propagation characteristics in the THz band (0.1–10 THz) are reviewed in the present chapter with focus on the impact of major propagation mechanisms such as partition loss, reflection, and diffuse scattering. The most important propagation and channel modeling approaches are presented and discussed, including empirical-statistical approaches, stochastic channel modeling, and deterministic approaches based on ray tracing.

10.1 Introduction

Indoor environments represent the most natural ground for THz applications. The high traffic density and high bitrate requirements, typical of indoor wireless communications scenarios, can be coped with at THz frequencies due to the large available bandwidth and to the possibility of using very directive transmission beams to increase the signal-to-interference ratio [1]. At the same time, short-link distances, low mobility, and the prevalent use of large terminals such as laptops, tablets, and appliances with high-gain antenna arrays and good power supplies can help to overcome the power-budget limitations of current THz technology. Information showers for museums and public spaces, high-definition holographic gaming, and infotainment for households, ultra-broadband mobile access for offices, high-throughput density cellular overlays for sport, show and conference venues, robotic control for industrial applications, centimeter-level localization, are among the many envisioned THz applications that will primarily take place indoors [2, 3].

Compared to outdoor or close-proximity THz propagation, indoor THz propagation shows specific characteristics. Typical indoor propagation distances of the order of a few meters to a few hundred meters are often not enough to guarantee the validity of far-field conditions and of plane-wave models [4]. Although rain and fog

V. Degli-Esposti (✉)
Department of Electrical Engineering "G. Marconi", University of Bologna, Bologna, Italy
e-mail: v.degliesposti@unibo.it

T. Kürner et al. (eds.), *THz Communications*, Springer Series in Optical Sciences 234, https://doi.org/10.1007/978-3-030-73738-2_10

attenuations are not present, molecular absorption peaks cannot be neglected above 300 GHz [4]. Moreover, unlike lower frequency signals, THz signals can hardly penetrate through walls, including dividing walls [3]. Therefore, typical THz indoor propagation environments are limited to one room and to the adjacent rooms at most. Techniques exploiting reflection from surfaces or from properly located mirrors have been proposed to extend coverage to NLoS locations and around corridors' corners [2].

As with lower frequencies, THz propagation and channel models can be roughly divided into stochastic and deterministic models. Stochastic channel models are usually derived by extracting from measurements – and/or calibrated ray tracing simulations - statistical moments of important propagation and channel parameters for a given kind of environments and are therefore defined *empirical-statistical models*. Stochastic models find one important application in channel simulation for link-level or system-level design. A complete coverage on this topic is given in Chap. 16.

Deterministic models are based on physics rather than on empirical observation and can therefore be defined *physical-deterministic models*. The most popular deterministic models for indoor environment are Ray-Based models such as Ray Tracing that naturally simulate the multipath propagation process and allow a physically sound, multidimensional propagation characterization. Their drawback is that they need an accurate description of the propagation environment and only yield site-specific results, i.e., specific to the input environment configuration only. Once properly validated and calibrated versus measurements, deterministic models are often used to complement or replace measurements in the derivation and parametrization of statistical channel models [5].

Unlike outdoor channel models, several indoor channel models for THz frequencies are available in the literature [2]. The rest of the present chapter features a review of state-of-the-art indoor THz channel measurements and empirical modeling in Sect. 10.2, while a survey on deterministic THz propagation modeling is provided in Sect. 10.3.

10.2 Indoor Measurements and Empirical Channel Modeling

Measurement-based studies on THz propagation in indoor environment can be roughly divided into two categories: (i) propagation characterization studies addressing specific aspects and (ii) channel sounding and modeling studies for real-life environments. In the present section only (ii) will be covered with some detail, while a few general considerations on indoor propagation characteristics will be briefly given below.

10.2.1 Indoor Propagation Characteristics

The indoor environment is characterized by space partitioning realized through either real walls or lighter structures such as glass, chipboard, or gypsum panels. While concrete walls' transmission losses are too high for THz waves to overcome them, lighter partition walls can be partially transparent, especially at sub-THz frequencies. If we model partition walls as uniform flat layers, partition loss *Lp* can be easily estimated using simple formulas. Since multiple reflections inside the material layer are usually negligible at THz frequencies due to the small wavelength with respect to thickness, *Lp* is dominated by (i) reflection attenuation on the air-material and material-air interfaces and (ii) exponential attenuation due to propagation inside the layer. An approximate partition loss estimate for normal incidence and low-loss materials is [6]:

$$L_p = e^{2\alpha w}\left(1 - |\Gamma|^2\right)^{-2} \quad ; \quad \alpha \simeq \frac{\sigma}{2\varepsilon_0 c\sqrt{\varepsilon_r{}'}} \tag{10.1}$$

where w is slab width, $|\Gamma|$ is the reflection coefficient modulus at either interfaces, σ is the conductivity [S/m], and α is the loss constant for low-loss dielectric materials of relative complex permittivity $\varepsilon_r = \varepsilon_r^{'} - j\varepsilon_r^{''}$ with $\varepsilon_r^{''} \ll \varepsilon_r^{'}$. Formula (10.1) can also be used to estimate α from reflection and transmission measurements. Since σ generally increases linearly with the frequency in a $\log(\sigma) - \log(\mathrm{f})$ scale [7], the same applies to α which is proportional to it. Additional transmission losses due to diffuse scattering should be added in case of unhomogeneous compound materials such as chipboard or some kinds of stone and wood. In case stratified materials, more complex formulations are necessary [8].

There is a vast literature on THz penetration through specific materials used for technology and devices, but a very limited number of *Lp* measurement results for indoor furnishing and partition materials have been published: most of them are summarized in Table 10.1. While *Lp* is generally low enough at sub-THz frequencies, it is very high at 300 GHz in most cases, except for gypsum, ceramic, and nylon slabs.

Due to the high *Lp* values, indoor NLoS links cannot rely on transmission. Therefore, one question that needs to be answered is whether NLoS links are possible using reflected paths (RNLoS paths). Experimental investigation in [11] at 100, 200, 300, and 400 GHz, using a simple setup with a CW wave modulated with a 1Gbit/s signal, demonstrated that NLoS connection is possible through up to two specular reflections: a BER of 10^{-8} can be achieved over a 2 m, 400 GHz link undergoing 1 reflection on a painted cinderblock wall using a Tx power of −8 dBm and 26 dB gain antennas. Similar results have been proved using a VNA setup [9]. Reflection from ceiling's acoustic panels can be a good way to replace the LoS in case of blocking, while reflection from wooden surfaces is generally too weak [12]. Specular reflection losses due to diffuse scattering are limited to a few dBs because the roughness of most indoor surfaces is still small compared to the wavelength at

Table 10.1 Measured mean penetration loss [dB] for some material slabs and slab-like objects at different frequencies. Frequency values refer to center-band as most measurements are wideband

Material	*w* [cm]	100 GHz	140 GHz	300 GHz	Ref.
Clay tile	1.5	9.3			[6]
Ceramic tile	0.7	4.7		12.9	[6]
Cement tile	2.5	~39			[2]
Granite	2.1	18.5			[6]
Gypsum	2.0			11.0	[6]
Plywood	1.8	17.1		38.1	[6]
Solid wood	2	20			[3]
	3.5			65.5	[3]
	4.0	41.6			[3]
Drywall	14.5		15.0		[10]
Clear glass	0.6		9		[10]
	2.5			86.7	[9]
Glass door	1.3		16.2		[10]
MDF door	3.5			65.5	[9]
Chipboard-1	1.6	16.5		46.9	[6]
Chipboard-2	1.6	11.9		45.3	[6]
Nylon	0.6	1.6		4.3	[6]

400 GHz. However, a stronger impact of diffuse scattering from surface roughness is expected in the higher range of THz frequencies.

10.2.2 Channel Sounding and Modeling

Proper characterization of indoor channels requires a dynamic range of at least 40 dB to allow the identification of multiple-bounce contributions, high time- and angle-resolution and a very large bandwidth, of at least several gigahertz to reflect the foreseen indoor applications' bandwidth. Over such large bandwidths, several propagation parameters can change, including molecular absorption - mostly due to water vapor – free-space loss, antenna characteristics, and material properties: this must be taken into account at both measurement and modeling stages. Although Doppler frequencies are limited to a few kilohertz due to low mobility, the high probability of human blockage can make nonetheless the indoor channel dynamic.

To compensate for the high isotropic losses, antennas with a directivity of at least 20–25 dBs must be used even for short indoor links. This fact makes beam alignment quite critical, but allows properly aligned links using horn antennas to be relatively immune from multipath, including RNLoS paths. Under these conditions, dB-Path-Loss (PL) versus link distance R in LoS links can be described using a one-slope (or Hata-like) formula:

$$PL = PL_0 + 10n \cdot Log(R) + \chi_\sigma \quad (10.2)$$

where PL_0 is path loss for R = 1 m and χ_σ is a zero-mean random variable of standard deviation σ to model beam misalignments. Path-loss exponent n is found to be very close to 2 (i.e., free space) at both 300 GHz and in the D-band (110–170 GHz) while σ is of only 1.44 dB and 0.17 dB (for $R < 1$ m links, with a Gaussian χ_σ), respectively [9, 13]. Reflection loss must be added up in (10.2) for RNLoS links. Moreover, an additional linear attenuation term $L_\alpha = R\alpha$, α being specific attenuation [dB/m], should be added for frequencies where molecular absorption is non negligible. A two-ray effect is observed in desktop links where the presence of reflection from the table surface, or from objects nearby, can generate frequency and space selectivity with deep fades [9, 13]. Reflections from the transmitter and receiver heads front faces can also affect the link. RMS delay spread is shown to be of a few picoseconds in LoS links with good antenna alignment and of a few tens of picoseconds in Obstructed LoS (OLoS) links with blocking due to small objects such as a glass or a ceramic mug [13].

Although measurements with fixed horn antennas are useful, rotating directive antennas at one or both link ends must be used to grasp the spatial characteristics of the channel. This kind of setup allows to achieve at the same time a high dynamic range and a complete channel characterization, including antenna-independent path-loss, time-dispersion parameters and directional characteristics of the channel. The knowledge of such characteristics is necessary to design future indoor THz system that will make use of MIMO or beam-switching arrays to cope with mobility and blockage of the main path and/or to increase channel performance. By rotating the antennas in the azimuth plane and summing up power contributions from all orientations, quasi-omnidirectional PL trends are derived in [14, 15] in the D-band for several indoor environments, including a shopping mall with link distances of up to 65 m. The PL trend is confirmed to follow Eq. (10.2), with a path loss exponent close to 2 and a σ of a few dBs, slightly higher than with fixed directive antennas, especially when partially obstructed LoS links are included [14]: in this case χ_σ can be used to accounts for shadowing effects rather than misalignment. Instead of summing power, by keeping track of the time-distribution or of the angle-distribution of the signal at the Rx, RMS delay-spread and RMS angle-spread can be computed, respectively. Differently from PL, the quasi-omnidirectional RMS delay-spread is much higher than the directional RMS delay-spread (with directive antennas): values between 3–10 ns in an office environment and up to 30 ns in the shopping mall are reported. Average azimuth spread is found to be of about 30° in all environments [14]. Interestingly, similar results were found for the 300 GHz band in a small office environment using a very high-resolution setup [16].

Equation (10.2) represents the simplest prototype of empirical-statistical propagation model. More complex statistical channel models that take into account also the frequency, angle, and space domains have been proposed in the last years for the simulation of THz channels. A good example is the stochastic channel model for THz MIMO links proposed in [5], which is based on the multidimensional

MIMO channel transfer function concept [17] properly modified to suit a broadband channel:

$$H_{m,n}(f) = \sum_{i=1}^{N_{rays}} H_i \cdot g_m\left(f, \phi_{AoD,i}, \theta_{AoD,i}\right) e^{j\mathbf{k}_i(f)\mathbf{r}_m} \cdot g_n\left(f, \phi_{AoA,i}, \theta_{AoA,i}\right) e^{j\mathbf{k}_i(f)\cdot\mathbf{r}_n} \tag{10.3}$$

with

$$\begin{aligned} H_i\left(f, \phi_{AoD}, \theta_{AoD}, \phi_{AoA}, \theta_{AoA}\right) = a_i \cdot e^{j\varphi_i} \cdot D_i(f) \cdot e^{-j2\pi\tau_i(f_i - f_0)} \\ \cdot \delta\left(\phi_{AoD} - \phi_{AoD,i}\right) \cdot \delta\left(\theta_{AoD} - \theta_{AoD,i}\right) \\ \cdot \delta\left(\phi_{AoA} - \phi_{AoA,i}\right) \cdot \delta\left(\theta_{AoA} - \theta_{AoA,i}\right) \end{aligned}$$

where $H_{m,n}(f)$ is the direction- and space-dependent transfer function between the m-th and n-th MIMO antenna elements of position vectors $\mathbf{r}_{\mathrm{m}}$ and $\mathbf{r}_{\mathrm{n}}$, and antenna characteristics g_m and g_n, respectively, and the footers AoD, AoA in the angle symbols stand for Angle of Departure and Angle of Arrival. All the remaining quantities have intuitive meaning: for a complete explanation of the model refer to Chap. 16. It is worth noting that factors a_i and $D_i(f)$ represent the i-th path-gain and its frequency dispersion due to the frequency variability of reflection coefficients and of free-space loss. In [5], the model is extended to the polarization domain by using a fully polarimetric channel matrix $\boldsymbol{H}_i$ – instead of a scalar transfer function H_i as in (10.3) – for the i-th path [5]. All the quantities in (10.3) – except for the characteristics of the antenna elements - are considered random variables that can be correlated with each other: for example amplitude a_i and delay τ_i of the i-th ray are correlated according to the Saleh-Valenzuela model [18]. The model generates realizations of all random variables according to statistical distributions derived through multiple ray tracing simulations in a given kind of environment (e.g., a small office in [5]), and therefore extracts realistic realizations of the transfer function random process that are representative of *that* kind of environment. Once the ray tracing tool is properly calibrated and the statistical distributions calculated, the method is 100 times faster than rerunning ray tracing for different positions over the reference environment to get the same number of channel realizations.

An alternative approach to channel simulation is possible through Geometric Stochastic Channel Models (GSCM) where the starting point is the generation of a distribution of "scatterers", according to a realistic statistical spatial distribution, around the radio terminals positions. The channel's transfer function realization is then generated using simplified ray tracing techniques in order to preserve spatial consistency with moving terminals. The GSCM approach, widely used for lower frequency bands and standardized at international level [19], is not of widespread use in THz propagation modeling.

10.3 Deterministic Propagation Modeling

Deterministic propagation models rely on the electromagnetic propagation theory rather than on formulas derived from measurements to describe the propagation process. Due to the small wavelength compared to the dimensional scale of the indoor propagation environment, electromagnetic methods that imply a domain discretization with a sub-wavelength mesh size such as Finite Difference Time-Domain (FDTD) methods, require huge computational resources at THz frequencies. Therefore, the most widely used deterministic models for THz frequencies are ray-based propagation models, namely Ray Tracing (RT) and its variations such as ray launching (also referred to as shooting and bouncing ray method [20]). RT being founded on the sound, albeit approximate, theory of Geometrical Optics (GO) [21] and its generalization to treat diffraction, the Uniform Theory of Diffraction [22], can be considered fairly reliable when the wavelength is small compared to the size of the obstacles and when the propagation environment is limited and therefore a detailed environment description (environment database), including electromagnetic material properties is possible. Both conditions are satisfied in indoor THz propagation.

The propagating rays are found using image theory and other geometric techniques applying GO rules [23]. Usually all rays undergoing a maximum number of interactions (or "bounces", i.e., reflections and diffractions), N_b, with walls and obstacles are taken into account. After geometric ray tracing, which is frequency-independent, field computation at the Rx position is performed. In order to simulate ultra-broadband propagation for THz applications, the so called "ultra-broadband RT" or "frequency-domain RT" approach is followed: the signal bandwidth is subdivided into N_f frequency components f_i. The field phasor for the j-th ray and for the frequency f_i is then computed:

$$\mathbf{E}^j\left(f_i\right) = \left(\prod_{n=1}^{N_b^j} \underline{\underline{C}}_n^j\left(f_i\right)\right) \cdot \mathbf{E}_0^j\left(f_i\right)\ A_{SF}^j\left(f_i\right)\ e^{-jk(f_i)r_{tot}^j} \tag{10.4}$$

where $\mathbf{E}_0^j$ is the field emitted by the Tx antenna at 1 m in the direction of departure of the j-ray, $\underline{\underline{C}}_n^j$ is the n-th interaction dyadic (a matrix that includes Fresnel's reflection coefficients or UTD diffraction coefficients [21]), A_{SF}^j is the spreading factor taking into account free space attenuation [21], r_{tot}^j is the total length of the j-th ray, and k is the wavenumber. Field computation (10.4) must be performed recursively for each one of the N_f frequency components f_i used to subdivide the total bandwidth. If the field $\mathbf{E}^j(f_i)$ of (10.4) corresponds to a signal of unit value at the Tx, the channel's transfer function can be computed using the following formulas [16]:

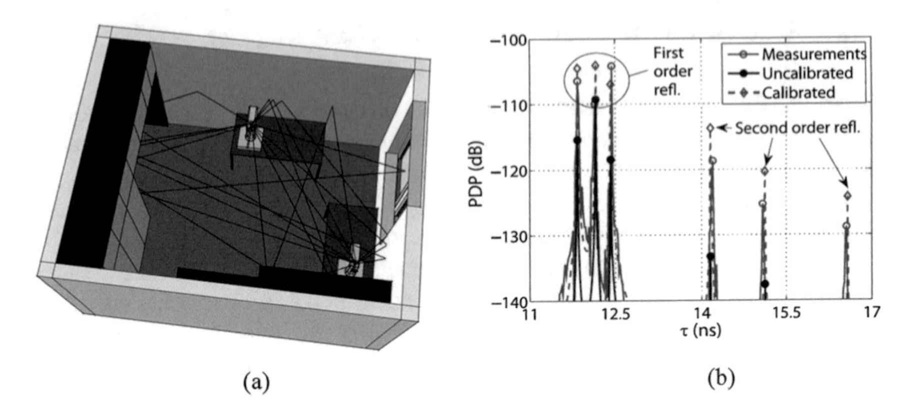

Fig. 10.1 (**a**) Small office scenario with traced rays; (**b**) measured, RT- predicted (uncalibrated) and RT-predicted (calibrated) power delay profile for environment depicted in (**a**) [16] (© 2013 KICS, reproduced with permission)

$$\begin{aligned} H_j(f) &= \sum_{i=1}^{N_f} \mathbf{E}^j\left(f_i\right) \cdot \hat{p}_{Rx}\left(\phi_j, \theta_j\right)\ \ell_{Rx}\left(f_i, \phi_j, \theta_j\right)\ \delta\left(f - f_i\right) \quad ; \\ H(f) &= \sum_{j=1}^{N_{rays}} H_j(f) \end{aligned} \tag{10.5}$$

where $\hat{p}_{Rx}$ is a unit phasor expressing the polarization characteristics of the Rx antenna, ℓ_{Rx} is its effective length, and (ϕ_j, θ_j) is the direction of arrival of the j-th ray.

The channel's impulse response, that takes into account both multipath dispersion and pulse broadening for each ray, can be derived by performing an inverse fast Fourier transform of the transfer function (10.5). It is worth noting that interaction coefficients can vary quite a lot with the frequency due to the effect of stratified material such as painted walls or lacquered surfaces. Therefore, the description of material properties fed to the RT model through the environment database is necessarily inaccurate and a calibration of interaction coefficients versus measurements is often necessary. If calibration is done properly, ultra-broadband RT can yield very accurate results. In Fig. 10.1, a predicted and measured power-delay profile for a small office environment is shown as an example [16].

According to high-resolution directional measurements presented in [16], only reflections (up to the fourth order) could be identified, therefore only reflections were considered in RT simulations. Diffraction is usually negligible at THz frequencies unless specific problems such as the characterization of blocking from humans and small objects must be addressed. Nevertheless, diffuse scattering from rough or irregular surfaces cannot be neglected if the standard deviation of surface roughness σ_h is comparable to the wavelength, which can happen for very rough surfaces or very high frequencies. Under these circumstances, reflection coefficients must be accordingly attenuated, for example using the Rayleigh factor [24] and new rays scattered by the whole surface, and therefore originating from every surface element Δs must be taken into account.

There are two main approaches for the computation of scattered rays' field. The first one is based on the Beckmann-Kirchhoff theory for scattering of incident plane waves from Gaussian rough surfaces [25]. The surface is divided into surface elements Δs in order for the incoming wave-front to be considered locally plane on Δs even for close Tx positions. Then a scattering coefficient that depends on the incident and scattering angles, as well as on the surface's roughness parameters (σ_h and correlation length) is applied to each tile to compute the intensity of the scattered field given the intensity of the incident field. Although derived for perfectly conducting surfaces, Kirchhoff's theory has been generalized to finite conductivity surfaces and successfully implemented into a RT tool [26]. The major limitations of Kirchhoff's theory are that roughness must be Gaussian and the correlation length must be much larger than the wavelength, which prevents application to very rough surfaces with dents and edges.

The second approach, called Effective Roughness (ER) approach, is semiheuristic and is based on a simple power-flow budget at the generic surface element Δs [27]. Differently from the Kirchhoff model, specular reflected wave and the scattered wave are treated as distinct waves where the attenuation of the former is due to part of its power being diverted into the latter. Therefore, the following power budget holds for a passive surface: $Pi = Pr + Ps + Pp$, where Pi, Pr, and Ps are the incident, reflected, and scattered power flows at the surface element and Pp is the remaining power that penetrates into it. The relative amplitude of the scattered wave versus the reflected wave is determined by an independent scattering parameter S, while a scattering pattern determines the directional properties of the scattered wave. The specular wave attenuation factor derives from S and from the power-budged equation. While the ER roughness model is not based on a rigorous electromagnetic theory and requires calibration of both S and the scattering pattern's parameters versus measurements, is does not suffer from the foregoing limitations of Kirchhoff's theory and has been applied with success to mm-wave as well as to THz frequencies [3].

References

1. Petrov, V., Kokkoniemi, J., Moltchanov, D., Lehtomaki, J., Koucheryavy, Y., & Juntti, M. (June 2018). Last meter indoor terahertz wireless access: Performance insights and implementation roadmap. *IEEE Communications Magazine, 56*(6), 158–165.
2. Elayan, H., Amin, O., Shihada, B., Shubair, R. M., & Alouini, M. (2020). Terahertz band: The last piece of RF spectrum puzzle for communication systems. *IEEE Open Journal of the Communications Society, 1*, 1–32.
3. Rappaport, T. S., et al. (2019). Wireless communications and applications above 100 GHz: Opportunities and challenges for 6G and beyond. *IEEE Access, 7*, 1.
4. Han, C., & Chen, Y. (2018, June). Propagation modeling for wireless communications in the Terahertz band. *IEEE Communications Magazine, 56*(6), 96–101.

5. Priebe, S., & Kürner, T. (2013, September). Stochastic modeling of THz indoor radio channels. *IEEE Transactions on Wireless Communications, 12*(9), 4445–4455.
6. Possenti, L., Pascual Garcia, J., Degli-Esposti, V., Lozano Guerrero, A., Barbiroli, M., Martinez-Inglés, M. T., Fuschini, F., Rodríguez, J. V., Vitucci, E. M., Molina-García-Pardo, J. M. (2020). Improved Fabry-Pérot electromagnetic material characterization: Application and results. Radio Science, Special Section "Radio channel modeling for 5G millimetre wave communications in built environments" 55, e2020RS007164.
7. *Effect of building materials and structures on radiowave propagation above about 100 MHz*, International Telecommunication Union Recommendation ITU-R P.2040-1, 2015.
8. Jansen, C., Piesiewicz, R., Mittleman, D., Kurner, T., & Koch, M. (2008, May). The impact of reflections from stratified building materials on the wave propagation in future indoor terahertz communication systems. *IEEE Transactions on Antennas and Propagation, 56*(5), 1413–1419.
9. Priebe, S., Jastrow, C., Jacob, M., Kleine-Ostmann, T., Schrader, T., & Kürner, T. (2011, May). Channel and propagation measurements at 300 GHz. *IEEE Transactions on Antennas and Propagation, 59*(5), 1688–1698.
10. Xing, Y., & Rappaport, T. S. (2018). Propagation measurement system and approach at 140 GHz-Moving to 6G and above 100 GHz. *2018 IEEE global communications conference (GLOBECOM), Abu Dhabi, United Arab Emirates*, 2018, pp. 1–6.
11. Ma, J., Shrestha, R., Moeller, L., & Mittleman, M. (2018, February). Channel performance for indoor and outdoor terahertz wireless links. *APL Photonics, 3*(5), 051601.
12. Khalid, N., & Akan, O. B. (2016). Wideband THz communication channel measurements for 5G indoor wireless networks. *2016 IEEE International Conference on Communications (ICC), Kuala Lumpur*, pp. 1–6, 2016.
13. Kim, S., Khan, W. T., Zajić, A., & Papapolymerou, J. (2015, July). D-band channel measurements and characterization for indoor applications. *IEEE Transactions on Antennas and Propagation, 63*(7), 3198–3207.
14. Nguyen, S. L. H., Järveläinen, J., Karttunen, A., Haneda, K., & Putkonen, J. (2018). Comparing radio propagation channels between 28 and 140 GHz bands in a shopping mall. *12th European Conference on Antennas and Propagation (EuCAP 2018)*, pp. 1–5, London, 9–13 April, 2018.
15. Pometcu, L., & D'Errico, R. (2018). Characterization of sub-THz and mm-wave propagation channel for indoor scenarios. *12th European Conference on Antennas and Propagation (EuCAP 2018)*, pp. 1–4, London, 9–13 April, 2018.
16. Priebe, S., Kannicht, M., Jacob, M., & Kürner, T. (2013, December). Ultra broadband indoor channel measurements and calibrated ray tracing propagation modeling at THz frequencies. *Journal of Communications and Networks, 15*(6), 547–558.
17. Clerkx, B., & Oestges, C. (2013). *MIMO wireless networks*. New York: Elsevier.
18. Saleh, A., & Valenzuela, R. (1987, February). A statistical model for indoor multipath propagation. *IEEE Journal on Selected Areas in Communications, 5*(2), 128–137.
19. 3GPP TR 25.996. (2009, December). *Spatial channel model for multiple input multiple output (MIMO) simulations*. [Online]. Available: http://www.3gpp.org/DynaReport/25996.htm
20. Lu, J. S., Vitucci, E. M., Degli-Esposti, V., Fuschini, F., Barbiroli, M., Blaha, J., & Bertoni, H. L. (2019, February). A discrete environment-driven GPU-based ray launching algorithm. *IEEE Transactions on Antennas and Propagation, 67*(02), 1558–2221.
21. Balanis, C. A. (1989). *Advanced engineering electromagnetics*. New York: Wiley.
22. Kouyoumjian, R. G., & Pathak, P. H. (1974, November). A uniform geometrical theory of diffraction for an edge in a perfectly conducting surface. *Proceedings of the IEEE, 62*(11), 1448–1461.
23. Fuschini, F., Vitucci, E. M., Barbiroli, M., Falciasecca, G., & Degli-Esposti, V. (2015, June). Ray tracing propagation modeling for future small-cell and indoor applications: A review of current techniques. *Radio Science, 50*(6), 469–485.
24. Ament, W. S. (1953, January). Toward a theory of reflection by a rough surface. *Proceedings of the IRE, 41*(1), 142–146.
25. Beckmann, P., & Spizzichino, A. (1987). *The scattering of electromagnetic waves from rough surfaces*. Norwood: Artech House.

26. Jansen, C., et al. (2011, November). Diffuse scattering from rough surfaces in THz communication channels. *IEEE Transactions on Terahertz Science and Technology, 1*(2), 462–472.
27. Degli-Esposti, V., Fuschini, F., Vitucci, E., & Falciasecca, G. (2007, January). Measurement and modelling of scattering from buildings. *IEEE Transactions on Antennas and Propagation, 55*(1), 143–153.

Chapter 11
Intra-device and Proximity Channel Modeling

Alenka Zajic and Danping He

Abstract Ultra-broadband terahertz (THz) communication systems are expected to help satisfy the ever-growing need for smaller devices that can offer higher-speed wireless communication anywhere and anytime. In the past years, it has become obvious that wireless data rates exceeding 10 Gbit/s will be required in several years from now. The opening up of carrier frequencies in the terahertz range is the most promising approach to provide sufficient bandwidth required for ultrafast and ultra-broadband data transmissions. This large bandwidth paired with higher-speed wireless links can open the door to a large number of novel applications such as intra-device wireless communications and device-to-device wireless communications. To design these new communication systems, the first step is to understand propagation effects in these environments at THz frequencies and develop channel models. This chapter summarizes advances in THz channel modeling for intra-device and proximity wireless communication systems.

11.1 Intra-device Channel Modeling

Communication between components, such as processor and memory within a computer system, currently relies on metal wires, and a transition to optical waveguides is expected in the future. In intra-device communications, optics promises much higher bandwidth (and thus improved computing performance). However, both wires and optics suffer from significant challenges in terms of assembly cost, airflow, service time, overall cost, etc. [1–3]. For system components, the

A. Zajic (✉)
School of Electrical and Computer Engineering, Georgia Institute of Technology, Atlanta, GA, USA
e-mail: alenka.zajic@ece.gatech.edu

D. He
State Key Laboratory of Rail Traffic Control and Safety, Beijing Jiaotong University, Beijing, China
e-mail: hedanping@bjtu.edu.cn

T. Kürner et al. (eds.), *THz Communications*, Springer Series in Optical Sciences 234, https://doi.org/10.1007/978-3-030-73738-2_11

number of pins or optical interfaces that a small chip package can have is limited, and sophisticated connections can also make component insertion (e.g., during assembly) and removal (e.g., to replace a failed component) more time-consuming and costly [4–6].

Wireless communication can alleviate such cable management, serviceability, and packaging constraints. Integration of wireless transceivers and antennas into the chip package would provide communication bandwidth without adding pins or fiber connectors to the chip package. A key challenge for wireless communication is that the required data rates in existing systems are already in the hundreds of gigabits per second for intra-system communication. For example, within a server computer system, data rates already exceed 500 Gbits/s, e.g., since late 2014, the Intel Core i7 Extreme processors and most Intel Xeon E5 v3 processors can communicate with the system's main memory using four DDR4-2133 channels, with a total throughput of 533 Gbits/s, and this is expected to soon increase to 800 Gbits/s when DRR4-3200 support is introduced.

Terahertz (THz) wireless communication has two key advantages that can be combined to achieve the required data rates. First, the usable frequency band around each frequency is much larger, so each channel can have a much higher data rate. This alone can increase data rates to several tens of Gbits/s, but multiple-input, multiple-output (MIMO) antennas are still needed to reach Tbits/s data rates. Fortunately, THz frequencies allow smaller antennas and antenna spacing, which provides for more MIMO channels within the same array aperture.

The first step to design communication systems for intra-device links is to understand propagation environment inside a metal box and derive statistical characteristics of this channel. At THz frequencies, there has been a large number of measurement campaigns that characterize indoor propagation environment including line-of-sight (LoS) propagation, non-line-of-sight (NLoS) propagation, angles of departure and arrival, shadowing effects, and reflection and diffraction from various materials [7–16]. Measurements have also been conducted to characterize waveguide-like structures with different dimensions at 60 and 300 GHz for intra-device communication [17]. Furthermore, on-board THz wireless communication measurements have been conducted by considering different possible scenarios, i.e., LoS, reflected non-line of sight (RNLoS), NLoS, and obstructed line of sight (OLoS) [9]. Finally, the channel measurements at 300 GHz inside a desktop size metal enclosure have been presented in [18].

Similarly, several channel models have been proposed for THz wireless channels in indoor environments [19–26]. The stochastic channel model based on ray-tracing (RT) has been proposed for the propagation channel of THz Kiosk download application in [19]. A scenario-specific stochastic model for THz indoor radio channels has been introduced in [20]. The stochastic model for distance-dependent angular and RMS delay spreads has been conducted in [21]. The statistical channel model for THz indoor multipath fading channels has been proposed in [22]. The performances of different large-scale path loss models at 30 GHz, 140 GHz, and 300 GHz have been compared in [23]. The models for characterization of reflections from the surfaces of different materials have been conducted in [24–26]. For signal propagation inside a waveguide structure, the channel model based on ray-tracing has been proposed in [17].

Compared to the indoor environment, THz propagation in metal enclosures experiences both traveling and resonant waves. This yields to larger number of multiple reflections as well as larger multipath spread [18]. Also, due to the resonant nature of the fields, the received power can vary with transceivers' positions. Based on these findings, a path loss model in an empty desktop size metal enclosure has been presented in [27]. To characterize path loss in metal enclosures, in addition to Friis equation terms, another term is required to describe the effect of the resonant modes. For the THz wireless channel in metal enclosure, the theoretical path loss $(PL)_{dB}$ can be calculated as

$$(PL)_{dB} = \overline{(PL)^t}_{dB} + 10log_{10}(|E|^2)^{-1} + X_\sigma, \tag{11.1}$$

where $\overline{(PL)^t}$ is the mean path loss of traveling wave and can be calculated by averaging Friis formula over the frequency band as

$$\overline{(PL)^t} = \frac{1}{\Delta f}\int_{\Delta f}\left(\frac{4\pi D^{\frac{\gamma}{2}} f}{c_0}\right)^2 df, \tag{11.2}$$

where D represents the signal traveled distance, γ is the path loss exponent, and c_0 is the speed of light.

The parameter $10\log_{10}(|E|^2)^{-1}$ which depends on the antenna height h_r represents the received power variation contributed by resonating modes and is defined as in [27]. Finally, X_σ is a zero-mean Gaussian random variable with standard deviation σ. It describes the random shadowing effects.

Based on the cavity environment and the statistical properties of the channel inside the metal cavity observed through measurements [18], a geometrical model which describes propagation in resonant cavity as a superposition of LoS, single-bounced (SB), double-bounced (DB), and multi-bounced (MB) rays is presented [28]. Figure 11.1 illustrates the propagation mechanisms inside the desktop size metal cavity with a geometry-based statistical model. Based on the geometrical model, a parametric reference model has been presented with the consideration of the signal propagation mechanisms in the excited cavity. The unique features of this model are (1) inclusion of resonant modes in the cavity into statistical channel model; (2) inclusion of path loss characterization that considers the losses due to antenna misalignment, which is required for THz wireless communication; and (3) coverage of large range of real-world scenarios, making it a very flexible and useful modeling tool. From the geometrical model, the input delay spread function of the Tx–Rx link can be expressed as the superposition of the LoS, SB, DB, and MB rays as

$$h(t,\tau) = h_{LoS}(t,\tau) + h_{SB}(t,\tau) + h_{DB}(t,\tau) + h_{MB}(t,\tau), \tag{11.3}$$

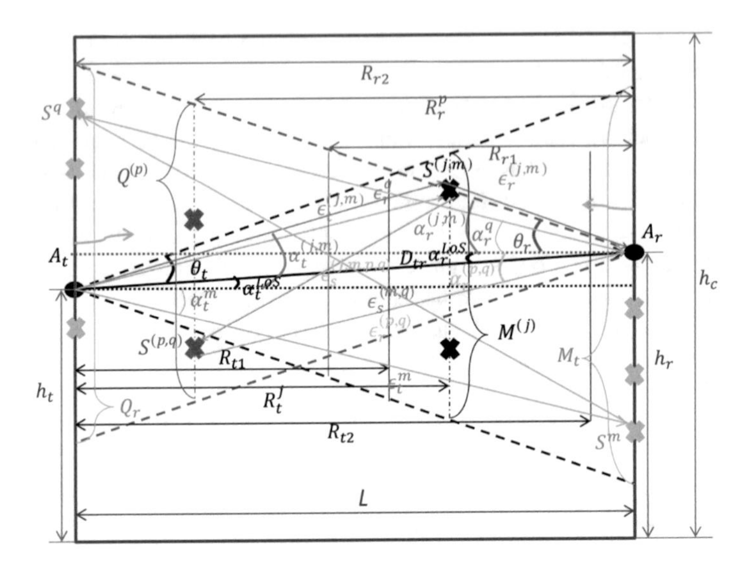

Fig. 11.1 The intact geometrical model for chip-to-chip wireless communication channel in a desktop size metal cavity [28]

where each channel impulse component is represented as a product of amplitude and Dirac impulse function that represents time delay. The amplitudes are calculated from the path loss, and time delays are obtained by calculating each ray's traveling distance and dividing it with the speed of light.

11.2 Proximity Channel Modeling

As a typical proximity communication, THz kiosk application offers ultrahigh downloads of digital information to users' handheld devices. The general configuration of a kiosk downloading application is shown in Fig. 11.2a. The Tx is deployed in a container and is surrounded by wave-absorbing material. The container has a front cover (window) that not only protects the Tx but also allows communication between Tx and receiver Rx. The front cover might be tilted with different angles in different configurations. The Rx is situated in a handheld device, and the distance between Rx and front cover may vary due to different user behaviors. The propagation environment is determined by the kiosk station design and the placements of the Tx and the Rx. Typical transmission ranges are in the order of centimeter, and the multiple paths between Tx and Rx have significant impact on the achievable data rates. The ray-tracing-based channel analysis and modeling of the THz kiosk downloading applications are first presented in [19].

In order to obtain the first-hand information of the kiosk downloading channel, channel measurements are conducted by using a vector network analyzer (VNA)

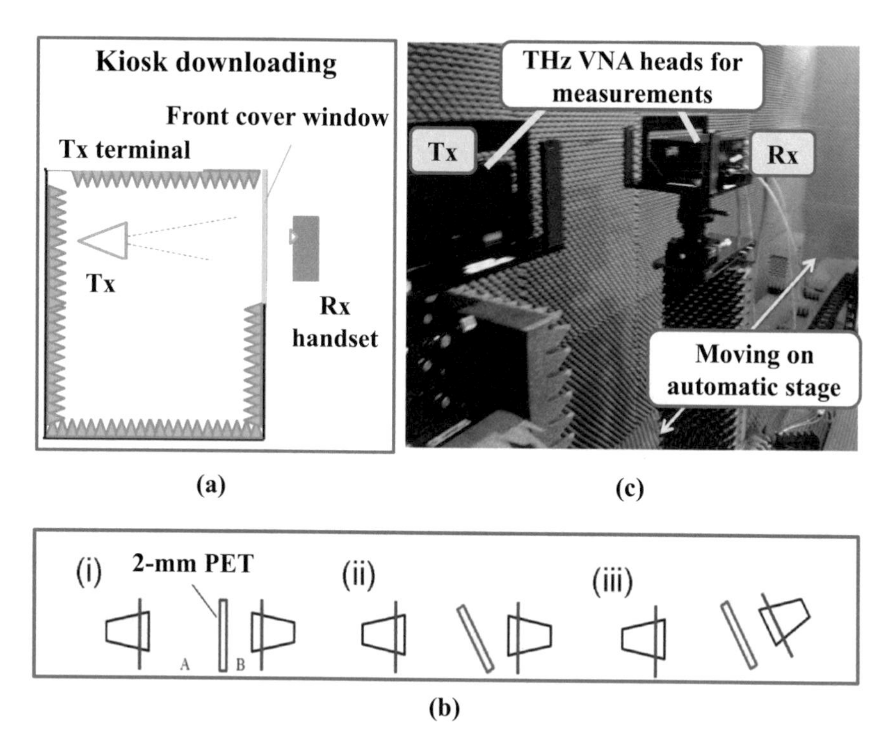

Fig. 11.2 (**a**) Overview of kiosk downloading communication system and the equivalent model [29], (**b**) three possible Kiosk downloading scenarios, (**c**) measurement campaign: VNA with automatic stage surrounded by wave absorber [29]

in an anechoic chamber between 220 GHz and 340 GHz [29]. Figure 11.2b shows three possible configurations of Tx, Rx, and window alignment that are suggested by IEEE 802.15 TG3d [29]. For all the three scenarios, both Tx and Rx have a metal plate surrounding the antenna, which represents possible hardware design, e.g., the radio unit inside the kiosk and the body of a mobile phone. In Scenario (i), the metal plates and the front cover plate are parallel. The front cover plate is tilted in Scenario (ii), while the metal plates at Tx and Rx are still parallel. In Scenario (iii), the front cover plate and Rx metal plate are tilted and parallel to each other. The measurement principle and corresponding devices are shown in Fig. 11.2c. In the measurements, a smooth polyethylene terephthalate (PET) plate with a thickness of 2 mm is used to represent the front cover. Commercially available 25-dBi horn antenna with $HPBW = 10°$ is used at both Tx and Rx. The radio channel is characterized by measurement of the S-parameters (S21) for varying distances between Tx and Rx using an automatic translation stage.

As the movements of Tx and Rx are constrained by the experimental equipment, a 3D RT simulator [30] is calibrated based on the measured data to explore more results for various deployments. Figure 11.3 shows the RT calibration result. The average errors of ray power, root-mean-square (RMS) delay spread, and Rician

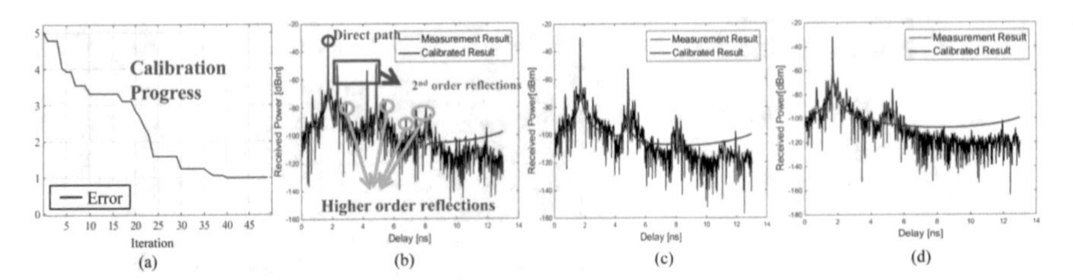

Fig. 11.3 Calibrated ray tracing results: (**a**) calibration progress of Scenario (i); (**b**) calibrated result of Scenario (i), (**c**) calibrated result of Scenario (ii); (**d**) calibrated result of Scenario (iii)

K-factor of the calibrated results are 1.0 dB, 0.02 ns, and 0.05 dB, respectively. Thereafter, key parameters of the kiosk downloading channel at all potential positions and angles can be accurately extracted and modeled via intensive calibrated RT simulations.

During an analysis of the significance of different ray types, it can be observed that the direct path and the second-order reflections contribute more than 97% of the received power. Due to the varying geometry alignments of the devices, the number of significant multipath components as well as the RMS delay spread decreases from Scenario (i) to Scenario (iii). For these, the amplitudes, time of arrival, phases, angles of departure and arrival, polarization ratios, and other key parameters of each scenario can be extracted from the RT result, and the stochastic channel model can be derived. The model is validated by an analysis of the Rician K-factors and RMS delay spreads in terms of rays, and the errors of which are less than 1.48 dB and 0.07 ns, respectively. The specific values of these channel parameters can be found in [19]. Provided that the thermal noise is not the bottleneck of the system performance, the symbol rate increases as the communication distance between Tx and Rx increases from 0.5 to 0.9 m, and the symbol rate increases as delay spread decreases. The maximum symbol rate reaches 3 GBd at 0.9 m for the evaluated system bandwidths. The example calculation proves that the data rate can reach 108 Gbit/s at 0.9 m with 60 GHz system bandwidth, when 64-QAM and carrier aggregation technologies are applied. As the average RMS delay spread decreases from Scenario (i) to (iii), the data rates of Scenario (ii) and (iii) are expected to be higher compared to Scenario (1). Therefore, the target data rate of 100 Gbit/s can be achieved for all the three scenarios with proper communication distance, system bandwidth, and technologies such as higher-order modulation and carrier aggregation. The developed channel model allows for the fast generation of channel transfer functions and can be used to guide close proximity communication system design at THz frequencies.

References

1. Chen, G., Chen, H., Haurylau, M., et al. (2006). On-chip copper-based vs. optical interconnects: Delay uncertainty, latency, power, and bandwidth density comparative predictions. In *Proceedings of the International Interconnect Technology Conference (IITC)*. https://doi.org/10.1109/IITC.2006.1648640
2. Srivastava, N., & Banerjee, K. (2005). Performance analysis of carbon nanotube interconnects for VLSI applications. In *Proceedings of the IEEE/ACM International Conference on Computer Aided Design (ICCAD)*. https://doi.org/10.1109/ICCAD.2005.1560098
3. Naeemi, A., Sarvari, R., & Meindl, J. (2006). On-chip interconnect networks at the end of the roadmap: Limits and nanotechnology opportunities. In *International Interconnect Technology Conference (IITC)*. https://doi.org/10.1109/IITC.2006.1648693
4. Banerjee, K., Souri, S., Kapur, P., et al. (2001). 3-D ICs: A novel chip design for improving deep-submicrometer interconnect performance and systems-on-chip integration. *Proceedings of the IEEE, 89*(5), 602–633. https://doi.org/10.1109/5.929647
5. Miller, D. A. B. (2000). Rationale and challenges for optical interconnects to electronic chip. *Proceedings of the IEEE, 88*(6), 728–749. https://doi.org/10.1109/5.867687
6. Havemann, R. H., & Hutchby, J. A. (2001). High-performance interconnects: An integration overview. *Proceedings of the IEEE, 89*(5), 586–601. https://doi.org/10.1109/5.929646
7. Khalid, N., & Akan, O. B. (2016). Wideband THz communication channel measurements for 5G indoor wireless networks. In *2016 IEEE International Conference on Communications (ICC)* (pp. 1–6). IEEE. https://doi.org/10.1109/ICC.2016.7511280
8. Kim, S., & Zajic, A. G. (2014). Statistical characterization of 300 GHz propagation on a desktop. *IEEE Transactions on Vehicular Technology, 64*(8), 3330–3338. https://doi.org/10.1109/TVT.2014.2358191
9. Kim, S., & Zajic, A. G. (2016). Characterization of 300 GHz wireless channel on a computer motherboard. *IEEE Transactions on Antennas and Propagation, 64*(12), 5411–5423. https://doi.org/10.1109/TAP.2016.2620598
10. Priebe, S., Jastrow, C., Jacob, M., et al. (2011). Channel and propagation measurements at 300 GHz. *IEEE Transactions on Antennas and Propagation, 59*(5), 1688–1698. https://doi.org/10.1109/TAP.2011.2122294
11. Priebe, S., Jacob, M., & Jastrow, C. (2010). A comparison of indoor channel measurements and ray tracing simulations at 300 GHz. In *35th International Conference on Infrared, Millimeter, and Terahertz Waves*. https://doi.org/10.1109/ICIMW.2010.5612330
12. Jacob, M., Priebe, S., Dickhoff, R., et al. (2012). Diffraction in mm and sub-mm wave indoor propagation channels. *IEEE Transactions on Microwave Theory and Techniques, 60*(3), 833–844. https://doi.org/10.1109/TMTT.2011.2178859
13. Kleine-Ostmann, T., Jastrow, C., Priebe, S., et al. (2012). Measurement of channel and propagation properties at 300 GHz. In *Conference on Precision Electromagnetic Measurements* https://doi.org/10.1109/CPEM.2012.6250900
14. Fricke, A., Rey, S., Achir, M., et al. (2013). Reflection and transmission properties of plastic materials at THz frequencies. In *38th International Conference on Infrared, Millimeter, and Terahertz Waves (IRMMW-THz)*. https://doi.org/10.1109/IRMMW-THz.2013.6665413
15. Fricke, A., Achir, M., Le Bars, P., et al. (2015). *Characterization of transmission scenarios for terahertz intra-device communications*. In *IEEE-APS Topical Conference on Antennas and Propagation in Wireless Communications (APWC)*. https://doi.org/10.1109/APWC.2015.7300195
16. Cheng, C., Kim, S., & Zajić, A. (2018). Study of diffraction at 30 GHz, 140 GHz, and 300 GHz. In *IEEE International Symposium on Antennas and Propagation USNC/URSI National Radio Science Meeting*. https://doi.org/10.1109/APUSNCURSINRSM.2018.8608803
17. Kürner, T., Fricke, A., Rey, S., et al. (2015). Measurements and modeling of basic propagation characteristics for intra-device communications at 60 GHz and 300 GHz. *Journal of Infrared, Millimeter, and Terahertz Waves, 36*(2), 144–158. https://doi.org/10.1007/s10762-014-0117-5

18. Fu, J., Juyal, P., & Zajić, A. (2019). THz channel characterization of chip-to-chip communication in desktop size metal enclosure. *IEEE Transactions on Antennas and Propagation, 67*(12), 7550–7560. https://doi.org/10.1109/TAP.2019.2934908
19. He, D., Guan, K., Ai, B., et al. (2017). Stochastic channel modeling for kiosk applications in the terahertz Band. *IEEE Transactions on Terahertz Science and Technology, 7*(5), 502–513. https://doi.org/10.1109/TTHZ.2017.2720962
20. Priebe, S., & Kürner, T. (2013). Stochastic modeling of THz indoor radio channels. *IEEE Transactions on Wireless Communications, 12*(9), 4445–4455. https://doi.org/10.1109/TWC.2013.072313.121581
21. Priebe, S., Jacob, M., & Kürner, T. (2014) Angular and rms delay spread modeling in view of THz indoor communication systems. *Radio Science, 49*(3), 242–251. https://doi.org/10.1002/2013RS005292
22. Kim, S., & Zajic, A. (2016). Statistical modeling and simulation of short-range device-to-device communication channels at sub-THz frequencies. *IEEE Transactions on Wireless Communications, 15*(9), 6423–6433. https://doi.org/10.1109/TWC.2016.2585103
23. Cheng, C., Kim, S., & Zajić, A. (2017). Comparison of path loss models for indoor 30 GHz, 140 GHz, and 300 GHz channels. In *11th European Conference on Antennas and Propagation (EUCAP)*. https://doi.org/10.23919/EuCAP.2017.7928124
24. Piesiewicz, R., Jansen, C., & Mittleman, D., et al. (2007). Scattering analysis for the modeling of THz communication systems. *IEEE Transactions on Antennas and Propagation, 55*(11), 3002–3009. https://doi.org/10.1109/TAP.2007.908559
25. Fricke, A., Kürner, T., Achir, M., et al. (2017). A model for the reflection of terahertz signals from printed circuit board surfaces. In *11th European Conference on Antennas and Propagation (EUCAP)*. https://doi.org/10.23919/EuCAP.2017.7928148
26. Priebe, S., Jacob, M., & Jansen, C. (2011). Non-specular scattering modeling for THz propagation simulations. In *Proceedings of the 5th European Conference on Antennas and Propagation (EUCAP)*, ISSN: 2164-3342.
27. Fu, J., Juyal, P., & Zajić, A. (2019). Path loss model as function of antenna height for 300 GHz chip-to-chip communications. In *IEEE International Symposium on Antennas and Propagation and USNC-URSI (APS)*. https://doi.org/10.1109/APUSNCURSINRSM.2019.8888326
28. Fu, J., Juyal, P., & Zajić, A. (2020). Modeling of 300 GHz chip-to-chip wireless channels in metal enclosures. *IEEE Transactions on Wireless Communications, 19*(5), 3214–3227. https://doi.org/10.1109/TWC.2020.2971206
29. Fricke, A., et al. (2016). Channel modelling document (CMD). IEEE 802.15 Plenary Meeting, Macau, 2016, DCN: 15-14-0310-19-003d.
30. He, D., Ai, B., Guan, K., et al. (2019). The design and applications of high-performance ray-tracing simulation platform for 5G and beyond wireless communications: A tutorial. *IEEE Communications Surveys & Tutorials, 21*(1), 10–27. https://doi.org/10.1109/COMST.2018.2865724

Chapter 12
Backhaul/Fronthaul Outdoor Links

Akihiko Hirata, Bo Kum Jung, and Petr Jurcik

Abstract One of the promising applications of THz wireless link is mobile backhaul/fronthaul. Terahertz mobile backhaul/fronthaul is outdoor fixed wireless link over a distance of from tens of metres to several kilometres. This chapter describes the system design, available link distance, technical challenges and field trials of backhaul/Fronthaul THz wireless links.

Fast growing network traffic and increasing demand for high speed connectivity over the next years require new wireless technologies to build communication systems supporting ultra-high data rates. The growth of data traffic, capacity and connectivity in radio access networks must also be reflected in wired or wireless transport networks (i.e. backhaul and fronthaul links). In general, wireless systems offer important advantages over optical fibres and free-space optical (FSO) alternatives not only for mobile and nomadic terminals, but also in numerous fixed communication scenarios. In fixed outdoor applications, the deployment of optical fibre (or any other wired alternative) is often prohibitively expensive, technically unfeasible or too time-consuming. In addition, for high data rate indoor applications, such as machine-to-machine/device-to-device communication in data centres, wireless local and personal area networks, smart offices and home theatres, the versatility of wireless communication systems is a major asset over any kind of wired solution (e.g. optical fibres). On the other hand, FSO communication using infrared (IR) laser light can avoid the aforementioned drawbacks of optical fibre, but compared to THz wireless communication, the IR signal is significantly more attenuated by

A. Hirata (✉)
Chiba Institute of Technology, Narashino, Japan
e-mail: hirata.akihiko@p.chibakoudai.jp

B. K. Jung
Institute for Communications Technology, Technische Universität Braunschweig, Braunschweig, Germany

P. Jurcik
Deutsche Telekom, Prague, Czech Republic

T. Kürner et al. (eds.), *THz Communications*, Springer Series in Optical Sciences 234, https://doi.org/10.1007/978-3-030-73738-2_12

the presence of dust in the air than the THz signal which undergoes almost no degradation [1]. In general, wireless systems profit from fast deployment, flexibility and easy reconfiguration, as well as lower deployment costs (CAPEX).

12.1 System Design of Wireless Backhaul/Fronthaul

Since the concept of wireless Backhaul/Fronthaul is to deliver aggregated data from UE to anchor point, the high capacity of wireless link is demanded. In this sense, the 5G access networks will extend the applied frequency spectrum above 6 GHz (e.g. to the 26/28 GHz, V-band) where wireless transport links are currently operated. Hence, the further improvement in data capacity of wireless transport networks will be limited by the availability of frequency spectrum below 100 GHz, which will be intensively used by 5G services. Significant allocation of higher frequency bands beyond 100 GHz is expected to be necessary for the next generation(s) of wireless transport networks. The characteristics of the most significant and promising frequency bands used in current or future multi-gigabit wireless transport networks (fixed services) are summarised in Table 12.1 [2].

The W-band and D-band have been already considered in the table of frequency allocation issued by the ITU-R Radio Regulations for fixed wireless services [3]. The frequency spectrum beyond 275 GHz has not yet been allocated, but it is a key agenda item for the ITU World Radio communication Conference (WRC) 2019 [4].

Table 12.1 Characteristics of the frequency bands under consideration [2]

	Frequency [GHz]	Total BW [GHz]	Type of licensing	Max link capacity [Gbps]	Max link length [km]
V-band	57–66	9 (14)	Unlicensed	1–2	<1
	(51–71)				
E-band	71–76	10 (5 + 5)	Lightly licensed	10–20	2-3 k
	81–86				
W-band	92–94	17.85	Lightly licensed	Expectation ~40	<1
	94.1–100				
	102–109.5				
	111.8–114.25				
D-band	130–134	31.8	NA	Expectation ~40	<1
	141–148.5				
	151.5–164				
	167–174.8				
THz-band	252–325	73	NA	Expectation >100–200	<1

The volume of data traffic consumed by the 5G and beyond 5G (B5G) use-cases, services and applications is expected to significantly grow in comparison to today's 4G/LTE generation. A factor of approximately 5–10× is foreseen. Evolving from 4G/LTE to 5G network architecture, the main change is that the original single-node baseband functions in 4G/LTE are split between Central Unit (CU), Distributed Unit(s) (DU) and Radio Unit(s) (RU) resulting in a so-called centralised network architecture with functional split.

The required throughput of the transport network depends greatly on the particular split option. The lower the split point, the higher the required throughput and the typical configurations of 5G macro BS require hundreds to thousands of Gbps of throughout. Such throughputs can be served by the future terahertz-based products targeting throughputs beyond 100 Gbps [5]. Table 12.2 shows the example specification of THz band wireless link [5]. Table 12.3 shows maximum achievable link distance where 100 Gbps is available [5]. The throughputs are calculated from bandwidth and modulation scheme (efficiency) appropriate to the realised CNR at each condition. The throughput is not proportional to the bandwidth since the bandwidth expansion causes SNR degradation. The receiver sensitivities have been derived by link level simulations and the SNR requirements for a BER of 10. The latter values have been derived by link level simulations and the SNR requirements for a BER of 10^{-12}. In case of single carrier system, bandwidth of 25.92 GHz (64-QAM) or 51.84 GHz (8-PSK, 16-QAM) is necessary for 100 Gbps data transmission. When four channels are used, bandwidth necessary for 100-Gbps throughput becomes 12.96 GHz (8-PSK, 16-QAM).

The maximum transmission distance is estimated under the further assumption of a 20 dB margin for atmospheric attenuation due to weather conditions. In case of single carrier system, the maximum available link distance becomes 168–352 m. In case of four channel system, the maximum available link distance increases to 470–704 m.

Table 12.2 Example of beyond 5G THz backhaul/fronthaul [5]

Parameter	Value	Remarks
RF frequency [GHz]	300	
Baud rate [Gbaud]	BW/1.2	Due to roll-off
NF [dB]	10	T = 300 K
TX power [dBm]	10	
Link distance [m]	1000	
Antenna gain [dBi]	50	Common for both TX and RX
Payload rate	0.9	Payload/frame length

Table 12.3 Maximum available link distance of THz link, where 100 Gbps is available [5]

Number of Carrier	Bandwidth (GHz)	Modulation	FEC Rate	Receiver Sensitivity (dBm)	Maximum Link Distance (m)
1	51.84	8-PSK	14/15	−43	352
1	51.84	16-QAM	14/15	−39	235
1	25.92	64-QAM	14/15	−36	168
2	25.92	8-PSK	14/15	−46	498
2	25.92	16-QAM	14/15	−42	332
2	12.96	64-QAM	14/15	−40	237
4	12.96	8-PSK	14/15	−49	704
4	12.96	16-QAM	14/15	−45	470

12.2 Challenges of Wireless Backhaul/Fronthaul

Researches about wireless Backhaul/Fronthaul are in general early stage and currently taking the first step especially into using of sub-THz frequency spectrum. Therefore, there are several facing challenges to overcome in order to utilise new technologies in live access networks. First of all, Backhaul/Fronthaul link should be principally extreme stable otherwise customers will experience bad quality of service since the function of Backhaul/Fronthaul is to transport accumulated packets from users to the core network or anchor points; therefore, unstable channel may high probably occur naturally high BER which results unsuccessful transmission of data. In this sense, wireless channel has flaw due to the nature property of fluctuate channel state occurred by discrete reasons. Herewith weather condition is counted. Since target service environment of wireless Backhaul/Fronthaul is outdoor usage, the impact of weather changes on communication channel is not avoidable. In general, attenuation of propagating signal cased by weather is more sensible at higher frequency spectrum [6–8]. In addition, an exemplar of attenuation under average weather condition of pertaining cities is provided in [5]. Out of the weather condition itself, antenna mismatching has huge influence on the wireless communication. Free space attenuation of radiate energy is highly relevant on the frequency. Consequently, using 300 GHz frequency spectrum encounters high attenuation, e.g. 122 dBm attenuation over 100 m at 300 GHz shown in Fig. 12.1.

As a result, communication range considering link budget is highly limited. The average propagation loss of the novel THz band is 6.6 dB higher than D-band, 10 dB higher than W-band, 12.4 dB higher than E-band and 13.1 dB higher than V-band for a link distance of 100 m. The free-space loss is the dominant factor which limits achievable link distance. Due to high atmospheric attenuation in V-band (up to 1.5 dB per 100 m), the propagation loss rises more steeply than for the other frequency bands. For example, the propagation loss in V-band is lower than E-band at the link distance below 150 m, but higher than in D-band at link distances above 600 m. Note that due to additional atmospheric attenuation at V-band, there is no significant advantage of V-band over the THz band at link distances above

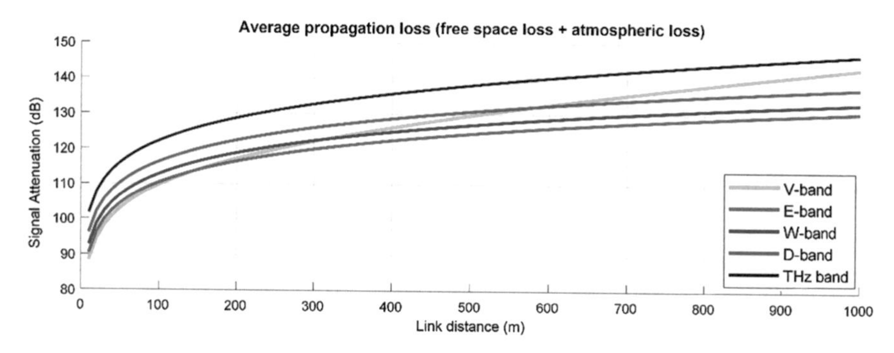

Fig. 12.1 Average propagation loss as a function of link distance

600 m. To overcome this high attenuation, high gain antenna even approximately 50 dBi antenna both of Tx and Rx is indispensable key for supporting a current wireless communication technique. However, as the gain of antenna is higher, main beam width of antenna becomes narrower. In other words, antenna should be pointing in right way unless received power of signal to wireless communicate will be decreased rapidly according to the mismatched angle [9]. By influence of the mismatch of antenna, SINR is likely to drop which limits potential feasible capacity of wireless channel in the sense of coding/modulation scheme and occurs further wireless link failure in extreme cases. Besides of propagation perspective, extreme real-time communication is one of the new application features of 5G according to NGMN 5G white paper [10]. In the 5G mobile network, generally under 10 ms within some specific cases under 1 ms of latency is aimed to be supported. Which means, entire computational time including coding/modulation of communication path between end user to end user should be guaranteed in order to support extreme low latency communication. In fact, latency is dominantly determined by more or less computational time of physical devices. This leads to the limited number of allowed relay (anchor) cell sites. Consequently, entire planning method of wireless Backhaul/Fronthaul is bounded by the computational speed of physical devices.

References

1. Su, K., Moeller, L., Lothar, R., Barant, B., & Federici, J. F. (2012). Experimental comparison of terahertz and infrared data signal attenuation in dust clouds. *Journal of the Optical Society of America A, 29*(11), 2360–2366.
2. *ThoR deliverable D2.1 requirements for B5G backhaul/fronthaul*. https://thorproject.eu/wp-content/uploads/2018/12/ThoR-D2.1-Requirements-for-B5G-back-fronthaul_NOT-YET-APPROVED.pdf
3. ITU-R. (2016). *Radio regulations*.
4. WRC. (2019). *ITU World Radio communication Conference 2019*. [Online]. Available: https://www.itu.int/en/ITU-R/conferences/wrc/2019

5. *ThoR deliverable D2.2 overall system design*. https://thorproject.eu/wp-content/uploads/2019/07/ThoR_SIKLU_190417_F_WP2-D2.2-Overall-System-Design.pdf
6. ITU-R P.676-12. (2019, August). *Attenuation by atmospheric gases and related effects*.
7. ITU-R P.838-3. (2005, March). *Specific attenuation model for rain for use in prediction methods*.
8. ITU-R P.840-8. (2019, August). *Attenuation due to clouds and fog*.
9. ITU-R WP3M Contribution 415. (2019). *Annex 01 – Liaison statement to Asia-Pacific Telecommunity – Task Group Fixed Wireless Systems – Working document towards a preliminary draft new APT[Recommendation/Report] on 'model[s] for FWS link performance degradation due to wind'*.
10. NGMN Alliance (2015, February). *NGMN 5G WHITE PAPER, v1.0*.

Chapter 13
Smart Rail Mobility

Ke Guan and Bo Ai

Abstract As a widely acknowledged efficient and green transportation model, rail traffic is expected to evolve into a new era of "smart rail mobility" where infrastructure, trains, and travelers will be interconnected to achieve optimized mobility, higher safety, and lower costs. Thus, a seamless high-data rate wireless connectivity with up to dozens of GHz bandwidth is required. Such a huge bandwidth requirement motivates the exploration of the underutilized millimeter (mm) wave and terahertz (THz) bands. In this chapter, the motivations of developing mmWave and THz communications for railway are clarified by first defining the applications and scenarios required for smart rail mobility. Then, the wireless channel in one "smart rail mobility" scenario—the intra-wagon scenario—is characterized through ultra-wideband (UWB) channel sounding and ray-tracing at mmWave and THz bands. Moreover, the train-to-infrastructure (T2I) inside-station channel is measured, simulated, and characterized at the THz band for the first time. All parameters are fed into and verified by the 3GPP-like quasi-deterministic radio channel generator (QuaDRiGa). This can provide the foundation for future work that aims to add the smart rail mobility scenario into the standard channel model families and, furthermore, provides a baseline for system design and evaluation of THz communications. Finally, we point out the future directions toward the full version of the smart rail mobility which will be powered by THz communications.

13.1 Introduction

Empowered by future communication technologies, such as the sixth-generation mobile communication system (6G), rail transport is expected to evolve into a new era of "smart rail mobility" where infrastructure, trains, passengers, and goods will be fully interconnected. In this vision, railway communications are required to

---

K. Guan (✉) · B. Ai
Beijing Jiaotong University, Beijing, China
e-mail: kguan@bjtu.edu.cn; boai@bjtu.edu.cn

T. Kürner et al. (eds.), *THz Communications*, Springer Series in Optical Sciences 234, https://doi.org/10.1007/978-3-030-73738-2_13

evolve from handling only the critical signaling applications to various high data rate applications: on-board and wayside high-definition (HD) video surveillance, on-board real-time high data rate connectivity, train operation information, real-time train dispatching HD video, and multimedia journey information [1, 2]. These applications should be realized in the five scenarios—train-to-infrastructure (T2I), inside-station, train-to-train (T2T), infrastructure-to-infrastructure (I2I), and intra-wagon [3]. Figure 13.1 illustrates the panorama of all the five communication scenarios of smart rail mobility.

According to the evaluation approach introduced by Guan et al. [3], the bandwidth requirements in both the inside station and I2I scenarios go from several hundred MHz to several GHz. For the T2T and intra-wagon scenarios, up to dozens of GHz bandwidths will be required, respectively. The biggest challenge is posed by the T2I scenario because this link needs to achieve very high data rates, low latencies, as well as close to 100% availability while traveling at speeds around 500 km/h (as of 2018, high-speed trains do not exceed 350 km/h in commercial travels, but the speed record is at 574 km/h since 2007). As the main interface between the network on-train and the fixed network, in the T2I scenario, an aggregated stream for the backhaul for both T2T and intra-wagon scenarios is transmitted. Therefore, a bandwidth of dozens of GHz (or even higher) is needed to accommodate up to 100 Gbps data rates. Such high data rate and huge bandwidth requirements form a strong driving force to exploit the the terahertz (THz) band, i.e., beyond 300 GHz, where the available spectrum is massively abundant [4, 5]. Hence, in order to effectively support the design, evaluation, and development of the THz communication-enabled smart rail mobility, the fundamental knowledge of the propagation channel characteristics is of paramount significance [6].

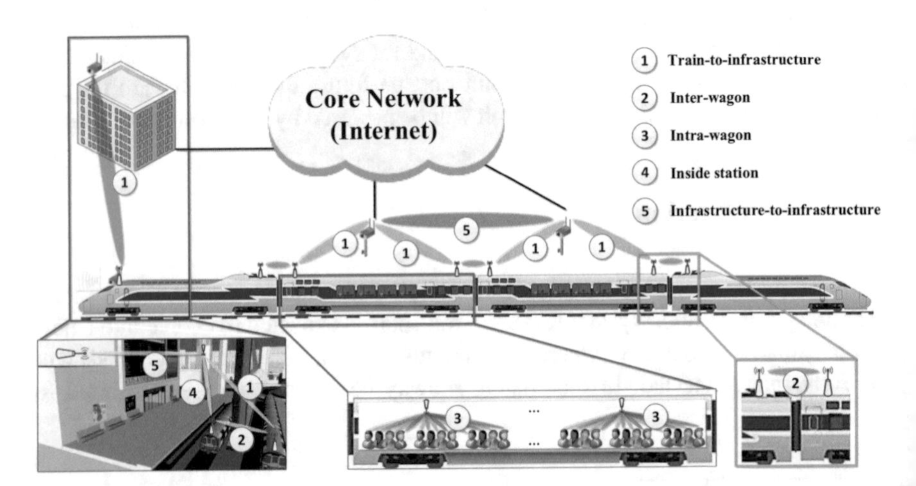

Fig. 13.1 Panorama of all the five communication scenarios of smart rail mobility [3] (Copyright © 2016 IEEE, reproduced with permission)

13.2 Train-to-Infrastructure Inside-Station Channel Characteristics at the Terahertz Band

The T2I inside-station channel at 300 GHz are characterized through measurement, simulation, and modeling for the first time in [7]. To begin with, channel measurements are conducted in a train test center at 304.2 GHz with 8 GHz bandwidth, using an M-sequence (a kind of pseudorandom sequence) correlation-based ultra-wideband (UWB) channel sounder. This channel sounder is composed of a controlling laptop, a base unit, sensor nodes, frequency extensions, and cables. The M-sequence is generated by the clock frequency of 9.22 GHz, which has a length of 4095 chips and duration of 444.14 ns, allowing a maximum path length of around 133 m. Correspondingly, the delay resolution is the chip duration of 108.5 ps which corresponds to a length of 3.25 cm. In the measurements, the channel impulse responses (CIRs) at 304.2 GHz with 8 GHz bandwidth are measured. This sounder has been reported and detailed in [8]. More information of doing measurements using this channel sounder can be found in [9]. As shown in Fig. 13.2, two groups of measurements emulating the T2I inside-station channel are done in a train test center with trains, tracks, and lampposts, highly similar to the T2I inside-station scenario.

The Rician K-factor and root-mean-square (RMS) delay spread extracted from measurements are 3.52–3.60 dB and 8.92–9.23 ns, respectively. The Rician K-factors are larger than the measured results (1–3 ns) in an open train station at 930 MHz [10]; the RMS delay spreads are smaller than the measured results (16 ns) in a railway depot at 60 GHz [11]. This implies that the multipath propagation in THz channel in a train test center and its similar scenarios such as T2I inside station is sparser than the lower-frequency channels.

With the purpose of physically interpreting the measurement results of channel sounding, an in-house-developed high-performance computing (HPC) and cloud-

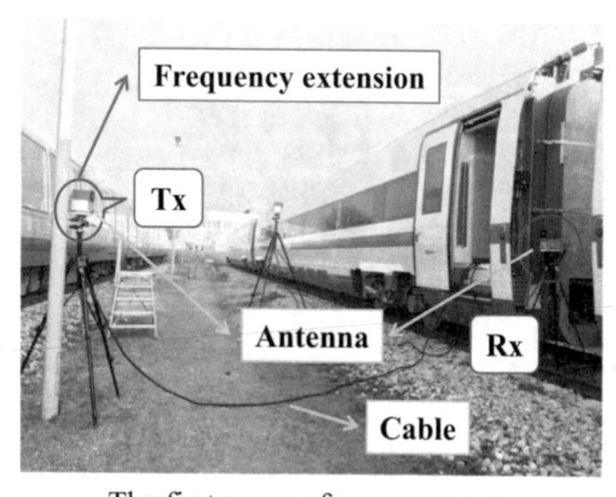

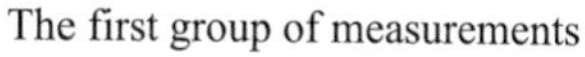
The first group of measurements

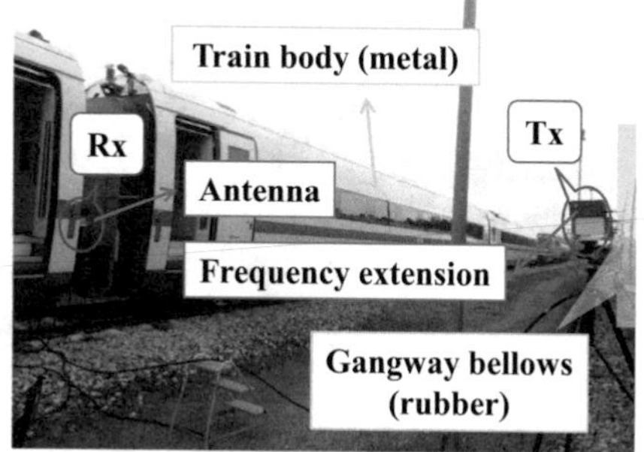

The second group of measurements

Fig. 13.2 Measurement campaign in a train test center. [7] (Copyright © 2019 IEEE, reproduced with permission)

based three-dimensional (3D) ray-tracing (RT) simulator is applied to the same scenario of measurement campaign. This RT simulator—CloudRT—is jointly developed by Beijing Jiaotong University and Technische Universität Braunschweig. It can trace rays corresponding to various propagation mechanisms, such as direct rays, reflected rays, scattered rays, etc., and is validated and calibrated by a large number of measurements at sub-6 GHz [12, 13] and THz band [14, 15]. More than ten properties of each ray can be output from RT results, such as reflection order, time of arrival, power, AoA, azimuth angle of departure (AoD), elevation angle of arrival (EoA), elevation angle of departure (EoD), and so on. More information on CloudRT can be found in tutorial [16] and http://www.raytracer.cloud/.

As shown in Fig. 13.3, the in-house-developed RT simulator confirms that the most important multipath in the first measurement is a second-order reflection even from the inner of the train wagon through the open door, while for the second measurement, even the fourth-order reflection can influence the channel. These findings reflect the fact that when intensive metallic reflectors exist, the THz propagation paths could be formed through multiple reflections. Thus, any metallic objects with smooth surface and dimensions obviously larger than the wavelength of THz wave should not be ignored.

Then, the measurement-validated RT simulator is utilized to do extensive simulations in a reconstructed realistic train station model for totally four cases of T2I inside-station channels defined by different Tx deployments and train conditions. Through analyzing the channel characteristics based on RT results, it is suggested to deploy the Tx on the catenary mast because it has smaller path loss and RMS delay spread. Moreover, it is found that when the Tx is on the catenary mast, the train on the adjacent track provides strong multipaths and enhances the waveguide effect, which decreases the path loss and suppresses the time dispersion of the channel. More details of the channel characteristics in this scenario can be found in [7], which provides the foundation for future work that aims to add the T2I scenario into the standard channel model families, providing a baseline for system design and evaluation of relevant THz railway communications.

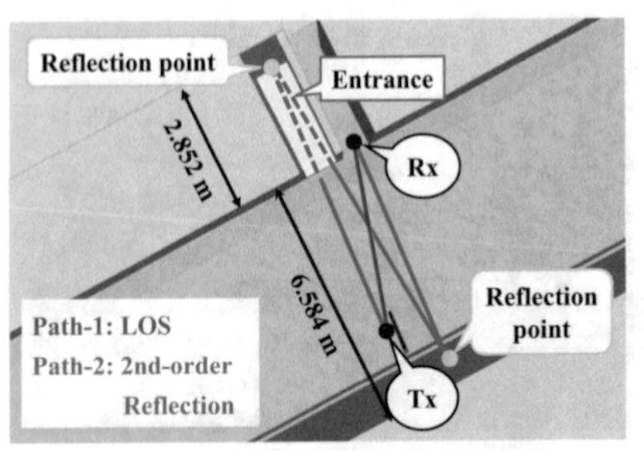

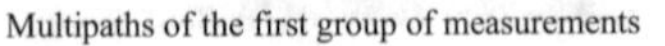
Multipaths of the first group of measurements

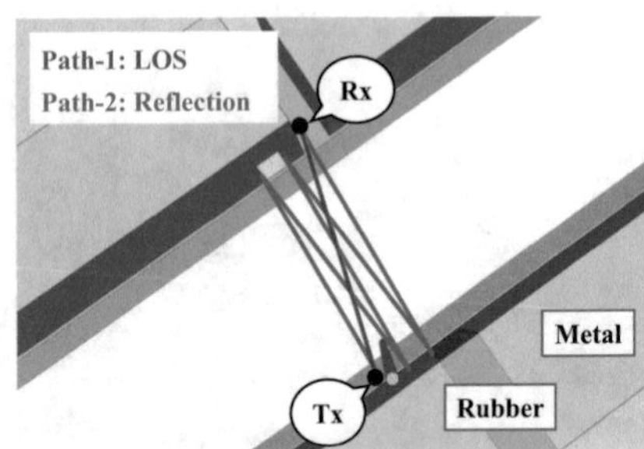

Multipaths of the second group of measurements

Fig. 13.3 Significant rays in the measurement campaign in a train test center. [7] (Copyright © 2019 IEEE, reproduced with permission)

13.3 Intra-wagon Channel Characteristics at the Terahertz Band

The intra-wagon channels at the 300 GHz band are characterized through measurement-validated RT simulations in [17]. As shown in Fig. 13.4, the CloudRT is calibrated and validated by a series of THz channel measurements inside a high-speed train wagon. With this measurement-validated RT simulator, extensive simulations with different transmitter (Tx) and receiver (Rx) deployments are conducted. At low frequencies, the channel is strongly influenced by the line of sight (LOS) and, therefore, is usually classified into LOS and non-LOS (NLOS) regions. However, the simulation results at 300 GHz bands show that the first-order reflection also imposes a significant impact on the channel characteristics. This motivates us to further classify the NLOS region into light NLOS (L-NLOS) and deep NLOS (D-NLOS) according to the existence of the first-order reflection. With this new classification, the promising NLOS regions can be distinguished from the real dark ones, providing a more precise baseline for channel modeling and system evaluation.

As can be seen in Fig. 13.5, the LOS, L-NLOS, and D-NLOS regions identified by RT simulations are marked by yellow, green, and dark blue, respectively. When only one Tx can be deployed, putting it at the ceiling in the center can offer 6.20 and 6.40% of the area ratios of LOS and L-NLOS higher than deploying it on the side. But still, even the Tx-center strategy can only achieve 49.50% of the LOS and L-NLOS regions in total. The rest 50.50% of the area will be hard to build the link. However, if we combine these two deployment strategies, the total area ratio of the LOS and L-NLOS regions can achieve 78.41%. Thus, the optimum solution is to deploy three transmitters per wagon, one at the ceiling in the center (Tx-center) and two at both end sides (Tx-side1 and Tx-side2). In this way, most of the users inside wagon receive either LOS or at least one strong reflection to build the THz links.

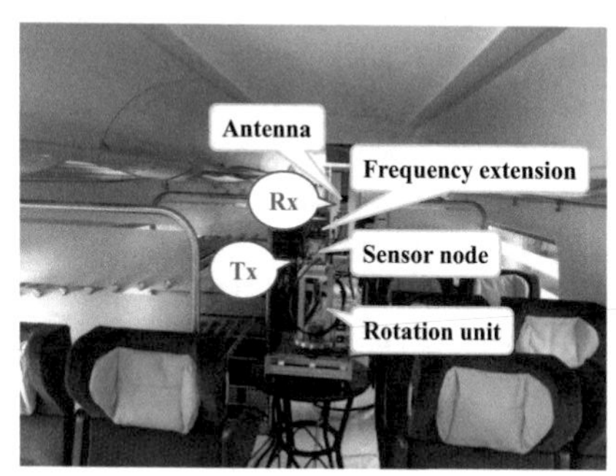

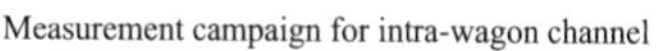
Measurement campaign for intra-wagon channel

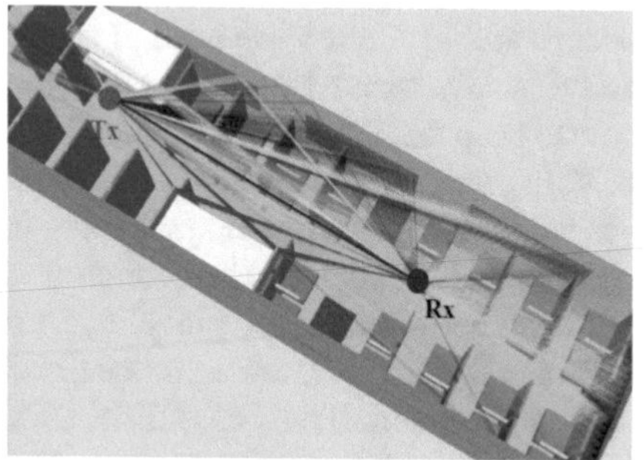

Ray-tracing simulations for intra-wagon channel

Fig. 13.4 Measurement campaign and ray-tracing simulations for intra-wagon channel. [17] (Copyright © 2019 IEEE, reproduced with permission)

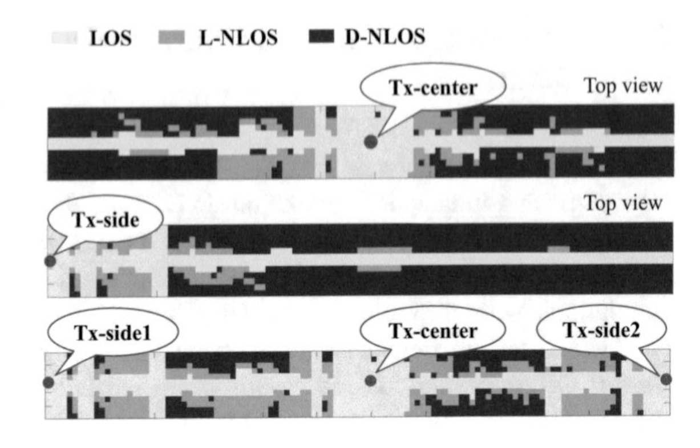

Fig. 13.5 LOS, L-NLOS, and D-NLOS regions identified by RT simulations. [17] (Copyright © 2019 IEEE, reproduced with permission)

Based on RT simulation results, a total of 12 cases (3 propagation regions with 2 Tx deployments at 2 frequencies) are characterized in terms of path loss, shadow fading, root-mean-square delay spread, Rician K-factor, azimuth/elevation angular spread of arrival/departure, cross-polarization ratio, and their cross-correlations. The values of all these parameters can be found in [17]. These results provide valuable insights into the system design and evaluation for intra-wagon communications.

13.4 Future Directions on Smart Rail Mobility

In this section, we draw the roadmap and discuss the future directions by defining two versions of the smart rail mobility paradigm—the basic version and the full version—in terms of data rate, bandwidth, carrier frequency, technical challenges, and implementation timeline.

In the basic version, considering the available bandwidths at mmWave band from 1 GHz (at 28 GHz) to 9 GHz (at 60 GHz), data rates of 1–10 Gbps could be achieved even with moderate spectral efficiencies. However, considering adaptive beam-forming at mmWave band, if using current beam training protocols involving an exhaustive search or prioritized sector search ordering, challenges will arise on how to keep the T2I link configuration time within the latency bounds of HD video or VoIP services. Moreover, novel handover scheme (e.g., efforts in [18]) and multi-beam concurrent transmission scheme (such as the heterogeneous multi-beam cloud radio access network architecture proposed in [19]) will be needed for improving handover performance and providing seamless mobility and coverage for mmWave network. To sum up, referring to the timeline of the standardization of 5G mmWave railway communications, e.g., IEEE 802.15 Interest Group HRRC (Interest Group High Rate Rail Communications), as well as the 5G standardization in 3GPP and ITU-R, mmWave communications are practical to enable the basic version of smart rail mobility in the next 2 years—2020–2025.

For the full version, taking advantage of the ultrahigh bandwidths beyond 20 GHz available in the terahertz (THz) frequency range, higher data rate transmission (i.e., beyond 100 Gbps) is possible to achieve. However, compared to the mmWave band, the beam width at THz frequencies is much narrower (even 1°) in order to compensate the higher path loss. Thus, it is more challenging to realize adaptive beam-forming in the T2I scenario. But if we can accurately predict the train location, it is possible to avoid (or at least greatly mitigate) the need of beam training, and then the aforementioned disadvantage of longer link configuration time at THz can be neglected. Note that the speed of a train can be time-variant, which will raise challenges to design competent localization algorithms. For inter-car and infrastructure-to-infrastructure scenarios, to overcome the performance loss from aerodynamics-induced beam misalignment, it is still necessary to develop efficient beam alignment techniques (like the new angle of arrival estimation algorithms in [20]), which are more challenging for THz due to the narrower beam widths. But from a long-term point of view, THz communications have broader prospects, because the very strong distance-frequency-dependent characteristics of the THz band can motivate the development of novel communication schemes, such as distance-aware multi-carrier modulation (DAMC), hybrid beam-forming scheme with DAMC transmission, etc. To summarize, according to the progress on THz communication standardization, for instance, IEEE 802.15.3d-2017 and IEEE 802.15 IGthz, the full version of smart rail mobility providing a seamless high data rate connectivity beyond 100 Gbps is expected to be enabled by THz communications between 2025 and 2030.

References

1. Moreno, J., Riera, J. M., Haro, L. D., & Rodriguez, C. (2015). A survey on future railway radio communications services: Challenges and opportunities. *IEEE Communications Magazine, 53*(10), 62–68. https://doi.org/10.1109/MCOM.2015.7295465
2. Ai, B., Guan, K., Rupp, M., et al. (2015). Future railway services-oriented mobile communications network. *IEEE Communications Magazine, 53*(10), 78–85. https://doi.org/10.1109/MCOM.2015.7295467
3. Guan, K., Li, G., Kürner, T., et al. (2016). On millimeter wave and THz mobile radio channel for smart rail mobility. *IEEE Transactions on Vehicular Technology, 66*(7), 5658–5674. https://doi.org/10.1109/TVT.2016.2624504
4. Rappaport, T. S., Xing, Y., Kanhere, O., et al. (2019). Wireless communications and applications above 100 GHz: Opportunities and challenges for 6G and beyond. *IEEE Access, 7*, 78729–78757. https://doi.org/10.1109/ACCESS.2019.2921522
5. Mumtaz, S., Jornet, J. M., Aulin, J., et al. (2017). Terahertz communication for vehicular networks. *IEEE Transactions on Vehicular Technology*. https://doi.org/10.1109/TVT.2017.2712878
6. Kürner, T., & Priebe, S. (2013). Towards THz communications—status in research, standardization and regulation. *Journal of Infrared, Millimeter, and Terahertz Waves, 35*(1), 53–62. https://doi.org/10.1007/s10762-013-0014-3
7. Guan, K., Peng, B., He, D., et al. (2019). Measurement, simulation, and characterization of train-to-infrastructure inside-station channel at the terahertz band. *IEEE Transactions on Terahertz Science and Technology, 9*(3), 291–306. https://doi.org/10.1109/TTHZ.2019.2909975

8. Rey, S, Eckhardt, J. M., Peng, B., et al. (2017). Channel sounding techniques for applications in THz communications: A first correlation based channel sounder for ultra-wideband dynamic channel measurements at 300 GHz. In *2017 9th International Congress on Ultra Modern Telecommunications and Control Systems and Workshops (ICUMT)*, Munich.
9. Vitucci, E. M., Zoli, M., Fuschini, M., et al. (2018). Tri-band mm-wave directional channel measurements in indoor environment. In *2018 IEEE 29th Annual International Symposium on Personal, Indoor and Mobile Radio Communications (PIMRC)*, Bologna.
10. Guan, K., Zhong, Z., Ai, B., et al. (2014). Propagation measurements and analysis for train stations of high-speed railway at 930 MHz. *IEEE Transactions on Vehicular Technology, 63*(8), 3499–3516. https://doi.org/10.1109/TVT.2014.2307917
11. Bulut, B., Barratt, T., Kong, D., et al. (2017). Millimeter wave channel measurements in a railway depot. In *2017 IEEE 28th Annual International Symposium on Personal, Indoor, and Mobile Radio Communications (PIMRC)*, Montreal, QC.
12. Guan, K., Zhong, Z., Ai, B., et al. (2013). Deterministic propagation modeling for the realistic high-speed railway environment. In *2013 IEEE 77th Vehicular Technology Conference (VTC Spring)*, Dresden.
13. Abbas, T., Nuckelt, J., Kürner, T, et al. (2015). Simulation and measurement-based vehicle-to-vehicle channel characterization: Accuracy and constraint analysis. *IEEE Transactions on Antennas and Propagation, 63*(7), 3208–3218. https://doi.org/10.1109/TAP.2015.2428280
14. Priebe, S., Kürner, T. (2013). Stochastic modeling of THz indoor radio channels. *IEEE Transactions on Wireless Communications, 12*(9), 4445-4455. https://doi.org/10.1109/TWC.2013.072313.121581
15. He, D., Guan, K., Fricke, A., et al. (2017). Stochastic channel modeling for kiosk applications in the terahertz band. *IEEE Transactions on Terahertz Science and Technology, 7*(5), 502–513. https://doi.org/10.1109/TTHZ.2017.2720962
16. He, D., Ai, B., Guan, K., et al. (2019). The design and applications of high-performance ray-tracing simulation platform for 5G and beyond wireless communications: A tutorial. *IEEE Communications Surveys & Tutorials, 21*(1), 10-27. https://doi.org/10.1109/COMST.2018.2865724
17. Guan, K., Peng, B., He, D., et al. (2019). Channel characterization for intra-wagon communication at 60 and 300 GHz bands. *IEEE Transactions on Vehicular Technology, 68*(6), 5193–5207. https://doi.org/10.1109/TVT.2019.2907606
18. Kim, J., Chung, H., Choi, S., et al. (2017). Mobile hotspot network enhancement system for high-speed railway communication. In *2017 11th European Conference on Antennas and Propagation (EUCAP)*, Paris.
19. Liu, Y., Fang, X., Xiao, M., et al. (2018). Decentralized beam pair selection in multi-beam millimeter-wave networks. *IEEE Transactions on Communications, 66*(6), 2722–2737. https://doi.org/10.1109/TCOMM.2018.2800756
20. Peng, B., & Kürner, T. (2017). Three-dimensional angle of arrival estimation in dynamic indoor terahertz channels using a forward-backward algorithm. *IEEE Transactions on Vehicular Technology, 66*(5), 3798–3811. https://doi.org/10.1109/TVT.2016.2599488

Chapter 14
Data Centers

Johannes M. Eckhardt and Tobias Doeker

Abstract The requirements on modern data centers grow continuously. An integration of wireless links at terahertz frequencies promise a high data rate and new flexibility. In this book chapter, we summarize the benefits and challenges of terahertz communication in a data center and present channel measurements for the top-of-rack and intra-rack use case. It is shown that multipath propagation in combination with high side lobes from beam-steerable antennas might cause intersymbol interference even in directional point-to-point communication.

14.1 Benefits and Challenges of THz Communication in a Data Center

Data centers experience a transformation from storage-driven applications such as managing huge data bases toward more complex and computationally intense tasks that handle processing and analysis requests [1]. Therefore, the new requirements request ultralow latency, high capacity, reliability, scalability, and flexibility [2]. The performance of present data center networks (DCNs) is limited by short bursts on links that are sparsely used on average, however, these bursts still lead to a drop of the quality of service [3].

The integration of wireless links into a data center (DC) brings along several benefits. In combination with beam steering, wireless links enable a full dynamic reconfiguration of the DCN. Modifying a deployed network is normally complex, lengthy, and costly. With wireless connections, the links are no longer restricted by physical cabling. The controller of a software defined network (SDN) will be able to change the network topology based on the current demands and traffic patterns [4].

J. M. Eckhardt (✉) · T. Doeker
Institute for Communications Technology, Technische Universität Braunschweig, Braunschweig, Germany
e-mail: eckhardt@ifn.ing.tu-bs.de; doeker@ifn.ing.tu-bs.de

T. Kürner et al. (eds.), *THz Communications*, Springer Series in Optical Sciences 234, https://doi.org/10.1007/978-3-030-73738-2_14

In addition, a direct communication between nodes that bypasses the aggregate and core switches is favorable for latency requirements.

Basically, the integration can be realized by a complete replacement of wired links [5]. Alternatively, the wireless links are added in addition and transform the DC into a wirelessly augmented DC [6]. In either way, a reduced cabling complexity leads to better airflow and improved cooling [7].

However, the wireless links are only beneficial for a data transmission if the provided data rate is comparable to the data rate of fiber connections. The terahertz (THz) communication meets this requirement aiming at high data rates up to several 100 Gbit/s. In addition, the highly directional antennas that compensate the high free-space path loss (FSPL) at THz frequencies reduce the interference. Hence, more links can operate in parallel.

The DC application case was early identified [8] and already tackled [9] by the research community. Nonetheless, the THz communications were made possible only in the last years by the advances in the device development, and a long way is still to go. Devices with higher output power are needed, and beam-steerable antenna arrays with low side lobes and a sufficient gain and steering range have to be developed [10]. Also, higher-layer protocols have to be rethought or adapted [4].

To successfully build a new communication system operating at THz frequencies, the radio channel has to be carefully studied and modeled in order to adapt the physical layer and higher-layer protocols. Thus, channel models are needed for each specific application incorporating the propagation in the complex environment, the geometry, and the materials.

A first simulation-based model is presented in [11] that investigates the effect of water vapor absorption using a free-space propagation model up to 2 THz. A more detailed simulation-based approach is described in [12] that derives a stochastic channel model from ray-tracing simulations in an ideal DC model and provides simulated transfer functions of the radio channel in the band from 252 GHz to 325 GHz. A measurement series in a mock-up environment at 300 GHz is analyzed in [13], whereas the first measurement campaign in a real DC is reported in [14]. The latest modeling approach was presented in [15] showing a cluster-based model that is derived from measurements for medium-height rack-to-rack connections. The rest of this book chapter presents channel characteristics for inter-rack, more specifically top-of-rack (ToR) and intra-rack (IR) scenarios based on measurements from [14].

14.2 Environment Description and Setup

The general DC is characterized by a clear and regular arrangement of the server racks. Metal, glass, and building materials represent the most common materials in this environment. In a high number of cases, the propagation scenario is a static line-of-sight (LOS) scenario with fixed transmitter (TX) and receiver (RX) positions. In some cases, a first-order reflection via a reflector is considered if a direct link is not possible.

The DC under investigation is organized in rows, while the cabling is laid in the floor. On top of each row of racks, a plastic curtain separates the air into cold and hot regions. In order to efficiently cool the servers, the air from a cold aisle passes through the server rack, and the heated air in the hot aisle is aspirated from the ceiling. In doing so, the curtains prevent an inefficient mixing of the air. For a more detailed description of the environment, the reader is referred to [14].

The channel measurements are performed with a sub-mmWave channel sounder operating at 304.26 GHz with a bandwidth of 8 GHz. It measures the channel impulse response (CIR) by cross-correlating the transmitted M-sequence of order 12 and the received signal. In the process, the system is driven by a 9.22 GHz clock that allows a chip duration of 108.5 ps corresponding to a spatial resolution of 3.24 cm. The length of the test sequence enables a maximum excess delay of 444.14 ns that equate to a maximum measureable path length of 133.3 m. The RX uses a subsampling factor of 128 that results in a measurement rate of 17 590 CIR/s. By averaging 4096 measurements, a dynamic range of 60 dB can be reached [16].

The channel sounder is equipped with standard gain horn antennas with a gain of 26.4 dBi and a half power beam width (HPBW) of 8.5° for the ToR measurements and a gain of 15 dBi and an HPBW of 30° for the IR measurements. Measurements are performed with both vertical and horizontal polarizations. To obtain omnidirectional channel characteristics in the ToR scenario from the directional channel measurement, TX and RX are mounted on top of automated rotating units that scan the horizontal plan with a programmable step size. The omnidirectional CIR is then created in a post-processing step. For IR measurements, height-adjustable tripods are used.

14.3 Inter-rack Measurements

Inter-rack communication, also called rack-to-rack communication, is a promising use case of wireless data transmission in a DC. Here, the inter-rack links can be divided into two regimes. Links at medium height will face a different propagation environment than ToR links that are located above the rack. The data traffic within a rack is managed by the ToR switch. Thus, the positioning of the THz interface on top of the rack is intuitive. In addition, the geometry of the DC allows for a communication over a longer distance and multiple rows in a likely LOS scenario. Measurements are performed for TX and RX being located in the same aisle, the adjacent aisle, and the next but one aisle. Figure 14.1 illustrates the measurement setup.

In all ToR measurements, a strong direct path is detected. The path loss can be modeled with the FSPL whereby each row of curtains adds a transmission loss of 7 dB. In the CIR, two parts can be identified. The first part contains a high density of multipath components (MPCs). After a certain delay threshold, only lone MPCs are present. This threshold is observed at an excess delay of 83 ns, 30 ns, and 40 ns

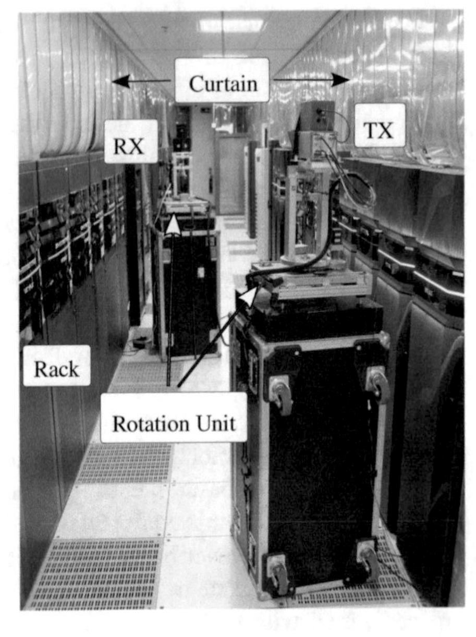

Fig. 14.1 ToR measurement setup

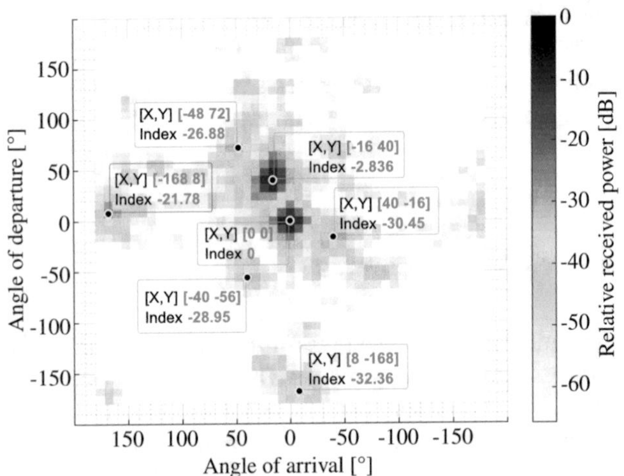

Fig. 14.2 PAP of the ToR scenario in the same aisle

for the same aisle, adjacent aisle, and next but one aisle, respectively. The intensity of this characteristic decreases with the number of rows between TX and RX.

The investigation of the multipath propagation in the DC is of particular importance. The scenario in the same aisle shows a strong MPC from a first-order reflection at the curtains with an additional path loss of only 2.8 dB at an excess delay of 1.42 ns. The corresponding power angular profile (PAP) is presented in Fig. 14.2. In combination with a side lobe of an antenna array, this MPC can lead to inter-symbol interference (ISI) and limit the data rate drastically.

In the adjacent aisle scenario, two striking MPCs are present beside the direct path. They are measured with an additional path loss of 16.0 dB and 14.1 dB at an excess delay of 21.01 ns and 55.50 ns, respectively. All other MPCs show an additional path loss of at least 20 dB. The next but one aisle scenario is characterized by a couple of MPCs that have an additional path loss between 27.3 dB and 32.4 dB and occur up to an excess delay of 40 ns.

The polarization also effects the propagation. In horizontal polarization, some of the rather weak MPCs can no longer be detected. All other paths can easily be matched whereby the difference stays within an interval of 5 dB. The direct path is not affected by the polarization in our measurements.

In terms of root mean square (RMS) delay spread, first an increase is observed from 6.6 ns to 10.4 ns for the same aisle and adjacent aisle, respectively. Then, the delay spread decreases to 3.1 ns in the next but one aisle. This behavior can be explained by the fact that in the same aisle scenario, the powerful MPCs with a short delay dominate. In the adjacent aisle setup, rather MPCs with medium power at larger delays exist, and all paths are affected by the transmission loss of the curtain. Finally, the number of paths with a reasonable impact decreases for the next but one aisle setup. In contrast, the angular spread monotonically decreases from 22.0° to 11.8° and to 8.7°, respectively, which also reflects the decrease in MPCs.

14.4 Intra-rack Measurements

To gain adaptability, reconfigurability, and less cabling on a smaller scale, wireless links can be added within a rack. Here, the metal housing poses a challenge since it can lead to strong multipath propagation. The IR channel is investigated by a measurement setup that emulates a likely IR link in a server rack with a height of 1.95 m, a width of 0.603 m, and a depth of 1.09 m. TX and RX are placed in the bottom and top corner inside the rack facing toward each other. Three different transmission distances and a different number of servers in the rack are evaluated. The cabling is omitted assuming a proper cable management and a high degree of wireless integration. Figure 14.3 shows an exemplary setup. It should be noted that the doors are closed for the measurement.

All measurements are placed in a LOS scenario whereby no particular influence of the number of server blades on the path loss of the direct path is visible. However, in the presence of servers, multiple clusters of MPCs are detected but no direct relation between the number of servers and the number of clusters is observed. Figure 14.4 compares the CIRs of the measurement with the largest distance and six servers (see Fig. 14.3) and the measurement with the largest distance without any servers. Additionally, the ground strut is covered with absorbers in the latter case. Clusters belonging to the server blades are visible from 7 ns to 9 ns, 10 ns to 12 ns, 13 ns to 15 ns, and 22 ns to 24 ns. The two peaks at a delay of 16.13 ns and 16.33 ns with an additional path loss of 19.7 dB and 19.2 dB, respectively, are linked to the ground strut of the rack. In conclusion, the influence of the rack housing is stronger than the impact of the server blades. To reduce its impact, the rack could be covered with a thin foil absorbing THz frequencies.

Fig. 14.3 IR measur. setup

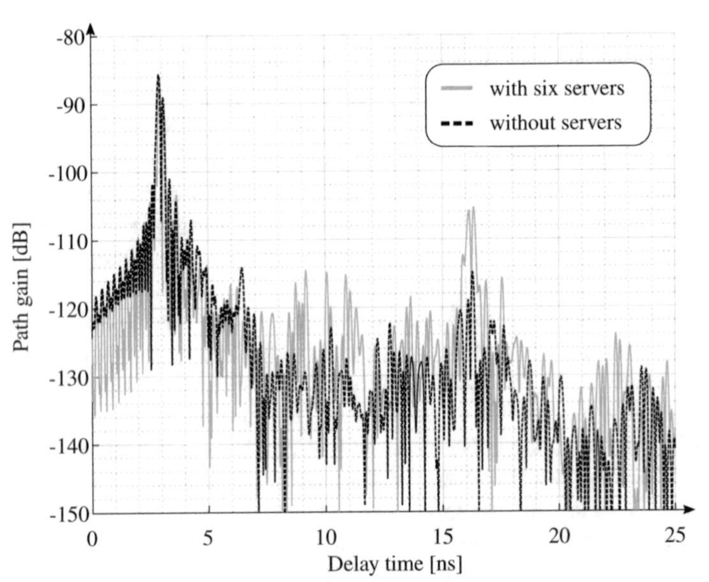

Fig. 14.4 CIR of the IR scenario.

14.5 Conclusion

Wireless links at THz frequencies in DCs show multiple promising benefits. Especially the ToR use case will feature higher flexibility and reconfigurability at runtime. ToR links are characterized by a static LOS scenario with known and fixed positions of TX and RX. Although the open space over the rack top offers the most favorable condition for wireless communication, reflecting objects such as plastic curtains or columns cause strong multipath propagation. In combination with high side lobes of beam-steerable antenna arrays, this might lead to ISI even in a directional point-to-point link. In the IR scenario, the metal housing causes reflections that have to be considered for the link design.

References

1. Rommel, S., Raddo, T. R., & Monroy, I. T. (2018). Data center connectivity by 6G wireless systems. In *Proceedings of IEEE Conference on Photonics in Switching and Computing (PSC 2018)*, Limassol, Cyprus.
2. Hamza. A. S., Deogun, J. S., & Alexander, D. R. (2016). Wireless communication in data centers: A survey. In *IEEE Communications Surveys & Tutorials, 18*(3), 1572–1595.
3. Benson, T., Akella, A., & Maltz, D. A. (2010). Network traffic characteristics of data centers in the wild. In *Proceedings of the 10th ACM SIGCOMM Conference on Internet Measurement*, New York.
4. Ahearne, S. (2019). Integrating THz wireless links in a data center network. In *Proceedings of IEEE 2nd 5G World Forum (5GWF)*, Dresden.
5. Shin, J., Sirer, E. G., Weatherspoon, H., & Kirovski, D. (2013). On the feasibility of completely wireless datacenters. *IEEE/ACM Transactions on Networking, 21*(5), 1666–1679.
6. Halperin, D., Kandula, S., Padhye, J., Bahl, P., & Wetherall, D. (2011). Augmented data center networks with multi-gigabit wireless links. In *Proceedings of the 11th ACM SIGCOMM Conference on Internet Measurement*, Toronto.
7. Rosenkrantz, E., & Arnon, S. (2016). Reducing energy consumption of data centers using optical wireless links. In *Proceedings of IEEE International Conference on Wireless for Space and Extreme Environments (WiSEE 2016)*, Aachen.
8. Kürner, T. (2015). TG3d applications requirements document. In *IEEE P802.15 Working Group for Wireless Personal Area Networks (WPANs)*. Retrieved October 28, 2020, from https://mentor.ieee.org/802.15/dcn/14/15-14-0304-16-003d-applications-requirement-document-ard.docx
9. Davy, A. S., Pessoa, L., Renaud, C., et al. (2017). Building an end user focused THz based ultra high bandwidth wireless access network: The TERAPOD approach. In *Proceedings of the 9th International Congress on Ultra Modern Telecommunications and Control Systems and Workshops (ICUMT)*, Munich (pp. 454–459).
10. Katayama, Y., Takano, K., Kohda, Y., Ohba, N., & Nakano, D. (2011). Wireless data center networking with steered-beam mmwave links. In *Proceedings of IEEE Wireless Communications and Networking Conference (WCNC 2011)*, Cancun.
11. Mollahasani, S., & Onur, E. (2016). Evaluation of Terahertz channel in data centers. In *Proceedings of IEEE/IFIP Network Operations and Management Symposium (NOMS 2016)*, Istanbul.
12. Fricke, A. (2016). Channel modeling document. In *IEEE P802.15 Working Group for Wireless Personal Area Networks (WPANs)*. https://mentor.ieee.org/802.15/dcn/14/15-14-0310-19-003d-channel-modeling-document.docx

13. Cheng, C., & Zajić, A. (2020). Characterization of propagation phenomena relevant for 300 GHz wireless data center links. *IEEE Transactions on Antennas and Propagation, 68*(2), 1074–1087.
14. Eckhardt, J. M., Doeker, T., Rey, S., & Kürner, T. (2019). Measurements in a real data center at 300 GHz and recent results. In *Proceedings of 13th European Conference on Antennas and Propagation (EuCAP 2019)*, Krakow.
15. Cheng, C., Sangodoyin, S., & Zajić, A. (2020). THz cluster-based modeling and propagation characterization in a data center environment. *IEEE Access, 8*, 56544–56558.
16. Rey, S., Eckhardt, J. M., Peng, B., Guan, K., & Kürner, T. (2017). Channel sounding techniques for applications in THz communications: A first correlation based channel sounder for ultra-wideband dynamic channel measurements at 300 GHz. In *Proceedings of 9th International Congress on Ultra Modern Telecommunications and Control Systems and Workshops (ICUMT)*, Munich (pp. 449–453).

Chapter 15
Vehicular Environments

Vitaly Petrov, Johannes M. Eckhardt, Dmitri Moltchanov, Yevgeni Koucheryavy, and Thomas Kürner

Abstract High-rate data exchange between the prospective (semi)autonomous vehicles is one of the promising usage scenarios for THz band communication systems. However, the vehicular scenarios have certain specifics to be accounted for when developing THz vehicular communication technologies. Particularly, the propagation of the THz signal in vehicular setups differs from the one in conventional indoor or outdoor cellular deployments. This chapter summarizes the preliminary findings in the area of THz communications in vehicular environments.

15.1 Motivation and Specifics of THz Vehicular Communications

The prospective dynamic networks formed by smart connected vehicles are expected to generate and exchange massive amounts of data coming from heterogeneous sensors: up to 1 TB/h generated by a single vehicle, according to [1]. Such high volumes cannot be delivered with state-of-the-art communication technologies. Therefore, the research community is actively exploring the adaptation of extremely high-rate THz communication systems to vehicle-centric scenarios [2]. At the same time, the successful adoption of THz communications in vehicular deployments requires a better understanding of the major vehicle-specific propagation effects, as outlined further in this chapter.

Following prior chapters, the characteristics of the THz channel highly depend on the scenario geometry and the materials in the environment. Even minor changes, such as road signs, may notably affect the performance [3]. There are not many

V. Petrov · D. Moltchanov · Y. Koucheryavy
Unit of Electrical Engineering, Tampere University, Tampere, Finland

J. M. Eckhardt (✉) · T. Kürner
Institute for Communications Technology, Technische Universität Braunschweig, Braunschweig, Germany
e-mail: eckhardt@ifn.ing.tu-bs.de; t.kuerner@tu-bs.de

T. Kürner et al. (eds.), *THz Communications*, Springer Series in Optical Sciences 234, https://doi.org/10.1007/978-3-030-73738-2_15

measurement campaigns reported to date for THz or sub-THz frequencies [4–6]. When extrapolating the results available for the millimeter wave bands [7–11], we expect the nonnegligible impact of the reflections and scattering from the vehicle bodies, the road infrastructure, the roadbed itself, and even the plants surrounding the road. Admitting the importance of all the listed effects, in this chapter, we focus on the two phenomena that are crucial for the design and evaluation of the prospective THz vehicular communication systems, namely, *vehicular blockage*, detailed in Sect. 15.2, and *vehicle-specific interference*, discussed in Sect. 15.3.

15.2 Vehicle-Body Blockage

15.2.1 Measuring the Impact of Blockage

To characterize the impact of vehicular blockage in THz communications, a measurement campaign has been conducted and described in [4]. The authors explored the setup, where the signal propagates through the vehicle body (see Fig. 15.1). For this study, the low-THz (sub-THz) channel sounder manufactured by Ilmsens was used. The measurements were conducted in a sub-THz frequency range: from 300.2 GHz to 308.2 GHz. Both the transmitter (Tx) and receiver (Rx) sides were equipped with horn antennas, each featured by approximately $\approx 8^{\circ}$ beamwidth and $\approx$26 dBi gain.

The selected results of this campaign are reported in Table 15.1. As can be observed from this table, the vehicle-body blockage has a notable effect. Additional attenuation from 28 dB to 50 dB is introduced on top of already high propagation losses, discussed in prior chapters. The exact penetration losses depend primarily on the height of the vehicle-to-vehicle V2V link. Particularly, the propagation at

Fig. 15.1 Measuring vehicle-body penetration loss at sub-THz frequencies

Table 15.1 Penetration losses in vehicular communications at 300 GHz

Measured path	Loss
Bumper level, $h \in [0.9\,\text{m} \ldots\ 1.1\,\text{m}]$	45 dB
Engine level, $h \in [1.1\,\text{m} \ldots\ 1.2\,\text{m}]$	50 dB
Front and rear windows, $h \in [1.4\,\text{m} \ldots\ 1.6\,\text{m}]$	40 dB
Near the rooftop, $h = 1.6\,\text{m}$	20 dB
Front side windows, $h \in [1.25\,\text{m} \ldots\ 1.4\,\text{m}]$	33 dB
Rear side windows, $h \in [1.25\,\text{m} \ldots\ 1.4\,\text{m}]$	28 dB

the engine-level, $h \in [1.1\,\text{m} \ldots\ 1.2\,\text{m}]$, is affected the most, as the signal has to penetrate through or diffract around many metal constructions. In contrast, the weakest attenuation is observed at the window level, $h \in [1.4\,\text{m} \ldots\ 1.6\,\text{m}]$, as the signal penetrates only two narrow pieces of glass. Similar observations are made for side propagation.

Summarizing, the following conclusions are made:

1. **Vehicle-body penetration loss is highly *height-selective*.** Therefore, in contrast to human-body blockage, where the obstacle can be modeled as a homogeneous cylinder (e.g., as in [12, 13]), the vehicle-body blockage calls for the development of more fine-graded models. Particularly, the height of the THz signal and its direction, when propagating through the vehicle body, must be accounted for.
2. **Vehicle-body is a *nonnegligible* blocker for THz communication systems.** The introduction of $\geq$28 dB loss in Table 15.1 may change the signal-to-noise (SNR) level at the Rx from, e.g., 10 dB (reliable data exchange) to -18 dB (outage with most of the modulation and coding schemes). Therefore, additional techniques should be applied for blockage mitigation in THz vehicular communication systems.

15.2.2 Blockage Mitigation

The vehicle-body blockage has a major impact on THz vehicular communications. In this subsection, we introduce two possible approaches for blockage mitigation.

Under-vehicle Propagation As detailed in [4], THz signals reflect from the road with moderate losses and thus can propagate under the vehicle body. This approach is particularly beneficial for direct V2V THz communications in case Tx and Rx antennas are located at the bumper level (close to the road surface). Hence, a notable part of the THz beam is not blocked. In addition, the angle of incidence when reflecting from the road is close to 90°, thus leading to moderate reflection losses.

Multi-connectivity Another possible solution to mitigate the unexpected blockage in THz vehicular communications is by utilizing the *multi-connectivity* technique. The approach suggests the target Tx and Rx devices stay connected via other vehicles in proximity. When the primary path (e.g., a direct link) is blocked, the nodes can utilize one of the alternative options. The ubiquitous use of this approach leads to the appearance of directional vehicular mesh networks [14]. However, the constant maintenance of many connectivity paths increases system complexity. Therefore, the number of used paths, termed as *degree of multi-connectivity*, must be properly selected, balancing the target reliability level and the associated overheads [15].

15.3 Directional Interference in Vehicular Setups

15.3.1 Measuring the Impact of Interference

To study the multipath interference in THz vehicular communication systems, the impact of side reflections from the neighboring vehicles has been measured in [4]. In this measurement setup, the separation distance between the Tx and Rx vehicles varied from 2 m to 12 m, while the car on the side lane was always equally distant from both sides (see Fig. 15.2). In addition, custom rotation units were used, allowing to obtain the power angular profile (PAP) in all the considered configurations.

The selected results of this study are presented in Fig. 15.3, illustrating the PAP for the Tx-Rx distances of 2 m and 6 m. The presented results are normalized to the corresponding values for the LoS path (top left corners of the plots). We observe that the side reflection effect is visible for both distances. However, the difference in terms of the power and the angles is lower for the longer distances: 3.5 dB and ≈24° for $d = 6$ m vs. 12 dB and ≈40° for $d = 2$ m. Hence, the reflections from the neighboring vehicles contribute to the multipath interference.

Fig. 15.2 Measuring the vehicle side reflection for the multipath interference modeling at THz frequencies

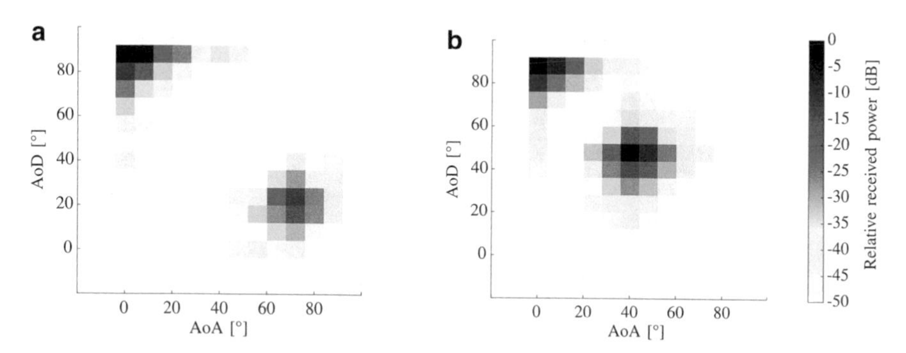

Fig. 15.3 Power angular profile for the side reflection at 300 GHz. (**a**) $d = 2$ m. (**b**) $d = 6$ m

The THz vehicular communications may be affected not only by the multipath interference but also by the direct interference coming from the neighboring vehicles. To analyze this effect, a detailed study has been conducted in [5]. For this purpose, mathematical and simulation-based frameworks were developed for two typical deployments: (1) *highway*, characterized by large separation distances between the vehicles (tens of meters), and (2) *urban*, characterized by short and random inter-vehicle distance (a few meters). It has been observed that the impact of interference is nonnegligible in both deployment options. In addition, the average interference power grows fast with the beamwidth of the employed radiation patterns and slowly decreases with the distance between the vehicles on the neighboring lanes.

Summarizing, the following conclusions are made:

1. **The impact of the multipath interference is *distance-dependent*.** As illustrated in both [4] and [5], the interference caused by the multipath reflections from the vehicles on the neighboring lanes is nonnegligible. Particularly, the power of the reflected signal may be just 3 dB to 6 dB lower than one of the direct line of sight (LoS) paths. At the same time, the actual impact of the interference greatly depends on the separation distance between the Tx and Rx vehicles. When the distance is short, beams cannot reach the vehicles on other lanes and thus do not contribute to the interference level. In contrast, for longer distances, the length of the reflected path becomes comparable to one of the LoS paths, while the reflection losses also decrease as the reflection angle of incident approaches 90°.
2. **The impact of the inter-vehicle interference is *density-dependent*.** A high density of vehicular traffic (e.g., in urban cities) leads to low separation distances between the Tx and Rx vehicles. Hence, most of the interference coming from the neighboring lanes is blocked by the vehicle bodies. In contrast, a low density of vehicular traffic (e.g., in empty highways) leads to a low quantity of vehicles interfering with the target link. Finally, as discussed in [6], the most profound impact of inter-vehicle interference is observed for moderate densities of traffic ($\approx$10 m to 20 m between the vehicles). For such a setup, the density is high enough to cause collided transmissions, while the scenario geometry still allows

most of the interference to reach the target Rx (low probability of blockage by the vehicles).

15.3.2 Interference Mitigation

Both the multipath interference and direct interference from the neighboring vehicles play an important role in THz vehicular communications. In this subsection, we present two possible approaches for interference mitigation in vehicular setups.

Narrow THz Beams A straightforward approach to mitigate directional interference is to reduce the beamwidth of the employed antenna radiation patterns. Following [5], the average interference power (both the multipath self-interference and the interference from other vehicles) decreases notably when narrowing the beams. Particularly, the utilization of the antenna radiation patterns of less than 5° beamwidth leads to favorable conditions for THz band vehicular communications in a highway scenario. However, the utilization of extremely directional radiation patterns challenges the beam steering and beam tracking in the presence of mobile nodes [16].

Directional CSMA Another possible approach to mitigate the interference in THz vehicular communications is by applying the *directional carrier sense multiple access (CSMA) approach*, as detailed in [6] and [17]. The measurements reported in [4] and [6] illustrate that the THz signal reflects from the front and rear of the vehicle with moderate losses. Therefore, almost any vehicle that may potentially interfere with the target THz link may sense the channel before sending any data and postpone its transmission in case an active data exchange is detected. Particularly, the use of robust physical layer preambles is suggested in [6] to facilitate the reliable detection of active communication. With the presented approach, both the range and the capacity of THz vehicular communications are improved by over 30% each.

15.4 Conclusions

Vehicle-centric scenarios have multiple specific features that must be considered when developing the THz vehicular communication systems. In this chapter, we discussed the two major relevant features: (1) vehicle-body blockage and (2) vehicle-centric interference. The research on THz vehicular systems is still at an early stage, and many questions remain open. Nevertheless, the presented first-order performance estimations, together with the listed approaches to mitigate these effects, will facilitate the development of reliable and efficient THz vehicular communication systems.

References

1. Choi, J., Va, V., Gonzalez-Prelcic, N., et al. (2016). Millimeter-wave vehicular communication to support massive automotive sensing. *IEEE Communications Magazine, 54*(12), 160–167.
2. Mumtaz, S., Jornet, J. M., Aulin, J., et al. (2017). Terahertz communication for vehicular networks. *IEEE Transactions on Vehicular Technology, 66*(7), 5617–5625.
3. Guan, K., Ai, B., Nicolas, M. L., et al. (2016). On the influence of scattering from traffic signs in vehicle-to-X communications. *IEEE Transactions on Vehicular Technology, 65*(8), 5835–5849.
4. Eckhardt, J. M., Petrov, V., Moltchanov, D., Koucheryavy, Y., & Kurner, T. (June 2021). Channel measurements and modeling for Low-Terahertz band vehicular communications. *IEEE Journal on Selected Areas in Communications, 39*(6), 1590–1603.
5. Petrov, V., Kokkoniemi, J., Moltchanov, D., et al. (2018). The impact of interference from the side lanes on mmwave/THz band V2V communication systems with directional antennas. *IEEE Transactions on Vehicular Technology, 67*(6), 5028–5041.
6. Petrov, V., Fodor, G., Kokkoniemi, J., et al. (2019). On unified vehicular communications and radar sensing in millimeter-wave and low terahertz bands. *IEEE Wireless Communications, 26*(3), 146–153.
7. Solomitckii, D., et al. (2020). Characterizing radio wave propagation in urban street canyon with vehicular blockage at 28 GHz. *IEEE Transactions on Vehicular Technology, 69*(2), 1227–1236.
8. Sato, K., Fujise, M., Tachita, R., et al. (2001) Propagation in ROF road-vehicle communication system using millimeter wave. In *Proc. of IEEE IVEC* (pp. 131–135)
9. Kato, A., Sato, K., Fujise, M., & Kawakami, S. (2001). Propagation characteristics of 60-GHz millimeter waves for its inter-vehicle communications, *IEICE Transactions on Communications, E84-B*, 2530–2539.
10. Ben-Dor, E., Rappaport, T. S., Qiao, Y., & Lauffenburger, S. J. (2011). Millimeter-wave 60 GHz outdoor and vehicle AOA propagation measurements using a broadband channel sounder. In *Proc. of IEEE GLOBECOM* (pp. 1–6).
11. Schneider, R., Didascalou, D., & Wiesbeck, W. (2000). Impact of road surfaces on millimeter-wave propagation. *IEEE Transactions on Vehicular Technology, 49*, 1314–1320.
12. Venugopal, K., Valenti, M. C., & Heath, R. W. (2016). Device-to-device millimeter wave communications: Interference, coverage, rate, and finite topologies. *IEEE Transactions on Wireless Communications, 15*(9), 6175–6188.
13. Petrov, V., Solomitckii, D., Samuylov, A., et al. (2017). Dynamic multi-connectivity performance in ultra-dense urban mmwave deployments. *IEEE JSAC, 35*(9), 2038–2055.
14. Andreev, S., et al. (2019). Dense moving fog for intelligent IoT: Key challenges and opportunities. *IEEE Communications Magazine, 57*(5), 34–41.
15. Gapeyenko, M., et al. (2019). On the degree of multi-connectivity in 5G millimeter-wave cellular urban deployments. *IEEE Transactions on Vehicular Technology, 68*(2), 1973–1978.
16. Petrov, V., et al. (2020). Capacity and outage of terahertz communications with user micro-mobility and beam misalignment. *IEEE Transactions on Vehicular Technology, 69*(6), 6822–6827.
17. Petrov, V., Fodor, G., Andreev, S., et al. (2019). V2X connectivity: From LTE to joint millimeter wave vehicular communications and radar sensing. In *Proc. of ASILOMAR* (pp. 1120–1124).

Chapter 16
Stochastic Channel Models

Danping He and Zhangdui Zhong

Abstract Considering the fact that stochastic channel models are indispensable for THz system design and performance evaluation for nonspecific sites, practical and efficient models covering various types of complex environments are urgently needed. This chapter offers the in-depth look at the modeling approaches for the aforementioned environments and summarizes the general way to derive THz stochastic channel models.

16.1 Review of THz Stochastic Channel Models

High data rate THz communications are envisioned to bring benefits to indoor, wireless data center, backhaul/fronthaul outdoor, intra-device, close proximity, railway, and vehicular scenarios. As accurate and efficient channel models are critical for THz system design and network planning, the features of propagation environments, the channel measurement, and channel characteristics have been studied by many researchers and standardization organizations. The state-of-the-art works have been introduced in Chaps. 10–15 for the potential environments.

According to the current studies, channel modeling approaches can be classified as stochastic and deterministic channel models. Detailed classifications of the models are shown in Fig. 16.1. Stochastic channel models include cluster-based, ray-based approaches, and so on. Ray-tracing (RT) and map-based approaches are typical deterministic models. Fairly speaking, the stochastic models with summarized statistics are suitable for scenario-specific but not site-specific overall system design and statistical performance evaluation, while the deterministic approaches are more suitable for evaluating design performance and network planning in specific sites. As a roadmap to obtain accurate deterministic models is much clearer

D. He (✉) · Z. Zhong
Beijing Jiaotong University, Beijing, China
e-mail: hedanping@bjtu.edu.cn; zhdzhong@bjtu.edu.cn

T. Kürner et al. (eds.), *THz Communications*, Springer Series in Optical Sciences 234, https://doi.org/10.1007/978-3-030-73738-2_16

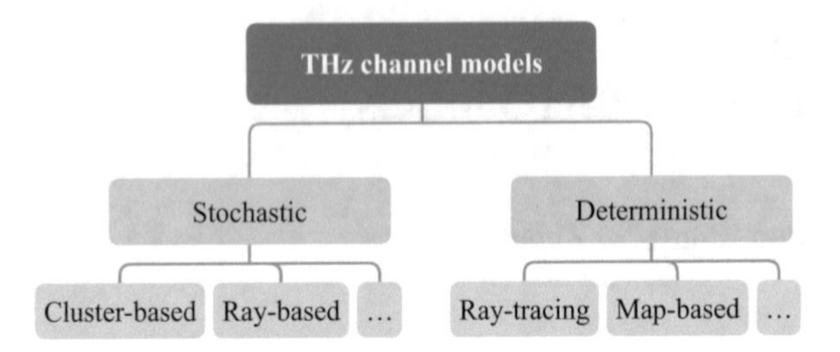

Fig. 16.1 Categories of THz channel models

than the stochastic models, we focus on discussing the way to derive THz stochastic channel models for various environments.

16.1.1 Cluster-Based Stochastic Channel Modeling

Clusters are sets of multipath that have similar propagation characteristics in delay domain and/or spatial domain. In data center [1], indoor [2], and train-to-infrastructure (T2I) inside-station [3] communication scenarios, the cluster-based models were developed and validated. The K-power means the clustering algorithm is applied in the three works to group the multipath components into a reasonable number of clusters.

Extended Saleh-Valenzuela (S-V) models are proposed to use in data center and indoor environments. Typical S-V model can be expressed as:

$$h(t) = \sum_{l=0}^{L} \sum_{k=0}^{K_l} \beta_{kl} e^{j\Phi_{kl}} \cdot \delta\left(t - T_l - \tau_{kl}\right) \tag{16.1}$$

where l and k are the indexes of cluster and path and L and K_l are the number of the cluster and the sub-path in cluster "l," respectively. T_l is the arrival time of the lth cluster, while β_{kl}, Φ_{kl}, and τ_{kl} are the amplitude, phase, and delay of the kth sub-path in the lth cluster, respectively. The path gain is modeled as:

$$\overline{\beta_{kl}^2} = \overline{\beta^2\left(T_l, \tau_{kl}\right)} = \overline{\beta^2(0,0)} e^{-\frac{T_l}{\Gamma}} e^{-\frac{\tau_{kl}}{\gamma}} \tag{16.2}$$

where $\beta^2(0,0)$ is the average power of the first sub-path of the first cluster and Γ and γ are the power-delay time constant of the cluster and sub-paths, respectively.

The authors of [1] extended the S-V model by considering the angular domain information, whereas the authors of [1] found that Γ could not precisely capture the attenuation of the multipath clusters in the THz data center environment. Therefore, S-V model is modified, and Γ is expressed into two sections as a function of T:

$$\Gamma(T)\begin{cases}\Gamma_1, & 0 < T < \tau_{th}\\ \Gamma_2, & \tau_{th} \leq T < \tau_{max}\end{cases} \tag{16.3}$$

where τ_{th} is a delay threshold value serving as a breakpoint for Γ and can be selected based on the distribution of multipath clusters. Γ_1 and Γ_2 can be determined through linear regression of cluster peak powers in dB and the associated delays in nanoseconds.

The 3GPP-like QuaDRiGa model is employed and validated in the T2I scenario [3]. Compared with Saleh-Valenzuela model, QuaDRiGa supports time-variant channel. However, QuaDRiGa requires more statistical features derived from key channel parameters, including PL, shadow fading, RMS delay spread, Rician K-factor, azimuth/elevation angular spread of arrival/departure, cross-polarization ratio, and their cross correlations.

16.1.2 Ray-Based Stochastic Channel Modeling

In addition to the cluster-based stochastic models, ray-based stochastic models are also popular in THz scenarios like metal enclosures [4], kiosk downloading [5], and indoor office environment [6]. By taking advantage of the high-resolution channel measurement in the delay and spatial domain, the multipath components can be distinguished better in THz band than the sub-6 GHz and mmWave frequency bands. Moreover, as the geometric structures of these environments are much less complex than the data center, integrated indoor, and inside-station environments, fewer MPCs and clearer propagation phenomena are observed. Therefore, significant rays are extracted and categorized into different types according to the propagation mechanisms, and they are modeled individually. The CTF $H_i(f)$ of each ray is a function of time, frequency, and angle:

$$\begin{aligned} H_i(f) = {} & a_{i,f_0} e^{j\varphi_i} D_{f_0}(f) e^{-j\pi\tau_i(f-f_0)} \\ & \cdot \delta(\phi_{AOD} - \phi_{AOD,i})\delta(\theta_{AOD} - \theta_{AOD,i}) \\ & \cdot \delta(\phi_{AOA} - \phi_{AOA,i})\delta(\theta_{AOA} - \theta_{AOA,i}) \end{aligned} \tag{16.4}$$

where a_{i,f_0} is the amplitude of ray i at reference frequency f_0, φ_i is the phase of ray i at f_0, $D_{f_0}(f)$ is the frequency dependency function of f at f_0, τ_i is the delay of ray i, and AOA and AOD are the angles of arrival and departure, respectively. The multipath amplitudes in a_i can be further decomposed into their co- and cross-polarized components $a_{i,co}$ and $a_{i,cross}$:

$$a_i = \begin{pmatrix} a_{i,co,\theta} & a_{i,cross,\theta} \\ a_{i,cross,\phi} & a_{i,co,\phi} \end{pmatrix} \tag{16.5}$$

In order to obtain the CTF of the entire channel, the path-specific CTFs of all the rays in (16.4) should be summarized and weighted with the polarimetric Tx and Rx antenna radiation patterns G_{tx} and G_{rx}:

$$H = \sum_{i=1}^{N_{Rays}} G'_{tx} H_i G_{rx} \tag{16.6}$$

Compared with the existing cluster-based methods, ray-based method is more suitable for the environments with less objects so that the propagation of multi-path components can be well explained.

16.2 Work Flow of Deriving Stochastic Channel Modeling

The key features that a THz stochastic channel model should represent include, but are not limited to, the path loss (PL), delay spread (DS), K-factor (KF), 3D angular spreads (AS) in azimuth and elevation domain, cross-polarization ratio (XPR), and number of multi-path components (MPCs). Extracting the key channel parameters and analyzing their statistical features are very important to stochastic channel models.

For example, "A-B" model is one of the typical ways to model PL:

$$PL\,(dB) = A \cdot \log_{10}\left(\frac{d}{d_0}\right) + B + X_\sigma \tag{16.7}$$

where d is the distance between the Tx and the Rx in meter; d_0 is the reference distance, which is 1 m; A is the slope; B is the intercept; and the X_σ is the shadow fading (SF), which can be expressed as a Gaussian variable with zero mean and a standard deviation σ_{SF}.

The Rician K-factor is defined as the ratio of the power of the strongest component to the power of the sum of the remaining components in the received signal. Traditionally, the Rician K-factor is calculated from the narrow-band channel sounding results by using a moment-based method. However, the UWB channel sounding results have high resolution in the time domain. Thus, the Rician K-factor can be calculated according to its definition:

$$KF = 10 \cdot \log_{10}\left(\frac{P_0}{\sum P_{\mathrm{remaining}}}\right) \tag{16.8}$$

where KF denotes the Rician K-factor and P_0 and $P_{\mathrm{remaining}}$ denote the power of the first component and the power of each of the remaining components, respectively.

RMS delay spread is used to quantify the dispersion effect of the multipath channel. It is defined as the square root of the second central moment of the PDP:

$$\sigma_\tau = \sqrt{\frac{\sum\limits_{n=1}^{N} {\tau_n}^2 \cdot P_n}{\sum\limits_{n=1}^{N} P_n} - \left(\frac{\sum\limits_{n=1}^{N} \tau_n \cdot P_n}{\sum\limits_{n=1}^{N} P_n}\right)^2} \tag{16.9}$$

where σ_τ denotes the RMS delay spread and P_n and τ_n denote the power and the excess delay of the nth multipath, respectively.

The angular spreads are calculated by the 3GPP definition:

$$\sigma_{\mathrm{AS}} = \sqrt{\frac{\sum\limits_{n=1}^{N} \left(\theta_{n,\mu}\right)^2 \cdot P_n}{\sum\limits_{n=1}^{N} P_n}} \tag{16.10}$$

where σ_{AS} denotes the angular spread (AS), P_n denotes the power of the nth multipath, and $\theta_{n,\mu}$ is defined by:

$$\theta_{n,\mu} = \mathrm{mod}\left(\theta_n - \mu_\theta + \pi, 2\pi\right) - \pi \tag{16.11}$$

where θ_n is the AoA/AoD/EoA/EoD of the nth ray. μ_n is:

$$\mu_\theta = \frac{\sum_{n=1}^{N} \theta_n \cdot P_n}{\sum_{n=1}^{N} P_n} \tag{16.12}$$

The calculations of the channel parameters strongly depend on the accuracy of the extracted clusters/rays. However, it is always challenging to identify clusters/rays from the measurements. Traditional approaches (see Fig. 16.2) are realized by postprocessing channel measurement results directly according to identification algorithms (e.g., SAGE, MUSIC) and vision inspection. However, there are unexpected user-specified parameters in the identification algorithms to be manually adjusted, which makes the extracted result of the traditional approach arbitrary and lacks rigorous explanation.

To tackle the aforementioned issues, RT, which can trace rays for direct reflection, scattering, transmission, and diffraction propagations, has been successfully applied in the recent works to dig the influence of the objects in the environment and identify the traveling paths of rays.

As shown in Fig. 16.2, if channel measurement and environment model are available, RT can be used to calibrate the geometries of the 3D environment model and the electromagnetic (EM) parameters of materials. Then, the reliable RT deterministic channel simulator is obtained. The comprehensive channel parameters can be derived from the detailed extracted ray information, and the key coefficients of the stochastic channel model can be tuned as well. The works [5, 6] are

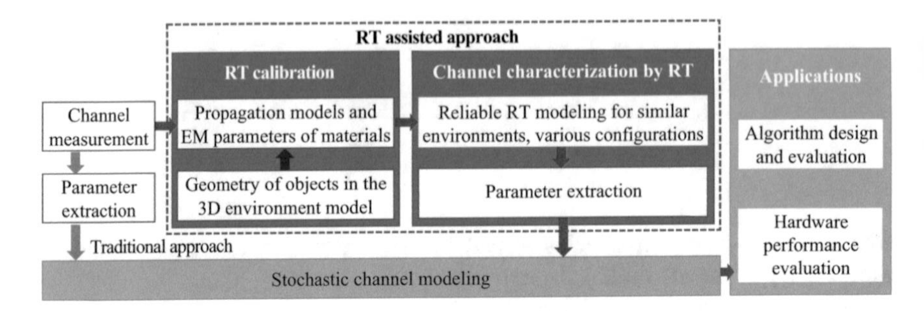

Fig. 16.2 Work flow

representatives of this approach. Based on RT calibration, the material parameters including penetration loss, dielectric parameters, and scattering coefficients can be tuned to match the power level of each traced ray. Practical RT simulations can be conducted for different simulation configurations in similar scenarios. This approach shows clear advantages to overcome the constraints of measurement: not only the same measured environment but also different ones with similar objects and materials, as well as various transceiver configurations, can be explored.

References

1. Cheng, C. L., Sangodoyin, S., & Zajić, A. (2020). THz cluster-based modeling and propagation characterization in a data center environment. *IEEE Access, 8*, 56544–56558.
2. Pometcu, L., & D'Errico, R. (2020). An indoor channel model for high data-rate communications in D-band. *IEEE Access, 8*, 9420–9433.
3. Guan, K., Peng, B., He, D., et al. (2019). Measurement, simulation, and characterization of train-to-infrastructure inside-station channel at the terahertz band. *IEEE Transactions on Terahertz Science and Technology, 9*(3), 291–306. https://doi.org/10.1109/TTHZ.2019.2909975
4. Fu, J., Juyal, P., & Zajić, A. (2020). Modeling of 300 GHz chip-to-chip wireless channels in metal enclosures. *Transactions on Wireless Communications, 19*(5), 3214-3227. https://doi.org/10.1109/TWC.2020.2971206
5. He, D., Guan, K., Fricke, A. et al. (2017). Stochastic channel modeling for kiosk applications in the terahertz band. *IEEE Transactions on Terahertz Science and Technology, 7*(5), 502–513. https://doi.org/10.1109/TTHZ.2017.2720962
6. Priebe, S., & Kürner, T. (2013, September). Stochastic modeling of THz indoor radio channels. *IEEE Transactions on Wireless Communications, 12*(9), 4445–4455.

Part IV
Antenna Concepts and Realization

Chapter 17
High-Gain Antennas

Akihiko Hirata and Jiro Hirokawa

Abstract High-gain antennas are necessary for terahertz wireless communications in order to achieve high C/N ratio required for high data rate, and to compensate for low transmitter power and low receiver sensitivity. However, it is difficult to build high-gain antennas at THz band. This chapter describes the nonplanar and planar antenna technologies in order to achieve high-gain at THz band.

17.1 Introduction

In THz communications, antennas tend to require higher realized gain. The required realized gain to get an equal receiving power to a lower frequency antenna should be increased when the operating frequency of an antenna is higher. The realized gain is degraded by the reflection of the antenna from the gain. The gain is reduced by the loss of the antenna material from the directivity. The loss of the antenna material includes not only that of the material itself such as the loss tangent of a dielectric but also the conductivity of a metal. The conductivity of a metal depends on the fabrication process of metal. For an example, the roughness of a metal reduces the conductivity effectively.

Antennas are classified to nonplanar antennas and planar antennas. A nonplanar antenna increases the directivity almost proportionally by enlarging the size while a planar antenna not. A planar antenna can increase the directivity with keeping the height unchanged. Figure 17.1 shows the relationship between the antenna gain and frequency of the antennas shown in this chapter.

A. Hirata
Chiba Institute of Technology, Narashino, Japan

J. Hirokawa (✉)
Tokyo Institute of Technology, Tokyo, Japan
e-mail: jiro@ee.e.titech.ac.jp

T. Kürner et al. (eds.), *THz Communications*, Springer Series in Optical Sciences 234, https://doi.org/10.1007/978-3-030-73738-2_17

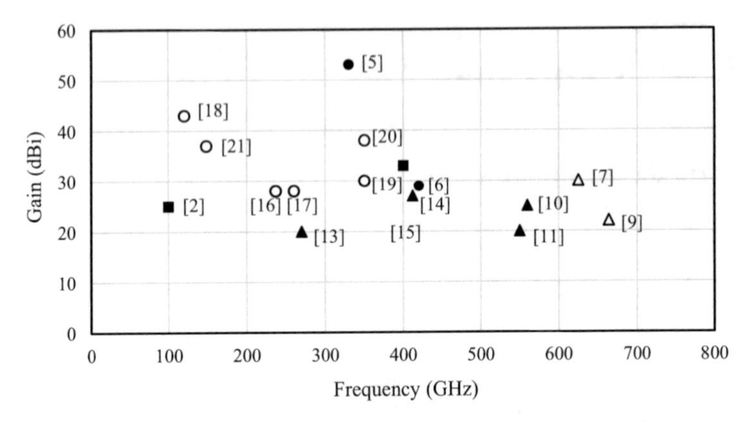

Fig. 17.1 Relationship between the antenna gain and frequency

17.2 Nonplanar Antennas

Nonplanar antennas are categorized mainly into reflector antennas, horn antennas, and lens antennas. One of the major reflector antennas at THz band is a Cassegrain antenna. Cassegrain antennas are already commercially available up to 325 GHz, and their maximum gain exceeds 50 dBi. These antennas are used for outdoor wireless data transmission experiments over a distance of >100 m. One of the problems for reflector antenna is the processing accuracy. As the operation frequency increases, the roughness and spherical accuracy of parabolic reflector and position of secondary reflector severely affect the efficiency of the antenna. P. Nayeri proposed a 3D printed dielectric reflectarrays at 100 GHz [1]. These reflectarrays were fabricated by Polymer-jetting 3-D printing technology, which made it possible to achieve rapid prototyping at a low-cost. A 400-GHz single-layered folded reflectarray antenna with a gain of 33.7 dBi was presented by Z.-W. Miao et al.[2]. It was composed of a feed source, a single-layered reflectarray using a lithography process on quartz, and a wire-grid polarizer implemented by the printed-circuit-board technology.

Another problem of high gain reflector antennas is the accurate measurement of the radiation pattern, because the output power of THz transmitters and the sensitivity of THz receivers are limited, which reduce the dynamic range of the measurement system. Radiation patterns of reflector antennas are defined by Recommendation ITU-R P.1245 [3], and they are used for sharing studies with other services, such as earth exploration satellite service or radio astronomy. However, this recommendation covers up to 86 GHz, and there is no recommendation that can be applied for THz band high-gain antenna. Therefore, it is necessary to measure the accurate radiation pattern of THz-band high-gain antenna and to make new antenna models at this band. Near field measurement of high-gain antenna by EO probe and calculation of far field from the measured near field is one of the promising solution [4].

Offset reflector antennas are also commercially available at a frequency of up to 400 GHz. H. Wang et al. present the terahertz high-gain offset reflector antennas [5]. They employed silicon carbide material and thermally stable carbon fiber-reinforced plastic, and they achieved antenna gain of 55.3 dB and 57.2 dB, respectively. K. Fan demonstrated a quasi-planar reflector antenna at 325–500 GHz. The antenna was composed of a feeding horn, quasi-planar reflectors, choke slots, and an E-plane flared horn. The choke slots were used to suppress the reflected wave from the sidewall and reduce the impact on the radiation performance, and its maximum gain was 32 dBi at 500 GHz [6].

Various types of horn antennas, such as rectangular horn, circular horn, are commercially available at up to 500 GHz. The gain of these antennas is about 25 dBi, and these antennas are used for indoor wireless communications experiments up to 10 m. At the research and development level, horn antennas operating at higher frequencies have been reported. B. Maffei reported WR-1.5 corrugated horn whose directivity was over 30 dBi at 625 GHz [7]. N. Chahat et al. demonstrated a multiflare angle horn with a directivity of 31.7 dBi at 1.9 THz [8]. A spline-profile diagonal horn at 664 GHz was presented by H. J. Gibson et al.[9]. The gain of the antenna was 22 dBi, and this was achieved by applying a spline curve to the vertical cutting axis of the "V" cutting tool, while retaining the ease of direct machining into a split block.

Si lens antennas are commonly used to collimate the output of semiconductor THz source, such as photoconductor, photodiode, resonant tunneling diode, transistors, and so on, by integrating Si lens on the backside of the semiconductor substrate. A 500 GHz-band Si lens antennas with a gain of over 20 dBi had been reported [10, 11]. A. Neto et al. firstly demonstrated the applicability of the leaky lens antenna concept at THz frequencies [12]. The antenna is integrated with a Kinetic Inductance Detector, so that the two of them function as an ultra sensitive detector over a bandwidth ranging from 0.15 to 1.5 THz. In order to increase the antenna gain, or in order to reduce the fabrication cost at THz band, cavity-backed Fresnel zone plate lens [13] and metallic lens [14] are reported.

17.3 Planar Antennas

Planar antennas are categorized mainly to leaky antennas with continuous radiation and array antennas with discretized radiations. In an array antenna, the radiating antenna elements should be connected with the feeding point by transmission lines. Not only the antenna efficiency of the radiating elements but also the loss of the transmission line should be considered. The losses of the transmission line are categorized to the conductor loss if a metal is used, the dielectric loss if a dielectric is used, and the radiation loss if an open transmission loss is used. As functions of the height of the transmission line, the conductor loss is almost antiproportional, the dielectric loss is almost unchanged, and the radiation loss is almost proportional. If a dielectric is used, the dielectric loss is not avoidable. In a closed transmission line, the height should be large, but needs not to excite higher modes. In an open transmission line, the height should be chosen properly to minimize the overall loss.

A. J. Alazemi et al. presented millimeter-wave and terahertz double bow-tie slot antennas on a synthesized elliptical silicon lens to cover 0.1 THz–0.3 THz and 0.2 THz–0.6 THz, respectively [15]. The double bow-tie slot antenna resulted in a wide impedance bandwidth and 78–97% Gaussian coupling efficiency over a 3:1 frequency range. A wideband coplanar-waveguide low-pass filter is designed using slow-wave techniques, and the measured filter response shows an $S_{21} < -25$ dB over a 3:1 frequency range.

K. Sarabandi et al. designed and developed a frequency beam-scanning array antenna operating at Y-band (220–325 GHz) [16]. The antenna is a traveling-wave antenna with a meandered rectangular waveguide structure. The array generates a beamwidth of about 2.5° with ±25° beam steering capability as the frequency sweeps from 230 GHz to 245 GHz. The antenna consists of over 600 patch elements and is confined within a volume of 45 mm × 8.5 mm × 1.25 mm and weighs less than 4.5 g. A prototype of the antenna is fabricated using silicon micromachining with a high level of accuracy. The antenna exhibits a measured scanning range of over 48° with a gain of over 28.5 dBi.

A. Gomez-Torrent et al. designed a dielectric filled parallel-plate waveguide (PPW) leaky-wave antenna fed by a pillbox [17]. The pillbox, a two-level PPW structure, has an integrated parabolic reflector to generate a planar wave front. The operation bandwidth of the antenna spans from 220 GHz to 300 GHz providing a simulated field of view of 56°. The micromachined low-loss PPW structure resulted in a measured average radiation efficiency of −1 dB and a maximum gain of 28.5 dBi with an input reflection below −10 dB. The overall frequency beam steering frontend is 24 mm × 24 mm × 0.9 mm.

A corporate-feed slotted waveguide array antenna is developed with broadband characteristics in terms of gain and reflection by using diffusion bonding of thin laminated plates in the 120 GHz and 350 GHz bands. In the 120 GHz band antennas, copper plates with chemical etching can be used [18]. When the number of the slots becomes large such as 32x32 or more, long and narrow metal parts exist in a thin plate and they could be easily bended in handling. Double layer structure is introduced in the feeding circuit. A 32x32-element array antenna shows more than a 38 dBi antenna gain with over 60% antenna efficiency and 15 GHz bandwidth (119.0–134.0 GHz) and a 64x64-element array shows a higher than 43 dBi antenna gain with over 50% antenna efficiency and 14.5 GHz bandwidth (118.5–133.0 GHz), respectively. In the 350 GHz band antennas, silicon wafers are etched by deep reactive ion etching process with accuracy lower than ±5 μm [19]. They are gold plated and then bonded with the diffusion bonding process. A 16x16-element array antenna is designed and fabricated. The 3 dB down gain bandwidth is 50.8 GHz in simulation and is 44.6 GHz in measurement. This antenna is improved by optimizing the designs to the process requirements of the Silicon-On-Insulator

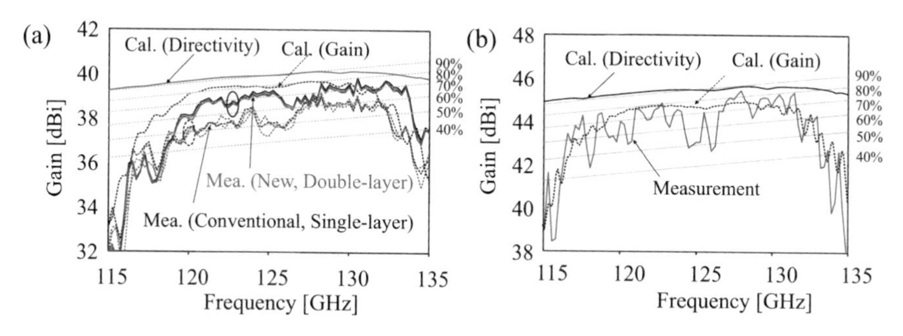

Fig. 17.2 Calculated and measured gains and efficiencies of array antennas with the new feeding structure (**a**) 32x32-element array and (**b**) 64x64-element array antenna [18] (@ 2014 IEEE, reproduced with permission)

micromachining technology in order to achieve the lower measured loss and the larger bandwidth [20]. Two antennas have measured gains of 32.8 dBi and 38.0 dBi and consist of a 16x16-element array and a 32x32-element array, respectively. The measured operation bandwidth for the both antennas is 80 GHz (22% fractional bandwidth), and the total measured efficiency is above −2.5 dB and above −3.5 dB for the two designs in the whole bandwidth.

Gap waveguide supplies contactless characteristics to avoid the good electrical contact between the different metallic layers. A series feed array antenna and a full corporate feed array antenna are fabricated and compared in D-band (110 GHz–170 GHz) [21]. In a 24x16-slot series feed array antenna, the measured gain is around 30 dBi to 31.5 dBi from 141 GHz to 149 GHz and the antenna efficiency is between 40% and 60%. In a 32x32-slot corporate feed array antenna, the measured gain is about 37 dBi and the antenna efficiency is around 50% from 138 GHz to 148 GHz (Fig. 17.2).

References

1. Nayeri, P., Liang, M., Sabory-Garcıa, R. A., Tuo, M., Yang, F., Gehm, M., Xin, H., & Elsherbeni, A. Z. (2014, April). 3D printed dielectric reflectarrays: Low-cost high-gain antennas at sub-millimeter waves. *IEEE Transactions on Antennas and Propagation, 62*(4), 2000–2008.
2. Miao, Z.-W., Hao, Z.-C., Wang, Y., Jin, B.-B., Wu, J.-B., & Hong, W. (2019). A 400-GHz high-gain quartz-based single-layered folded reflectarray antenna for terahertz applications. *IEEE Transactions on Terahertz Science and Technology, 9*(1), 78–88.
3. https://www.itu.int/rec/R-REC-F.1245/en
4. Huy, H., Pham, N., Hisatake, S., & Nagatsuma, T. (2014). *Far-field antenna characterization in the sub-THz region based on electrooptic near-field measurements*. 2014 International Topical Meeting on Microwave Photonics (MWP), WC-7.
5. Wang, H., Dong, X., Yi, M., Xue, F., Liu, Y., & Liu, G. (2017). Terahertz high-gain offset reflector antennas using SiC and CFRP material. *IEEE Transactions on Antennas and Propagation, 65*(9), 4443–4451.

6. Fan, K., Hao, Z.-C., Yuan, Q., & Hong, W. (2017, July). Development of a high gain 325-500 GHz antenna using quasi-planar reflectors. *IEEE Transactions on Antennas and Propagation, 65*(7), 3384–3391.
7. Maffei, B., von Bieren, A., de Rijk, E., Ansermet, J.-P., Pisano, G., Legg, S., & Macor, A. (2014). High performance WR-1.5 corrugated horn based on stacked rings. *Proceedings of SPIE, 9153.*
8. Chahat, N., et al. (2015, November). 1.9-THz multiflare angle horn optimization for space instruments. *IEEE Transactions on Terahertz Science and Technology, 5*(6), 914–921.
9. Gibson, H. J., Thomas, B., Rolo, L., Wiedner, M. C., Maestrini, A. E., & de Maagt, P. (2017). A novel spline-profile diagonal horn suitable for integration into the split-block components. *IEEE Transactions on Terahertz Science and Technology, 7*(6), 657–663.
10. Alonso-DelPino, M., Llombart, N., Chattopadhyay, G., Lee, C., Jung-Kubiak, C., Jofre, L., & Mehdi, I. (2013). Design guidelines for a terahertz silicon micro-lens antenna. *IEEE Antennas and Wireless Propagation Letters, 12*, 84–87.
11. Llombart, N., Lee, C., Alonso-delPino, M., Chattopadhyay, G., Jung-Kubiak, C., Jofre, L., & Mehdi, I. (2013, September). Silicon micromachined lens antenna for THz integrated heterodyne arrays. *IEEE Transactions on Terahertz Science and Technology, 3*(5), 515–522.
12. Neto, A., Llombart, N., Baselmans, J. J. A., Baryshev, A., & Yates, S. J. C. (2014, January). Demonstration of the leaky lens antenna at submillimeter wavelengths. *IEEE Transactions on Terahertz Science and Technology, 4*(1), 26–32.
13. Xu, J., Chen, Z. N., & Qing, X. (2013, April). 270-GHz LTCC-integrated high gain cavity-backed fresnel zone plate lens antenna. *IEEE Transactions on Antennas and Propagation, 61*(4), 1679–1687.
14. Hao, Z.-C., Wang, J., Yuan, Q., & Hong, W. (2017). Development of a low-cost THz metallic lens antenna. *IEEE Antennas and Wireless Propagation Letters, 16*, 1751–1754.
15. Alazemi, A. J., Yang, H. H., & Rebeiz, G. M. (2016, May). Double bow-tie slot antennas for wideband millimeter-wave and terahertz applications. *IEEE Transactions on Terahertz Science and Technology, 6*(5), 682–689.
16. Sarabandi, K., Jam, A., Vahidpour, M., & East, J. (2018, November). A novel frequency beam-steering antenna array for submillimeter-wave applications. *IEEE Transactions on Terahertz Science and Technology, 8*, 654–665.
17. Gomez-Torrent, A., Garcia-Vigueras, M., Le Coq, L., Mahmoud, A., Ettorre, M., Sauleau, R., & Oberhammer, J. (2020, February). A low-profile and high-gain frequency beam steering sub-THz antenna enabled by silicon micromachining. *IEEE Transactions on Antennas and Propagation, 68*(2), 672–682.
18. Kim, D., Hirokawa, J., Ando, M., Takeuchi, J., & Hirata, A. (2014, March). 64x64-element and 32x32-element slot Array antennas using double-layer hollow-waveguide corporate-feed in the 120 GHz band. *IEEE Transactions on Antennas and Propagation, 62*(3), 1507–1512.
19. Tekkouk, K., Hirokawa, J., Oogimoto, K., Nagatsuma, T., Seto, H., Inoue, Y., & Saito, M. (2017, January). Corporate-feed slotted waveguide array antenna in the 350-GHz band by silicon process. *IEEE Transactions on Antennas and Propagation, 65*(1), 217–225.
20. Gomez-Torrent, A., Tomura, T., Kuramoto, W., Hirokawa, J., Watanabe, I., Kasamatsu, A., & Oberhammer, J. (2020, June). A 38 dBi gain, low-loss, flat array antenna for 320 GHz to 400 GHz enabled by silicon-on-insulator micromachining. *IEEE Transactions on Antennas and Propagation, 68*(6), 4450–4458.
21. Liu, J., Zaman, A. U., & Yang, J. (2019, September). Two types of high gain slot array antennas based on ridge gap waveguide in the D-band. *IEEE-APS Topical Conference on Antennas and Propagation in Wireless Communications*, 75–78.

Chapter 18
Antenna Arrays for Beamforming

Muhsin Ali, Alejandro Rivera-Lavado, Álvaro José Pascual-Garcia, David González-Ovejero, Ronan Sauleau, Luis Enrique García-Muñoz, and Guillermo Carpintero

Abstract This chapter presents the antenna arrays for beamforming and beam-steering in the mmW and THz frequency ranges based on phased arrays and leaky-wave antennas. The common feature of the presented approaches is that the antenna emitters (AE) are fed by high-speed photomixers. This enables photonic control of the phase in the antennas and introducing the delays in optical domain for true time-delay steering. Among the beamforming approaches presented, the first one uses dielectric rod waveguide technology for coupling the radiation out of the substrate with wideband performance, while the other two are planar antennas.

18.1 Introduction

The development of next-generation mobile communication networks (e.g., 5G, 6G) is addressing the bandwidth challenge, leaping from 20 MHz currently available at microwave bands below 6 GHz to tens of GHz in the low-loss windows at millimeter (mmW, 30 GHz to 300 GHz) and Terahertz (THz, 300 GHz to 3 THz) wave frequency ranges. However, at these frequency ranges, the free-space propagation losses and atmospheric absorption are of primary concern, mainly due to the low RF output power generated by the currently available THz signal sources, without a clear winner technology. It is a formidable system design challenge to minimize losses at the generation side to deliver enough power at the receiving end. To address the challenge of boosting the radiated power at high frequencies, beamforming is a promising technology to consider [1], taking advantage of the highly directive nature of THz antennas with small physical aperture.

M. Ali (✉) · A. Rivera-Lavado · L. E. García-Muñoz · G. Carpintero
Universidad Carlos III de Madrid, Getafe, Spain
e-mail: muali@ing.uc3m.es

Á. J. Pascual-Garcia · D. González-Ovejero · R. Sauleau
Université de Rennes-l, Rennes, France

T. Kürner et al. (eds.), *THz Communications*, Springer Series in Optical Sciences 234, https://doi.org/10.1007/978-3-030-73738-2_18

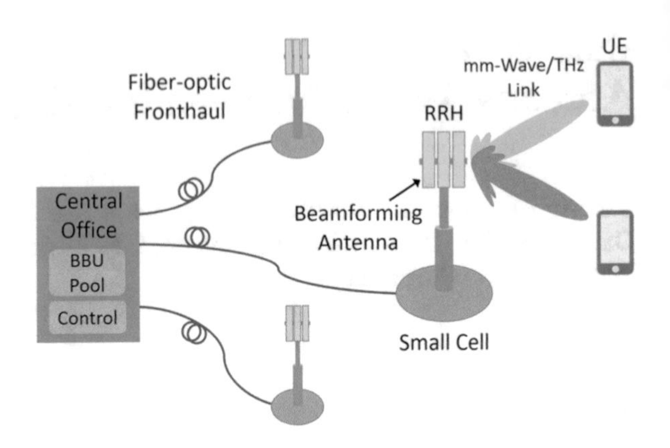

Fig. 18.1 An illustration of next-generation communication system utilizing mmW/THz beamforming

Beamforming is defined as the art to control the shape, directivity, and radiation direction of the beam. In this chapter, we present an overview of mmW and THz antenna array structures enabling dynamic beamforming and beam-steering. These will be needed in next-generation small-cell wireless access networks as illustrated in Fig. 18.1. In a distributed architecture, the antenna sites contain remote radio head (RRH) units that need to be controlled from the base band unit (BBU). The connection between BBU and RRH is established through low-loss fiber-optic fronthaul, allowing the centralization of resources and processing at central office to reduce the RRH complexity. Complex antenna arrays with beamforming are advantageous for the directional coverage to the mobile user equipment (UE), especially if we have control of the radiation direction in the optical domain.

The mmW/THz beamforming has sparked an interest in developing antenna arrays. Attempts have been made to develop linear horn antennas for phased array systems [2, 3], using complex fabrication, complicating the 2D scaling, see also Chap. 33. CMOS technology has proven to offer densely integrated 2D THz phased arrays on silicon [4], albeit with very limiting frequency tuning capability due to narrowband nature of electronic phase shifters. In [5], the authors presented fast frequency-steerable antennas for the application in THz radar.

This chapter describes three breakthrough approaches to integrated antenna arrays for beamforming and beam-steering in the mmW and THz frequency ranges based on phased arrays and leaky-wave antennas. The common feature of the presented approaches is that the antenna emitters (AE) are fed by high-speed photomixers, either photoconductors (PC) or photodiodes (PD). This enables controlling the phase of the signals in the optical domain for true time-delay (TTD) steering. Among the approaches, we have selected key technologies using planar antennas that can be adapted to the THz bands. The first one to be discussed uses dielectric rod waveguide technology for coupling the radiation out of the substrate.

18.2 Dielectric Rod Waveguide Antennas

Optoelectronic THz sources are grown on high-permittivity substrates, such as InP or GaAs. When placed in the gap of a printed AE, most of the power is scattered into the substrate [6]. The use of dielectric structures, such as silicon lenses, dielectric horns [7], and dielectric rod waveguides (DRW) [8], is required to achieve efficient radiation of the generated signal. DRW antennas, Fig. 18.2a and b, are a cost-affordable, low weight, and compact alternative to electrically large lenses that efficiently radiate the power into a single beam as shown in Fig. 18.2c. This occurs for frequencies on which most of the power is radiated close to the DRW antenna tip [8], that is, from 100 GHz to 200 GHz for the device shown in Fig. 18.2b.

Figure 18.2e and f shows the simulated E-field amplitude on a DRW antenna with an embedded planar lens. The planar lens is defined by reducing the permittivity of the DRW antenna around it by etching electrically small holes in the substrate. This dielectric structure extends the range of frequencies on which most of the power is radiated in a single area close to the tip. It is possible to achieve a single-lobe radiation pattern from 100 GHz up to 800 GHz [9] when the AE is placed in the lens focus. The most relevant radiation zones are marked in Fig. 18.2f. As can be seen, there are radiation zones that lead to secondary lobes in the radiation pattern at 800 GHz. Since most of the power is still radiated close to the tip, the interference between different zones is mitigated, which avoids radiation nulls in the endfire direction.

An array of DRW antennas can effectively combine the power of an arbitrarily large number of mm- and sub-mm wave sources in a single output beam. This is not possible with an electrically large silicon lens since the power generated in the out-of-focus elements is not combined into the main beam [11]. Arrays with one lens per element lead to sparse configurations and/or smaller efficiencies [12]. Since DRW antennas are compact, it is still possible to place one element per AE in a dense configuration with good scanning properties.

To reduce the experimental complexity, first tests of DRW antenna arrays were conducted with rectangular waveguide feeds. An early 2 × 1 DRW antenna array prototype was presented in [13]. Figure 18.3 shows a 4 × 4 silicon DRW antenna array [14]. It is designed for a frequency of 100 GHz. It consists of 16 E-tapered DRW antennas of length 15 mm, width 1 mm, and thickness 0.5 mm (Fig. 18.3a) fed by a WR-10 power divider (Fig. 18.3b). An E-plane pitch of $2\lambda_0/3$ and an H-plane pitch of λ_0 was used, with λ_0 being wavelength at center frequency.

A total gain of 23.5 dBi and a cross-polar discrimination (XPD) of 26 dB are measured at 100 GHz, shown in Fig. 18.3c. According to our simulation results, the coupling between the adjacent elements both in E- and H-plane is smaller than 25 dB.

Figure 18.4a shows a 1 × 4 array of E-plane tapered DRW antennas fed by four antenna emitters. They work from 100 GHz to 200 GHz. A pitch of 1 mm between the elements is chosen since a spacing of $\lambda_0/2$ at a central frequency of 150 GHz is

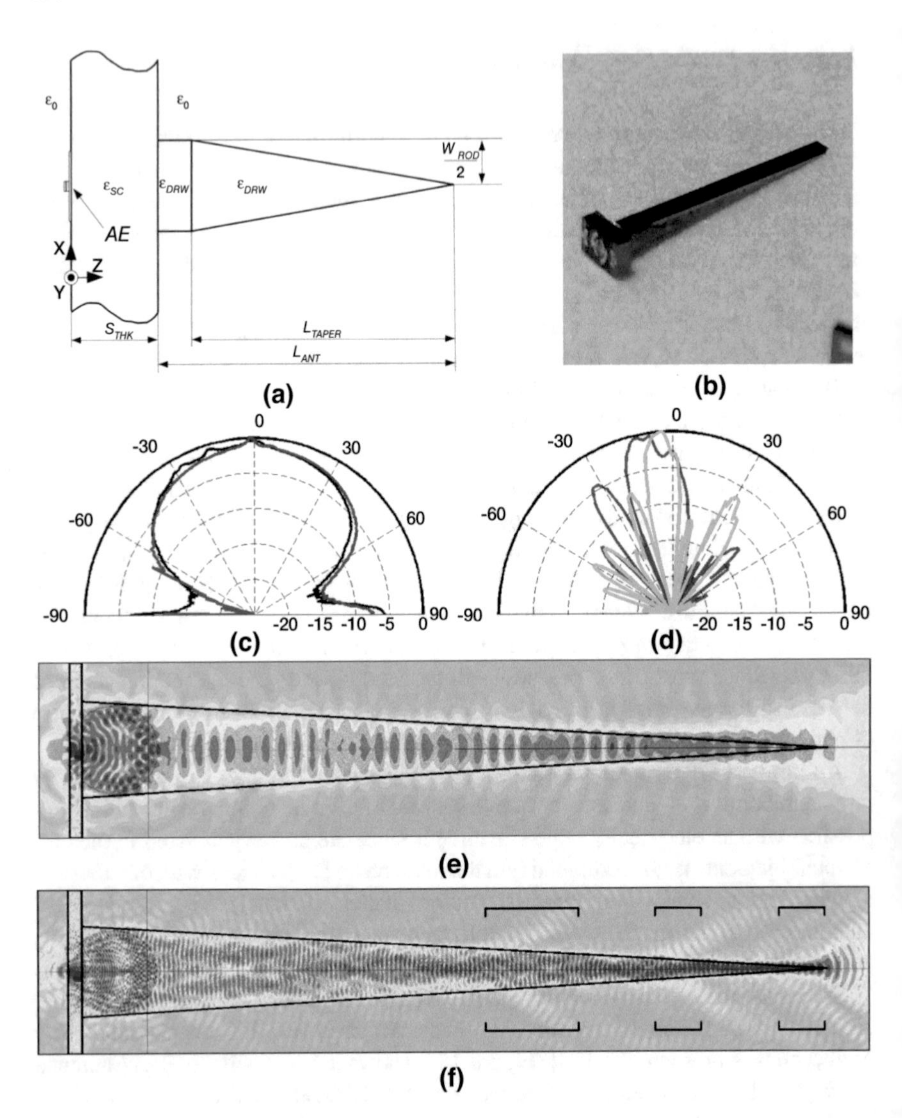

Fig. 18.2 (**a**) Sketch of a DRW antenna placed at the backside of the antenna emitter wafer. (**b**) DRW antenna assembled to a n-i-pn-i-p AE. The rod length, L_{TAPER}, is 8 mm, width W_{ROD}, equal to 0.5 mm, and the thickness is 0.5 mm. (**c**) Measured radiation pattern at 150 GHz. Co-polar component for both E-plane ($\phi = 0°$, black) and H-plane ($\phi = 90°$, red) is shown. (**d**) Measured radiation pattern of a *TeraScan1550* quasi-optical terahertz source [10] at 100 GHz (red) and 150 GHz (green). Only H-plane co-polar component is shown. Simulated E-field amplitude distribution on a DRW antenna with an embedded planar lens at (**e**) 300 GHz and (**f**) 800 GHz. [8] © 2015 IEEE, [9] © 2019 Springer nature, reproduced with permission

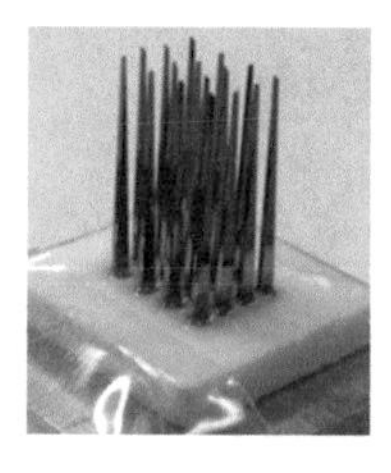

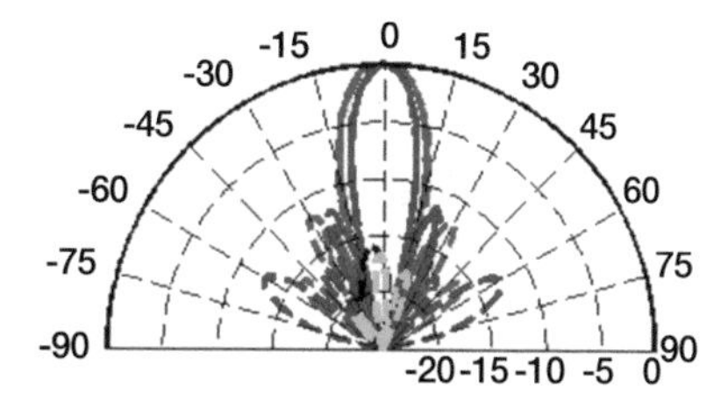

Fig. 18.3 (**a**) 4 × 4 DRW antenna array. (**b**) WR-10 power divider for feeding the 16 antennas. (**c**) Simulated (dashed) and measured (solid) radiation pattern. Both E-plane ($\phi = 0°$, red) and H-plane ($\phi = 90°$, blue) are shown. The cross-polar component (H-plane black and E-plane green) is also displayed. [14] © 2019 Springer nature, reproduced with permission

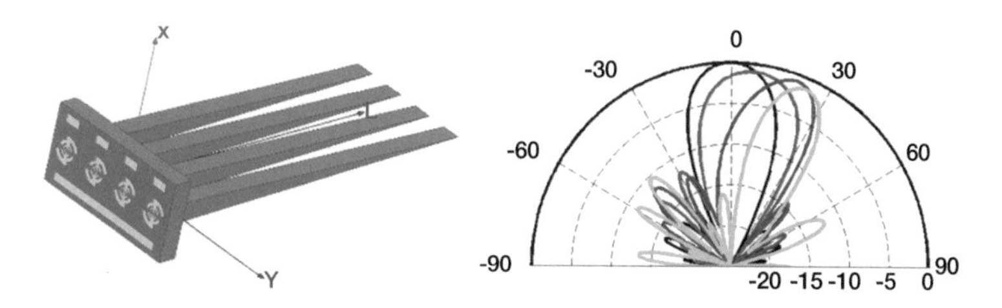

Fig. 18.4 (**a**) 1 × 4 DRW antenna array fed by four log-periodic AEs. (**b**) Scanning on the E-Plane. A progressive phase shift β of 0° (black), 30° (red), 60° (blue), and 90° (green) is shown

suitable for broadside scanning arrays [15], as confirmed by the simulation results given in Fig. 18.4b.

Next, we consider the mm and sub-mm signal phase shifting. Since the DRW is an open transmission line, it is possible to modify the propagation constant by modifying the conditions in the waveguide walls, so the phase shifter can be easily integrated with the DRW antenna. Some technologies, such as microelectromechanical systems (MEMS), have been proved suitable for this purpose [16].

Carbon nanotubes (CNT) can be used for implementing a cost-affordable DRW-based phase shifter. Figure 18.5 shows an E-plane tapered sapphire DRW with CNTs deposited on top. They are illuminated by a tungsten lamp. Two W85104A WR-10 frequency-extension heads allow us to characterize the DRW S_{21} parameter for different illumination conditions.

Three power densities were considered: $P_0 = 0$ mW/mm^2, $P_1 = 0.63$ mW/mm^2 and $P_2 = 1.26$ mW/mm^2. Figure 18.6 shows the amplitude and the phase of the S_{21} parameter. To test repeatability in the measurements, two samples of P_0 and three of P_1 were taken alternatively. All the plots have been normalized to the P_0 measurement. As can be seen, there is no noticeable change in the S_{21} amplitude. A maximum phase shift of 33° was achieved.

At this point, the mechanism of this effect is not fully understood. With the results presented in Fig. 18.6, it is not possible to determine whether there is optical interaction between light and nanotubes or there is heating due to light absorption in the CNT layer. Any possible effect on the DRW substrate in these results has been

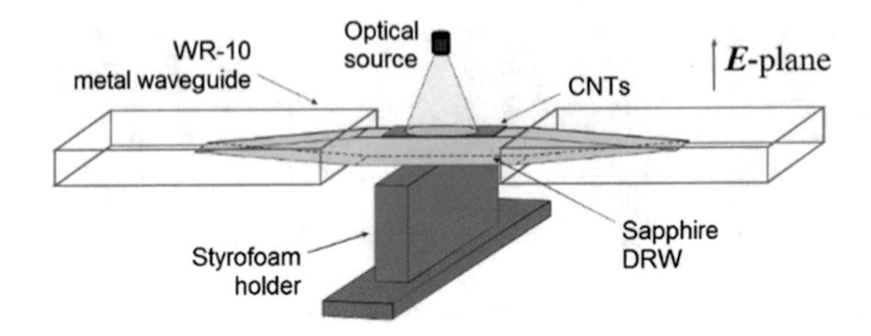

Fig. 18.5 Sketch of a CNT-based phase shifter

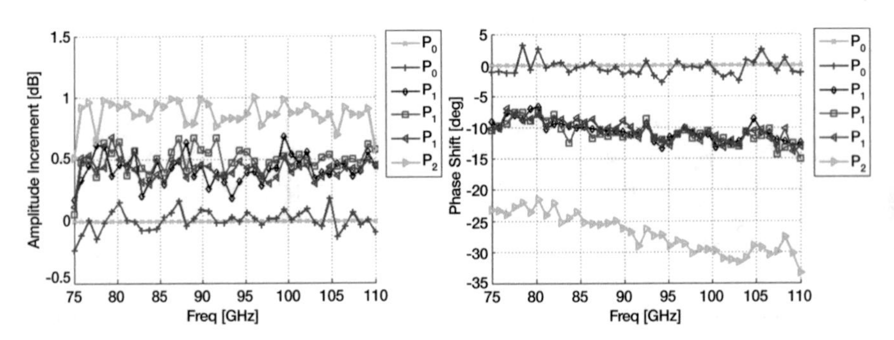

Fig. 18.6 (**a**) Amplitude and (**b**) phase of the measured S_{21} parameter for three different light intensities: $P_0 = 0$ mW/mm^2, $P_1 = 0.63$ mW/mm^2, and $P_2 = 1.26$ mW/mm^2

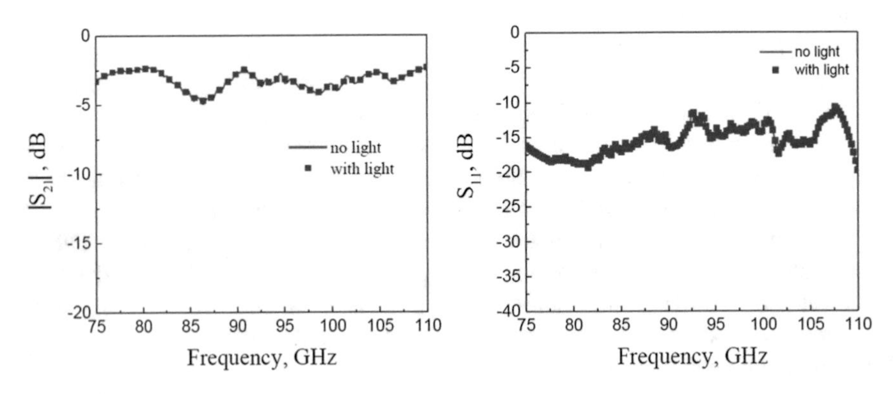

Fig. 18.7 (**a**) Amplitude and (**b**) phase of the measured S_{21} of the unloaded sapphire DRW without and with 2.52 mW/mm^2 light intensity illumination

discarded by characterizing the same DRW with no CNTs and two illumination conditions: no power and 2.52 mW/mm^2, with the results shown in Fig. 18.7. No difference between the two conditions is noticed.

18.3 Leaky-Wave Antenna

One of the simplest forms to steer the beam in 1D is frequency-selective beamforming using a leaky-wave antenna (LWA). In this configuration, a traveling wave encounters periodic perturbations in one direction, so that it leaks progressively. If the electrical path between successive perturbations is such that the traveling wave encounters them in phase, the antenna will radiate a beam at or near broadside, as is clear from the array factor. Conversely, as frequency increases, the electrical path between elements will increase too, and the array factor will produce a beam moving with frequency from the backward to the forward quadrant [17]. The design of the LWA is motivated by the necessity of increasing the emitted power and or directivity by a scalable layout. This feature is advantageous with respect to classical mmW antennas. A proof of concept for E-band frequencies has been reported in [18], and it is shown in Fig. 18.8.

The geometrical features have been printed on a low-loss grounded Duroid 5880 substrate ($\varepsilon_r = 2.24$, $\tan\delta = 0.004$ at 60 GHz), suitable for high-frequency applications. The antenna is excited by a 2×1 array of high-speed uni-traveling carrier photodiodes (UTC-PD) grown on InP [19]. Each UTC-PD presents a 50-Ω grounded coplanar waveguide output. By controlling the number of rows and their length, one can easily accomplish symmetrical patterns (at E- and H-planes) or increase the directivity or the emitted power (more PDs can be included in the direction perpendicular to the rows).

The antenna is comprised of three main sections: A feed network, a radiating section including four rows, and an RF choke plus DC pads for PD polarization, as indicated in Fig. 18.8. The choke is composed of seven cascaded T-sections and shows an isolation level better than 35 dB for the range of operation of the antenna (75 GHz to 85 GHz). The feed network contains a transition between the PD output to the microstrip line in the PCB substrate, and a power splitter so that each PD feeds two columns of the array for a more symmetric radiation pattern. The rows in the radiating section are mirrored to decrease the level of cross-polarization. Each row is periodically loaded by groups of $\lambda/4$ stubs that progressively radiate the input power. The stubs are carefully designed so that the LWA does not present the open-stopband effect common in LWAs at frequency of broadside radiation [17]. For experimental

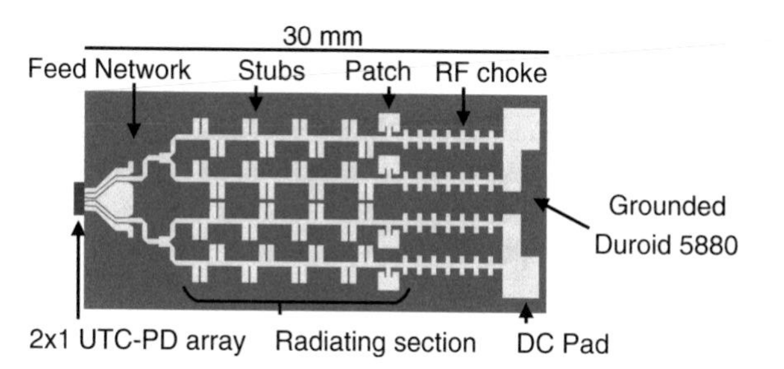

Fig. 18.8 Layout of the LWA with the different sections indicated [18]

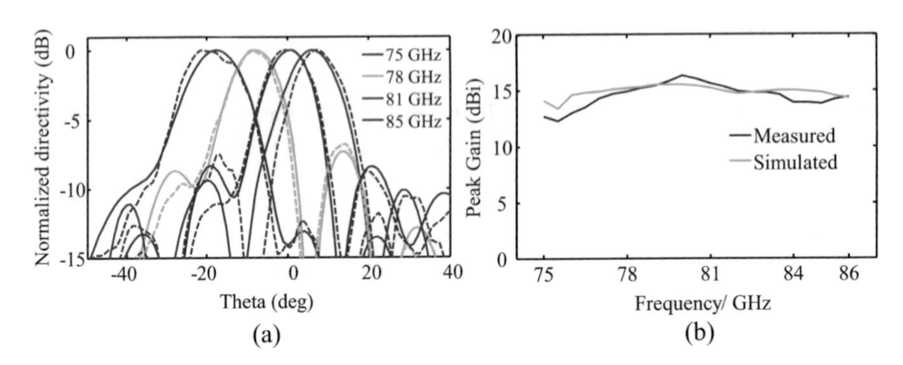

Fig. 18.9 (**a**) Normalized co-polar directivity at H-plane at different frequencies for the two-row prototype adapted for GSG probe measurements. Continuous line: simulation, dashed line: measurements. (**b**) Simulated and measured peak gain [18]

verification with quasi-optical elements, the antenna is truncated in a matched patch that radiates the remaining power.

A two-row prototype, adapted for excitation with a GSG probe, has been manufactured to examine frequency-selective beam steering along the plane of the row (H-plane). The prototype was measured at University of Nice. The results, shown in Fig. 18.9a, show that beam-steering in 1D is accomplished for a scanning range of 22°. Note that side-lobes are substantially high because the row is truncated by a patch, and the patterns are optimized for the prototype excited by the PDs. Figure 18.9b shows the antenna gain at different frequencies. The LWA does not present a reduced level of gain at the frequency of broadside operation (around 81 GHz); thus, the scanning range is extended from the backward quadrant ($\theta < 0$) to broadside and the forward quadrant ($\theta > 0$). Finally, it is worth mentioning that extending the row to 19 periods leads to a half-power beam-width (HPBW) of 4° at broadside (18.5 dBi gain for one row), while maintaining good input matching characteristics. The prototype, integrated with one active UTC-PD, shows a 3-dB power drop bandwidth between 75 GHz and 85 GHz, and is capable of error-free transmission at 2.15 Gbps for a 25 cm link.

18.3.1 Beam-Switching with a Patch Array

An attractive variation of beam steering is beam switching, as illustrated in Fig. 18.10a. In the case shown, two antennas (upper and lower) are located at the back focal plane of a lens (F), a distance s out of the optical axis. From geometrical optics, the beam tilt for each antenna is given by:

$$\tan\theta = \pm\frac{s}{F}. \tag{18.1}$$

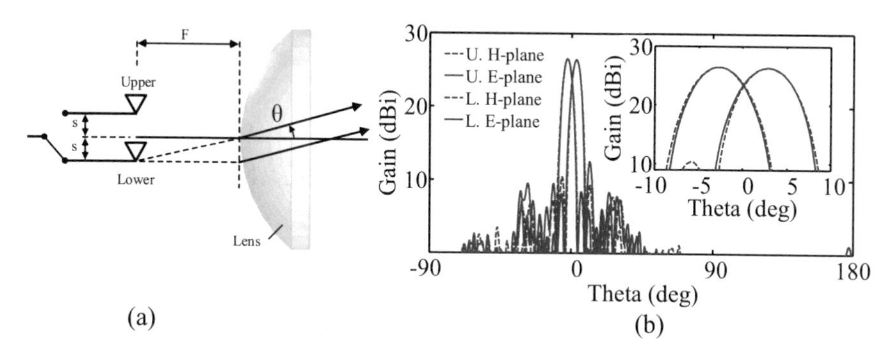

Fig. 18.10 (**a**) Schematic of beam switching, (**b**) simulation results of beam switching using two 2 × 2 patch arrays (upper, U. and lower, L.) at the focal plane of a PTFE lens (51 mm diameter)

Only one antenna is active at a time; thus, the beam can be switched among different directions governed by the previous equation. In [20], we have reported a prototype to demonstrate beam-switching using a photonic transmitter. It consists of two 2 × 2 patch arrays excited by one PD each. The arrays feature an impedance bandwidth >20%, which is enough to cover the E-band. They are located at the focal of a PTFE lens, and their separation, 2 *s*, is 5 mm. Figure 18.10b shows the simulated gain for E- and H-planes. They cross at −2.9 dB, with a peak gain of 27 dBi, and point angles equal ±2.7°, stable over the E-band.

Beam switching for photonic transmitters has been first proposed in [20], it presents several trade-offs, discussed below, that make it an extremely interesting route to explore. Compared to a classical phased array approach, beam switching can achieve high gain (due to the lens) using a reduced number of elements. For instance, a square array over a ground plane of λ/2 dipoles, spaced $\lambda_0/2$, presents a gain of 19.1 dBi for 25 elements and 25.2 dBi for 100 elements. On the downside, beam switching allows discrete scanning directions only and the pattern degradation when *s* is comparable to *F* or lens diameter limits the scan range.

Besides, the grating lobe condition of an array disappears, so the element spacing *s* can be in principle larger than $\lambda_0/2$. This is a crucial characteristic for photonic transmitters where efficient optical illumination of the PD is of utmost importance. It can be challenging to integrate a number of fibers if the spacing between elements is only 1.5 mm or 0.5 mm ($\lambda_0/2$ for 100 GHz and 300 GHz respectively). For beam switching, the spacing is taken as a design variable. A larger separation between elements also eases the integration of amplifiers or matching circuits compared to phased arrays. On the other hand, the power combining of a phased array does not occur for a transmitter featuring beam switching.

Finally, beam switching is especially well suited to photonic transmitters since it can be performed using low insertion loss and low power-consuming optical switches with a switching time in the order of tens of nanoseconds. This approach would help alleviate the stringent alignment conditions required for long-distance THz wireless links.

18.4 Tapered Slot Antenna and Photonic TDD Beam-Steering

The last beamforming antenna example uses a tapered slot antenna (TSA) array, aided by photonics-enabled true TDD technique. In the traditional method of packaging the photomixer emitters, there is little room for integrating custom antennas to achieve the desired performance and functionality [7]. This limit is primarily imposed by the nature of lens and horn elements. We overcome this problem by developing a low-profile and low-cost planar antenna array leading to a simpler integration of AE in a photonic phased array (PAA).

The antenna forms an important element in the mmW/THz sources, and its complexity is even higher when realizing phased arrays. Two corrugated TSAs in 2 × 1 configuration, with grounded coplanar input, are designed to work in the E-band frequency. The overall structure of the antennas, and passive circuit, is shown in Fig. 18.11a. A proof-of-concept TSA array AE for power scaling has been reported earlier [21]. The 2 × 1 TSAs are implemented on a commercial low-loss substrate Duroid 5880. The spacing d between antennas is kept to λ_0 (4.2 mm) at f_0 of 70 GHz. To ease the complexity of fiber array coupling, the number of antenna elements was limited to only two. However, good performance of beam-steering was achieved, as will be shown. The PAA emitter prototype is shown in Fig. 18.11b, packaged in a custom housing. It mainly consists of the fabricated passive circuit, which includes the antennas and bias network (bias-T), and a pair of UTC-PD chips [19]. A picture of the assembly can be seen in the inset. The pitch of passive optical waveguides (POW) in the PD chip is 500 μm. The active PDs are optically excited using a 2 × 1 lensed fiber array of equal pitch.

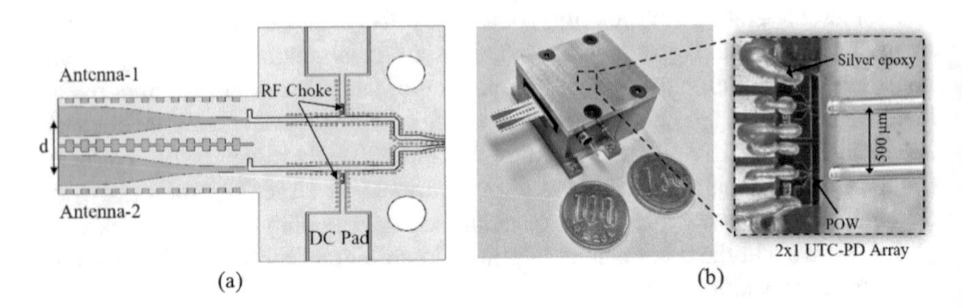

Fig. 18.11 (**a**) Layout of the 2 × 1 tapered slot antenna array with the passive circuit board including bias network, and (**b**) the packaged photonic PAA emitter prototype with antenna and photodiode chips, inset: microscopic view of the 2 × 1 UTC-PD array bonded to the antenna, and excited by lensed fiber array. [21] © 2019 IEEE, reproduced with permission

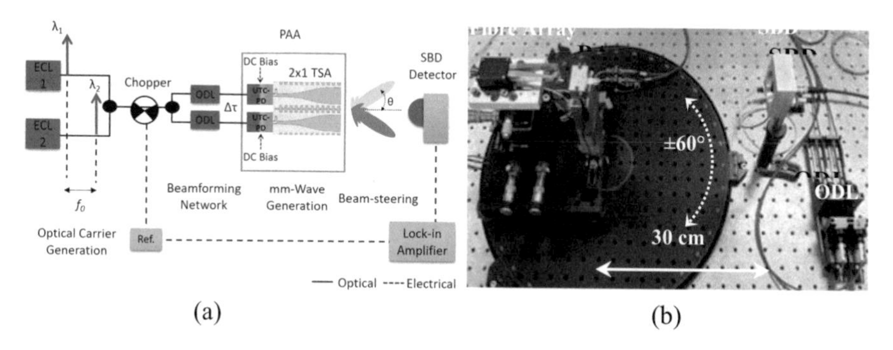

Fig. 18.12 (**a**) Schematic diagram of the experimental arrangement for photonic-enabled beam-steering, and (**b**) picture showing the benchtop setup, with the emitter module mounted on a rotation stage, used for radiation pattern measurements

Next, we evaluate the performance of the antenna itself and mmW beam-steering. The performance of beamforming antennas particularly depends on the structure feeding them, called beamforming network (BFN). Unlike the microwave systems, the phase-tuning BFNs in the mmW range and beyond often lead to beam-squint problem due to narrowband nature of phase-shifters. A TTD based BFN on the other hand is a promising alternative, due to its ability to provide near-constant delays over whole bandwidth of interest [22]. While electronic delay lines do not meet the bandwidth requirements and often difficult to implement, the photonic TTDs [23] offer a significant advantage due to broadband nature of optical components.

Figure 18.12a shows the schematic of the setup to characterize the TSA array emitter, utilizing optical heterodyne technique for the generation of mmW signals. Two external cavity lasers (ECL) providing wavelengths λ_1 and λ_2 are separated at a frequency difference of f_0. An optical chopper then introduces the modulation at reference frequency of 30 Hz for lock-in detection, before splitting into two optical paths. Thereon, tunable optical delay lines (ODL) are used in each of the two paths, providing optical TTD. The delay $\Delta\tau$ required to steer the radiated beam by an angle θ from the boresight is calculated using (2), where d is the spacing between the antennas mentioned above, and c is the speed of light. The optical signals are then fed into the PD-integrated antennas, converted into RF by photomixing and radiated into free-space. Finally, a Schottky barrier diode (SBD) module detects the mmW signals and the signal intensity is measured with a lock-in amplifier. A picture of benchtop setup used to measure radiation patterns is given in Fig. 18.12b.

$$\sin\theta = \frac{\Delta\tau}{d}c \tag{18.2}$$

First, the simulated gain of 2 × 1 TSA array is provided in Fig. 18.13a, to observe radiation performance of the antenna array. The gain estimated is better than 14 dBi over the frequency range of 70–100 GHz, with a peak value of 15.9 dBi at 86 GHz. An undesired drop around 67 GHz is caused by reflections at the antenna input.

We then measured the relative frequency response for the integrated emitter module, shown in Fig. 18.13b, in a broad range between 53 GHz and 120 GHz. We can see that the variations are within the 3-dB limits over much of the frequency range, providing a bandwidth of up to 45 GHz, translating to a fractional bandwidth of 45%.

Finally, the beam-steering behavior of the TSA array is evaluated by obtaining the 1D far-field patterns in azimuth (E-) plane. The angular resolution is set to 3°. The steering is performed at f_0 of 60 GHz and 70 GHz. Figure 18.14 shows the measured beam patterns at each of the carrier frequencies for the steering angles of 0° and ±10°. Note that the side-lobes of less than −10 dB are maintained, when

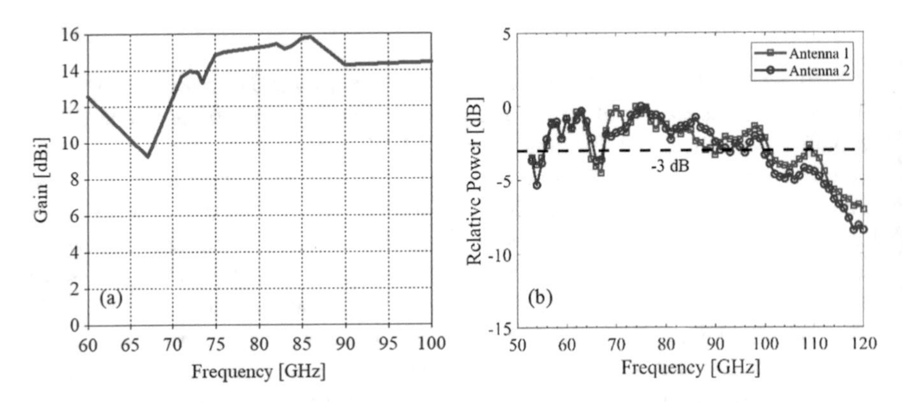

Fig. 18.13 (**a**) Simulated gain of 2x1 TSA array over frequency and (**b**) measured frequency response of the PAA emitter. © 2019 IEEE, reproduced with permission

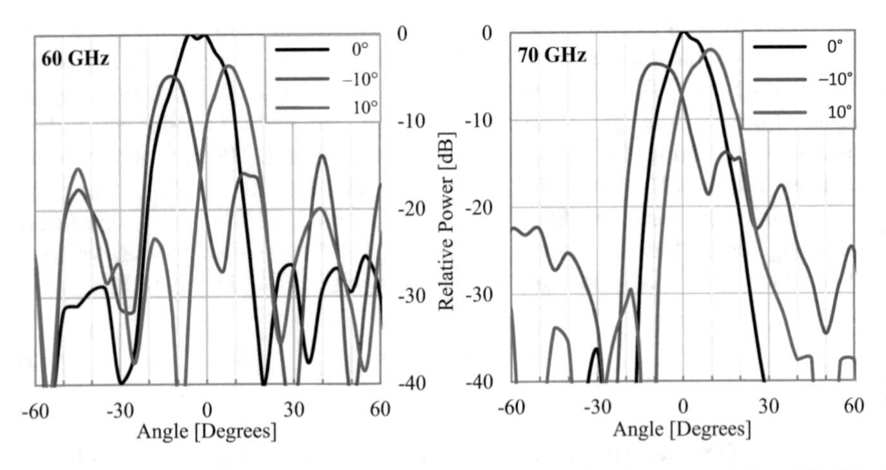

Fig. 18.14 Beam-steering radiation patterns measured at the carrier frequencies of 60 and 70 GHz for three different scanning angles each

performing the steering, indicating the possibility of further steering of the beam. As a result, more than 20° scan range can be achieved. Steering also results in a power reduction of 2 dB to 4 dB in the main lobe due to an increase in the levels of grating lobes. To achieve the steering, the amount of delay $\Delta\tau$ between antenna elements for both the frequencies was set to 0 ps and 2.4 ps for 10°, and 2.4 ps and 0 ps for $-10°$. The HPBW is measured to be around 20°.

References

1. Headland, D., Monnai, Y., Abbott, D., Fumeaux, D., & Withayachumnankul, W. (2018). Tutorial: Terahertz beamforming, from concepts to realizations. *APL Photonics, 3*(5), 051101–051103.
2. Liu, Y., Isaac, B., Kalkavage, J., Adles, E., Clark, T., & Klamkin, J. (2019). *Programmable integrated microwave photonics beamforming networks for millimeter wave communications and phased arrays*. In Proceedings 2019 IEEE avionics and vehicle fiber-optics and photonics conference (AVFOP), Arlington, VA, USA, pp. 1–2.
3. Rey, S., Merkle, T., Tessmann, A., & Kürner, T. (2016). *A phased array antenna with horn elements for 300 GHz communications*. 2016 International Symposium on Antennas and Propagation (ISAP), Okinawa, pp. 122–123.
4. Tousi, Y., & Afshari, E. (2015, February). A high-power and scalable 2-D phased array for terahertz CMOS integrated systems. *IEEE Journal of Solid-State Circuits, 50*(2), 597–609.
5. Murano, K., et al. (2017, January). Low-profile terahertz radar based on broadband leaky-wave beam steering. *IEEE Transactions on Terahertz Science and Technology, 7*(1), 60–69.
6. Dohler, G. H., Garcia-Munoz, L. E., Preu, S., Malzer, S., Bauerschmidt, S., Montero-de-Paz, J., Ugarte-Munoz, E., Rivera-Lavado, A., Gonzalez-Posadas, V., & Segovia-Vargas, D. (2013, September). From arrays of THz antennas to large-area emitters. *IEEE Transactions on Terahertz Science and Technology, 3*(5), 532–544.
7. Andres-Garcia, B., Garcia-Munoz, E., Bauerschmidt, S., Preu, S., Malzer, S., Döhler, G. H., Wang, L. J., & Segovia-Vargas, D. (2011, September). Gain enhancement by dielectric horns in the terahertz band. *IEEE Transactions on Antennas and Propagation, 59*(9), 3164–3170.
8. Rivera-Lavado, A., et al. (2015). Dielectric rod waveguide antenna as THz emitter for photomixing devices. *IEEE Transactions on Antennas and Propagation, 63*(3), 882–890.
9. Alejandro Rivera-Lavado, et al. (2019). Planar lens-based ultra-wideband dielectric rod waveguide antenna for tunable THz and sub-THz photomixer sources. *Journal of Infrared, Millimeter, and Terahertz Waves, 40*(8), 838–855.
10. Toptica Photonics AG. (2020, April). [Online]. Available: https://www.toptica.com/products/terahertz-systems/frequency-domain/terascan/
11. Preu, S., et al. (2017). Fiber-coupled 2D n-i-pn-i-p superlattice photomixer array. *IEEE Transactions on Antennas and Propagation, 65*(7), 3474–3480.
12. Rivera-Lavado, A., García-Muñoz, L.-E., Lioubtchenko, D., Preu, S., Abdalmalak, K. A., Santamaría-Botello, G., Segovia-Vargas, D., & Räisänen, A. (2019, August). Planar lens-based ultra-wideband dielectric rod waveguide antenna for tunable THz and sub-THz photomixer sources. *Journal of Infrared, Millimeter, and Terahertz Waves, 40*(8), 838–855.
13. Pousi, J. P., Lioubtchenko, D. V., Dudorov, S. N., & Räisänen, A. V. (2010). High permittivity dielectric rod waveguide as an antenna array element for millimeter waves. *IEEE Transactions on Antennas and Propagation, 58*(3), 714–719.
14. Rivera-Lavado, A., García-Muñoz, L. E., Generalov, A., Lioubtchenko, D., Abdalmalak, K.-A., Llorente-Romano, S., Garcia-Lamperez, A., Segovia-Vargas, D., & Räisänen, A. V. (2016). Design of a dielectric rod waveguide antenna array for millimeter waves. *Journal of Infrared, Millimeter and Terahertz Waves*.

15. Balanis, C. A. (2005). *Antenna theory: Analysis and design*. New York: Wiley.
16. Chicherin, D., Sterner, M., Oberhammer, J., Dudorov, S., Lioubtchenko, D., Niskanen, A. J., Ovchinnikov, V., & Räisänen, A. V. (2010). *MEMS based high-impedance surface for millimetre wave dielectric rod waveguide phase shifter*. In European Microwave Conference (EuMC), pp. 950–953.
17. Jackson, D. R., & Oliner, A. A. (2008). Leaky-wave antennas. In C. A. Balanis (Ed.), *Modern antenna handbook* (1st ed., p. 458). Hoboken: Wiley.
18. Pascual, A. J., Ali, M., Carpintero, G., Ferrero, F., Brochier, L., Sauleau, R., García-Muñoz, L. E., & González-Ovejero, D. (2020). A photonically-excited leaky-wave antenna array at e-band for 1-D beam steering. *Applied Sciences, 10*(10), 3474.
19. Rouvalis, E., et al. (2012). High-speed photodiodes for InP-based photonic integrated circuits. *Optics Express, 20*, 9172–9177.
20. Pascual, A. J., Batté, T., de Sagazan, O., Carpintero, G., Sauleau, R., & González-Ovejero, D. (2020, November). *A photonic transmitter for beam switching in mm-wave wireless links*. Submitted to the 45th International Conference on Infrared, Millimeter, and Terahertz Waves (IRMMW-THz), Buffalo, US.
21. Ali, M., Jankowski, A., Guzmán, R. C., Dijk, F. v., García-Muñoz, L. E., & Carpintero, G. (2019). *A broadband millimeter-wave photomixing emitter array employing UTC-PD and planar antenna*. 2019 44th international conference on Infrared, Millimeter, and Terahertz Waves (IRMMW-THz), Paris, France, pp. 1–2.
22. Mailloux, R. J. (2005). *Phased array antenna handbook*. Boston: Artech House.
23. Nanzer, J. A., Wichman, A., Klamkin, J., McKenna, T. P., & Clark, T. R., Jr. Millimeter-wave photonics for communications and phased arrays. *Fiber and Integrated Optics, 34*(4), 159–174.

Chapter 19
Algorithms for Multiple Antennas

Bile Peng

Abstract As described in previous chapters, the THz channels are subject to unfavorable propagation conditions. In order to establish a stable wireless connection, advanced antenna technologies are expected to compensate for the propagation conditions. Compared to a single high-gain antenna (e.g., a horn antenna), a multi-antenna system can provide higher performance and more flexibility. This chapter discusses the application of multi-antenna systems for the THz communications with focus on the channel estimation and signal processing algorithms. We first introduce three different types of multi-antenna systems, including phased array with the ability to adjust the main lobe direction, multiple-input-multiple-output (MIMO) array with complete signal processing ability, and hybrid antenna arrays, i.e., an array of phased arrays. Since the AoA and channel estimations are essential to the signal processing, the estimation techniques are discussed in the next two sections. Finally, signal processing for massive MIMO is presented in the last section.

19.1 The Multiple Antenna Systems

The multiple antenna systems are a major approach to develop advanced signal processing methods in order to improve the system performance and/or the stability. Given the unfavorable propagation conditions of the THz wireless channels described in the previous chapters, the multiple antenna systems can play an important role to establish a stable and fast THz link [1].

B. Peng (✉)
Institute for Communications Technology, Technische Universität Braunschweig, Braunschweig, Germany
e-mail: peng@ifn.ing.tu-bs.de

T. Kürner et al. (eds.), *THz Communications*, Springer Series in Optical Sciences 234, https://doi.org/10.1007/978-3-030-73738-2_19

In this chapter, we discuss the following three major categories of multiple antenna systems: phased array, full MIMO array, and hybrid array. The purposes of this section are to provide a mathematical description how these multiple antenna systems work and discuss their advantages, disadvantages, and domains of applications.

19.1.1 Phased Array

The phased arrays have compact array sizes, and the distance between the antenna elements is small enough compared to the wavelength (this distance is usually assumed to be less than half of the wavelength), such that we can assume the only difference between the channel impulse responses (CIRs) of the antennas elements are the phases (delays if the signal is broadband, see Remark 19.1) of the complex channel gains. All other channel parameters, including angle of departure (AoD), angle of arrival (AoA), path loss, and Doppler shift, are the same for all antenna elements. In this case, we can adjust the phases of the signals in order to enhance or reduce the antenna gain in a given direction by constructively or destructively overlapping signals from different antenna elements, respectively.

Consider the phased array in Fig. 19.1, where the black dots represent antenna elements. Due to the geometric displacement between antenna elements 0 and i, the propagation distance difference (i.e., the distance difference to reach the same wavefront of antenna element 0, where the wavefront is the plane perpendicular to the propagation direction if we apply the plane wave assumption) for direction $\mathbf{d}$ is $\mathbf{h}_i$. We denote the vector from antenna element 0 to antenna element i as $\mathbf{l}_i$, the unit direction vector that we would like to transmit the signal as $\mathbf{d}$ and the angle between $\mathbf{l}_i$ and $\mathbf{d}$ as ϕ (ϕ is the same for all antenna elements because of the above-described assumption). The length of $\mathbf{h}_i$ is the projection length of $\mathbf{l}_i$ on $\mathbf{d}$, namely, $|\mathbf{d}||\mathbf{l}_i|\cos\phi = |\mathbf{d}\mathbf{l}_i|$. Therefore, the phase difference between antenna element 0 and antenna element i is

$$\delta_i = \frac{2\pi|\mathbf{d}\mathbf{l}_i|}{\lambda} = \frac{2\pi f|\mathbf{d}\mathbf{l}_i|}{c} \tag{19.1}$$

where λ is the wavelength, f is the frequency, and c is the speed of light. We define the *steering vector* $\mathbf{t}(\mathbf{d}) = (t_0, \ldots, t_N)^T$, where the ith element $t_i(\mathbf{d}) = \exp\left(-j2\pi f|\mathbf{d}\mathbf{l}_i|/c\right)$ and N is the number of antenna elements.

In a two-dimensional space, the unit direction vector $\mathbf{d} = (\cos\phi, \sin\phi)$ as shown in Fig. 19.1. In a three-dimensional space, we can apply the spherical coordinate system to define a direction with its elevation θ and azimuth ϕ. In this case, $\mathbf{d}(\theta, \phi) = (\cos\theta\cos\phi, \cos\theta\sin\phi, \sin\theta)$.

We can shift signal phases before transmission in order to strengthen the signal power in a given direction. Let us denote the phase shift of antenna element i as φ_i, as shown in Fig. 19.1, and the *beamforming vector* $\mathbf{b} = (b_0, \ldots, b_N)^T$, where the

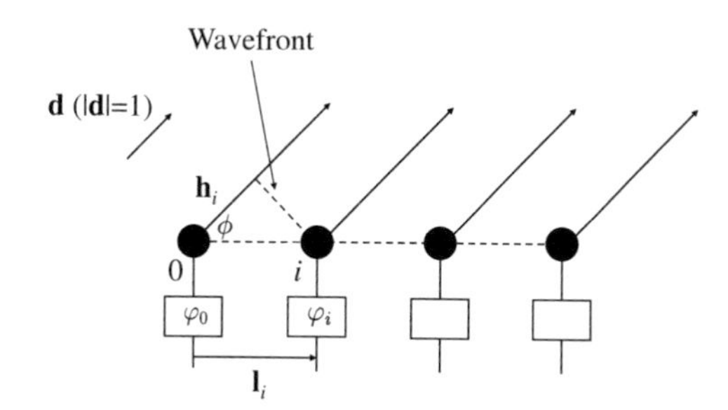

Fig. 19.1 The phased array

ith element $b_i = \exp(-j\varphi_i)$. The transmitted signal s is the sum of signals from all antenna elements, whose phases are first shifted artificially with the beamforming vector $\mathbf{b}$ and then shifted with the steering vector $\mathbf{t}$. Namely

$$s(\mathbf{d}) = \mathbf{t}^T(\mathbf{d})\mathbf{b}s_0. \tag{19.2}$$

As a transmitting antenna array, the radiation power in all directions must be normalized to the total transmission power. Therefore, we define the *antenna directivity* as the portion of the transmitted power in direction $\mathbf{d}$ in the total transmission power, i.e.

$$D(\mathbf{d}) = \frac{4\pi |s(\mathbf{d})|^2}{\int_{\mathbf{d}} |s(\mathbf{d})|^2 d\mathbf{d}} \tag{19.3}$$

If we use the spherical coordinate system (θ, ϕ), (19.3) can be rewritten as

$$D(\theta, \phi) = \frac{4\pi |s(\mathbf{d}(\theta, \phi))|^2}{\int_0^{\pi} \int_0^{2\pi} |s(\mathbf{d}(\theta, \phi))|^2 \sin\theta d\phi d\theta}. \tag{19.4}$$

Intuitively, a high antenna directivity suggests that the transmitted power is concentrated in the given direction.

Remark 19.1 It is to note from (19.1) that the phase shift depends on the frequency. For a broadband signal, which is usually the case for the THz communication, the frequency cannot be considered constant; the phase shift is therefore a function of the frequency. In this case, we can shift the delay of the signal (e.g., via varying the waveguide length) instead of the phase in order to compensate for the propagation length difference shown as $\mathbf{h}_i$ in Fig. 19.1.

Remark 19.2 The above discussion considers the phased array as a transmitter antenna. If it is applied as the receiver antenna, the phases of the received signals of each antenna element are adjusted in a similar manner as Fig. 19.1 shows, where the phase shifts are characterized by the *combining vector* $\mathbf{c}$, where the ith element $c_i = \exp(j\varphi_i)$ with φ_i being the phase shift of signal from antenna element i. The shifted signals are then added up to the output signal s. If we use the column vector

$\mathbf{s}$ to denote the received signals, where the ith element s_i is the signal from the ith antenna element, then $s = \mathbf{s}^T\mathbf{c}$.

The derivation of beamforming and combining vectors given the desired AoD and AoA is presented in Sect. 19.2.

In a phased array, all antenna elements share the same radio frequency (RF) frontend. The signal processing is limited to phase (delay) shifting. The simple structure is a major advantage of the phased array. Besides, only the AoD or the AoA is required to compute the beamforming or combining vector. The signal processing is therefore simple and robust. However, the size of the phased array is limited by the wavelength, which is particularly small in THz communications. This poses a constraint on the antenna gain. In addition, the antenna gain is only a function of the angle. As a result, the phased array cannot differentiate signals to or from similar directions.

19.1.2 Full MIMO Array

Compared to the phased array, the size of the full MIMO array does not have to be compact, and there is no requirement on the CIR similarity of the antenna elements. Correspondingly, signal processing is not restricted to phase shifting. Each antenna element has a complete RF frontend such that spatial multiplexing can be implemented. This type of array realizes higher performance at the cost of complexity.

One special property about the broadband THz channel is the frequency selectivity due to the broad bandwidth. Thus, the channel can only be characterized by the vector CIR rather than a scalar channel gain as in ordinary narrowband channels. Although we can use a multi-carrier system such as orthogonal frequency-division multiplexing (OFDM) to make the subchannels narrowband, it is not likely that the subchannel can be assumed flat given the very broad bandwidth of THz communications (up to 50 GHz) and the limited number of subcarriers due to constraints such as hardware complexity, sensitivity to frequency shift (e.g., Doppler shift), and peak-to-average power ratio (PAPR). Therefore, we assume the channel to be frequency-selective and use the CIR rather than a scalar channel gain in the following derivation.

For transmitting antennas, the signal is processed before transmission and therefore is named *precoding*. For receiving antennas, the transmitted symbol is detected from the noisy received signals in all antennas, and the signal processing for receiver is referred to as *detection*. If one communication participant (e.g., the access point (AP)) has higher computation power than the other (e.g., the user equipment (UE)), the more powerful participant can be responsible for the precoding and detection such that the signal processing of other participants can be minimized.

19.1.2.1 Precoding

Let us assume the CIR between the mth transmitting antenna and kth receiving antenna has L taps, where L is the upper bound of the lengths of all CIRs in the considered system, and denote the CIR from transmitter antenna m to receiver antenna k as the $L \times 1$ vector

$$\mathbf{h}_{k,m} = \left[h_{k,m}(0), h_{k,m}(1), \ldots, h_{k,m}(L-1)\right]^T \in \mathbb{C}^L. \tag{19.5}$$

Data symbol stream for the kth receiving antenna is precoded for all M transmitting antennas using a bank of finite impulse response (FIR) filters. We assume that all FIR filters of the transmitter have the same length of N. The transmit filter that precodes the kth symbol stream for the mth antenna is denoted by

$$\mathbf{p}_{m,k} = \left[p_{m,k}(0), p_{m,k}(1), \ldots, p_{m,k}(N-1)\right]^T \in \mathbb{C}^N. \tag{19.6}$$

The symbol stream of the kth data symbol stream at time instant $i \in \mathbb{Z}$ is represented by the $(N+L-1) \times 1$ vector

$$\mathbf{s}_{ki} = \left[s_k(i), s_k(i-1), \ldots, s_k(i-N-L+2)\right]^T \in \mathbb{C}^{N+L-1}. \tag{19.7}$$

For the transmit symbols, we assume the cross correlation

$$\mathrm{E}\{s_k(i)s_l^*(i-\Delta)\} = \delta_{k-l} \cdot \begin{cases} r_s(\Delta) & \text{for } \Delta \geq 0 \\ r_s^*(|\Delta|) & \text{otherwise} \end{cases}, \forall k, l, i \tag{19.8}$$

yielding the cross-correlation matrix

$$\mathbf{R}_{\mathbf{s}}^{N+L-1} = \mathrm{E}\{\mathbf{s}_{ki}\mathbf{s}_{ki}^H\} = \begin{bmatrix} r_s(0) & r_s(1) & \ldots\ r_s(N+L-2) \\ r_s^*(1) & r_s(0) & \ldots\ r_s(N+L-1) \\ \vdots & \vdots & \ddots \quad \vdots \\ r_s^*(N+L-2) & r_s^*(N+L-1) & \ldots \quad r_s(0) \end{bmatrix}. \tag{19.9}$$

The expectation calculation depends on the modulation scheme.

Furthermore, we define the submatrix $\mathbf{R}_{\mathbf{s}}^N$ as first N rows and columns of $\mathbf{R}_{\mathbf{s}}^{N+L-1}$.

For notational convenience, we also introduce the following notational definitions. For the (k, m)th channel link, we define the $(N+L-1) \times N$ Toeplitz matrix as

$$\mathbf{H}_{k,m} = \begin{bmatrix} h_{k,m}(0) & h_{k,m}(1) & \dots & h_{k,m}(L-1) & 0 & \dots & 0 \\ 0 & h_{k,m}(0) & \dots & h_{k,m}(L-2) & h_{k,m}(L-1) & \dots & 0 \\ \vdots & \vdots & & \vdots & \vdots & \ddots & \vdots \\ 0 & 0 & \dots & 0 & 0 & \dots & h_{k,m}(L-1) \end{bmatrix}^T. \tag{19.10}$$

We stack all the symbol stream vectors, yielding the $K(N+L-1) \times 1$ vector

$$\mathbf{s}_i = \left[\mathbf{s}_{0,i}^T, \mathbf{s}_{1,i}^T, \dots, \mathbf{s}_{K-1,i}^T\right]^T \in \mathbb{C}^{K(N+L-1)}. \tag{19.11}$$

We stack the transmit filter coefficients of all streams and transmitting antennas, yielding the $KMN \times 1$ vector

$$\mathbf{p} = \left[\mathbf{p}_0^T, \mathbf{p}_1^T, \dots, \mathbf{p}_{M-1}^T\right]^T \in \mathbb{C}^{KMN} \tag{19.12}$$

where $\mathbf{p}_m = \left[\mathbf{p}_{m,0}^T, \mathbf{p}_{m,1}^T, \dots, \mathbf{p}_{m,K-1}^T\right]^T \in \mathbb{C}^{KN}$.

At the receiver side, we assume that the kth UE multiplies its received signal by a quasi-constant factor $\sqrt{\alpha_k}$ (known by both transmitter and receiver) such that a predefined receive signal level is ensured. The received signal vector $\mathbf{r}(i) = \left[r_0(i), \dots, r_{K-1}(i)\right]^T$ for the K UEs at time instant i is then given by

$$\begin{aligned} \mathbf{r}(i) &= \mathbf{A}^{\frac{1}{2}} \left[\mathbf{I}_K \otimes \mathbf{s}_i\right]^T \\ &\quad \cdot \begin{bmatrix} \mathbf{I}_K \otimes \mathbf{H}_{0,0} & \dots & \mathbf{I}_K \otimes \mathbf{H}_{0,M-1} \\ \vdots & \ddots & \vdots \\ \mathbf{I}_K \otimes \mathbf{H}_{K-1,0} & \dots & \mathbf{I}_K \otimes \mathbf{H}_{K-1,M-1} \end{bmatrix} \mathbf{p} \\ &\quad + \mathbf{A}^{\frac{1}{2}} \mathbf{n}(i) \\ &= \bar{\mathbf{S}}^T \bar{\mathbf{H}} \mathbf{p} + \mathbf{A}^{\frac{1}{2}} \mathbf{n}(i) \end{aligned} \tag{19.13}$$

with $\mathbf{A} = \text{diag}(\alpha_0, \dots, \alpha_{K-1})$, and $\mathbf{n}(i) = \left[n_0(i), \dots, n_{K-1}(i)\right]^T$ where $n_k(i)$ denotes the additive white Gaussian noise at receiver k for the ith time instant. For Eq. (19.13), we introduce the substitutions

$$\bar{\mathbf{H}} = \begin{bmatrix} \mathbf{I}_K \otimes \bar{\mathbf{H}}_{0,0} & \dots & \mathbf{I}_K \otimes \bar{\mathbf{H}}_{0,M-1} \\ \vdots & \ddots & \vdots \\ \mathbf{I}_K \otimes \bar{\mathbf{H}}_{K-1,0} & \dots & \mathbf{I}_K \otimes \bar{\mathbf{H}}_{K-1,M-1} \end{bmatrix} \in \mathbb{C}^{K^2(N+L-1)\times KMN} \tag{19.14}$$

$$\bar{\mathbf{H}}_{k,m} = \alpha_k^{\frac{1}{2}} \mathbf{H}_{k,m} \in \mathbb{C}^{(N+L-1)\times N}$$

$$\bar{\mathbf{S}} = \left[\mathbf{I}_K \otimes \mathbf{s}_i\right] \in \mathbb{C}^{K^2(N+L-1)\times K}.$$

The precoding vector $\mathbf{p}$ defined in (19.12) depends on the estimated CIRs and will be derived in Sect. 19.3.

19.1.2.2 Detection

As explained above, signal detection refers to recovering the transmitted symbols from the received noisy signals in all receiving antennas. Similar to precoding, the single-input-single-output (SISO) transmission from transmitting antenna k to receiving antenna m in the signal detection problem can be represented as

$$\begin{aligned} \mathbf{r}_m &= \begin{bmatrix} h_{k,m}(L-1) & h_{k,m}(L-2) & \dots & 0 \\ 0 & h_{k,m}(L-1) & \dots & 0 \\ \dots & \dots & \dots & \dots \\ 0 & 0 & \dots & h_{k,m}(0) \end{bmatrix} \mathbf{s}_k + \mathbf{n} \\ &= \mathbf{H}_{k,m}\mathbf{s}_k + \mathbf{n} \end{aligned} \tag{19.15}$$

where $\mathbf{H}_{k,m}$ is the uplink channel matrix from transmitting antenna k to receiving antenna m of size $N \times (L+N-1)$ and $\mathbf{s}_k$ is the vector of transmitted symbols. The uplink MIMO channel matrix is constructed as

$$\mathbf{H} = \begin{bmatrix} \mathbf{H}_{0,0} & \mathbf{H}_{0,1} & \dots & \mathbf{H}_{0,M-1} \\ \mathbf{H}_{1,0} & \mathbf{H}_{1,1} & \dots & \mathbf{H}_{1,M-1} \\ \dots & \dots & \dots & \dots \\ \mathbf{H}_{K-1,0} & \mathbf{H}_{K-1,1} & \dots & \mathbf{H}_{K-1,M-1} \end{bmatrix}. \tag{19.16}$$

The total MIMO transmission is described as

$$\mathbf{r} = \mathbf{H}\mathbf{s} + \mathbf{n}. \tag{19.17}$$

The purpose of the detection is to use a detection $\mathbf{p}$ such that the mean square error (MSE) is minimized, i.e.,

$$\min_{\mathbf{p}} \left(\left| \mathbf{p}^T \mathbf{r} - s_0 \right|^2 \right) \tag{19.18}$$

where s_0 is the transmitted symbol to be detected.

The readers are encouraged to refer to [2] for more details.

In the full MIMO array, each antenna is equipped with an RF frontend and is able to carry out full-signal processing (i.e., amplification and delay shifting). As a result, the full MIMO array can significantly outperform the phased array and different data streams can be separated even if the AoAs or AoAs are very similar (this is what the phased array cannot do). However, the high complexity and the requirement for knowledge of full CIRs may limit its application.

The detection vector **p** defined in (19.18) depends on the estimated CIRs and will be derived in Sect. 19.3.

19.1.3 Hybrid MIMO Array

As described above, the phased array has a simple structure, requiring only the angular information for signal processing, but its performance is limited by the small array size. On the contrary, the full MIMO array could achieve a significantly higher performance, but its complexity and requirement for CIR estimates of all channels make it difficult to apply.

A compromise between performance and complexity might be achieved with the so-called hybrid MIMO array, i.e., an array of several phased arrays [3]. Antenna elements in one phased array share the same RF frontend, and therefore, the complexity can be reduced significantly compared to the full MIMO array. Full-signal processing can be carried out between the subarrays, where the subarrays can be understood as directive antennas with the ability to change the main lobe direction with phase (delay) shifting within a subarray. In the application, the main lobes of the phased arrays are adjusted toward the other participant of the communication via beamforming or combining vectors, and the precoding or detection vectors are computed to achieve the maximum performance.

19.2 Angle-of-Arrival Estimation and Signal Processing in Phased Arrays

As explained in Sect. 19.1.1, the computation of optimal beamforming vector and combining vectors relies on the AoD/AoA estimation. In practice, AoD estimation is essentially identical to AoA estimation by exchanging transmitter and receiver. Therefore, we only discuss AoA estimation, i.e., the transmitter sends a signal, and the receiver estimates from which direction the signal comes from.

19.2.1 Static AoA Estimation

In this section, we consider the scenario where transmitter and receiver remain stationary and the AoA stays constant and therefore needs to be estimated only once.

In a general sense, the AoA estimation is a well-studied topic with different solutions. For example, power angular spectrum (PAS)-based AoA estimations measure the PAS and select the local maxima as the AoAs. A further possibility is to use the maximum likelihood (ML) estimation to compute the most probable values of parameters. Various algorithms are applied for the ML estimation, e.g., Newton-Raphson method, expectation-maximization (EM), space alternating generalized expectation-maximization (SAGE) algorithms, and sparse Bayesian learning (SBL). The ML estimation achieves better precision at the cost of higher computational complexity. Therefore, its application is mainly limited to data processing of channel measurement and is difficult to be applied in real-time applications. On the other hand, eigenvalue-based algorithms, such as multiple signal classification (MUSIC) and estimation of signal parameters via rotational invariance technique (ESPRIT), assume uncorrelated signal sources and are difficult to be applied for multipath propagation. Based on the above considerations, we apply the PAS-based AoA estimation in order to achieve an optimal compromise between simplicity, requirement, and performance.

An AoA estimation algorithm is proposed in [4]. This method focuses on improving efficiency of exhaustive search. Due to the extremely high propagation path loss, the antenna gain must be very high to realize a reasonable signal-to-noise ratio (SNR), which implies a narrow half power beam width (HPBW) and a long scanning time with an exhaustive manner. This problem becomes worse if the received signal strength is stronger than the noise power only if both transmitter and receiver have found the correct AoD and AoA, respectively, because the number of candidates is the product of numbers of possible AoAs and AoAs.

As a consequence, an exhaustive search is prohibitively time-consuming. For this problem, the IEEE 802.15.3c standard [5] has provided a solution: the efficiency can be improved by introducing a preliminary estimation with low precision. Namely, an antenna with a wide HPBW is applied for a fast but less accurate AoA estimate (first step). After that, another antenna with a narrow HPBW is applied to estimate the precise AoA within the range of the rough estimate determined in the first step (second step). By means of this estimation with two steps, main lobe directions far from the correct AoA can be excluded quickly in the first step, and therefore, the efficiency is significantly improved compared to the original exhaustive search.

Unfortunately, this method cannot be applied to the terahertz (THz) communication because a wide HPBW implies a low antenna gain. Due to the high propagation path loss and the strong thermal noise of the THz communication, the received signal strength would be below the noise level. However, it is proven that PAS in different frequencies are highly correlated [4], i.e., the AoA at a lower frequency

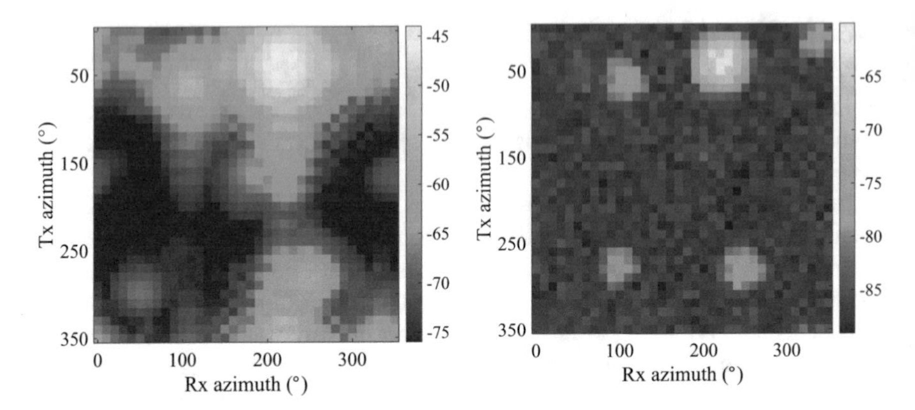

Fig. 19.2 PAS of different frequencies under the same measurement setup. The similarity of AoDs and AoAs of the strongest propagation paths is unambiguous (unit: dB) [4]. (**a**) 10 GHz. (**b**) 300 GHz. (©2019 IEEE, reproduced with permission)

is similar to the AoA at the THz frequency. This can be seen in Fig. 19.2, which is from a measurement in an indoor environment with multiple different frequencies.

According to the current system concept, the THz communication device is equipped with a low-frequency RF frontend as fallback solution. Therefore, we can design a two-step estimation, in which we estimate the angle with low resolution at the lower frequency in step 1 and estimate the angle with high resolution and at the THz frequency in step 2 within the rough range determined in previous step. Since the propagation path found at the lower frequency appears in similar angular domain at the THz frequency, it is possible to use the rough estimate (due to the broader antenna main lobe) at the lower frequency in step 1 as a confined search range in step 2.

According to the Friis' transmission equation, the propagation path loss at the lower frequency is significantly lower than at the THz frequency. This results in another advantage of estimation at lower frequency in step 1. We can use a quasi-omnidirectional antenna for the transmitter and a directional antenna for the receiver and hence decompose estimation of AoD and AoA. In this way, the number of candidates can be further reduced.

The two-step AoA estimation is therefore formulated as Algorithm 1.

Figure 19.3 shows a measurement result in an AoA estimation demonstration. The solid line represents the received power strength at the frequencies centered at 8 GHz known as ultra-wide band (UWB), whereas the dashed line is the received power strength of at the THz band. The dots represent the actual measurement. It can be observed that in step 1, bigger step size is taken to quickly approach the true AoA and in step 2, smaller step size is used for precise search. In this way, the number of steps is reduced significantly compared to the primitive exhaustive search.

Algorithm 1 Two-step AoA estimation

```
#1 transmits a pilot signal in all directions at low frequency
for every section of #2 do
    if #2 detects a signal in current section then
        #2 transmits in the current section a pilot signal at low frequency
        for every section of #1 do
            if #1 detects a signal then
                a pair of (AoD, AoA) on section level is found
            end if
        end for
    end if
end for
for every pair of (AoD, AoA) on section level do
    for every beam of #1 within current section do
        #1 transmit in current beam at THz frequency
        for every beam of #2 within current section do
            if #2 detects signal then
                a pair of (AoD, AoA) on beam level is found
            end if
        end for
    end for
end for
```

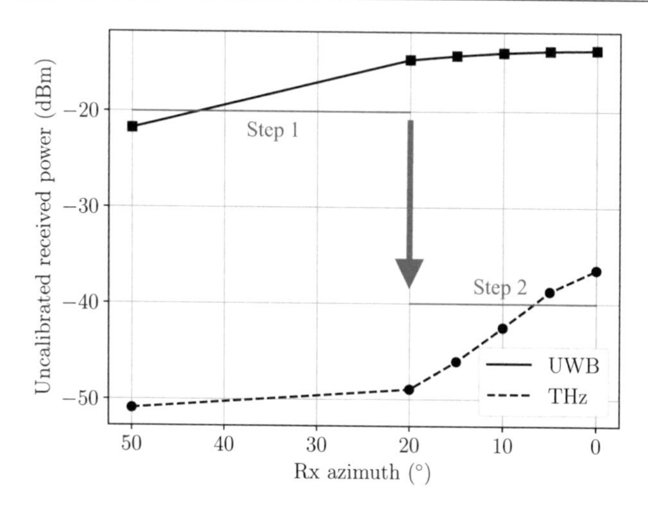

Fig. 19.3 Recorded received power strength in an experimental validation [4] (©2019 IEEE, reproduced with permission)

19.2.2 *Dynamic AoA Estimation*

Section 19.2.1 describes the static AoA estimation. In practice, we are interested in dynamic AoA estimation besides the static one. If transmitter or receiver moves during the data transmission, the AoA is dynamic and must be periodically

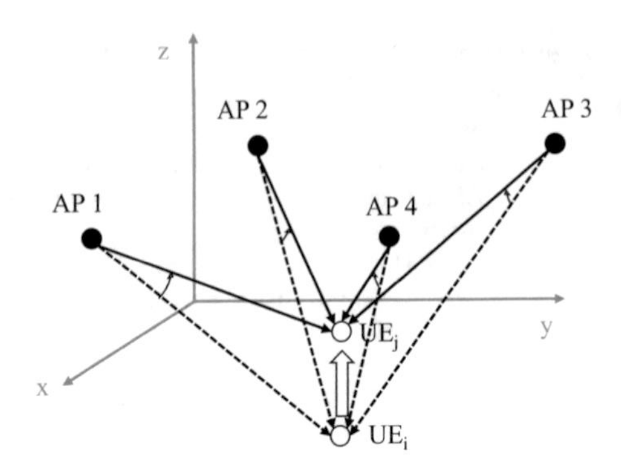

Fig. 19.4 Spatial correlation of AoA change [6] (©2018 IEEE, reproduced with permission)

estimated. In this case, the Bayesian inference is a useful tool to improve the estimation precision since the AoA change is spatially and temporally correlated.

Consider a scenario with four APs and one UE, as shown in Fig. 19.4. For THz channels, the LoS path is the preferable propagation path. If the UE moves upward between time instants t_0 and t_1, the AoA changes are temporally correlated because they are caused by the continuous UE movement. Furthermore, AoA azimuth of all four APs will remain constant, and the AoA elevation will decrease. This is an intuitive example of the space-time correlation: because the UE displacement is continuous, the AoAs are temporally correlated; since the mutual reason for AoA changes (denoted as ΔAoA) of all APs, if we use the estimates of other APs as extrinsic information, we can utilize this correlation to further improve the estimation accuracy.

The Bayesian inference is carried out with a hidden Markov model (HMM) along time within a single antenna and a message passing mechanism between the antennas, as shown in Fig. 19.5, where the likelihood is denoted as L, prior (inference from the antenna's own history) is denoted as P, and external message from other antennas is denoted as M. Details about the HMM and the message passing are available in [7] and [6], respectively.

Figure 19.6 shows the considered scenario with four AP antennas as red dots and a moving UE as the blue curve.

Figure 19.7 shows the realized effective antenna gains with different algorithms. It can be observed that without inference (w.o. inference, the blue curve), the AoA estimation precision is poor due to the strong thermal noise. With temporal forward inference (f. inference, the orange curve), the estimation precision is improved significantly, and the effective antenna gain is higher. With both forward inference and spatial correlation (SC), the stability of the estimation is improved further.

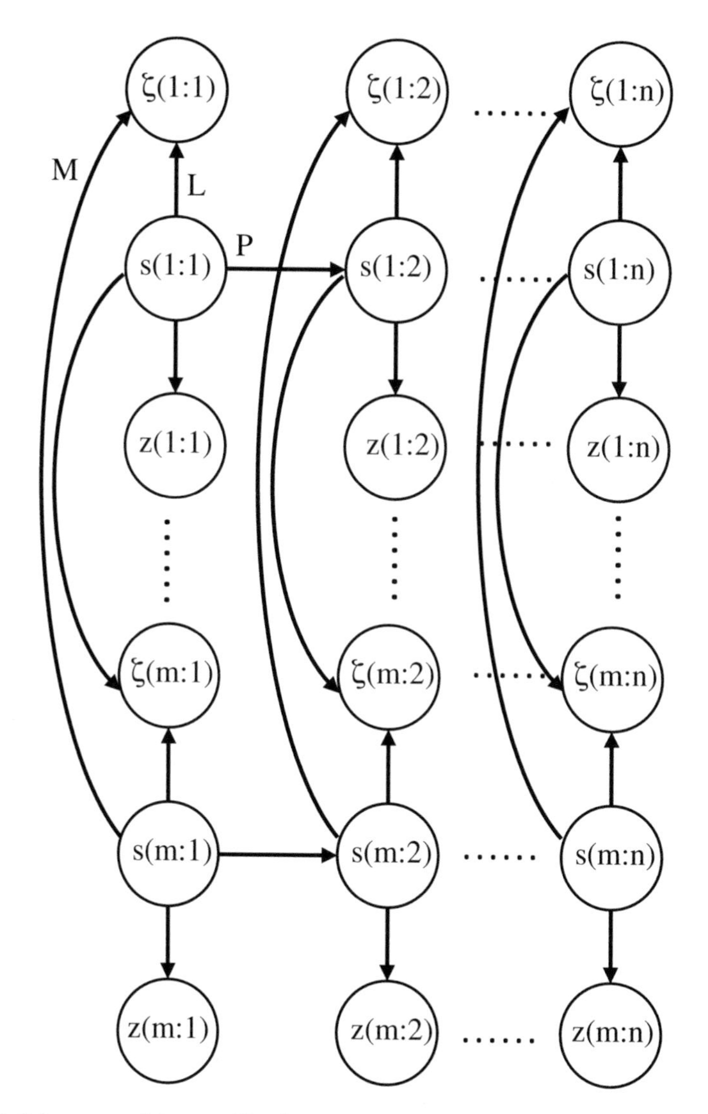

Fig. 19.5 Inference model to consider the space-time correlation [6] (©2018 IEEE, reproduced with permission)

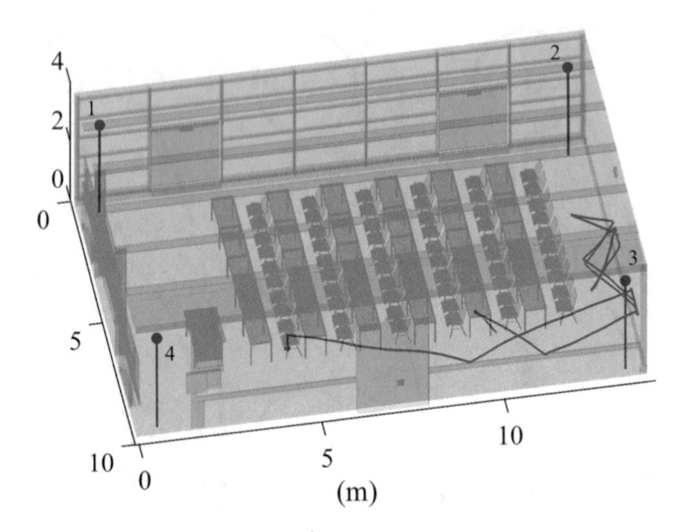

Fig. 19.6 The considered scenario with four directive AP antennas and one moving UE [6] (©2018 IEEE, reproduced with permission)

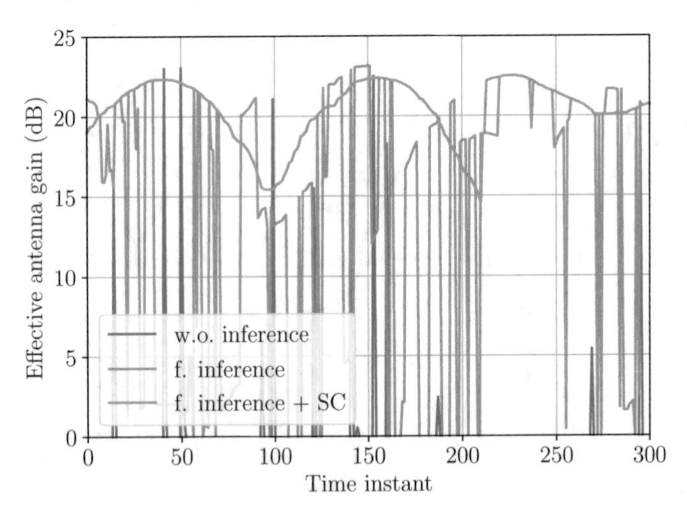

Fig. 19.7 Realized effective antenna gain with different algorithms [6] (©2018 IEEE, reproduced with permission)

19.3 Channel Estimation and Signal Processing in Massive MIMO

19.3.1 MMSE Channel Estimation

Besides the AoA estimation, other channel parameters, such as complex channel gain and delay, need to be estimated as well. The estimated channel state is defined as channel state information (CSI) and is the foundation of many advanced signal processing algorithms. In channel estimation, the transmitter sends a pilot signal, which is known to the receiver. The receiver compares the received signal with the original pilot signal and estimates the channel parameters.

We consider the minimum mean square error (MMSE) channel estimation in this section, which minimizes the MSE between the received signal and the signal model parameterized by channel gain and delay:

$$MSE = \mathbb{E}\left(\left\| \mathbf{y} - \sum_i \mathbf{s}_i(g_i, \tau_i) \right\|_2^2 \right) \tag{19.19}$$

where $\mathbf{y}$ is the vector of the measured received signals with each element as the received signal at each antenna element and $\mathbf{s}_i$ is the signal from ith path with complex channel gain g_i and delay τ_i.

One of the most important difficulties of the channel estimation is the strong thermal noise. In order to obtain a more accurate estimate, we can use the discrete Fourner transform (DTF) to transfer $\mathbf{y}$ to the *beamspace domain*:

$$\mathbf{h} = \mathbf{F}\mathbf{y} \tag{19.20}$$

where $\mathbf{F}$ is the DTF matrix. In the beamspace domain, each element of $\mathbf{h}$ corresponds to a certain AoA. Since the THz channel is sparse, i.e., the signal propagates via certain paths and arrives at the receiver in specular AoAs [8], $\mathbf{h}$ must be sparse as well. Therefore, we can apply the least absolute shrinkage and selection operator (LASSO) denoising method [9], which adds a one norm of $\mathbf{h}$ as a penalty term:

$$\hat{\mathbf{h}}^* = \arg\min \frac{1}{2}||\mathbf{h} - \hat{\mathbf{h}}'||_2^2 + \gamma||\hat{\mathbf{h}}'||_1 \tag{19.21}$$

where γ is a denoising parameter. It is shown [10] that Stein's unbiased risk estimate (SURE) is an unbiased estimator for the MSE solution. Using SURE, the optimal γ^* can be iteratively computed, and the channel parameters can be calculated with the denoised sparse vector $\hat{\mathbf{h}}^*$.

With the estimated CSI, multi-antenna algorithms can be carried out to improve the communication quality, including the application of massive MIMO, which will be described in Sect. 19.3.2.

19.3.2 Massive MIMO Precoding and Detection

With the estimated CSI, the signal processing can be formulated as linear optimization problem. We assume that the future UE can be portable, e.g., a smartphone and

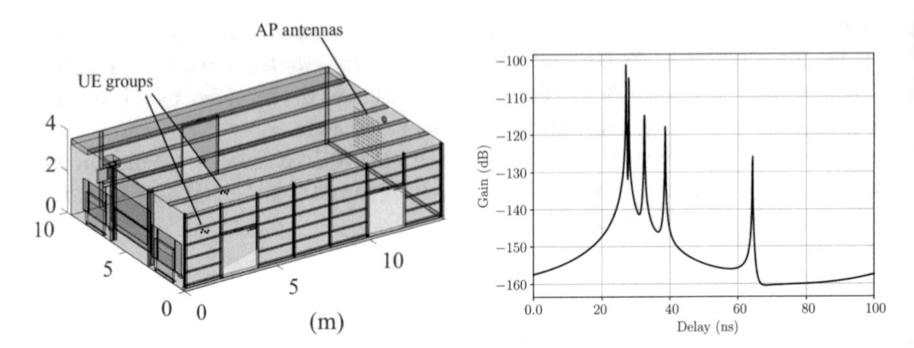

Fig. 19.8 The considered scenario and an example CIR of a SISO channel [2]. (**a**) The scenario. (**b**) A SISO CIR (©2019 IEEE, reproduced with permission)

a smartwatch, and the AP has more computation power. Therefore, we formulate the massive MIMO signal processing problem as downlink precoding (the signal is processed before transmission) and uplink detection (the signal is processed after receiving), such that the AP carries out signal processing in both downlink and uplink.

A lecture room (Fig. 19.8a) is chosen as application scenario, where the red dots are the AP antennas and the black dots are UEs. They are placed close to each other to represent the worst-case scenario. Figure 19.8b shows an example CIR of a SISO channel. It can be observed that low channel gain and strong multipath components (MPCs) are the major characteristics of the indoor THz channel.

19.3.2.1 Downlink Precoding

The objective for the downlink precoding is to minimize the sum MSE between the received symbol vector

$$\mathbf{r}(i) = \left[r_0(i), \ldots, r_{K-1}(i)\right]^T \tag{19.22}$$

and a delayed transmit symbol vector

$$\mathbf{s}(i-\Delta) = \left[s_0(i-\Delta), \ldots, s_{K-1}(i-\Delta)\right]^T \tag{19.23}$$

where the delay $\Delta \geq 0$ is a design parameter. In order to account for a transmit (sum) power constraint P_{TX} of the AP, we introduce the scaling factor β^{-1} for the received signal. The precoding weight vector that minimizes the sum MSE is obtained by

$$\mathbf{p}^0 = \arg\min_{\beta,\mathbf{p}} \epsilon(\beta, \mathbf{p}) \quad \text{s.t.} \quad \mathbf{p}^H \left(\mathbf{I}_{MK} \otimes \mathbf{R}_{\mathbf{s}}^N\right) \mathbf{p} = P_{\text{TX}} \tag{19.24}$$

with

$$\begin{aligned}
\epsilon(\beta, \mathbf{p}) &= \mathrm{E}\left\|\beta^{-1}\mathbf{r}(i) - \mathbf{s}(i-\Delta)\right\|^2 \\
&= \beta^{-2}\mathbf{p}^H\bar{\mathbf{H}}^H\mathrm{E}\bar{\mathbf{S}}^*\bar{\mathbf{S}}^T\bar{\mathbf{H}}\mathbf{p} \\
&- 2\beta^{-1}\Re\left\{\mathbf{p}^H\bar{\mathbf{H}}^H\mathrm{E}\bar{\mathbf{S}}^*\mathbf{s}(i-\Delta)\right\} \\
&\quad + \beta^{-2}\mathrm{tr}(\mathbf{A}\mathbf{R_n}) + K\cdot r_s(0) \\
&= \beta^{-2}\mathbf{p}^H\bar{\mathbf{H}}^H\left(\mathbf{I}_{K^2}\otimes\mathbf{R}_{\mathbf{s}}^{N+L-1}\right)\bar{\mathbf{H}}\mathbf{p} \\
&- \beta^{-1}\mathbf{p}^H\bar{\mathbf{H}}^H\mathbf{d}_\Delta^* \\
&\quad - \beta^{-1}\mathbf{d}_\Delta^T\bar{\mathbf{H}}\mathbf{p} + \beta^{-2}\mathrm{tr}(\mathbf{A}\mathbf{R_n}) + K\cdot r_s(0)
\end{aligned} \tag{19.25}$$

where

$$\begin{aligned}
\mathbf{d}_\Delta &= \mathrm{E}\left[\bar{\mathbf{S}}\mathbf{s}(i-\Delta)^*\right] \\
&= \left[(\mathbf{e}_K^0\otimes\mathbf{r}_s(\Delta))^T, \ldots, (\mathbf{e}_K^{K-1}\otimes\mathbf{r}_s(\Delta))^T\right]^T
\end{aligned} \tag{19.26}$$

and $\mathbf{r}_s(\Delta) = \left[r_s(\Delta), r_s(\Delta+1), \ldots, r_s(\Delta+N+L-2)\right]^T$.

The following theorem is derived to solve the above optimization problem:

Theorem 19.1 *The optimal MMSE precoding vector is given by*

$$\mathbf{p}^o = \beta^o\bar{\mathbf{p}} \tag{19.27}$$

where

$$\begin{aligned}
\bar{\mathbf{p}} = &\left(\bar{\mathbf{H}}^H\left[\mathbf{I}_{K^2}\otimes\mathbf{R}_{\mathbf{s}}^{N+L-1}\right]\bar{\mathbf{H}}\right. \\
&\left. + tr(\mathbf{A}\mathbf{R_n})/P_{TX}\mathbf{I}_{MK}\otimes\mathbf{R}_{\mathbf{s}}^N\right)^{-1}\bar{\mathbf{H}}^H\mathbf{d}_\Delta^*
\end{aligned} \tag{19.28}$$

and

$$\beta^o = \sqrt{\frac{P_{TX}}{\bar{\mathbf{p}}^H\left[\mathbf{I}_{MK}\otimes\mathbf{R}_{\mathbf{s}}^N\right]\bar{\mathbf{p}}}}. \tag{19.29}$$

Specifically, for binary phase-shift keying (BPSK) and quadrature phase-shift keying (QPSK), for which the cross-correlation matrix is a unit matrix, the expression is simplified as

$$\bar{\mathbf{p}} = \left(\bar{\mathbf{H}}^H\bar{\mathbf{H}} + tr(\mathbf{A}\mathbf{R}_\mathbf{n})/P_{TX}\mathbf{I}_{KMN}\right)^{-1}\bar{\mathbf{H}}^H\mathbf{d}_\Delta \tag{19.30}$$

and

$$\beta^o = \sqrt{\frac{P_{TX}}{\sigma_s^2\left\|\bar{\mathbf{p}}\right\|^2}}. \tag{19.31}$$

The proof is available in [2].

19.3.2.2 Uplink Detection

The objective of the uplink detection is to estimate the transmitted symbol by linear combination of the received symbols with the detection vector **p** such that the MSE is minimized, i.e.

$$\min_{\mathbf{p}} \mathrm{E}\left(\left|\mathbf{p}^T\mathbf{r} - s_0\right|^2\right). \tag{19.32}$$

Compared to the downlink precoding, the uplink detection is not subject to the transmission power constraint, and the noise power is higher with more receiving antennas.

To solve this problem, we have the following theorem:

Theorem 19.2 *The optimal MMSE detection vector is given by*

$$\mathbf{p}^* = \mathbf{M}^{-1}\hat{\mathbf{H}}\mathbf{c}_{s_0} \tag{19.33}$$

where

$$\mathbf{M} = \hat{\mathbf{H}}\mathbf{R}_\mathbf{s}\hat{\mathbf{H}}^H\mathbf{I} + \mathbf{R}_\mathbf{n} \tag{19.34}$$

for on-off-keying (OOK) and

$$\mathbf{M} = \hat{\mathbf{H}}\hat{\mathbf{H}}^H + \mathbf{R}_\mathbf{n} \tag{19.35}$$

for BPSK and QPSK.

The proof of the theorem is available in [2] as well.

Index of s_0 and coding length depend largely on the delay spread of the CIRs. According to previous THz channel measurement [11, 12], the line-of-sight (LoS) path contributes the most energy of the whole CIR. The choice of s_0 index and $\mathbf{r}_u$ length should ensure that every block (i.e., every AP antenna) in (19.16) with the column index corresponding to the index of s_0 has contribution of the LoS path. Subject to this condition, the length of $\mathbf{r}_u$ should be as short as possible because the more symbols are taken into consideration, the stronger the noise and interference would be. Other than that, the computational effort would also be higher with a larger channel matrix.

Considering these facts, the length of $\mathbf{p}$ is suggested to be at least the difference between the maximum and minimum LoS delays of all channels and the index of s_0 corresponds to the minimum LoS delay.

After the derivation in the previous sections, the problems of precoding and detection become solving (19.27) and (19.33), both of which are large linear equation systems due to the big number of antennas and length of the CIR. An efficient algorithm is required because the problems must be solved in real time.

The least square QR (LSQR) [13] algorithm is a conjugate gradient (CG)-based iterative algorithm to solve large and sparse linear equation systems. Compared to the original CG algorithm, it has a better numerical stability and is especially efficient when a large portion of the matrix elements are 0, which is true in our case because the broadband CIR is indeed sparse (see Fig. 19.8b). Based on the CIR sparsity, a sparse channel matrix $\mathbf{H}$ can be obtained by reducing channel gains below a given threshold (defined as 30 dB weaker than the LoS gain in this paper) to 0 without significantly distorting the CIRs.

Applying the LSQR algorithm, the precoding and detection problems can be solved using the LSQR algorithm. Its output $\mathbf{p}$ is the optimal precoding or detection vector.

19.3.2.3 Simulation Results

Simulation results are presented in this section. The simulation parameters are summarized in Table 19.1.

Figure 19.9 shows the equivalent downlink CIR with five UEs in the middle. The target user (user 3) is in the middle, while the other users (1, 2, 4, and 5) are around user 3. That is to say, we are considering the worst case where the target user is surrounded by other users.

Compare Figs. 19.8b and 19.9, it can be observed that

- The channel gain of the intended UE is improved from less than -100 dB to more than -60 dB. The gain is roughly the factor of antenna number ($20 \cdot \log_{10}(120) \approx 41.6$ dB), indicating that all the channel gains have constructively overlapped at the desired delay.

Table 19.1 Simulation parameters

Parameter	Value
Considered frequency range	300–310 GHz
Uniform antenna array size	12×10
Hybrid antenna array size	4×3
Elements per subarray in hybrid antenna array	10
Distance between AP antennas	0.2 m
Distance between UEs	0.2 m
Number of UEs	1, 2, 5
Transmission power	1 mW per antenna for uniform array and per subparray for hybrid array (if not specified otherwise)
Length of FIR filter	50
Truncation quantile to reduce PAPR	0.9999
Threshold to set channel gain to 0 for sparsity	30 dB lower than LoS gain

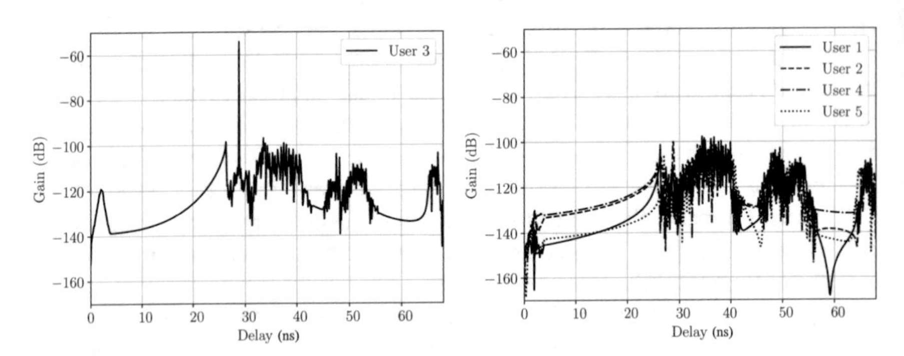

Fig. 19.9 Realized equivalent downlink CIR [2]. (**a**) CIR of target user. (**b**) CIR of interfered users. (©2019 IEEE, reproduced with permission)

- The multipath componentss (MPCs) have been significantly reduced compared to the LoS path. Therefore, the inter-symbol interference can be reduced significantly.
- The maximum gains of other UEs are at least 40 dB lower than the intended UE, resulting in a very low inter-user interference.

We assume a Gaussian antenna diagram [14] for Rx antenna. Figure 19.10 shows the relationship between UE antenna gain and bit error rate (BER) in downlink. All three modulation schemes and both uniform and hybrid AP antenna (an array of phased arrays) configurations are considered.

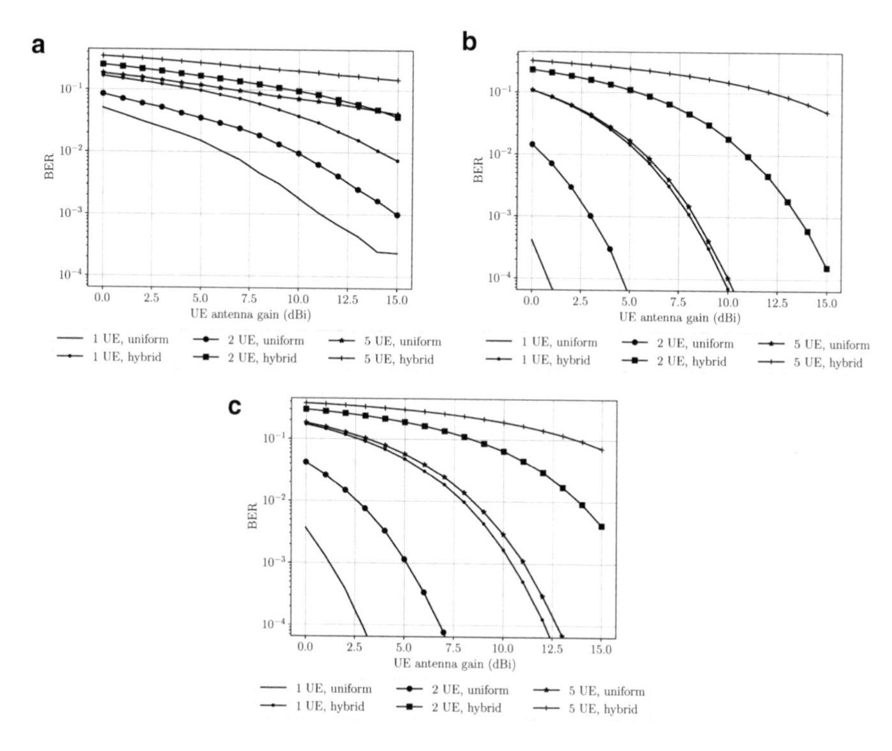

Fig. 19.10 Realized BER in downlink [2]. (**a**) OOK. (**b**) BPSK. (**c**) QPSK. (©2019 IEEE, reproduced with permission)

Simulation results of uplink detection are shown in Fig. 19.11. Compared to downlink, performances with different antenna configurations and numbers of UEs are similar in uplink. This is because the transmission power in uplink does not depend on these two factors, proving that the performance constraint is the transmission power rather than signal processing. With BPSK, up to two UEs and hybrid array, a BER of 10^{-3} is achievable with an ordinary UE antenna gain of 11 dB.

To conclude, the proposed downlink precoding and uplink detection algorithms can effectively compensate for the unfavorable propagation condition and realize a satisfactory THz link performance.

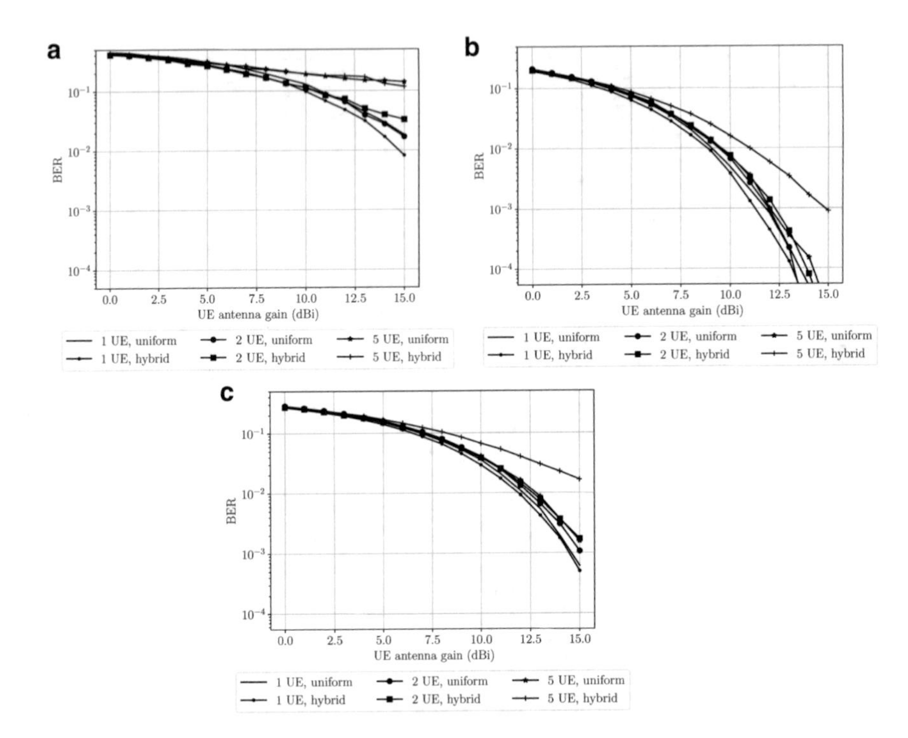

Fig. 19.11 Realized BER in uplink [2]. (**a**) OOK. (**b**) BPSK. (**c**) QPSK. (©2019 IEEE, reproduced with permission)

References

1. Lin, C., & Li, G. Y. (2015). Indoor terahertz communications: How many antenna arrays are needed? *IEEE Transactions on Wireless Communications, 14*(6), 3097–3107.
2. Peng, B., Wesemann, S., Guan, K., Templ, W., & Kürner, T. (2019). Precoding and detection for broadband single carrier Terahertz massive MIMO systems using LSQR algorithm. *IEEE Transactions on Wireless Communications, 18*(2), 1026–1040.
3. Lin, C., & Li, G. Y. L. (2016). Terahertz communications: An array-of-subarrays solution. *IEEE Communications Magazine, 54*(12), 124–131.
4. Peng, B., Guan, K., Rey, S., & Kürner, T. (2019). Power-angular spectra correlation based two step angle of arrival estimation for future indoor Terahertz communications. *IEEE Transactions on Antennas and Propagation, 67*(11), 7097–7105.
5. (2016) IEEE 802.15.3-2016, Standard for Wireless High-Rate Multi-Media Networks
6. Peng, B., Guan, K., & Kürner, T. (2018). Cooperative dynamic angle of arrival estimation considering space–time correlations for Terahertz communications. *IEEE Transactions on Wireless Communications, 17*(9), 6029–6041.
7. Peng, B., & Kürner, T. (2017). Three dimensional angle of arrival estimation in dynamic indoor terahertz channels using forward-backward algorithm. *IEEE Transactions on Vehicular Technology, 66*(5), 3798–3811. https://doi.org/10.1109/TVT.2016.2599488
8. Molisch, A. F. (2005). Ultrawideband propagation channels-theory, measurement, and modeling. *IEEE Transactions on Vehicular Technology, 54*(5), 1528–1545.

9. Tibshirani, R. (1996). Regression shrinkage and selection via the lasso. *Journal of the Royal Statistical Society: Series B (Methodological), 58*(1), 267–288.
10. Ghods, R., Gallyas-Sanhueza, A., Mirfarshbafan, S. H., & Studer, C. (2019). Beaches: Beamspace channel estimation for multi-antenna mmwave systems and beyond. In *2019 IEEE 20th International Workshop on Signal Processing Advances in Wireless Communications (SPAWC)* (pp. 1–5). IEEE.
11. Priebe, S., Jastrow, C., Jacob, M., Kleine-Ostmann, T., Schrader, T., & Kürner, T. (2011). Channel and propagation measurements at 300 GHz. *IEEE Transactions on Antennas and Propagation, 59*(5), 1688–1698.
12. Peng, B., Rey, S., & Kürner, T. (2016). Channel characteristics study for future indoor millimeter and submillimeter wireless communications. In *2016 10th European Conference on Antennas and Propagation (EuCAP)* (pp. 1–5). IEEE.
13. Paige, C. C., & Saunders, M. A. (1982). LSQR: An algorithm for sparse linear equations and sparse least squares. *ACM transactions on Mathematical Software, 8*(1), 43–71.
14. Priebe, S., Jacob, M., & Kürner, T. (2012). Affection of THz indoor communication links by antenna misalignment. In *Sixth European Conference on Antennas and Propagation (EUCAP)* (pp. 483–487). IEEE.

Part V
Transceiver Technologies 1: Silicon-based Electronics

Chapter 20
SiGe HBTs

Jae-Sung Rieh and Omeed Momeni

Abstract SiGe HBTs (heterojunction bipolar transistors) are a type of Si-based BJTs (bipolar junction transistors) that employ a SiGe alloy layer as a part of the device, typically as the base layer. Since its first demonstration in 1987, SiGe HBTs have become a mainstream Si bipolar transistor and are now widely employed for various high-frequency applications. In this chapter, a brief overview of SiGe HBTs is provided. Firstly, the DC characteristics are described in terms of the collector current and current gain, with focus on the comparison with Si BJTs and the origin of the advantages. A review on the definition of various breakdown voltages will be also presented along with device parameters that affect the breakdown voltages. Secondly, the RF characteristics of SiGe HBTs will be discussed in terms of f_T (cutoff frequency) and f_{max} (maximum oscillation frequency). Additional discussion will be made on high-frequency noise and $1/f$ noise of SiGe HBTs. Lastly, a couple of exemplary SiGe HBTs and circuits will be introduced to provide the practical aspect of modern high-performance SiGe HBTs.

20.1 Introduction

Silicon (Si) is, without question, the most prevailing semiconductor today; its popularity is arising from the highly favorable properties such as the availability of high-quality oxides, mechanical stability, high thermal conductivity, matured processing technology, and low production cost. Looking back, however, germanium (Ge) was once the semiconductor of choice in the early days of semiconductor industry. As the readers may be aware, the very first transistor in the world developed at Bell Lab in 1947 was a Ge transistor. Until Ge transistors gave way to Si

---

J.-S. Rieh (✉)
Korea University, Seoul, South Korea
e-mail: jsrieh@korea.ac.kr

O. Momeni
University of California Davis, Davis, CA, USA
e-mail: omomeni@ucdavis.edu

T. Kürner et al. (eds.), *THz Communications*, Springer Series in Optical Sciences 234, https://doi.org/10.1007/978-3-030-73738-2_20

transistors due to the apparent shortcomings of Ge such as the lack of the stable oxide and the performance instability at high temperatures, Ge was widely embraced as a useful semiconductor material. In view of this background, it is quite natural that the alloy of these two major elementary semiconductors, SiGe, attracts attention as a viable candidate for various device innovations.

At least a couple of attractive features of SiGe alloys can be mentioned from a device application perspective. First, their bandgap can be adjusted by controlling the Ge composition, which enables the bandgap engineering of Si-based devices. It is known that the bandgap of SiGe layers grown on Si substrates is reduced by ~80 meV for every 10% increase in the Ge content for a practical range of Ge composition. This enables heterostructure junctions for Si-based devices between two semiconductors with different bandgaps, which serves as an extremely useful tool for the device performance enhancement. Further, it allows the grading of bandgap in a certain region of the device, which also can improve the local carrier transport and resultant performance boost. Second, SiGe alloy layers exhibit enhanced carrier mobilities over Si layers under certain conditions of transport and material parameters, such as the carrier flow direction, Ge composition, doping concentration, and carrier polarity. For example, SiGe layers grown on top of Si substrates improve the hole mobility for both in-plane and out-of-plane directions of the carrier movement. The out-of-plane electron mobility is enhanced with SiGe alloys for the practical range of Ge composition (< 0.3) with high doping concentration, although the in-plane mobility will be degraded in the same condition.

These attractive properties gave rise to the application of SiGe layers for various types of devices, both optical and electrical. However, the most successful device application of SiGe has been with SiGe HBTs. The first SiGe HBT was demonstrated by IBM in 1987, which was based on MBE-grown SiGe/Si epitaxial layers [1]. This was soon followed by the first report on the RF characteristics of SiGe HBT by Stanford and Hewlett-Packard, which exhibited f_T of 28 GHz [2]. The first SiGe HBT with f_T higher than 100 GHz was reported early 1990s [3]. However, it was not until 1996 that the world's first commercial SiGe HBT technology (in fact a BiCMOS technology) was released by IBM, which exhibited f_T and f_{max} of 47 GHz and 65 GHz, respectively. Continued scaling and structural innovation have led to SiGe HBTs with further improved operation frequencies, notably those with $f_T/f_{max} = 350/160$ GHz in 2003 by IBM [4], $f_T/f_{max} = 300/500$ GHz in 2010 by IHP [5], and 505/720 GHz in 2016 by IHP [6]. Note that the early efforts on RF SiGe HBTs were made mainly to improve f_T, while the more recent works are focusing on higher f_{max}, the latter being now widely accepted as more relevant for most RF circuit designs.

20.2 DC Characteristics

20.2.1 Collector Current

The most distinct feature of SiGe HBTs from Si BJTs in terms of DC characteristics is the higher collector current level for a given base bias. As the collector current dictates the current gain and the transconductance in bipolar transistors, an overview of the collector current of SiGe HBTs in comparison with Si BJTs would be useful. Before jumping into the current analysis, it would help readers to review the energy band diagram of SiGe HBTs in comparison with that of Si BJTs, which are shown in Fig. 20.1 for *npn* devices in the forward active mode operation. Typical SiGe HBTs employ a SiGe layer as the base region while the emitter and the collector are based on Si, whereas the entire region of Si BJTs is made of Si. For the base SiGe layer, Ge composition is usually maintained below ~20% to avoid dislocations that may be caused by lattice mismatch with larger Ge compositions. Also, the grading of Ge composition across the base is very widely practiced, in which case Ge composition gradually increases from the emitter side toward the collector side. Called the retrograded Ge composition, it results in a gradual reduction of the bandgap toward the collector side as shown in Fig. 20.1a, which creates a slope in the conduction band. A quasi-electric field for electrons is generated due to the slope, which helps to improve the electron transport across the base and a consequent speed enhancement as will be discussed in the next subsection.

For the DC characteristics, the major effect of SiGe arises from the overall reduction in the bandgap of the base region. To understand the effect, a quantitative analysis would be helpful. The general expression for the collector current density of bipolar transistors in the forward active mode is given by Moll-Ross relation as follows [7]:

$$J_C = \frac{q e^{qV_{BE}/kT}}{\int_0^{W_B} \frac{N_B(x)}{D_{nB}(x) n_{iB}^2(x)} dx} \tag{20.1}$$

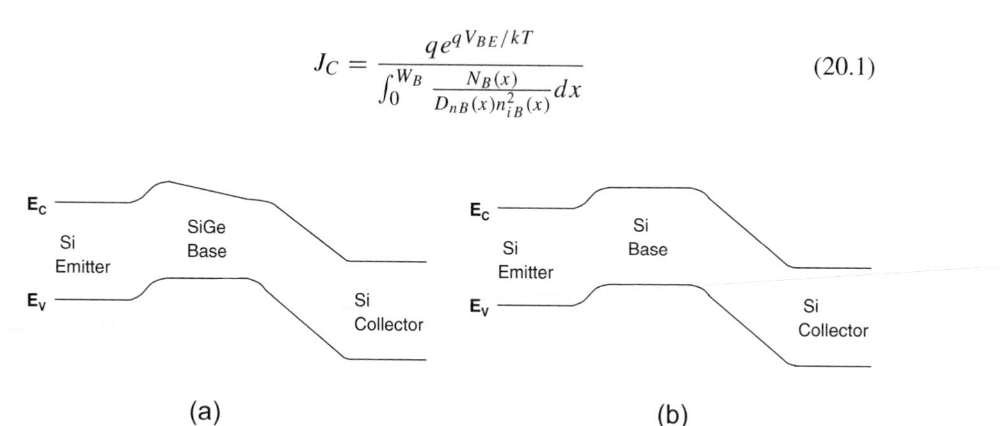

Fig. 20.1 Energy band diagram of *npn* type transistors in the forward active mode: (**a**) SiGe HBT. (**b**) Si BJT

where N_B, D_{nB}, and n_{iB} are the doping concentration, electron diffusion constant, and intrinsic carrier concentration, respectively, in the neutral base region (base region outside the depletion regions) and W_B is the neutral base width at bias V_{BE}. It is interesting to note that the collector current level is determined by the physical parameters in the *base* region, with no contribution from the emitter or collector regions. This is related to the fact that bipolar transistors are minority carrier devices, leading to *npn* device performance dictated by the minority electrons in the *p*-type base.

Two cases will be discussed separately. First, the case with a uniform Ge composition throughout the base region will be considered, which would provide a more insightful view with a simpler analysis when compared to the more popular case with graded Ge composition. With additional assumptions of uniform doping concentration and uniform diffusion constant across the base region, Eq. (20.1) becomes:

$$J_{C,SiGe,unif} = \frac{qn_{iB,SiGe}^2 D_{nB,SiGe}}{N_B W_B} e^{qV_{BE}/kT} \tag{20.2}$$

The readers may remember from the basic semiconductor device classes that the general expression for $n_i{}^2$ is given as:

$$n_i^2 = (N_C N_V)\, e^{-E_g/kT} \tag{20.3}$$

where E_g is the bandgap; N_C and N_V are the effective density of states at conduction and valence bands, respectively. Then, $n_{iB}{}^2$ of the SiGe base region with uniform Ge composition can be expressed as:

$$n_{iB,SiGe,unif}^2 = (N_C N_V)_{SiGe} e^{-(E_{g,Si} - \Delta E_{g,SiGe})/kT} \tag{20.4}$$

where $\Delta E_{g,SiGe}$ represents the bandgap reduction due to the addition of Ge into the base region. From Eqs. (20.2) and (20.4), the collector current density of the SiGe HBT is given as follows [8]:

$$J_{C,SiGe,unif} = \frac{q D_{nB,SiGe} (N_C N_V)_{SiGe}}{N_B W_B} e^{-(E_{g,Si} - \Delta E_{g,SiGe})/kT} e^{qV_{BE}/kT} \tag{20.5}$$

In comparison, the current density for Si BJTs is given by:

$$J_{C,Si} = \frac{q D_{nB,Si} (N_C N_V)_{Si}}{N_B W_B} e^{-E_{g,Si}/kT} e^{qV_{BE}/kT} \tag{20.6}$$

Hence, the collector current of SiGe HBTs is increased by a factor of:

$$\frac{J_{C,SiGe,unif}}{J_{C,Si}} = \frac{(N_C N_V)_{SiGe}}{(N_C N_V)_{Si}} \frac{D_{nB,SiGe}}{D_{nB,Si}} e^{\Delta E_{g,SiGe}/kT} = \frac{(N_C N_V)_{SiGe}}{(N_C N_V)_{Si}} \frac{\mu_{nB,SiGe}}{\mu_{nB,Si}} e^{\Delta E_{g,SiGe}/kT} \tag{20.7}$$

where Einstein relation of $\mu_n kT/q = D_n$ is used. While there are differences in the effective density of states and the mobilities between SiGe and Si, the dominating factor in Eq. (20.7) is still the exponential bandgap reduction factor, or $e^{\Delta E_{gB,SiGe}/kT}$. With the typical range of Ge composition of 10–20%, the enhancement factor can range from ~35 to as high as ~900 at room temperature. The number can be even greater at lower temperatures.

Second, the case with the retro-graded Ge composition in the base is considered. While an analytic derivation of the collector current density can be obtained if the grading is linear, only the result will be presented here for the sake of briefness. The detailed derivation can be found elsewhere [8, 9]. For the linearly retro-graded base, the collector current gain is approximated as:

$$J_{C,SiGe,grade} = \frac{q\tilde{D}_{nB,SiGe}\left(\tilde{N}_C \tilde{N}_V\right)_{SiGe}}{N_B W_B} e^{\Delta E_{gB,SiGe}(0)/kT} \cdot \frac{\Delta E_{gB,SiGe}(grade)}{kT} e^{qV_{BE}/kT} \tag{20.8}$$

where $\Delta E_{gB,SiGe}(0)$ is the bandgap reduction at the emitter-side edge of the neutral base region and $\Delta E_{gB,SiGe}(grade)$ is the total grading of the energy bandgap across the neutral base region. The tildes on some parameters indicate the average value of the corresponding parameters in the base region, as the values vary across the base region due to the position dependence of the Ge composition. With the given Ge composition profile, the enhancement factor becomes:

$$\frac{J_{C,SiGe,grade}}{J_{C,Si}} = \frac{\left(\tilde{N}_C \tilde{N}_V\right)_{SiGe}}{(N_C N_V)_{Si}} \cdot \frac{\tilde{\mu}_{nB,SiGe}}{\mu_{nB,Si}} e^{\Delta E_{gB,SiGe}(0)/kT} \frac{\Delta E_{gB,SiGe}(grade)}{kT} \tag{20.9}$$

Hence, the dominant enhancement factors are the exponential bandgap reduction factor, $e^{\Delta E_{gB,SiGe}(0)/kT}$, and the linear bandgap grading factor, $\Delta E_{gB,SiGe}(grade)/kT$. As the magnitude of the linear factor is limited, the design of the Ge profile needs to maintain the Ge composition at the emitter-side edge of the neutral base large enough to achieve a large enhancement in the collector current density. With a proper base Ge profile design, the enhancement factor of several tens can be easily obtained.

In a qualitative term, the main factor for the raised collector current level of SiGe HBTs over that of Si BJTs is the increased minority electron concentration in the base region, which stems from the increased intrinsic carrier concentration n_i in the base region as a result of the reduced bandgap of SiGe. As n_i has an exponential relation with the bandgap, a slight reduction in the bandgap induced by the addition of a small amount of Ge in the base region results in a remarkable increase in the collector current level.

20.2.2 Current Gain

For bipolar transistors, the current gain usually refers to the common-emitter current gain, which is the ratio between the collector current and the base current. Although the DC current gain and AC current gain are technically different, the current gain in this subsection will refer to the DC current gain, while it will not be much different from the AC current gain. The base current in *npn* transistors, which is basically a hole current, is composed of the recombined holes in the emitter-base depletion region and in the neutral base region, and holes injected into the neutral emitter region. For modern bipolar transistors with very thin neutral base layer and low defect density, the hole current injected into the emitter dominates the total base current. Assuming a thin emitter region with dominant hole recombination at the emitter contact, which is the typically the case, the base current density can be expressed as follows [10]:

$$J_B = \frac{q e^{qV_{BE}/kT}}{\int_0^{W_E} \frac{N_E(x)}{D_{pE}(x) n_{iE}^2(x)} dx + \frac{N_E(W_E)}{n_{iE}^2(x) S_p}} \tag{20.10}$$

where N_E, D_{pE}, and n_{iE} are the emitter doping concentration, hole diffusion constant, intrinsic carrier concentration, respectively, in the neutral emitter region. W_E is the neutral emitter width at bias V_{BE}, and S_p is the hole recombination velocity at the emitter contact. If we assume uniform doping concentration and diffusion constant across the emitter region, Eq. (20.10) is reduced to:

$$J_B = \frac{q e^{qV_{BE}/kT}}{\frac{N_E W_E}{D_{pE} n_{i,Si}^2} + \frac{N_E}{n_{i,Si}^2 S_p}} = \frac{q n_{i,Si}^2 D_{pE}}{N_E W_E \left(1 + \frac{D_{pE}}{W_E S_p}\right)} e^{qV_{BE}/kT} \tag{20.11}$$

It is obvious from the expression that the base current is determined by the parameters in the emitter region, similar to the collector current being dependent on base region parameters as mentioned earlier. This indicates that there is no difference in the base current density for the two base profiles being considered, one with uniform Ge composition and the other with graded Ge composition. By the same token, the base current will be identical for SiGe HBTs and Si BJTs as well, since the emitter region of typical SiGe HBTs is made of Si, as is the case for Si BJTs. Therefore, the enhancement factor for the current gain will be identical to that of the collector current density. Based on this observation, the current gain for the two types of SiGe HBTs is given as:

$$\beta_{SiGe,unif} = e^{\Delta E_{gB,SiGe}/kT} \cdot \frac{(N_C N_V)_{SiGe} D_{nB,SiGe} N_E W_E}{(N_C N_V)_{Si} D_{pE} N_B W_B} \left(1 + \frac{D_{pE}}{W_E S_p}\right) \tag{20.12}$$

$$\beta_{SiGe,grade} = e^{\Delta E_{gB,SiGe}(0)/kT} \frac{\Delta E_{gB,SiGe}(grade)}{kT} \cdot \frac{\left(\tilde{N}_C \tilde{N}_V\right)_{SiGe} \tilde{D}_{nB,SiGe} N_E W_E}{(N_C N_V)_{Si} D_{pE} N_B W_B} \left(1 + \frac{D_{pE}}{W_E S_p}\right) \tag{20.13}$$

The enhancement factors for the current gain of SiGe HBTs over Si BJTs, which is the same as those of the collector current density, are not be repeated here. Recalling the level of enhancement in the collector current density with the addition of Ge in the base, the enhancement in the current gain can range from several tens up to several hundreds with typical Ge composition profiles in the base region. The enormous current gain in SiGe HBTs is one of the most outstanding features of SiGe HBTs that distinguish them from Si BJTs. Again, the higher current gain of SiGe HBTs eventually originates from the increased minority electron carrier concentration in the base region with the smaller bandgap of SiGe, which results in the increased collector current while the base current remains the same for SiGe HBTs and Si BJTs. This can be compared to the cases of HBTs with a large emitter bandgap, often found in III-V HBTs, in which the increased current gain is mainly due to the reduction of the base current.

While it is true that higher current gain is a desired aspect of bipolar transistors, an excessively large current gain is not necessarily favored as it can lead to degradation in the breakdown voltage without clear benefits for actual circuit applications. A rule of thumb is that the current gain of about 100 is sufficient for most circuit applications. Any extra current gain larger than this level will not be desired. However, the real advantage of the high current gain with SiGe HBTs lies in the fact that the extra current gain can be *traded* for other performances of SiGe HBTs. One key trade-off is to increase the base doping concentration N_B in sacrifice of the current gain. Thus, increased base doping will benefit various aspects of device operation, including the reduction of base resistance, the availability of thinner base layer, and the prevention of the emitter-collector punch-through. These aspects are highly beneficial to the RF performance of the device as will be detailed later. For conventional Si BJTs, the base doping cannot be increased too much since it will result in insufficient current gain, while such constraint is eliminated with SiGe HBTs.

20.2.3 Breakdown Voltages

For bipolar transistors, various breakdown voltages can be defined. First of all, a couple of *p-n* junction breakdown voltages can be defined: the emitter-base breakdown voltage with collector open (denoted as BV_{EBO}) and the collector-base breakdown voltage with emitter open (denoted as BV_{CBO}). The breakdown voltage between collector and emitter is more intricate, since the two electrodes are separated by the base region and thus the breakdown strongly depends on the condition at the base terminal. In this situation, we can consider three different cases regarding the base terminal condition: base opened, base shorted to emitter, and base connected to emitter through a resistor. The collector to emitter breakdown voltage corresponding to each of these cases is denoted as BV_{CEO}, BV_{CES}, and BV_{CER}, respectively. Figure 20.2 illustrates the connection schematics for these five breakdown voltages. Following discussion will be focused on the collector to emitter breakdown voltages. To better understand the breakdown in each of the three conditions at the base terminal, let's take a look at the mechanisms of avalanche breakdown across the emitter and collector when the bias across the two terminals is near the breakdown voltage.

When the breakdown occurs across the base-collector junction, the generated holes by the breakdown are drifted toward the base region. Once they enter the base region, they will either exit the device through the base contact or recombine with electrons injected from the emitter into the base. For the latter case, a single such recombination event requires hundreds of additional electrons to be injected from the emitter, or, to be more exact, as many electrons as the current gain. The increased injection level of the electrons causes increase in the emitter current level, which further promotes the breakdown in the base-collector junction and thus increases the emitter current level again. This positive feedback mechanism, which is originated by the current gain action of bipolar transistors, tends to lower the breakdown voltages. Interestingly, the strength of such positive feedback depends on how much portion of the total breakdown-generated holes eventually end up triggering an emitter current rise, instead of exiting the device. The portion depends on the condition at the base terminal. The larger the impedance looking out from the base region at the base terminal, the larger portion of the breakdown-generated

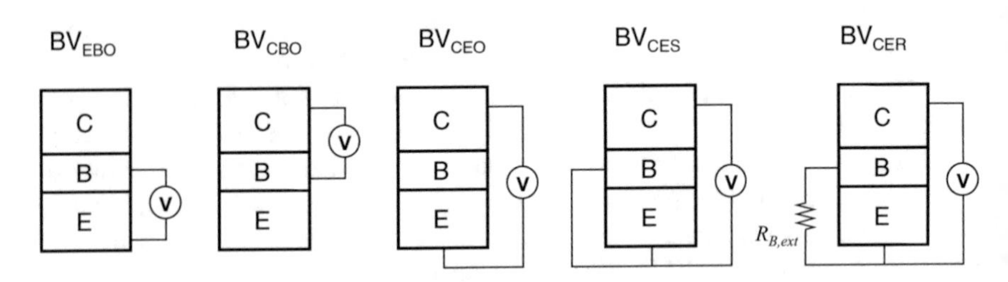

Fig. 20.2 Connection schematics of various breakdown voltages of bipolar transistors

holes will recombine with electrons, since the large impedance tends to block the holes from exiting the device. This would result in smaller breakdown voltage.

Based on the mechanism above, the relative magnitude of BV_{CEO}, BV_{CES}, and BV_{CER} is now discussed. For devices with the open base, the entire breakdown-generated holes will recombine with electrons injected from the emitter, thus severely triggering the emitter current level increase and thus the positive feedback. This worst scenario would lead to the lowest breakdown voltage possible with the given device, which is BV_{CEO} by definition. Although BV_{CEO} is commonly quoted as the breakdown voltage of bipolar transistors, open-base condition is in fact rarely faced in actual operation of the device in circuit applications. For the devices with the base shorted to emitter, the vast majority of the breakdown-generated holes will exit the device through the base. This is the most optimistic situation and will lead to the largest breakdown voltage possible across the emitter and collector, which is BV_{CES} by definition. Its value will be virtually equal to BV_{CBO} as the breakdown does not involve the positive feedback. Positioned between the two extreme cases is the device with the base terminal connected to the emitter through an external resistor. In this case, naturally, the corresponding breakdown voltage, BV_{CER}, will take on values somewhere between BV_{CEO} and BV_{CES}, depending on the values of the external resistor. In fact, this is the most practical situation in the actual circuit applications, and thus, BV_{CER} should be interpreted as the representative upper voltage limit allowed between the emitter and collector of a bipolar transistor. It is known that, for practical values of $R_{B,ext}$ ($< 1\ \text{k}\Omega$) and moderate emitter size, BV_{CER} is very close to BV_{CES}. Hence, from a practical point of view, the maximum allowed voltage between collector and emitter is not far from BV_{CBO}.

It is noted that a larger current gain would enhance the positive feedback of the breakdown, hence lowering the breakdown voltage. This is one main reason why the current gain should be maintained within a reasonable level in bipolar transistors. Also, it will be useful to remember that the ratio between BV_{CBO} and BV_{CEO} is around 2–4 for typical SiGe HBTs. For the most advanced high-frequency SiGe HBTs of today, BV_{CBO} is around 3–5 V, while BV_{CEO} ranges over 1.4–1.8 V. As for the comparison between SiGe HBTs and Si BJTs, there is no dominating factor that would make the breakdown voltages of the two devices different. The larger current gain and smaller base bandgap of SiGe HBTs would reduce the breakdown voltages of SiGe HBTs, but the effect is only moderate. While the reported breakdown voltages of SiGe HBTs are usually lower than those of Si BJTs, it is mainly due to the higher collector doping concentration of SiGe HBTs rather than the SiGe layer in the base. A similar level of collector doping concentration with Si BJTs would make a similar result, while such an aggressive collector doping for increased operation frequency was not routinely practiced for Si BJTs.

20.3 RF Characteristics

20.3.1 Cutoff Frequency f_T

Before jumping into the discussion on the cutoff frequency f_T, it will be useful to briefly overview the device operation speed and its measures. The operation speed of transistors is widely represented by two parameters. One is the cutoff frequency f_T, which is the frequency where the current gain becomes unity, and the other is the maximum oscillation frequency f_{max}, which is the frequency where the power gain (more specifically, the unilateral power gain U) drops down to unity. While the relative importance of the two parameters is still under discussion [11], there is a wide-spread notion that f_{max} is more relevant, at least for analog and RF circuit applications as briefly mentioned earlier in this chapter. On the other hand, f_T is less sensitive to the device layout and thus is more representative of the process technology rather than the actual devices employed for circuits. For this reason, f_T still retains its status as a key parameter widely quoted in evaluating semiconductor technologies. In this subsection, both parameters will be discussed. Because a better insight can be obtained with f_T, a more detailed description will be first given for f_T, from which the discussion on f_{max} will be derived in the following subsection.

The cutoff frequency f_T of bipolar transistors can be expressed in terms of the emitter-to-collector delay time, τ_{EC}, which is defined as the inverse of f_T with a factor of 2π:

$$\begin{aligned}\tau_{EC} &= \frac{1}{2\pi f_T} = \tau_E + \tau_C + \tau_B + \tau_{CSCL} \\ &= \frac{kT}{qI_C}C_{EB} + \left(\frac{kT}{qI_C} + R_C + R_E\right)C_{CB} + \frac{{W_B}^2}{\gamma D_{nB}} + \frac{W_{CSCL}}{2v_{SAT}}\end{aligned} \quad (20.14)$$

where k is the Boltzmann constant, C_{EB} and C_{CB} are emitter-base and collector-base capacitances, R_C and R_E are collector and emitter resistances, W_{CSCL} is the base-collector space charge region (SCR) width, v_{SAT} is the saturation velocity, and γ is the field factor that is a measure of the quasi-electric field established in the base due to the base bandgap grading.

The most effective strategy of bipolar transistor performance improvement is the scaling [12]. This strategy works not only for SiGe HBTs but also for Si BJTs. In fact, the scaling is a generic strategy that works for any semiconductor devices. Smaller device dimension tends to reduce the transit time of carriers, as well as the *RC* time delays in the device interior if proper optimization is made. As we narrow our focus down to SiGe HBTs, we can expect both vertical and lateral scalings to improve the device speed based on the delay time reduction as expected by Eq. (20.14). For SiGe HBTs, the vertical scaling will impose the primary impact as they are typically a vertical device. The lateral scaling affects the device speed through the charging time as dictated by the resistance and capacitance. Both scaling approaches are discussed below, starting with the vertical scaling.

The vertical scaling involves both collector scaling and base scaling. Let's bring out attention to the collector scaling first. The major effect of the collector scaling is the reduction of W_{CSCL}, which is usually realized with increased collector doping concentration N_C. The increased collector doping reduces the base-collector space charge layer width (depletion width) W_{CSCL}, leading to a smaller τ_{CSCL}. The reduced W_{CSCL}, however, necessarily involves an increase in C_{CB}, leading to an increased charging time and thus a larger collector delay τ_C. Therefore, the collector scaling requires a precise balance between these two competing factors. A widely practiced approach for SiGe HBTs to suppress the C_{CB} increase due to the increased N_C is *selectively* implanting the intrinsic region of the collector just below the emitter area. In this way, only the (laterally) central region of the collector will be highly doped, while the edge region maintains a relatively low doping concentration. This selectively implanted collector (SIC) effectively reduces the electron traveling time across the space charge layer while suppressing the extrinsic component of C_{CB}.

Another advantage of the collector scaling with increased N_C comes from the increase in the Kirk current I_K, or the collector current at which the Kirk effect takes place. The Kirk effect refers to a phenomenon in which a rather abrupt widening of the base width occurs when the collector current becomes excessively large. The Kirk effect kicks in when the mobile carrier concentration in the base-collector space charge region becomes comparable to the ionized dopant concentration in that region, which is almost the same as the background doping concentration. Based on this, I_K can be expressed as follows [13]:

$$I_K = qA_C v_{SAT} N_C \left(1 + \frac{2\varepsilon_s V_{CB}}{qN_C W_C^2}\right) \approx qA_C v_{SAT} N_C \tag{20.15}$$

where V_{CB} is the voltage applied between collector and base, ε_s is the dielectric constant of Si, A_C is the collector cross-sectional area, and W_C is the lateral width of the collector layer. Eq. (20.15) indicates that a higher doping concentration effectively increases I_K and thus delays the onset of the Kirk effect. This will allow a higher collector current level, leading to a reduction in the minimum available values of τ_E and τ_C (see Eq. (20.14)), which is basically a consequence of the reduced minimum charging time with the increased current. In addition, the collector scaling effectively increases the average velocity across the base-to-collector space charge region due to enhanced ballistic transport [14], which also contributes to reducing τ_{CSCL}.

As for the base scaling, its main effect is obviously the reduction of W_B and, thus, τ_B. It may sound straightforward, but the actual implementation is not so simple. Aggressively reduced neutral base width may trigger the troublesome emitter-to-collector punch-through, in which the emitter-base depletion region and the base-collector depletion regions merge. This causes the neutral base region to disappear, preventing the normal operation of bipolar transistors. Even if the punch-through is barely averted, the narrow neutral base width would significantly increase the base resistance R_B as the base current is supplied through the lateral path along the base layer. In either case, the most effective solution is to increase the base doping concentration. It prevents the depletion regions from encroaching too deep into the base region and, at the same time, reduces the sheet resistance of the base region. There are a couple of issues related with the heavily doped base, though. As mentioned earlier, a high base doping would reduce the current gain. With SiGe base, however, this issue becomes mostly irrelevant, since a sufficiently high current gain is already available, which can be traded for higher base doping concentration. The other issue, which is rather related to the material aspect of the device fabrication, is the outdiffusion of the base dopant. Typically employed for base doping is boron (B), and the diffusion of B atoms out of the base region during high-temperature cycles in device fabrication results in the widening of the base layer. This will counteract the scaling and degrade device speed with increased τ_B. Interestingly, this issue, a general problem for Si bipolar transistors, can be partly mitigated by employing SiGe for the base, as the added Ge atoms tend to suppress the diffusion of B atoms. To further suppress the B outdiffusion, carbon (C) atoms can be additionally included in the base region, which is widely employed in modern SiGe HBTs [15, 16]. While such carbon-doped SiGe HBTs are sometimes specifically referred to as SiGeC HBTs, you can assume that most modern SiGe HBTs include C in the base as a routine process.

There is another factor of the base scaling that enhances the device speed. As mentioned earlier, the quasi-electric field is established across the base when the retro-graded Ge composition profile is adopted for the base region. It will accelerate the electrons traveling across the base, leading to a reduction in the base transit time τ_B. This effect is represented by the field factor γ, which appears in the denominator of τ_B in Eq. (20.14). It can be explicitly expressed as a function of quasi-electric field strength E_0 as follows [17]:

$$\gamma = 2\left[1 + \left(\frac{qE_0 W_B}{kT}\right)^{\frac{2}{3}}\right] \tag{20.16}$$

E_0 is proportional to the slope of conduction band edge (E_C in Fig. 20.1a), which will increase with reduced W_B for a given total grading to Ge composition across W_B. Hence, the base scaling with the same peak Ge composition will enhance the impact of the quasi-electric field on the base transit time reduction. This effect, together with the reduced W_B itself, will improve the device speed, a result of the base scaling. One advantage of the base scaling over the collector scaling is the fact that it does not degrade the base-collector breakdown voltage, which renders the base scaling a more favored option for the speed enhancement particularly in power devices.

In the case of the lateral scaling, its effect on f_T is not as pronounced as the vertical scaling. However, the lateral scaling may help when performed together with the vertical scaling. For example, the increase in C_{CB} and C_{EB} due to the vertical scaling can be effectively compensated for by reducing the capacitance area with the lateral scaling [18]. Such compensation applies to C_{EB}, too. In the case of resistance, however, the contact area and the cross section of the vertical current paths are reduced with the lateral scaling. Hence, resistances such as R_E and R_C may be degraded as a result of the laterally pinched vertical current paths, if not significantly. However, R_B is dominated by the lateral path along the thin base layer and thus much helped by the lateral scaling. R_B does not apparently affect f_T as revealed by Eq. (20.14), though, while it does affect f_{max} as will be discussed shortly below. Also, the fringing current component at the emitter-base junction, which degrades the device speed due to the effectively increased current path across the base, may be more pronounced with the emitter stripe width reduction with the lateral scaling.

20.3.2 Maximum Oscillation Frequency f_{max}

For bipolar transistors, f_{max} can be approximated as:

$$f_{\max} = \sqrt{\frac{f_T}{8\pi R_B C_{CB}}} \tag{20.17}$$

Hence, it is closely related to f_T, but is additionally affected by R_B and C_{CB} (note that for field-effect transistors such as MOSFETs, gate resistance R_G and gate-drain capacitance C_{GD} affect f_{max} in the same way). As was the case for f_T, both vertical and lateral scalings affect f_{max} of SiGe HBTs. We have seen the effects of the vertical scaling on f_T in various ways above. As f_{max} increases with f_T, it inherits the overall advantages of the vertical scaling imposed on f_T, although the impact is diminished due to the square-root relation between f_{max} and f_T as shown in Eq. (20.17). However, the additional factors that appear in the f_{max} relation, R_B and C_{CB}, are also affected by the vertical scaling. In fact, both R_B and C_{CB} are adversely affected by the vertical scaling. The increased sheet resistance due to the thinned base layer results in an increase in R_B, and the reduced base-collector depletion width due to raised collector doping will lead to increase in C_{CB}. Therefore, it is fair to say that there are competing effects of the vertical scaling on f_T, R_B, and C_{CB}, and the overall effect may turn out either beneficial or detrimental, depending on structural details of the given device.

The lateral scaling, which has only limited influence on f_T, plays a major role on f_{max}. Base current is supplied laterally to the intrinsic region of SiGe HBTs, unlike emitter or collector currents that mostly flow through vertical paths. Consequently, the lateral scaling reduces the length of the lateral path for the base current, particularly through the region just beneath the emitter with a high sheet resistance, resulting in a reduction in R_B. The base-collector junction area decreases with the lateral scaling, too, leading to a reduction in C_{CB}. As f_{max} has a direct correlation with R_B and C_{CB} as shown in Eq. (20.17), it greatly benefits from such effects of the lateral scaling on these parameters. As mentioned earlier, most circuit performances are closely related to f_{max} of the device. Overall, the SiGe layer employed for the base region enables the aggressive vertical scaling of the base region and base resistance reduction through the high base doping concentration. This has led to the high f_{max} values achieved with modern SiGe HBTs, which has reached beyond 700 GHz [6].

20.3.3 Noise Characteristics

The noise characteristics of transistors can be treated in terms of the high-frequency noise and low-frequency noise (or 1/*f* noise). This sub-section will provide a brief overview of these two types of noise for SiGe HBTs.

For bipolar transistors, the high-frequency noise arises from the shot noise and the thermal noise, which arise from the discrete nature of carriers crossing *p-n* junctions and the finite resistance along the carrier paths, respectively. The high-frequency noise can be represented by four noise parameters: minimum noise figure (F_{min}), noise resistance (R_n), and the real and imaginary components of the source impedance match for lowest F_{min} (Γ_{opt}). Out of the four noise parameters, F_{min} is regarded as the most critical one, as it represents the lowest noise figure level available with the optimal noise matching. By assuming a simple common-emitter compact model, F_{min} of a bipolar transistor can be expressed as follows [12]:

$$F_{\min} \simeq 1 + \sqrt{\frac{2I_C}{V_T} R_B \left(\frac{f^2}{f_T^2} + \frac{1}{\beta} \right)} \tag{20.18}$$

where V_T is the thermal voltage given by kT/q.

In the lower frequency regime at which $f << f_T/\beta^{0.5}$, F_{min} can be reduced to following frequency-independent form:

$$F_{\min} \simeq 1 + \sqrt{\frac{2I_C}{V_T \beta} R_B} \tag{20.19}$$

It can be easily seen that a smaller R_B and a large β would help to reduce F_{min}. In the higher frequency regime where $f >> f_T/\beta^{0.5}$, F_{min} climbs up linearly with frequency:

$$F_{\min} \simeq 1 + f \sqrt{\frac{2}{V_T} R_B \left(\frac{I_C}{f_T^2} \right)} \tag{20.20}$$

In this regime, reducing R_B as well as increasing f_T is required for the reduction of F_{min}.

From these expressions for F_{min} in terms of the device parameters, it is clear SiGe HBTs will have obvious advantages over Si BJTs. Although its impact is minimal, current gain is typically larger for SiGe HBTs as mentioned earlier, which helps to bring down F_{min} at frequencies much lower than f_T. More importantly, a small R_B and a large f_T, which are readily available with SiGe HBTs as observed earlier, will greatly improve the noise characteristics. This clearly indicates that, for RF applications, SiGe HBTs are favored over Si BJTs for both high-speed and low-noise performances.

The low-frequency noise of transistors also affects RF characteristics of the circuits although it occurs at far lower frequency than the typical RF operation frequency, as it can be frequency up-converted with various mixing operation of the actual circuit. The low-frequency noise of semiconductor devices is often dominated by $1/f$ noise, which mostly arises from the trapping and de-trapping processes of carriers that travel inside the device. In MOSFETs, the carriers travel along the channel in close proximity of the oxide-semiconductor interface and hence they are highly vulnerable to trapping and de-trapping processes, leading to a significant $1/f$ noise. On the contrary, the current path of the typical (vertical) bipolar transistor is formed away from such interfaces or surfaces, for which bipolar transistors tend to exhibit a much lower $1/f$ than MOSFETs. However, $1/f$ noise cannot be completely eliminated, and it does affect the performance of bipolar transistors and circuits.

It is known that $1/f$ noise of the modern polysilicon emitter SiGe HBTs operating in the common-emitter configuration is primarily generated in the base current [19, 20]. At medium or high bias conditions, carriers are trapped and released by defects that are typically spread across emitter area A_E, near the thin interfacial oxide layer that is formed at the polysilicon-to-silicon emitter interface. Devices with a relatively large emitter area contain a large number of traps, leading to a rather smooth $1/f$ spectrum and little inter-device variation. Smaller devices contain a smaller number of traps and thus show a more fluctuating noise profile over the frequency and also a larger inter-device variation. At very low biases, the dominant noise mechanism becomes recombination and generation along the emitter perimeter, which mostly arises from defects in the emitter-base depletion region. Assuming the noise is dominated by the area-related traps, the base current noise power of bipolar transistors becomes [21]:

$$S_{Ib} = K_f \frac{I_B^2}{A_E} \cdot \frac{1}{f} = K_f \frac{I_C^2}{\beta^2 A_E} \cdot \frac{1}{f} = K_f \frac{J_C^2 A_E}{\beta^2} \cdot \frac{1}{f} \tag{20.21}$$

where K_f is a SPICE model parameter that is dependent on a trap density and the strength of the carrier-trap interaction.

According to Eq. (20.21), the higher current gain of SiGe HBT would help to lower $1/f$ noise for a fixed current density. At the same time, the higher collector current, typically achieved with SiGe HBTs with collector scaling, would result in higher noise level. However, remember that the raised noise level with a larger bias current is rather a natural relation, which also occurs with high-frequency noise as shown in Eq. (20.18). The corner frequency, widely adopted as a measure of the $1/f$

noise strength, of SiGe HBTs typically falls on near 1 kHz [22], which is 1–3 orders of magnitude smaller compared with typical MOSFETs.

20.4 SiGe HBT Technology Examples

As briefly mentioned in the introduction of this chapter, the first demonstration of SiGe HBT in 1987 [1] was soon followed by the first reported RF characteristics in 1989 [2]. Since then, the performance of SiGe HBTs has been rapidly enhanced, leading to device operation frequencies well beyond a few hundred GHz. The high-frequency Si BJT technologies by major industrial players have been transformed to SiGe HBT technologies in the course of a decade or two. In this sub-section, a couple of SiGe HBT technologies are introduced as examples, one from IBM (now GlobalFoundries) and IHP.

The first is a SiGe HBT from IBM, with a device cross-sectional schematic illustrated in Fig. 20.3. On top of the heavily arsenic-doped subcollector buried layer, the collector region is formed with a selective implantation in the central region of a lightly doped collector epitaxial layer. The collector region is linked to the collector contact through a vertical reach-through region, also heavily doped for reduced collector resistance. A boron-doped SiGe base layer is grown above the collector region by UHVCVD (Ultra High Vacuum Chemical Vapor Deposition), which reaches the base contact through a self-aligned raised extrinsic base. On

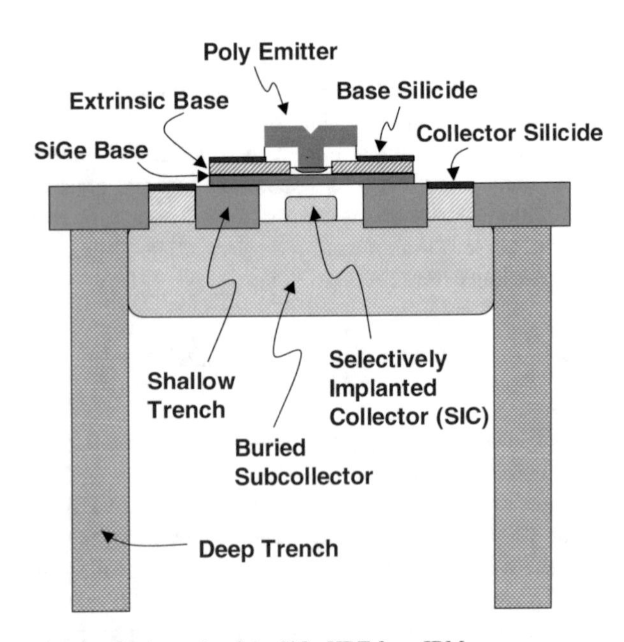

Fig. 20.3 Cross-sectional schematic of the SiGe HBT from IBM

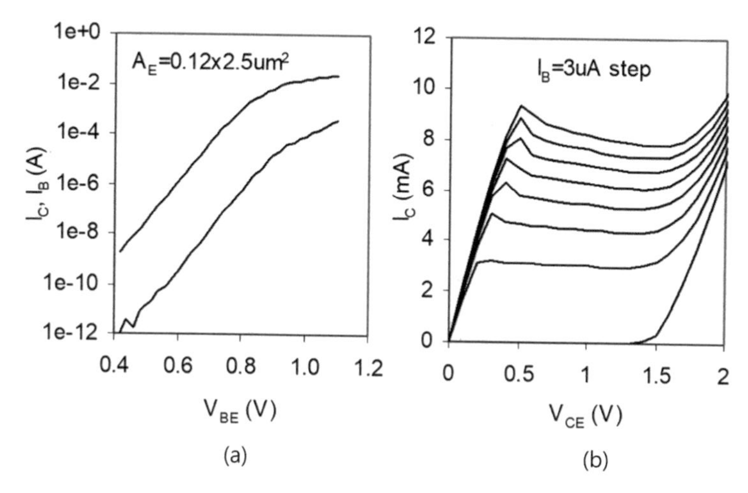

Fig. 20.4 DC characteristics of the IBM SiGe HBT with an emitter size of $A_E = 0.12 \times 2.5\ \mu\text{m}^2$: (**a**) Gummel plots. (**b**) Forced-I_B output characteristics. (© 2004 IEEE [22])

top of the intrinsic base region, a phosphorus-doped T-shaped polysilicon emitter is formed, which allows narrow emitter stripes close to ~0.1 μm without causing a significant increase in the emitter resistance. Contact resistance of electrodes is lowered with the silicidation of the exposed contact region. Oxide-based deep and shallow trenches provide device isolation without the use of mesa etching, leading to a planar topography and dense layout, an obvious difference from typical III-V HBTs. The completed devices are wired by Cu-based multi-level BEOL (Back End Of the Line) process, which offers low resistance and favorable electromigration characteristics for interconnect lines.

The DC characteristics of the device with an emitter size of $A_E = 0.12 \times 2.5\ \mu\text{m}^2$ are shown in Fig. 20.4. The Gummel plots show the near-ideal characteristics of both collector and base currents without a signature of base leakage down to $V_{BE} = 0.4$ V. The ideality factor is 1.05 and 1.08 for collector and base, respectively. The near-unity base ideality factor, an indication of the excellent control of surface and interface states, can be attributed to the planar structure and high-quality Si-oxide interface. The peak DC current gain is around 3500. BV_{CEO} and BV_{CBO} are 1.4 V and 5 V, respectively. A reduction of the excessive current gain is expected to increase BV_{CEO} with the given BV_{CBO}.

The RF characteristics are presented in Fig. 20.5, which shows measured f_T and f_{max} of the device for various layout options. The three layout variants, which are denoted as CBE, CBEB, and CBEBC following the order of the electrode contacts location, are also depicted in the figure with both cross-section and top view. The peak f_T and f_{max} values, 375 GHz and 210 GHz, respectively, are obtained with CBEBC option, which has dual contacts for both base and collector electrodes. Compared to the CBE case, the CBEB and CBEBC configuration improves f_{max} owing to the reduction of R_B, which arises from the fact that base currents are

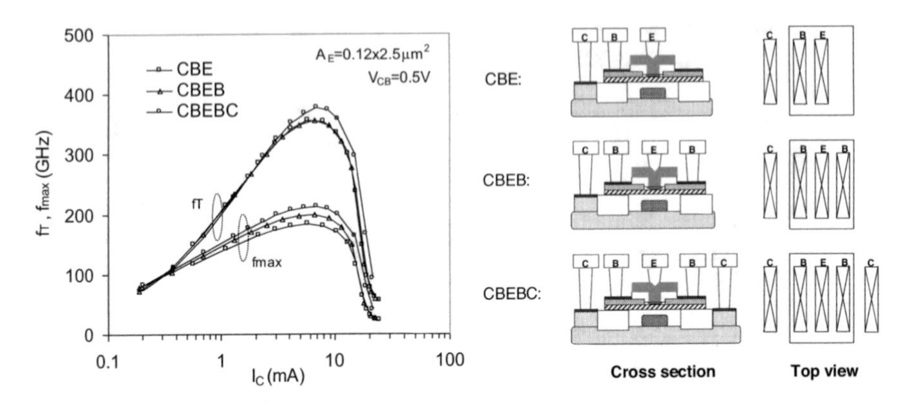

Fig. 20.5 f_T and f_{max} of the IBM SiGe HBT with various layout schemes. (© 2004 IEEE [22])

supplied from either side of the emitter contact, decreasing both contact and access resistance. Further, the transition from CBEB to CBEBC results in f_T improvement due to R_C reduction from the dual contact of the collector electrode. It is also helped by the symmetric spread of injected electrons at the collector region that effectively delays the onset of the Kirk effect. The resultant improvement in f_T also contributes to pushing up f_{max} slightly.

Additionally, the dependence of peak f_T and f_{max} on emitter length L_E and width W_E is depicted in Fig. 20.6, with CBE devices as an example. The most obvious effect is the increase in f_{max} with decreasing L_E. This can be ascribed to the decreasing R_B due to the shorter base current path to the other side of the emitter finger along the silicided extrinsic base. This effect is expected to be less pronounced for CBEB and CBEBC, though, as they have dual base contacts. f_T shows a weaker dependence on L_E compared with f_{max}, while it exhibits a moderate optimum point near $L_E = 2.5$ μm. This can be attributed to the competing scaling behavior of the resistance (R_E and R_C) and capacitance (C_{CB}) with L_E variation, which finds a balance point near this dimension. As for the W_E dependence, increased f_{max} with W_E reduction is obvious, which is from the reduced base access resistance underneath the emitter opening. On the other hand, f_T exhibits little dependence on the W_E variation, which indicates the balance established between reduced C_{CB} and C_{EB} and increased R_C.

Although the device described above is more inclined for f_T performance in sacrifice of f_{max}, it was afterward evolved into other variants with a performance shift toward f_{max}, including those that exhibited $f_T/f_{max} = 300/360$ GHz [23] and 285/475 GHz [24] based on a 90-nm BiCMOS technology.

As another example, a SiGe HBT developed from IHP is briefly introduced below, which has a record f_{max} performance of 720 GHz with associated f_T of 505 GHz [6]. The high f_T and f_{max} of the device can be contributed to the aggressive scaling in both base and collector. Tailoring of various process steps, such as the control of the annealing temperature and the avoidance of the collector implantation

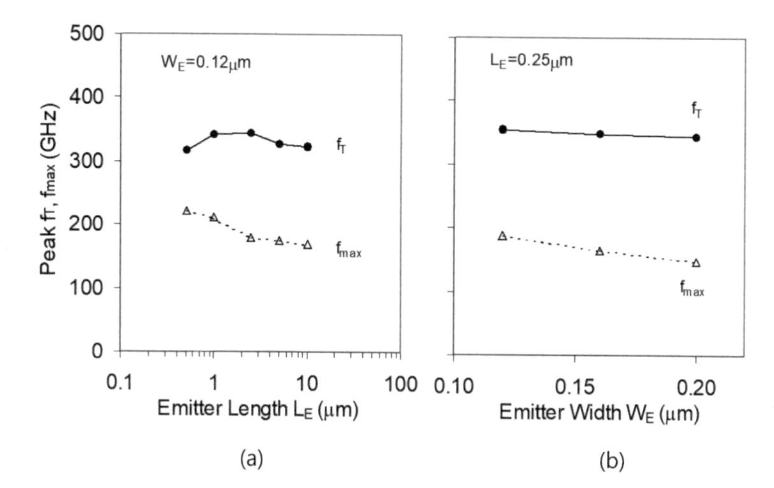

Fig. 20.6 Dependence of f_T and f_{max} of the IBM SiGe HBT on emitter dimension: (**a**) Dependence on emitter length L_E (**b**) Dependence on emitter width W_E. (© 2004 IEEE [22])

through the base layer, resulted in the optimized base and collector vertical scalings. In particular, extensive efforts were made to reduce R_B for improved f_{max}, which include the reduction of the emitter width, the optimization of the extrinsic base doping concentration and distribution, and the increase of the base silicide layer thickness. Additionally, the emitter sidewall spacer that determines the position of the self-aligned extrinsic base silicidation edge was also aggressively scaled, which would effectively reduce the base access resistance and thus enhance f_{max}.

The DC characteristics of the device are shown in Fig. 20.7. The device with 720-GHz f_{max} is denoted as D7bs in the plot, which is composed of 8 emitter fingers of $0.105 \times 1\ \mu m^2$. It exhibits breakdown voltages of $BV_{CEO} = 1.6$ V and $BV_{CES} = 3.2$ V. The peak current gain is estimated to be slightly lower than 1000 based on Fig. 20.7a. Extracted f_T and f_{max} are shown in Fig. 20.8. The peak f_T/f_{max} of 505/720 GHz is obtained at the collector current density of 34 mA/μm^2. This current density is higher than that of the SiGe HBT from IBM described above by a factor of nearly two. This suggests that a far more aggressive collector scaling with increased collector doping concentration was applied for this device, which led to a more effective delay of the onset of Kirk effect and thus improved peak f_T/f_{max} values. The relatively lower BV_{CES} ($\approx BV_{CBO}$) of this device (3.2 V vs 5 V) is also an apparent result of the higher collector doping concentration, while BV_{CEO} is in fact larger (1.6 V vs 1.4 V) as helped by the smaller current gain.

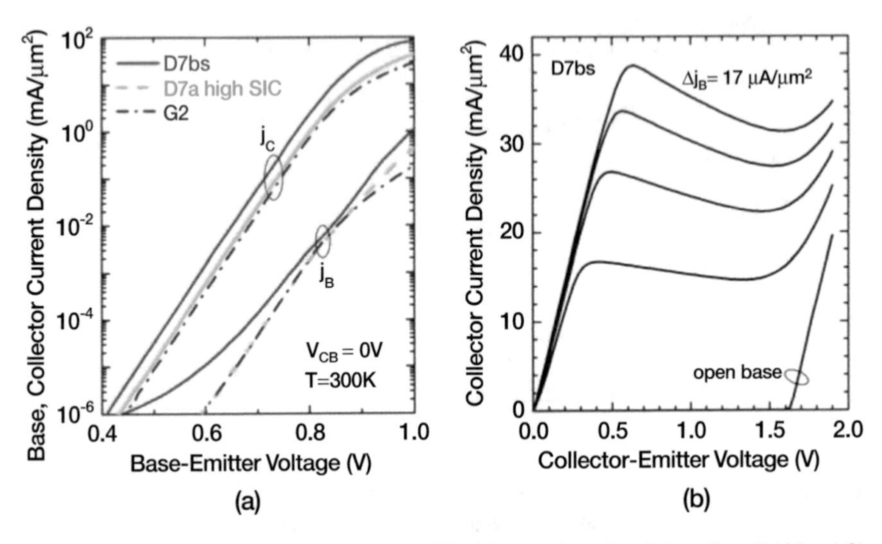

Fig. 20.7 DC characteristics of the IHP SiGe HBT with an emitter size of $A_E = 8 \times (0.105 \times 1.0)$ μm^2: (**a**) Gummel plot. (**b**) Forced-I_B output characteristics. (© 2016 IEEE [6])

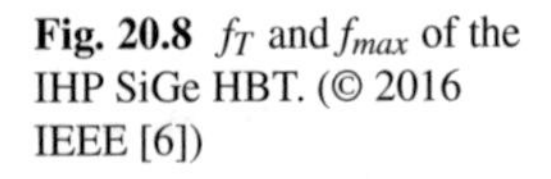

Fig. 20.8 f_T and f_{max} of the IHP SiGe HBT. (© 2016 IEEE [6])

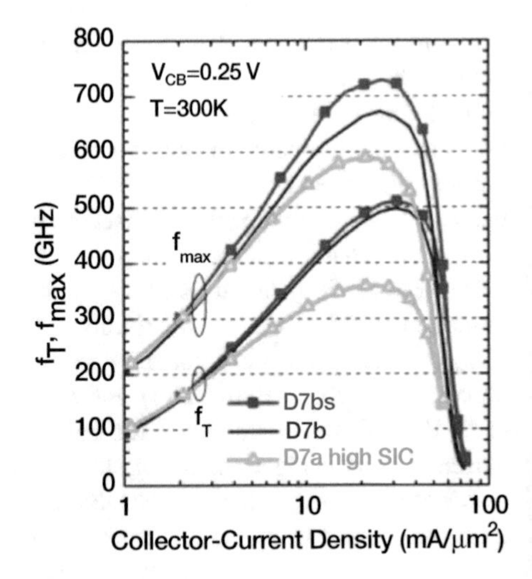

20.5 SiGe HBT Circuit Examples

There is a rapidly growing interest for implementing semiconductor-based electronic systems operating at hundreds of GHz, or the frequency range loosely defined as the THz band. The applications of such high-frequency systems include THz spectroscopy, THz imaging, THz radar, and, of course, THz communication, as this compiled book is intended to cover. Those systems can be implemented based on various semiconductor technology options. Among them, SiGe HBT technologies provide Si-based solution together with CMOS technologies. Compared to CMOS technologies, SiGe HBT technologies benefit from the fact that a more relaxed lithography node can be used for a similar level of performance, which would save lithography mask cost that will comprise a significant portion of the total chip cost especially with low-volume productions. Compatibility with digital circuitry, the chief advantage of CMOS technologies, is also supported by SiGe HBT technologies as they are mostly offered as a form of BiCMOS technology, although the available CMOS devices from BiCMOS technologies may not of the most scaled version.

20.5.1 Unit Block Circuit Examples

Virtually, all the key circuit components composing such THz systems have been demonstrated based on SiGe HBT (or BiCMOS) technologies. *Oscillators*, an indispensable component for signal generation at transmitters and local oscillation at receivers, have been extensively developed to operate at a few hundred GHz range [25–27], and oscillation frequency beyond 1 THz has been recently achieved with an oscillator array composed of 42 radiators [28]. For heterodyne systems, for both transmitters and receivers, *mixers* play a pivotal role together with local oscillation for frequency up- or down-conversion. While THz mixers have been within visibility for decades based on diodes or more exotic devices, such as SIS (superconductor-insulator-superconductor) or HEB (hot-electron bolometer) structures, integration-friendly transistor-based mixers are highly desired for semiconductor THz systems, which have also been widely realized with SiGe HBT technologies up to beyond 800 GHz [29]. *Frequency multipliers* are a critical component for THz systems to boost the signal frequency, a recent study having demonstrated a 0.92-THz frequency quadrupler [30]. While the circuit components mentioned so far can be operated at frequencies higher than the device cutoff frequency owing to the harmonic techniques, another core circuit component in electronics systems, *amplifiers*, is inherently based on the fundamental mode operation. Hence, its operation frequency is strictly limited by the device operation frequency limit. Such a restriction has limited the highest operation frequency of SiGe HBT amplifiers in the range of around 250 GHz to date [31, 32]. Below, a couple of circuit components based on SiGe HBT technologies will be briefly reviewed as example cases.

Let's first consider oscillators operating at 320 GHz as reported in [26] based on the IHP 130-nm SiGe HBT technology. Two circuits were compared. One is a push-push oscillator based on a common-base LC cross-coupled structure, in which the second harmonic signal is extracted at the collector common node of the differential core that oscillates at the fundamental frequency of 160 GHz. The other circuit is a 160-GHz fundamental-mode oscillator followed by a frequency doubler. The fundamental-mode oscillator also adopts a common-base LC cross-coupled structure, additionally employing a stacked common-base buffer to boost the oscillator output power. The frequency doubler takes on a G_m-boosting technique with a base-emitter cross-coupling in the core differential pair, with the output taken at the collector common node for second harmonic extraction. The circuit schematics and the chip photos are shown in Fig. 20.9. The push-push oscillator exhibited an output power of −6.3 dBm at 324 GHz. On the other hand, with the oscillator-doubler circuit, a higher output power of +1.6 dBm was obtained at 332 GHz. Of course, the latter consumed a larger DC power (101.2 mW vs 197.4 mW) and chip area (0.22 mm^2 vs 0.44 mm^2). Still, the overall efficiency was higher for the oscillator-doubler circuit (0.2% vs 0.7%), rendering this circuit a favored option, which can be ascribed to the fact independent optimizations are available for fundamental oscillation and second harmonic extraction.

As for an amplifier, a 260-GHz cascode differential amplifier based on the IHP 130-nm SiGe HBT technology is reviewed. One of the challenges in the design of circuits operating at frequencies beyond ~100 GHz is the uncertainty associated with the device model, which is typically not guaranteed for this high-frequency range. For amplifier designs, the model inaccuracy may cause not only the shift in the performance matrix, but also drift in the stability, which is more serious as it may result in the total failure of the amplifier operation. Particularly for bipolar transistors, it is known that the uncertainty in the base parasitic inductance value may aggravate the stability issue. Hence, for this 3-stage amplifier, a control unit is installed for each stage, which is basically a shunt transistor whose base bias is externally controlled with a tuning voltage (see Fig. 20.10a for the circuit schematic). By design, this control unit affects the gain of the amplifier as well, rendering the amplifier to operate as a variable gain amplifier. With the fabricated chip shown in Fig. 20.10b, a peak gain of 15 dB (with input and output balun losses of 1.5 dB compensated for) at 260 GHz with a 3-dB bandwidth of 16.5 GHz was achieved, for which a DC power of 112 mW was consumed. With the voltage tuning, the peak gain could be lowered down to 7 dB, which in turn improves the stability. Through the entire operation condition, the amplifier did not show any indication of instability, which imply the accuracy of the device model employed for the amplifier design.

20.5.2 Integrated System Examples

There have been an increasing number of reports on system-level integration of SiGe HBT circuits operating at THz frequencies, many of which are targeted at THz communication applications. A summary table for SiGe HBT transmitters and receivers intended for THz communication is presented in Table 20.1. It compiles SiGe HBT communication transmitters and receivers operating beyond 100 GHz, a good number of results have been reported recently [33–41]. The transmitter output power has reached beyond 8 dBm (at 240 GHz) [37], while the receiver noise figure as low as 10.7 dB (at 190 GHz, DSB (double side band)) has been reported [36]. Based on these transmitters and receivers, combined with various modulation

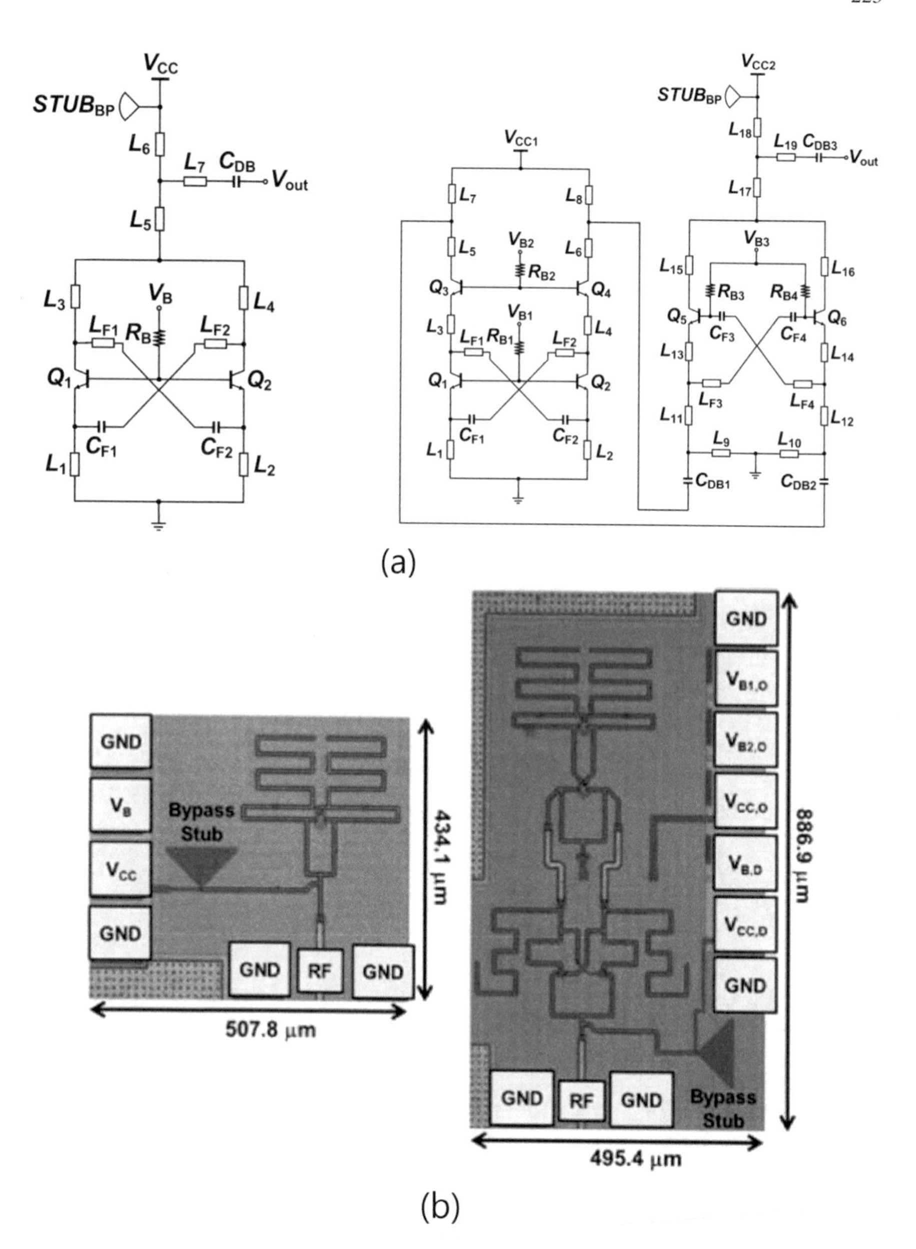

Fig. 20.9 (**a**) Circuit schematics of two 320-GHz signal sources (**b**) Chip photo of the fabricated signal sources. (left: push-push oscillator; right: integrated oscillator-doubler) (© 2015 IEEE [26])

schemes ranging over BPSK, QPSK, and QAM, data rate up to 100 Gbps have been reported with carrier frequency [40].

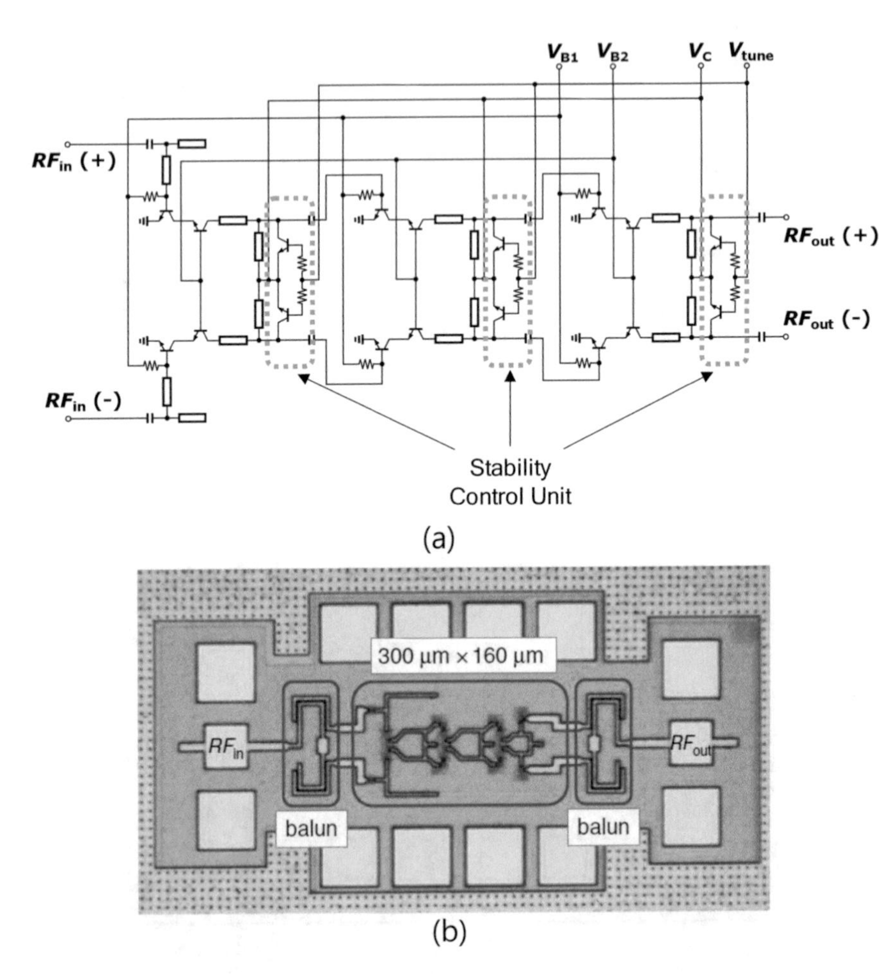

Fig. 20.10 (**a**) Circuit schematic of the 260-GHz SiGe HBT amplifier (**b**) Chip photo of the fabricated amplifier. (© The Institution of Engineering and Technology 2017 [31])

Other examples for a system-level integration include arrays for radiation and beam steering. Below are recent examples for such arrays in a 130-nm SiGe technology.

20.5.2.1 Scalable and Wideband Radiator Arrays

In spite of using performance-enhancing techniques for harmonic generation, there are still limitations on the maximum harmonic power and the number of power-boosted transistors that can be connected to a single antenna for THz signal generation and radiation. This notion entails the use of antenna array at these frequencies. Standing and traveling wave properties have been used in many

Table 20.1 SiGe HBT transmitters and receivers operating beyond 100 GHz for communication

	Frequency (GHz)	TX P_{out} (dBm)	TX EIRP (dBm)	Rx Gain (dB)	RX NF (dB)	Modulation	Data Rate (Gbps)
[33]							
[34]	210–275	−4.4	21.9	10.5	15	QPSK/64QAM	2.7
[35]	240	6	32.4	11	15	BPSK/QPSK	25
[36]	190	−6	−1	47	10.7	BPSK	50
[37]	220–260	8.3	-	24	12	32QAM	90
[38]	240	−0.8	13.2	32	13.4	BPSK	25
[39]	305–375	-	18.4	28	19.7*	–	–
[40]	220–255	5.5		8	14	16QAM	100
[41]	240	−8.2	5.8	44	13.4	16QAM	3.9

All data are based on 130-nm SiGe BiCMOS technology *calculated or simulated

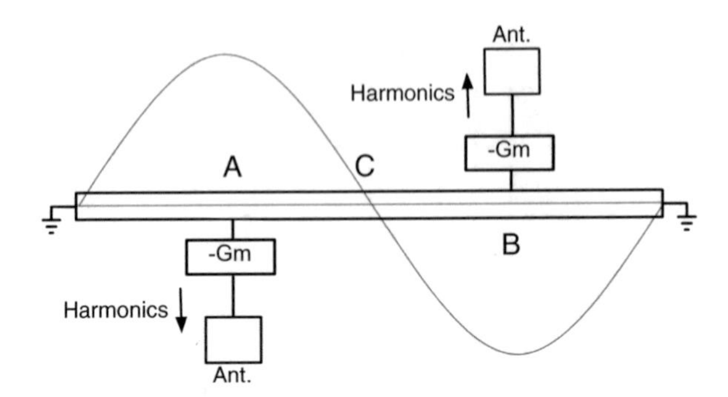

Fig. 20.11 Standing wave oscillator with negative resistance block and harmonic generation and radiation

electronic components such as oscillators, amplifiers, and frequency multipliers [42, 43]. Standing wave has also been used in clock distribution for digital circuits to reduce the clock skew [44]. Figure 20.11 shows a conceptual standing wave oscillator. This oscillator has a transmission line (TL) with electric length of one wavelength (λ) and both ends are shorted to ground. Active components ($-$Gm, negative resistance) are used to sustain oscillation. The negative resistance is usually placed at the voltage amplitude peaks of the standing wave, e.g., points A and B in Fig. 20.11. With a perfect short circuit termination at the transmission line's (TL) end, the signals at points A and B have the same amplitude and are 180° out of phase. Now if we can extract the even harmonics of the fundamental oscillation signal at points A and B and radiate it through antennas, the even harmonics would be in phase and their powers will add up constructively in space. Furthermore, all the odd harmonics will be canceled out, resulting in a cleaner spectrum.

The negative resistance block at points A and B will be used to generate the harmonic power and radiate it through an antenna. The length of the TL should be increased so that more negative resistance blocks and more antennas can be added. The problem here is the longer TL length will result in lower frequency modes as illustrated in Fig. 20.12b. For example, cross-coupled transistor pair is quite broadband and has even higher power at lower frequencies. So if the TL size is doubled in a cross-coupled based standing wave oscillator, the frequency of operation will indeed be reduced to half of the original frequency as shown in Fig. 20.12b. Therefore, it would not be possible to scale up the antenna array for higher radiated power.

A solution to this problem is a narrowband negative resistance that can only cancel the loss at and around the frequency of interest. As a result all the undesired modes are suppressed. By using this block in a standing wave oscillator, we can increase the length of the TL and the oscillation frequency stays the same as shown in Fig. 20.12c. Using this structure, we can easily scale the number of active components without changing the frequency and without using a lossy and power hungry coupling circuitry. This would result in a truly scalable structure for

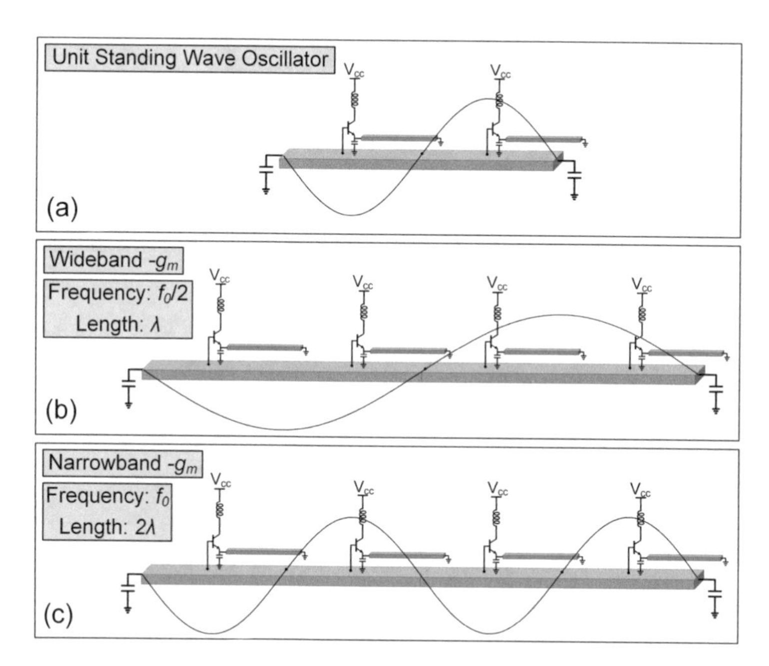

Fig. 20.12 Maintaining the operation mode of (**a**) unit standing wave oscillator (SWO) cell while scaling the array, (**b**) replicating the unit cell when there is negative transconductance at $f_o/2$ will force the oscillation to operate in an undesired mode, whereas (**c**) narrowband negative transconductance forces the oscillation to remain at the original frequency

high power radiation at THz frequencies. Moreover, this structure can be viewed as multiple coupled standing wave oscillators and therefore the same phase noise enhancement as in coupled oscillators applies here.

As a proof of concept, a 4-element radiator array at 340 GHz was implemented in a 130-nm BiCMOS process [45]. Figure 20.13 illustrates the structure. Two separate continuous standing waves are formed on the top and bottom horizontal TL with 4λ electrical lengths. These two waves oscillate out-of-phase, creating an electrical ground in the middle plane and producing in-phase fourth harmonics for coherent radiation in the far field. The fourth-harmonic power is fed to and radiated by integrated patch antennas. The TL at the collector and the capacitor at the emitter of the transistor are used to create a narrowband negative resistor at the base of the transistor. The TL at the collector is also used to extract and combine the fourth harmonic power from four adjacent transistors and feed it to a patch antenna. Rather than using lossy varactors, the base voltage is used to change the base capacitor and hence vary the frequency. The frequency can be tuned from 332.5 to 352.8 GHz which results in 5.9% tuning range at the center frequency of 343 GHz. The measured total radiated power of the array was measured to be -10.5 dBm at 342 GHz with maximum 5.5 dB variation across the tuning range. The measured and simulated radiation patterns of the array in the azimuth and elevation planes are shown in Fig. 20.14. The half-power beam widths are 30° and 80° in the azimuth

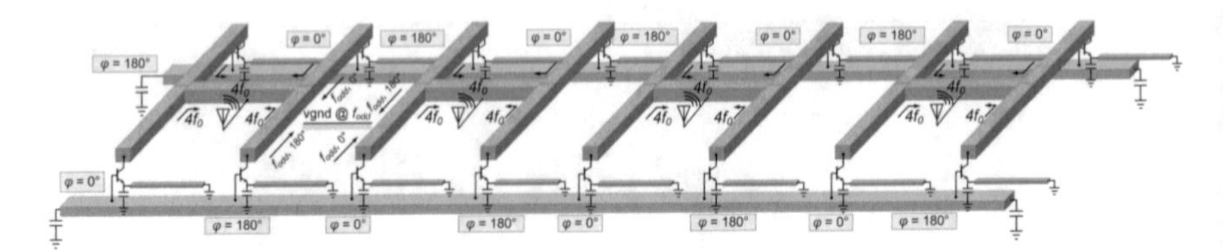

Fig. 20.13 Structure of the implemented four-element radiator array based on scaling the standing wave by replication of an oscillator unit cell

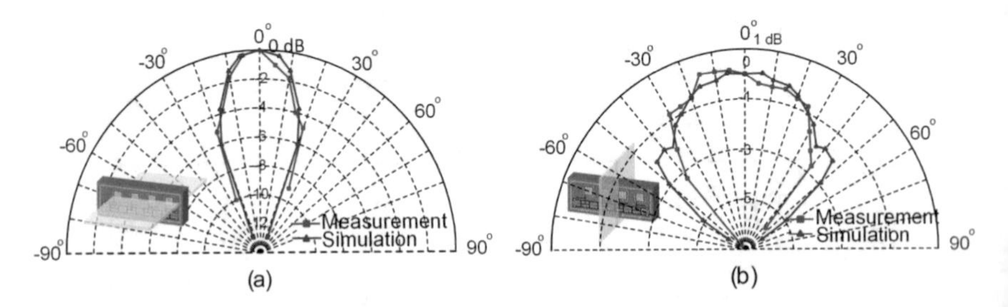

Fig. 20.14 Measured radiation patterns of the 340 GHz 1 × 4 array in (**a**) azimuth and (**b**) elevation planes

and elevation planes, respectively. This results in a measured maximum directivity of 11.3 dBi.

20.5.2.2 Wide-Angle Steering and Wideband Phased Arrays

As illustrated so far, standing wave oscillators have unique properties that make them ideal for coherent radiator arrays. However, if traveling wave is combined with standing wave, phase shift between transistors and hence beam steering is achievable. The circuit diagram in Fig. 20.15 shows a 2 × 2 standing/traveling wave phased array. The top and bottom rows are the familiar 2-element standing wave oscillator. The only difference is that the emitter TLs (TL_E) are connected to biasing voltages, Vst_1 to Vst_4, instead of ground. This enables an independent control over the base-emitter voltages of each element, and makes the beam steering possible. The voltage *Vfr* is the base voltage of all the transistors and is used to control the frequency of the array. The top and bottom oscillators are coupled to each other through TL_p lines and are phase/frequency locked through injection locking.

To understand beam steering in the *x*-direction consider the bottom half of one of the two-element oscillators as shown in Fig. 20.16. Beam steering in the *x*-direction is carried out by creating a voltage difference between control voltages *Vst* in a row. If $Vst_1 = Vst_2$, the balanced condition results in the presence of only a standing wave without phase difference and steering of the beam. This scenario is depicted in the ac equivalent half circuit of the row in Fig. 20.16a. However, a differential change in Vst_1 and Vst_2 (reducing Vst_1 and increasing Vst_2 with the same amount) results in

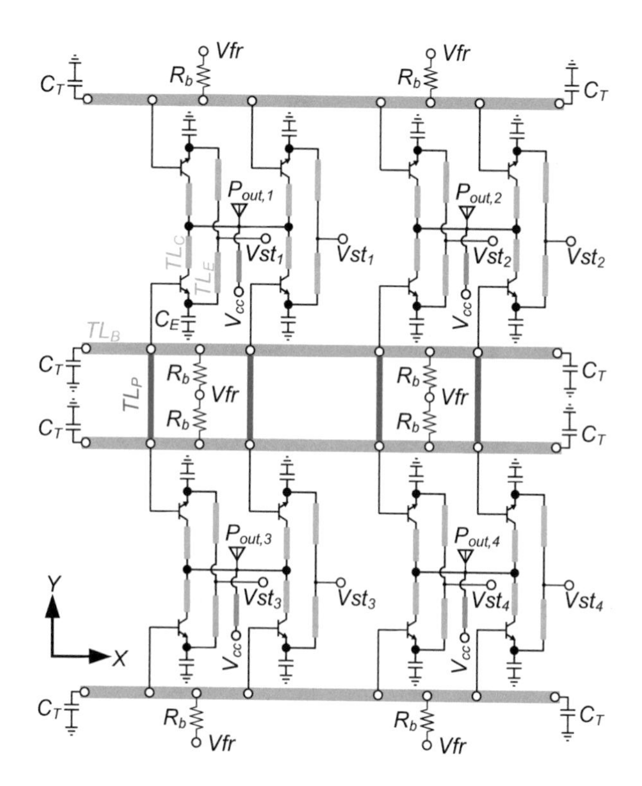

Fig. 20.15 Schematic of the implemented 2 × 2 standing/traveling wave phased array

the appearance of a residual traveling wave that creates a phase difference between transistors at fundamental frequency (f_0). This is shown in Fig. 20.16b. This phase difference is then translated and magnified to the desired fourth-harmonic radiated power and therefore steers the beam of radiation in the *x*-direction. In this case, the two transistors on the left side produce a stronger negative resistance than the others. As a result, the excess power generated by the stronger transistors is flown toward the weaker transistor to make up for its loss compensation deficiency. This flow of power can only occur through a traveling wave between the transistors. Unlike the *x*-direction, an injection mechanism is used in the y direction to create phase shift and steering in this direction.

As a proof of concept, a 2 × 2 phased array at 350 GHz was implemented in a 130-nm BiCMOS process [46, 47]. As shown in Fig. 20.17a, the frequency can be changed from 318 to 370 GHz that corresponds to 15.1% tuning range with 344-GHz center frequency. Figure 20.17b and c shows the measured beam-steering characteristic of the phased array in the *E*-plane (*x*-steering) and *H*-plane (*y*-steering), respectively, as well as the frequency change during this process. A total 128° and 53° scan angles are achieved in the *E*-plane and *H*-plane, respectively. This is a significant improvement compared to the phased arrays at this frequency

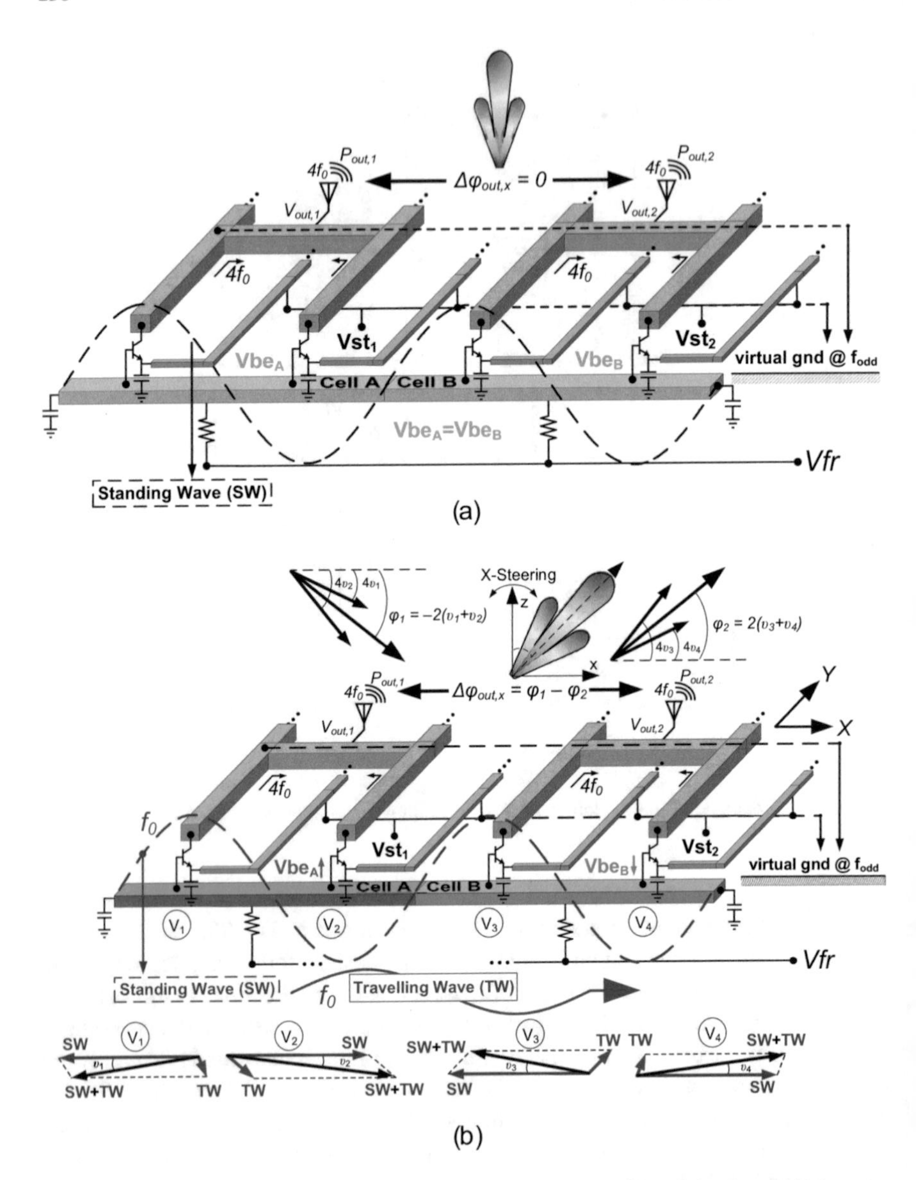

Fig. 20.16 (**a**) In-phase radiation when $Vst_1 = Vst_2$ and (**b**) generation of the phase shift between the radiated powers from the two antennas by differentially changing Vst_1 and Vst_2 and introducing a residual traveling wave

range. Figure 20.18 shows the measured radiation patterns of the phased array with and without steering in the *E*- and *H*-planes. EIRP and radiated power of the phased array have maximum values of 4.9 dBm and −6.8 dBm at 344 GHz, respectively.

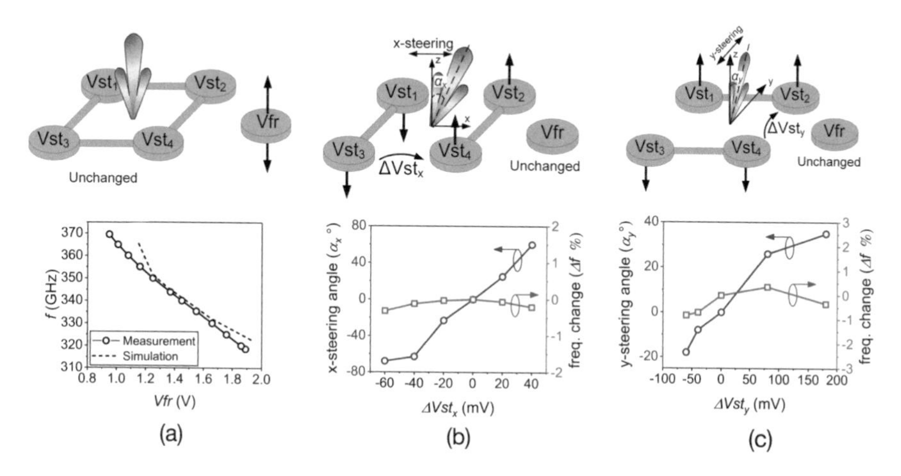

Fig. 20.17 Independent mechanisms used in measurement for (**a**) frequency tuning, (**b**) *x*-steering in the *E*-plane, and (**c**) *y*-steering in the *H*-plane, and the undesired frequency change during the process

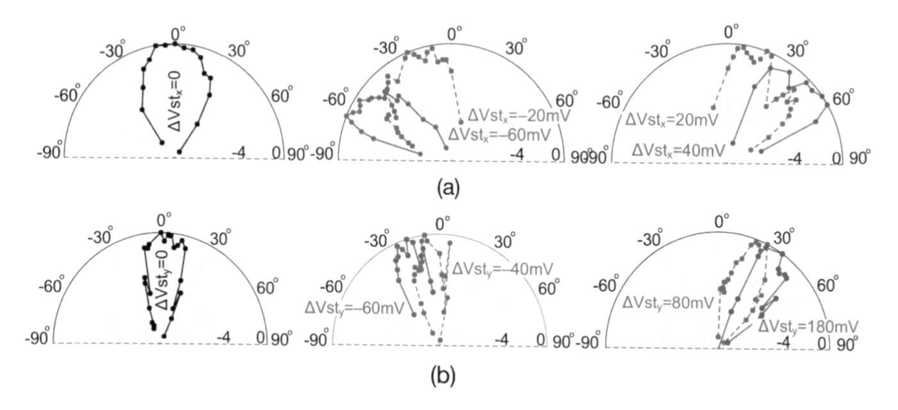

Fig. 20.18 Measured radiation patterns of the 2x2 phased array at 344 GHz in (**a**) *E*-plane and (**b**) *H*-plane

References

1. Iyer, S. S., Patton, G. L., Delage, S. S., Tiwari, S., & Stork, J. M. C. (1987). Silicon-Germanium base heterojunction bipolar transistors by molecular beam epitaxy. In *Technical digest of international electron device meeting*, pp. 874–876.
2. Kamins, T. I., et al. (1989). Small-geometry, high-performance, Si-$Si_{1-x}Ge_x$ heterojunction bipolar transistors. *IEEE Electron Device Letters, 10*(11), 503–505.
3. Crabbé, E. F., Meyerson, B. S., Stork, J. M. C., & Harame, D. L. (1993). Vertical profile optimization of very high frequency epitaxial Si and SiGe-base bipolar transistors. In *IEDM technical digest*, pp. 83–86.
4. Rieh, J.-S., et al. (2002). SiGe HBTs with cut-off frequency of 350 GHz. In *Technical digest of international electron devices meeting*, pp. 771–774.

5. Heinemann, B., et al. (2010). SiGe HBT technology with fT/fmax of 300GHz/500GHz and 2.0 ps CML gate delay. In *2010 International Electron Devices Meeting*, 6–8 Dec. 2010, pp. 30.5.1–30.5.4, https://doi.org/10.1109/IEDM.2010.5703452.
6. Heinemann, B., et al. (2016). SiGe HBT with fT/fmax of 505 GHz/720 GHz. In *2016 IEEE International Electron Devices Meeting (IEDM)*, pp. 3.1.1–3.1.4. https://doi.org/10.1109/IEDM.2016.7838335.
7. Kroemer, H. (1985). Two integral relations pertaining to the electron transport through a bipolar transistor with a nonuniform energy gap in the base region. *Solid-State Electronics, 28*(11), 1101–1103.
8. Harame, D. L., et al. (1995). Si/SiGe epitaxial-base transistors – Part I: Materials, physics, and circuits. *IEEE Transactions on Electron Devices, 42*(3), 455–468.
9. Rieh, J.-S. (2011). SiGe/Si heterojunction bipolar transistors and circuits. In *Comprehensive semiconductor science and technology*. Elsevier Inc., pp. 1–51.
10. Taur, Y., & Ning, T. H. (1998). *Fundamentals of modern VLSI devices*. Cambridge: Cambridge University Press.
11. Gosser, R. A., Foroudi, O., & Flanyak, S.(2002). New bipolar figure of merit "fo". In *Proceedings of Bipolar/BiCMOS circuits and technology meeting*, pp. 128–135.
12. Rieh, J.-S., Greenberg, D., Stricker, A., & Freeman, G. (2005). Scaling of SiGe heterojunction bipolar transistors. *Proceedings of the IEEE, 93*(9), 1522–1538.
13. Roulston, D. J. (1990). *Bipolar semiconductor devices*. New York: McGraw-Hill publishing Company, McGraw-Hill.
14. Ishibashi, T. (2001). Nonequilibrium electron transport in HBTs. *IEEE Transactions on Electron Devices, 48*(11), 2595–2605.
15. Lanzerotti, L., Stach, J. S., Hull, R., Buyuklimanli, T., & Magee, C. (1996). Suppression of boron out-diffusion in SiGe HBT's by carbon incorporation. In *Technical digest of international electron devices meeting*, pp. 249–252.
16. Osten, H. J., et al. (1997). The effect of carbon incorporation on SiGe heterobipolar transistor performance and process margin. In *Technical digest of international electron devices meeting*, pp. 803–806.
17. Sze, S. M. (1981). *Physics of semiconductor devices* (2nd ed.). New York: Wiley.
18. Rieh, J.-S., Khater, M., Jeseph, A., Freeman, G., & Ahlgren, D. (2006). Effect of collector lateral scaling on performance of high-speed SiGe HBTs with f_T > 300 GHz. *Electronics Letters, 42*(20), 1180–1181.
19. Deen, M. J., & Simoen, E. (2002). Low-frequency noise in polysilicon-emitter bipolar transistors. *IEE Proceedings of Circuits, Devices and Systems, 149*(1), 40–50.
20. Sanden, M., Marinov, O., Deen, M. J., & Ostling, M. (2002). A new model for the low-frequency noise and the noise level variation in polysilicon emitter BJTs. *IEEE Transactions on Electron Devices, 49*(3), 514–520.
21. Niu, G. (2005). Noise in SiGe HBT RF technology: Physics, modeling, and circuit implications. *Proceedings of the IEEE, 93*(9), 1583–1597.
22. Rieh, J.-S., et al. (2004). SiGe heterojunction bipolar transistors and circuits toward terahertz communication applications. *IEEE Transactions on Microwave Theory and Techniques, 52*(10), 2390–2408.
23. Pekarik, J. J., et al. (2014). A 90nm SiGe BiCMOS technology for mm-wave and high-performance analog applications. In *2014 IEEE Bipolar/BiCMOS circuits and technology meeting (BCTM)*, IEEE, pp. 92–95.
24. Liu, Q. Z., et al. (2014). SiGe HBTs in 90nm BiCMOS technology demonstrating fT/fMAX 285GHz/475GHz through simultaneous reduction of base resistance and extrinsic collector capacitance. *ECS Transactions, 64*(6), 285–294. https://doi.org/10.1149/06406.0285ecst.
25. Voinigescu, S. P., et al. (2013). A study of SiGe HBT signal sources in the 220–330-GHz range. *IEEE Journal of Solid-State Circuits, 48*(9), 2011–2021.
26. Yun, J., Yoon, D., Jung, S., Kaynak, M., Tillack, B., & Rieh, J.-S. (2015). Two 320 GHz signal sources based on SiGe HBT technology. *IEEE Microwave and Wireless Components Letters, 25*(3), 178–180. https://doi.org/10.1109/LMWC.2015.2391011.

27. Zeinolabedinzadeh, S., Song, P., Kaynak, M., Kamarei, M., Tillack, B., Cressler, J. D. (2014). Low phase noise and high output power 367 GHz and 154 GHz signal sources in 130 nm SiGe HBT technology. In *2014 IEEE MTT-S International Microwave Symposium (IMS2014)*, 1–6 June 2014, pp. 1–4, https://doi.org/10.1109/MWSYM.2014.6848559
28. Hu, Z., Kaynak, M., & Han, R. (2018). High-power radiation at 1 THz in silicon: A fully scalable Array using a multi-functional radiating mesh structure. *IEEE Journal of Solid-State Circuits, 53*(5), 1313–1327. https://doi.org/10.1109/JSSC.2017.2786682.
29. Ojefors, E., Grzyb, J., Zhao, Y., Heinemann, B., Tillack, B., & Pfeiffer, U. R. (2011). A 820GHz SiGe chipset for terahertz active imaging applications. In *IEEE international solid-state circuits conference*, pp. 224–226.
30. Aghasi, H., Cathelin, A., & Afshari, E. (2017). A 0.92-THz SiGe power radiator based on a nonlinear theory for harmonic generation. *IEEE Journal of Solid-State Circuits, 52*(2), 406–422.
31. Yoon, D., Seo, M., Song, K., Kaynak, M., Tillack, B., & Rieh, J. (2017). 260-GHz differential amplifier in SiGe heterojunction bipolar transistor technology. *Electronics Letters, 53*(3), 194–196. https://doi.org/10.1049/el.2016.3882.
32. Schmalz, K., Borngraber, J., Mao, Y., Rucker, H., & Weber, R. (2012). A 245 GHz LNA in SiGe Technology. *IEEE Microwave and Wireless Components Letters, 22*(10), 533–535. https://doi.org/10.1109/LMWC.2012.2218097.
33. Zeinolabedinzadeh, S. et al. (2014). A 314 GHz, fully-integrated SiGe transmitter and receiver with integrated antenna. In *2014 IEEE Radio Frequency Integrated Circuits Symposium*, IEEE, pp. 361–364.
34. Sarmah, N., et al. (2016). A fully integrated 240-GHz direct-conversion quadrature transmitter and receiver chipset in SiGe technology. *IEEE Transactions on Microwave Theory and Techniques, 64*(2), 562–574. https://doi.org/10.1109/TMTT.2015.2504930.
35. Sarmah, N., Vazquez, P., Grzyb, J., Foerster, W., Heinemann, B., & Pfeiffer, U. (2016). A wideband fully integrated SiGe chipset for high data rate communication at 240 GHz. In *2016 11th European Microwave Integrated Circuits Conference (EuMIC)*, IEEE, pp. 181–184.
36. Fritsche, D., Stärke, P., Carta, C., & Ellinger, F. (2017). A low-power SiGe BiCMOS 190-GHz transceiver chipset with demonstrated data rates up to 50 Gbit/s using on-chip antennas. *IEEE Transactions on Microwave Theory and Techniques, 65*(9), 3312–3323.
37. Rodríguez-Vázquez, P., J. Grzyb, Heinemann, B., & Pfeiffer, U. R. (2018). Performance evaluation of a 32-QAM 1-meter wireless link operating at 220–260 GHz with a data-rate of 90 Gbps. In *2018 Asia-Pacific Microwave Conference (APMC)*. IEEE, pp. 723–725.
38. Eissa, M. H., et al. (2018). Wideband 240-GHz transmitter and receiver in BiCMOS technology with 25-Gbit/s data rate. *IEEE Journal of Solid-State Circuits, 53*(9), 2532–2542.
39. Al-Eryani, J., Knapp, H., Kammerer, J., Aufinger, K., Li, H., & Maurer, L. (2018). Fully integrated single-chip 305–375-GHz transceiver with on-chip antennas in SiGe BiCMOS. *IEEE Transactions on Terahertz Science and Technology, 8*(3), 329–339.
40. Rodríguez-Vázquez, P., Grzyb, J., Heinemann, B., & Pfeiffer, U. R. (2019). A 16-QAM 100-Gb/s 1-M wireless link with an EVM of 17% at 230 GHz in an SiGe technology. *IEEE Microwave and Wireless Components Letters, 29*(4), 297–299.
41. Eissa, M. H., et al. (2020). Frequency interleaving IF transmitter and receiver for 240-GHz communication in SiGe: C BiCMOS. *IEEE Transactions on Microwave Theory and Techniques, 68*(1), 239–251.
42. Chen, Y.-J., & Chu, T.-S. (2013). 2-D direct-coupled standing-wave oscillator arrays. *IEEE Transactions on Microwave Theory and Techniques, 61*(12), 4472–4482.
43. Momeni, O., & Afshari, E. (2011). A broadband mm-wave and terahertz traveling-wave frequency multiplier on CMOS. *IEEE Journal of Solid-State Circuits, 46*(12), 2966–2976.
44. Sasaki, M. (2009). A high-frequency clock distribution network using inductively loaded standing-wave oscillators. *IEEE Journal of Solid-State Circuits, 44*(10), 2800–2807.
45. Jalili, H., & Momeni, O. (2017). A standing-wave architecture for scalable and wideband millimeter-wave and terahertz coherent radiator arrays. *IEEE Transactions on Microwave Theory and Techniques, 66*(3), 1597–1609.

46. Jalili, H., & Momeni, O. (2017). A 318-to-370GHz standing-wave 2D phased array in 0.13 μm BiCMOS. In *2017 IEEE International Solid-State Circuits Conference (ISSCC)*, 2017: IEEE, pp. 310–311.
47. Jalili, H., & Momeni, O. (2019). A 0.34-THz wideband wide-angle 2-D steering phased array in 0.13μm SiGe BiCMOS. *IEEE Journal of Solid-State Circuits, 54*(9), 2449–2461.

Chapter 21
Si-CMOS

Omeed Momeni and Minoru Fujishima

Abstract This chapter presents some important discussions on Si-CMOS implementations for THz applications. The technology's performance is compared with other platforms, the challenges are explained, and few important THz circuit blocks and systems are presented. In the first section, the nonlinearity mechanism in CMOS transistors and techniques to maximize harmonic generation is proposed. Moreover, a scalable oscillator array structure for high power and wideband THz radiation is presented. In the following section, a system of 300 GHz wireless transceiver technology using CMOS is introduced. This system includes many THz components including low-noise mixers, amplifiers, and power dividers.

21.1 Integrated Circuit Technologies for Signal Generation and Radiation

High power and wideband signal generation is one of the most challenging aspects of mm-wave and THz systems. Low-quality factor of the passive components including the varactors and the high operation frequency relative to the maximum oscillation frequency (f_{max}) of the active devices are the main reasons for inefficient signal generation, low signal power, and small-frequency tuning range for these systems. Compound semiconductor devices such as InP have relatively higher f_{max} and can certainly perform better in this regard. However, these devices are hard to integrate with low frequency and digital circuit blocks, have lower yields with higher number of transistors, and consequently are more costly especially for consumer electronics applications with high-quantity requirements. Silicon-based devices on the other hand are doing much better in terms of integration and cost but suffer from

---

O. Momeni
University of California, Davis, Davis, CA, USA

M. Fujishima (✉)
Hiroshima University, Hiroshima, Japan
e-mail: fuji@hiroshima-u.ac.jp

T. Kürner et al. (eds.), *THz Communications*, Springer Series in Optical Sciences 234, https://doi.org/10.1007/978-3-030-73738-2_21

low f_{max}. The fastest transistor in a BiCMOS technology has almost half the f_{max} (~600 GHz) of the fastest InP transistor (~1.2 THz) [1, 2]. The f_{max} is even worse (~400 GHz) for the fastest CMOS transistor [3, 4].

For these reasons silicon technologies are the preferred platform if they can deliver the required performances. Regardless of the technology, it is paramount to understand the operation of the device and to estimate its maximum power generation and the conditions to reach it at a specific frequency. Due to low f_{max} values in silicon technologies, low tuning range, and high phase noise in fundamental oscillators, harmonic signals are extracted at mm-wave and THz frequencies. To boost the output power beyond the maximum power of an individual transistor, on-chip and spatial power combining are employed [5–13]. Passive components such as transmission lines are used to combine the power on chip. As the number of transistors increases, the transmission lines in the power combiner become longer and the loss increases accordingly. Depending on the frequency, at some point the output power does not increase as the number of transistors increases. This is why spatial power combiners are necessary at mm-wave and THz frequencies. Multiple oscillators or frequency multipliers with radiating antennas are coupled together to achieve the same signal phase, and therefore their radiated powers add up in space [14–16]. Here the coupling circuitry is lossy, resulting in loss and lower radiated harmonic power. Moreover, in some cases the coupling dynamic is done in a loop structure, which can limit the scalability of the array. The power consumption of the coupling circuitry is also a concern and can considerably reduce the DC to RF efficiency.

In the next two subsections, we will first discuss the nonlinearity mechanism in CMOS transistors and propose signal conditions to maximize harmonic generation. We will then propose scalable oscillator array structures that eliminate the coupling circuitry to provide high power and wideband THz radiation.

21.1.1 Harmonic Generation Mechanisms and Limits in Transistors

The major nonlinearity mechanism used in small-signal regime to generate harmonic power is the first-order nonlinear relation between the gate-source voltage and the drain current in CMOS transistors. This is usually referred to as the square-law characteristic of a CMOS transistor and will result in even harmonic generation. Similarly there is an exponential relation between the base-emitter voltage and the collector current in BJT transistors. However, when the transistor node voltages deviate from small-signal approximation, the square-law or the exponential characteristics are not the major source of nonlinearity anymore. This is especially true in large signal circuit blocks such as oscillators and frequency multipliers, which are also the foremost blocks for high-frequency signal generation.

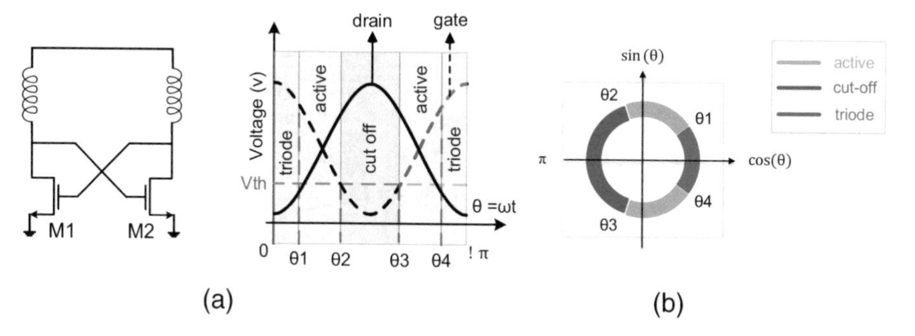

Fig. 21.1 Observing operation regions of transistors. (**a**) Circuit schematic of a conventional cross-coupled oscillator and its waveforms. (**b**) Operation map of one of the transistors in the oscillator

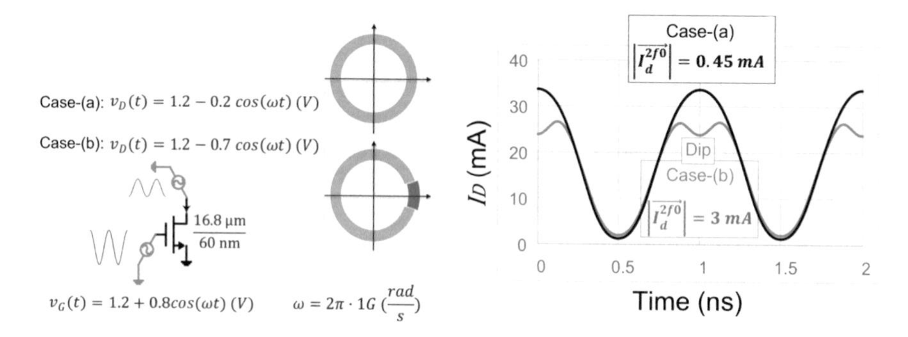

Fig. 21.2 Drain current waveform comparison. Case (**a**) the transistor stays in active. Case (**b**) effect of a short transition to triode region

Figure 21.1a shows the signal waveforms for the transistor nodes in a cross-coupled (push-push) oscillator. A transistor with large amplitudes on its terminals enters different regions of operation. For instance, in Fig. 21.1, transistors may enter all three major regions of operation. These transitions are also shown in a polar coordinate system in Fig. 21.1b which illustrates different operation regions of M1 for one complete phase of oscillation, 2π. Such a map can be determined for any oscillating transistor and is defined as its operation map [17]. In such a system, the change in the regions of operation is the main source of nonlinearity [17, 18]. For example, Fig. 21.2 shows the simulated harmonic generation of a transistor when its gate and drain node voltages are controlled to create specific operation maps. In case (a), the transistor is in the active region during the entire oscillation period. Therefore, second-harmonic generation is merely generated by the square-law characteristics of the transistor. On the other hand, in case (b), the transistor moderately enters the triode region for a short period. This figure clearly shows that the second harmonic drain current is almost seven times higher in case (b) compared to case (a).

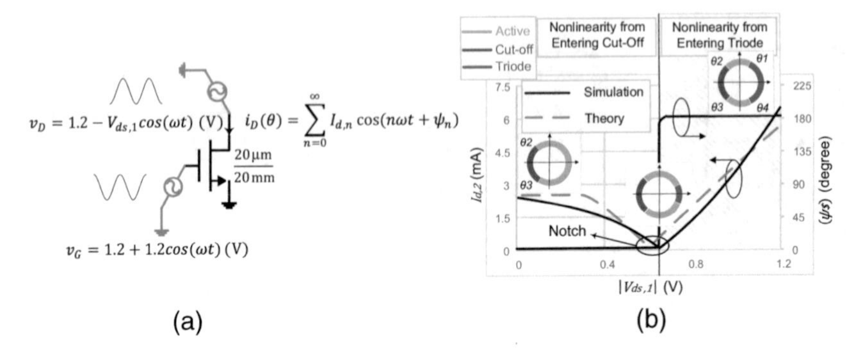

Fig. 21.3 (**a**) NFET transistor driven by 10 GHz tones at gate and drain. (**b**) Second harmonic amplitude of drain current from simulation and theory

This highlights the importance of operation map for any transistor used for harmonic generation. Different regions of operation in the operation map and the duration of each region in one period have significant effects on harmonic generation. For example, Fig. 21.3 shows the second harmonic generation of a CMOS transistor for different operation maps. The gate-source voltage is chosen so that the transistor is in the cutoff region for portion of the signal cycle. Now by changing the magnitude of the voltage swing at the drain (*Vds,1*), we can control the portion of the signal cycle during which the transistor is in the triode region. On the far left side of Fig. 21.3b, *Vds,1* is zero and transistor does not enter the triode region. Therefore almost all of the second harmonic drain current (*Id,2*) is generated by the change of operation region from active to cutoff and vice versa. As *Vds,1* increases the transistor starts to enter the triode region for portion of the signal cycle, and as a result *Id,2* starts to drop. This is because the portion of *Id,2* that is generated from entering the triode region is out-of-phase with the one generated from entering the cutoff region. At a specific *Vds,1* value, these two components become equal, and therefore, the total *Id,2* is zero (shown in Fig. 21.3 as a notch point). However, as *Vds,1* increases further, the duration of the triode region increases, and eventually *Id,2* is dominated by the portion generated from entering the triode region. This is evident from Fig. 21.3b as the phase of *Id,2* suddenly flips to 180° on the right side of the notch point.

The notch area in Fig. 21.3b should be avoided in designing blocks such as frequency multipliers or harmonic oscillators. In fact, for designing such blocks, designers should decide whether to deploy the nonlinearity from entering the cutoff region on the left side of Fig. 21.3b or using the nonlinearity from entering deep triode region on the right side of Fig. 21.3b. Traditional frequency doublers as shown in Fig. 21.4a exploit nonlinearity from the cutoff region [19]. In these circuits, drain nodes are usually virtual ground at the fundamental frequency. Thus, signal swing at drain nodes is not high and transistors barely enter triode. On the other hand, entering the triode region is inevitable in harmonic oscillators because oscillation amplitudes are fairly high. Figure 21.4b shows an example.

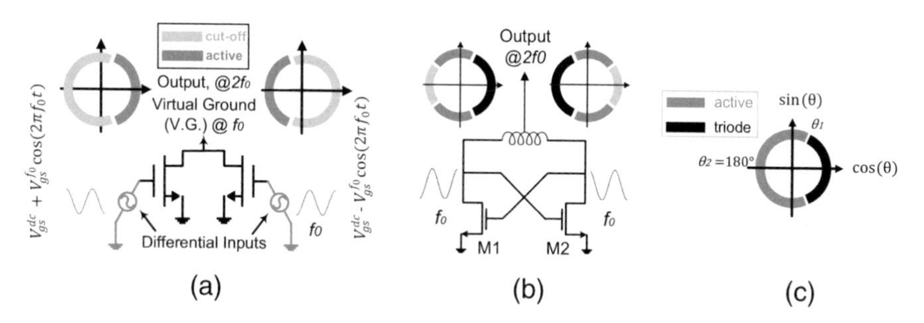

Fig. 21.4 (**a**) Traditional circuit topology for frequency doublers. (**b**) An example for a second harmonic oscillator. (**c**) Ideal operation map for maximum second harmonic power generation

Different harmonic components of *Id,2* can also be calculated as a function of the transistor's I-V curve and the operation map [17, 18]. These theoretical values are also shown in Fig. 21.3b and compare very well with the simulation results. That means we can use the same theoretical framework to calculate the harmonic power and maximize it with an optimum operation map. Since entering the triode region in harmonic oscillators is unavoidable, we can maximize the second harmonic generation by minimizing the cutoff region in the operation map and operate at the far right side of Fig. 21.3b. An example of this operation map is shown in Fig. 21.4c. Moreover, an optimum duration of the triode region (Θ_1) can be calculated to be $\sim 76°$ to maximize the second harmonic power generation. The full analysis is shown in [18].

Exploiting our new understanding for harmonic generation, we have implemented a 300 GHz voltage-controlled oscillator (VCO) with record-breaking output power (4.9 dBm) and DC-to-RF efficiency (3%) in a 65 nm CMOS process. The circuit and the measurement results are shown in Fig. 21.5.

21.1.2 Scalable and Wideband Radiator Arrays

An scalable standing wave architecture was already discussed in section 20.1.5.2 [20, 21]. This structure can be further scaled to more radiation elements simply by adding more replicas of the unit cell. However, in practice as the length of the TL increases, it becomes harder to keep all the transistors in phase. This is because the termination capacitors on the left and right side of the TL can never be a perfect ground and hence generate a reflected wave which is not exactly 180° out-of-phase with the incident wave. The effect of this phase error on in-phase radiation of the transistors is more significant as the length of the TL increases. To solve this problem and achieve scalability in a two-dimensional plane, the architecture shown in Figs. 21.6 and 21.7 is proposed. Here the unit cell is limited to a two-element standing wave oscillator. The phase relations in this unit cell remain balanced and accurate. By shifting the top part of the unit cell by one element, we can achieve

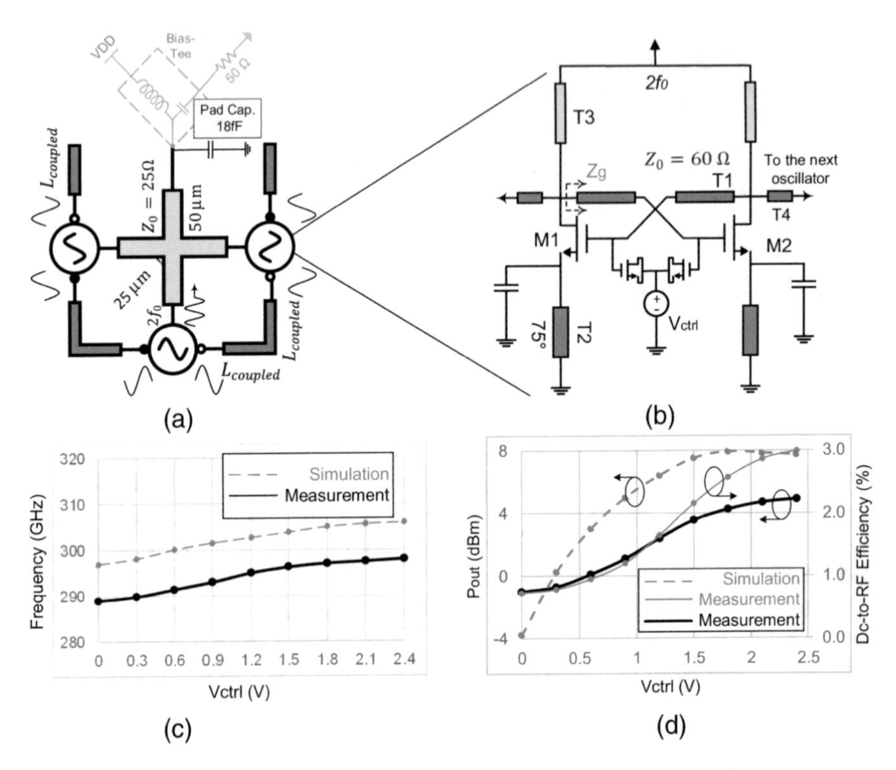

Fig. 21.5 (**a**) The implemented system of coupled oscillators, (**b**) individual oscillator schematic, (**c**) measured and simulated oscillation frequency as a function of varactor's control voltage (Vctrl), (**d**) measured and simulated output power and dc-to-RF efficiency

scalability in the x direction as shown in Fig. 21.6. In this configuration as many unit cells as needed can be used without altering the correct phase relation between the transistors. Figure 21.7 shows the technique to extent the array in the y direction. The "Hinge" cells at the end of each row is used to couple the transistors, synchronize the phase/frequency, and extend the array in two dimensions.

A 25-element radiator array at 450 GHz was implemented in a 65 nm CMOS process [22]. The negative resistor structure is similar to the one in the 1×4 array with BJT except that the sources of the transistors are directly connected to the ground. This is a more efficient way of generating negative resistors in a CMOS process. Folded slot antenna is used for each cell, and a silicon lens is implemented for an efficient backside (through the chip's substrate) radiation. Figure 21.8 shows the implemented structure and the measurement results using a lens with a radius, R_{lens}, of 5 and 12.5 mm. The frequency can be altered from 438.4 to 479.1 GHz with a 1.2 V supply which is equivalent to 8.9% tuning range. The equivalent isotropically radiated power (EIRP) of the circuit is 19.3 dBm, and the radiated power is −1.8 dBm at 448 GHz. Figure 21.8d shows a significant improvement in radiated power with respect to the state of the art.

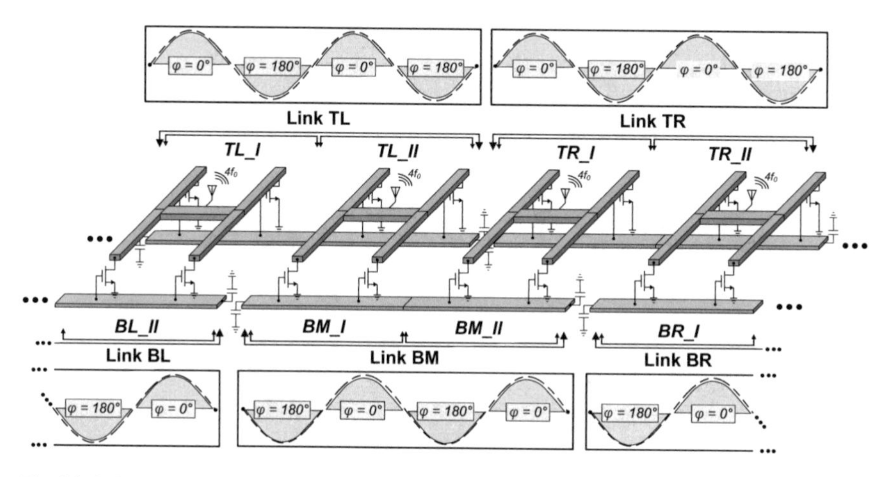

Fig. 21.6 Coupling scheme for scalable coupling in the x direction

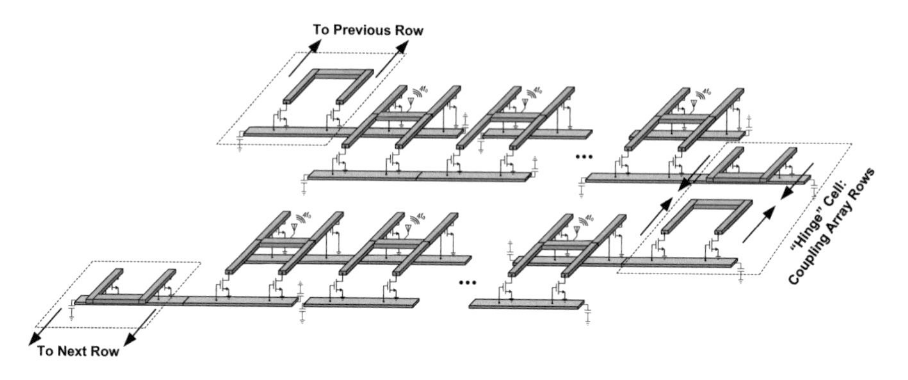

Fig. 21.7 Scaling of the array in two dimensions

21.2 300 GHz Band Communication Using Silicon Integrated Circuits

Below 275 GHz, frequencies that can be used for wireless communication are already assigned. Of these, the frequency range of 252–275 GHz can be used for wireless communication. Meanwhile, as shown in Fig. 21.9, its application to wireless communication in the frequency range of 275–450 GHz was discussed. As a result, frequencies such as 275–296 GHz were supported for use in land mobile and fixed service applications[23]. Therefore, it is expected that a continuous frequency band of 252–296 GHz can be used for wireless communication. This is the target frequency band for 300 GHz band communication.

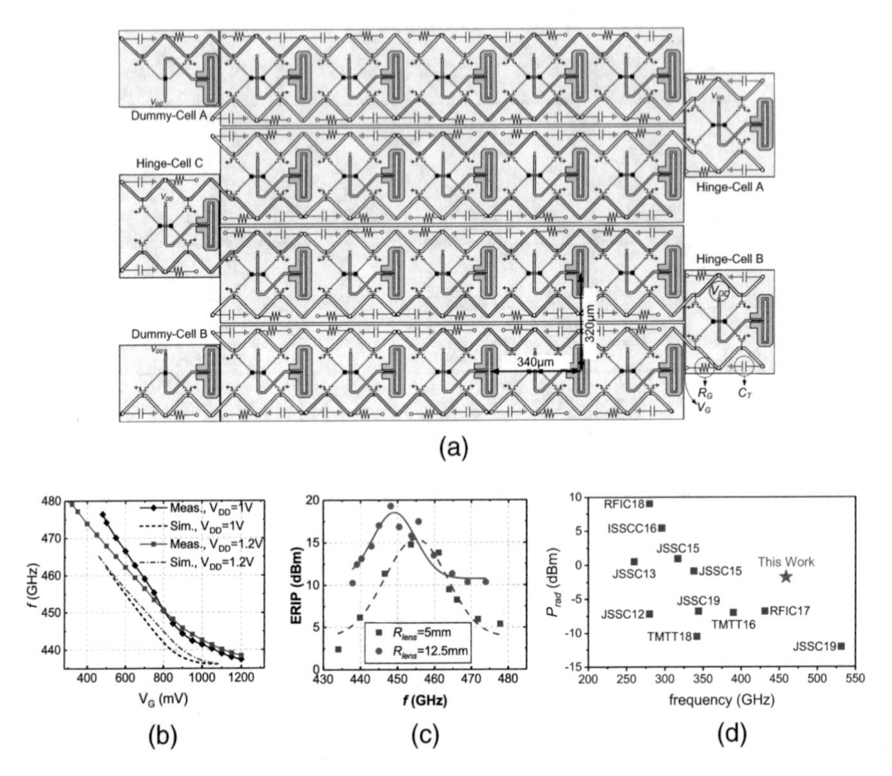

Fig. 21.8 (**a**) The full structure of the implemented 25-element array, (**b**) measured and simulated frequency of the source as a function of the gate voltage, (**c**) measured EIRP for $R_{lens} = 5$ mm and $R_{lens} = 12.5$ mm, (**d**) overview of the published radiated power levels among state of the art integrated coherent sources in silicon

Here, we will introduce the features of silicon integrated circuits while comparing them with compound semiconductor integrated circuits and introduce the technology of silicon integrated circuits that realize transceivers for 300 GHz band communication.

21.2.1 Integrated Circuit for Terahertz Communication

The frequency used in the 300 GHz band communication is 100 times higher than that of the widely used microwave communication. Generally, terahertz has a drawback that it is difficult to generate a high-power signal. Therefore, it is necessary to use a circuit that can output the necessary signal power for communication at ultra-high frequencies. In recent years, improvements in both device and circuit technology have helped solve this problem.

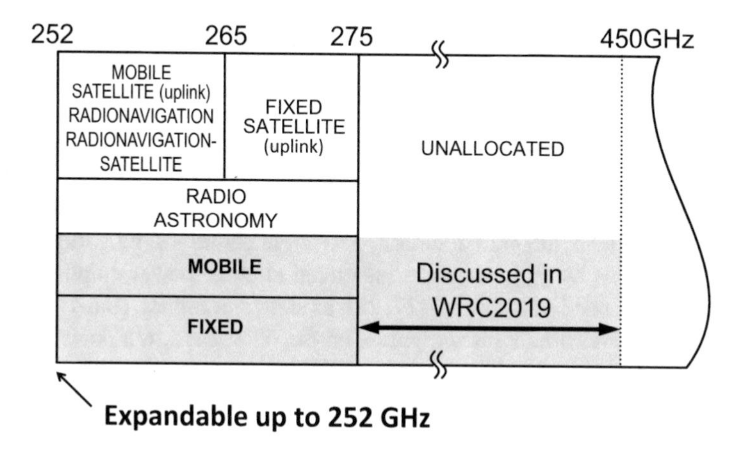

Fig. 21.9 Target frequency range for 300 GHz band wireless communication

The integrated circuits are broadly divided into compound semiconductor integrated circuits using indium phosphide (InP) and gallium arsenide (GaAs) substrates and silicon integrated circuits using silicon (Si) substrates. Silicon integrated circuits further include silicon germanium (SiGe) BiCMOS integrated circuits and CMOS integrated circuits. The difference between these integrated circuits is the maximum oscillation frequency f_{max} of the transistor and the degree of integration. The latest InP HEMT among compound semiconductor transistors has a f_{max} of 1.5 THz [24], and the SiGe bipolar transistor has a f_{max} of 720 GHz [25]. CMOS integrated circuits were considered unsuitable for terahertz communication because the high-frequency performance of silicon transistors was lower than that of compound semiconductor transistors. Among the MOSFETs in CMOS integrated circuits, the f_{max} of transistors using the SOI process is 410 GHz [26]. In an inexpensive bulk CMOS process, f_{max} is around 300 GHz. Here, wireless circuits used in mass-produced mobile phones and personal computers (PCs) are now made with cheap CMOS integrated circuits due to improvements in circuit and device technology. Advanced signal processing is essential for infrastructure devices that span satellites and their attached terminals. BiCMOS integrated circuits and CMOS integrated circuits can integrate digital circuits, but compound semiconductor integrated circuits cannot integrate digital circuits. In particular, CMOS integrated circuits, which are becoming more miniaturized than Bi-CMOS circuits, can incorporate large-scale digital circuits, so there is a possibility that a single-chip transceiver device can be realized in the future. Although the performance is inferior to that of compound semiconductors and SiGe integrated circuits, if terahertz communication can be realized with CMOS integrated circuits, the range of applications will be expanded to mobile devices. CMOS integrated circuits are important for realizing terahertz communication and advanced signal processing on a single chip at low cost.

21.2.2 300 GHz Band Transmitter Using Silicon Integrated Circuit

If the circuit technology can overcome the performance of silicon transistors, CMOS integrated circuits can be selected for the 300 GHz band transceiver. This chapter introduces the 300 GHz band wireless transmission technology using CMOS. Even for the same silicon integrated circuit, the architecture of the 300 GHz band transmitter depends on f_{max}. In SiGe integrated circuits, power amplifiers can be used for 300 GHz band transmitters [27, 28], as shown in Fig. 21.10a, as in ordinary transceivers. In CMOS integrated circuits, on the other hand, the power amplifiers normally included in transceivers are not available.

Here, a method of generating a terahertz signal using a photonics device will be introduced. It is difficult for photonics devices to directly generate low-frequency (long wavelength) signals such as terahertz. Of course, photonics devices cannot amplify terahertz signals. Therefore, a method called photomixing is used to generate the terahertz signal. In the case of photomixing, a signal in the 300 GHz band is generated by mixing two signals separated by a frequency of 300 GHz near a wavelength of 1.5 μm. On the other hand, in CMOS, it is difficult to directly generate a high-frequency (short wavelength) signal such as terahertz. There, a terahertz signal is generated by multiplying the low signal. For example, the frequency of 150 GHz is doubled, or the frequency of 100 GHz is tripled to generate the 300 GHz band.

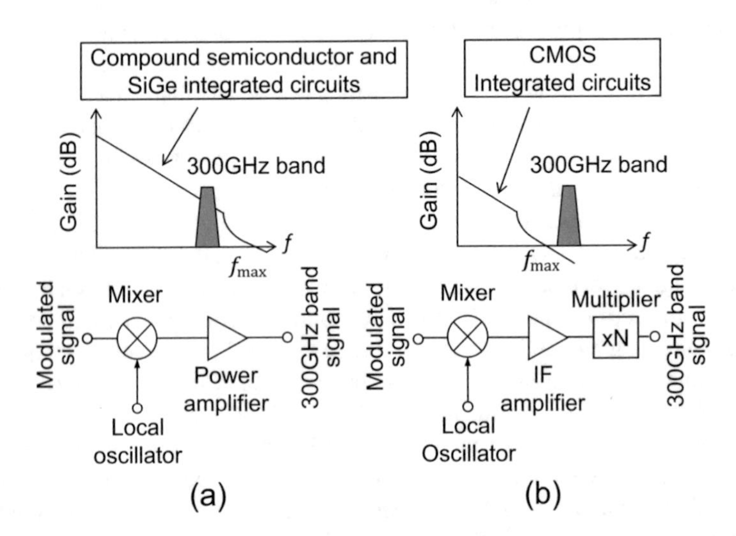

Fig. 21.10 Comparison of 300 GHz band transmitters. (**a**) Since the maximum oscillating frequency (f_{max}) of the transistors used in the most advanced compound semiconductor integrated circuits and SiGe integrated circuits exceeds 300 GHz, signals in the 300 GHz band can be amplified. (**b**) f_{max} of CMOS integrated circuit transistor is near 300 GHz or less, so power amplifier cannot be used. Therefore, a 300 GHz band signal will be generated using a multiplier

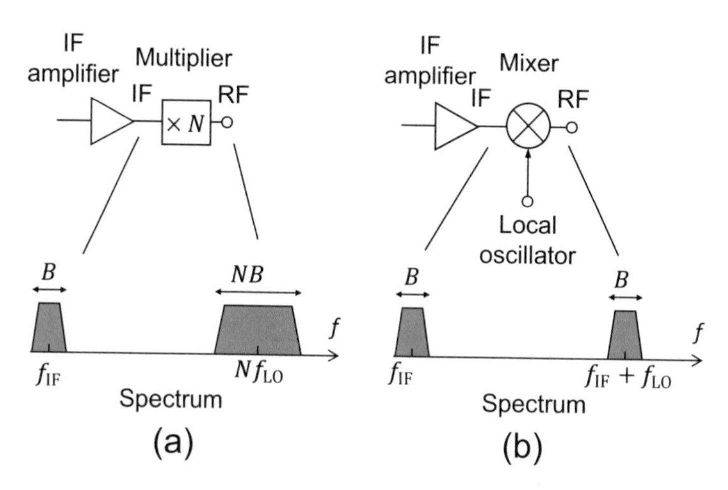

Fig. 21.11 Comparison between using a multiplier and a mixer in the final stage. (**a**) When the $\times N$ multiplier is used, not only the center frequency is multiplied by N, but the frequency band B is also multiplied by N. (**b**) When a mixer is used, the frequency shifts up (upconverted) by the local oscillation frequency f_{LO}

In order to realize a 300 GHz band transmitter with a CMOS integrated circuit, a method has been proposed in which the intermediate frequency of 80 GHz is tripled in the final stage to generate a 240 GHz signal as shown in Fig. 21.10b [29]. However, in the multiplier, as shown in Fig. 21.11a, not only the center frequency but also the frequency band is tripled, so the wideband available in the 300 GHz band cannot be fully utilized. To fully utilize the frequency band, it is necessary to use a mixer in the final stage of the transmitter that upconverts the frequency without changing the band, as shown in Fig. 21.11b. However, since the f_{max} of the MOSFET in the CMOS integrated circuit is low, the conversion gain of the mixer is 0 dB or less, and the output power cannot be increased.

If the 300 GHz band signal cannot be amplified, the output power of the mixer must be increased. In general, the output power can be increased by increasing the power of the local oscillator (LO) signal and the modulated signal that are the two mixer inputs. Of the LO signal and the modulated signal, at least the LO signal is in the millimeter wave band. In the millimeter wave band, the gain of the CMOS amplifier per stage is small, so the amplifier must be multistage. When using an intermediate frequency for the modulation signal, the amplifier must be prepared for both the LO signal and the modulation signal. In this case, not only the chip area increases, but also the power consumption increases. Here, if the mixer outputs are connected in parallel as shown in Fig. 21.12, the signal in the 300 GHz band can be increased even without a power amplifier [30, 31]. However, in order to increase the output power by connecting many mixers in parallel, many cross-wirings are created. Here, in the same frequency band, both signals can be amplified by the same amplifier by superimposing the signals. The terahertz signal is output when

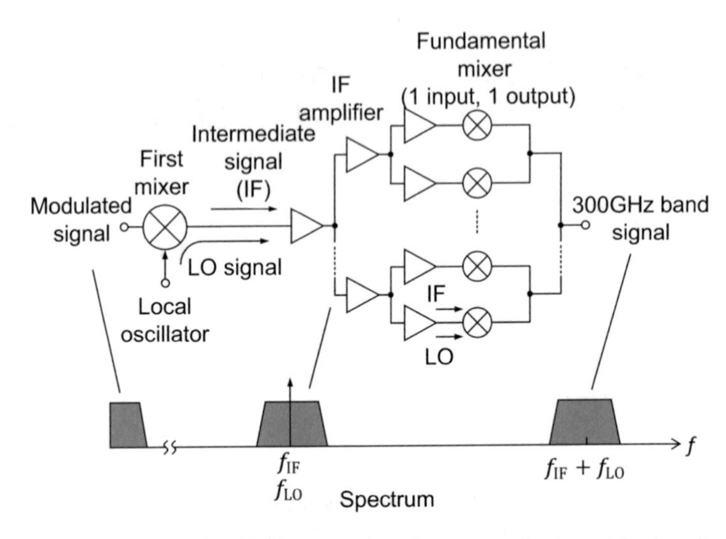

Fig. 21.12 Block diagram of a CMOS transmitter that uses a fundamental mixer in the final stage. The IF signal and LO signal are superimposed and input to the fundamental mixer at the final stage

the superimposed signal is amplified and input to the final stage mixer that utilizes the nonlinear characteristics of the transistor.

As shown in Fig. 21.12, a fundamental mixer that superimposes two input signals on one signal line is used at the final stage of the transmitter [32]. If the frequencies of the two superposed input signals are close to each other, the two signals can be amplified simultaneously by one amplifier. Then you can split and combine with a binary tree without complicating the layout. In the fundamental mixer used in the final stage, the input signal is squared as in the frequency doubler. Therefore, the sum of the superimposed IF and LO signals is squared. The expanded term contains the desired signal, which is the product of the two superimposed signals. On the other hand, in addition to the desired signal, an unwanted signal consisting of the square of the IF signal and the square of the LO signal is also generated. Since the frequencies of these desired and unwanted signals are close to each other, they cannot be removed using a bandpass filter. Therefore, as shown in Fig. 21.13, it is divided into two paths, a square mixer for inputting the $v_{IF} - v_{LO}$ signal and a square mixer for inputting the $v_{IF} + v_{LO}$ signal, and both outputs are input to the balun. Then, by extracting the differential signal from the balun, only the desired signal remains, and unnecessary signals can be canceled.

As shown in Fig. 21.13, a fundamental mixer that superimposes two input signals on one signal line is used at the final stage of the transmitter [32]. If the frequencies of the two superposed signals are close to each other, the two signals can be amplified by the same amplifier. Then, mixers can be connected in parallel by splitting and combining with a binary tree without complicating the layout. In the fundamental mixer used in the final stage, the input signal is squared as in the

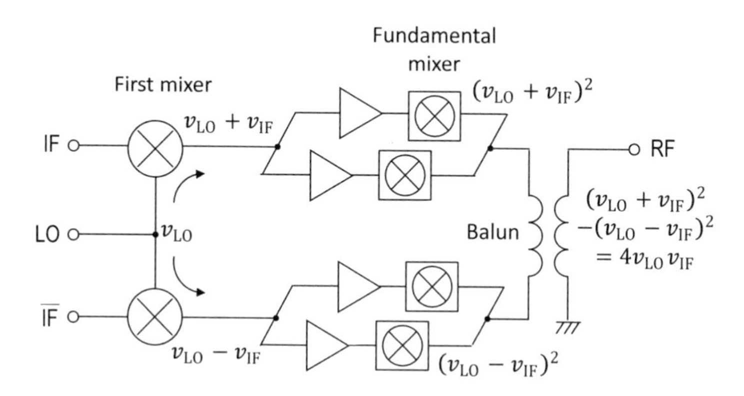

Fig. 21.13 Suppression of unnecessary signals generated by the fundamental mixer. The fundamental mixer outputs the squared IF signal and LO signal. Therefore, the unnecessary signal is canceled by using the two signal paths and extracting the difference between the output signals

frequency doubler. Therefore, the sum of the superimposed IF and LO signals is squared. The expanded term contains the desired signal ($v_{LO}v_{IF}$), which is the product of the two superimposed signals. On the other hand, in addition to the desired signal, an unwanted signal consisting of the square of the IF signal (v_{IF}^2) and the square of the LO signal (v_{LO}^2) is also generated at the same time. Since the frequencies of these desired and unwanted signals are close to each other, they cannot be removed using a bandpass filter. Therefore, as shown in Fig. 21.13, the differential IF signal ($\pm v_{IF}$ signal) is used. The differential IF signal is superimposed on the LO signal to create two ($v_{LO} + v_{IF}$) signals and ($v_{LO} - v_{IF}$) signals. These signals are amplified by the multistage amplifier and sent to the fundamental mixer. The fundamental mixer squares these signals to produce $(v_{LO} \pm v_{IF})^2$. Next,$(v_{LO} + v_{IF})^2$ and $(v_{LO} - v_{IF})^2$ are input to the balun, and only the differential mode signal is extracted. As a result, only the desired signal ($4v_{LO}v_{IF}$) remains, and unnecessary signals (v_{LO}^2, v_{IF}^2) can be canceled.

21.2.3 300 GHz Band Receiver Using Silicon Integrated Circuit

On the other hand, in the receiver, as shown in Fig. 21.14a, a receiver of SiGe integrated circuit can use a low-noise amplifier [27, 28]. Furthermore, in recent SiGe integrated circuit receivers, a technique has been reported that uses a driver amplifier operating in the 300 GHz band for the LO signal to improve the conversion gain of the fundamental mixer and to omit the low-noise amplifier. As a result, a conversion gain of +7 dB and a noise figure of 17 dB are realized while omitting

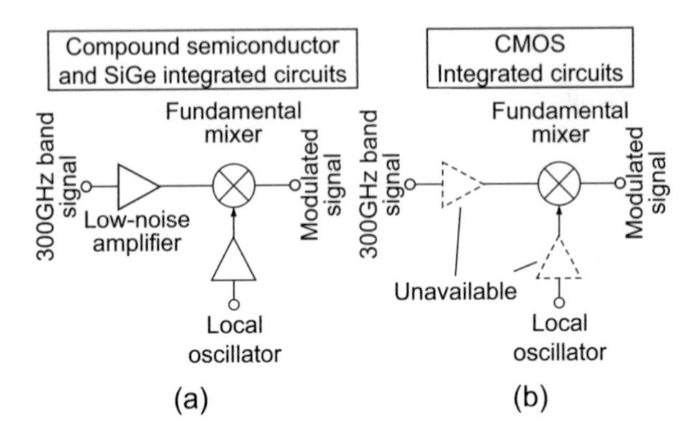

Fig. 21.14 Comparison of 300 GHz band receivers. (**a**) In the compound semiconductor integrated circuit and the SiGe integrated circuit, the received signal is amplified by the low-noise amplifier and then converted into the low-frequency signal by the down-conversion mixer, so that the noise figure of the mixer has a small influence on the receiver performance. (**b**) In a CMOS integrated circuit, since a low-noise amplifier cannot be used, the fundamental mixer is the first stage of the receiver, which has a large effect on the performance of the receiver. If a fundamental mixer with a small conversion loss is used, it cannot be amplified at the frequency of the local oscillation signal

the low-noise amplifier [33]. In a CMOS integrated circuit receiver, a low-noise amplifier cannot be used as shown in Fig. 21.14b. In this case, the down-conversion mixer must be used in the first stage. In general, harmonic mixers are often used in ultra-high frequency down-conversion mixers to lower the LO frequency. However, using a harmonic mixer with high conversion loss and high noise figure degrades the receiver performance. It is necessary to use a fundamental wave mixer in order to suppress the performance deterioration of the receiver by using the mixer in the first stage. In the fundamental mixer, the LO signal frequency is the same as the input 300 GHz band. On the other hand, a LO signal of sufficiently high power is required to suppress conversion loss, but a CMOS integrated circuit cannot make an LO signal driver. Therefore, in the 300 GHz band CMOS receiver, as shown in Fig. 21.15, the LO signal with the power required for the fundamental mixer is realized by using parallel connection as in the transmitter [34, 35].

21.2.4 One-Chip Transceiver

The combination of these 300 GHz transmitter technology and 300 GHz receiver technology will eventually enable CMOS transceiver chips that target the IEEE 802.15.3d channels 49–50 and 66 frequency bands [36] as shown in Fig. 21.16. Although IEEE 802.15.3d is expected to enable single-chip transceivers with QAM modulation, this is an important technology, especially for mass-produced transceivers for the general consumer. The one-chip transceiver uses an architecture

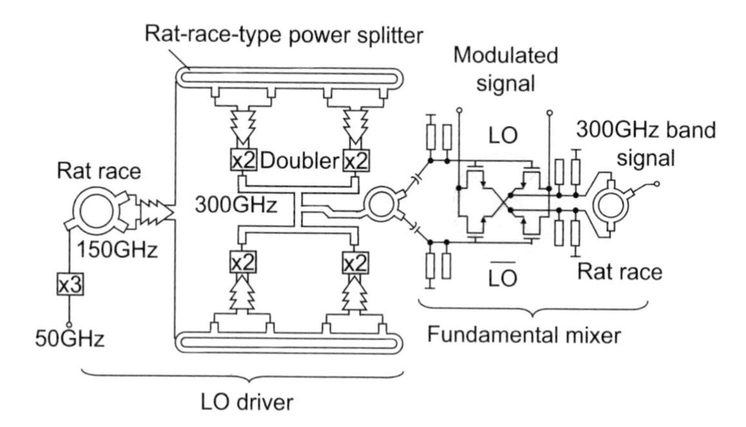

Fig. 21.15 300 GHz band CMOS receiver using the fundamental mixer in the first stage. To use a fundamental mixer with low conversion loss, the local oscillation frequency needs to be 300 GHz band. Since the 300 GHz band signal cannot be amplified by the CMOS circuit, the output of the doubler is connected in parallel to generate the 300 GHz signal of the power required for the fundamental mixer

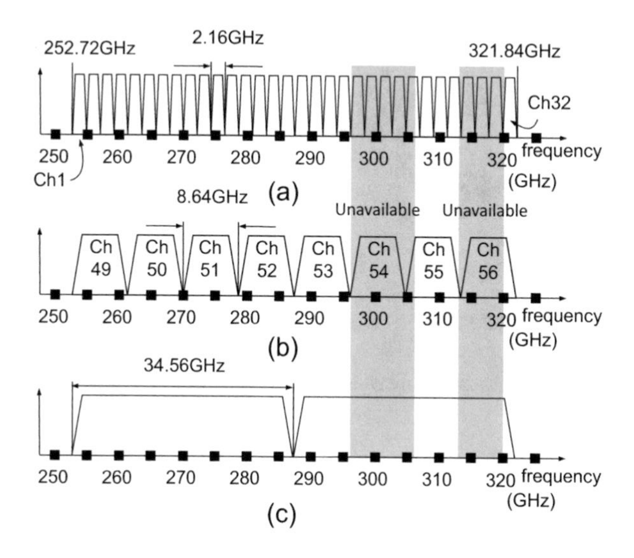

Fig. 21.16 In IEEE 802.15.3d, channels of various bands were assigned to the frequency band of 252.72 GHz to 321.84 GHz. However, it includes frequency bands that cannot be used above 296 GHz. (**a**) and (**b**) are examples of channels assigned by IEEE 802.15.3d. (**c**) Assigning a channel in the 34.56 GHz band enables connection with the wired communication interface CEI-28G

without a power amplifier for transmitter and a low-noise amplifier for receiver. A schematic diagram of the transceiver is shown in Fig. 21.17. It operates in transmit or receive mode as shown in Fig. 21.18. In either mode, the transmitter part is shared, and in receive mode it functions as a LO frequency multiplier. In transmit

Fig. 21.17 300 GHz band CMOS one-chip transceiver architecture

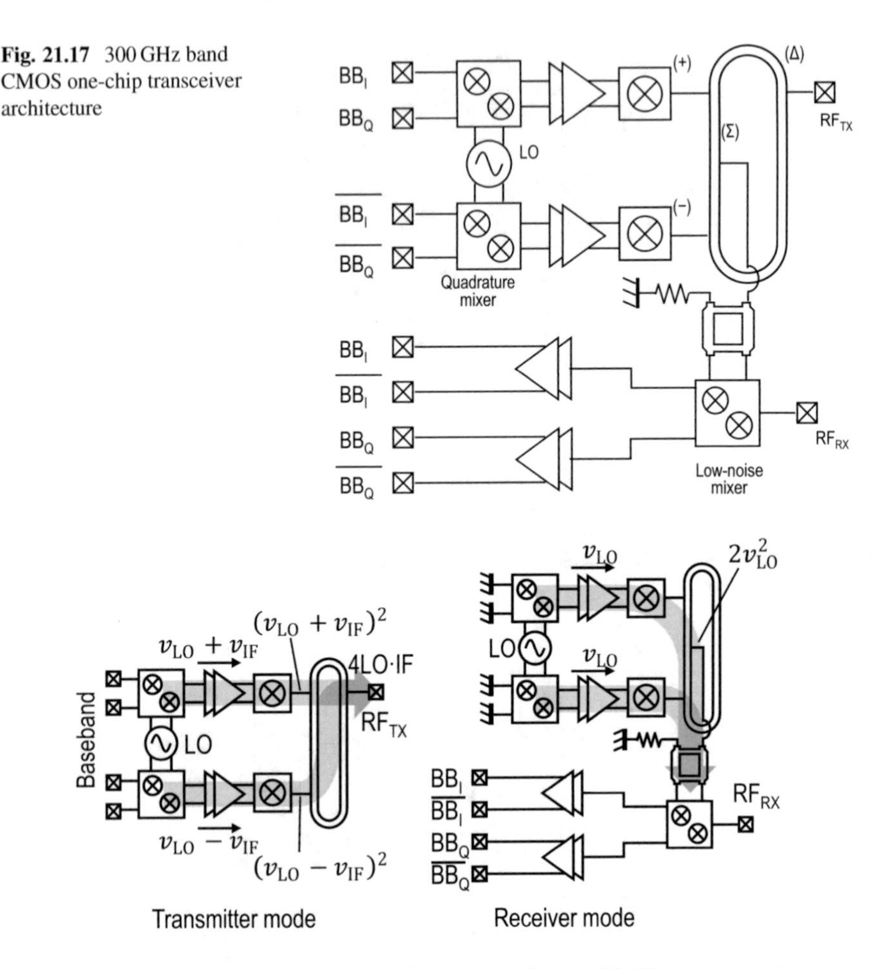

Fig. 21.18 Signal flow in transmitter mode and receiver mode in a 300 GHz band transceiver

mode, the incoming baseband signal is upconverted to IF by a quadrature mixer. The IF is superimposed with the second LO signal (~133 GHz) by the mixer. Two IF signals with different signs ($\pm v_{IF}$ signals) superimposed on the LO signal ($v_{LO} \pm v_{IF}$) are squared by the frequency doubler (fundamental mixer [31]) to generate$(v_{LO} \pm v_{IF})^2$. Next, connect $(v_{LO} + v_{IF})^2$ and $(v_{LO} - v_{IF})^2$ to the positive (+) and negative (−) ports of the double rat race circuit [31] that can input two pairs of differential signals, respectively. As a result, the desired transmitter output signal 4LOIF (about 266 GHz) is output from the differential (Δ) port. In receiver mode, it terminates the baseband input port for the transmitter, and the mixer only acts as an LO buffer. The doubler generates LO2 and supplies 2LO2 (~266 GHz) to the receiver mixer from the composite (Σ) port of the double rat race circuit. The mixer-first direct conversion receiver section consists of a low-noise mixer and baseband amplifier. A transceiver was prototyped using the 40 nm CMOS process. The size is

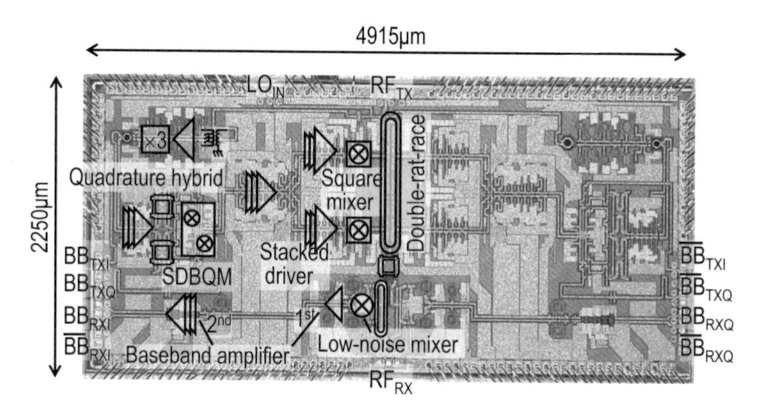

Fig. 21.19 Chip micrograph of 300 GHz CMOS transceiver

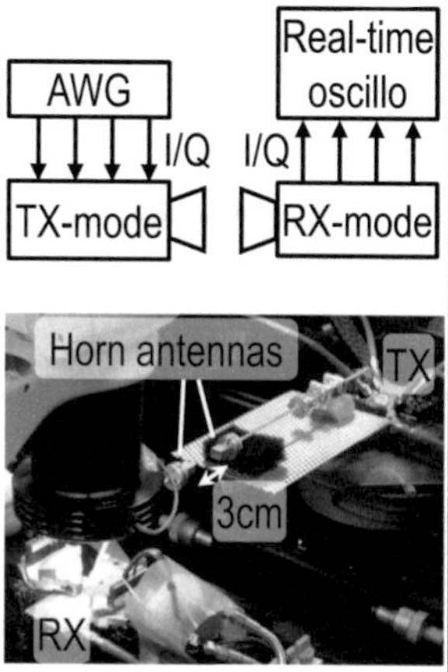

AWG
Real-time oscillo
I/Q I/Q
TX-mode
RX-mode
Horn antennas
TX
3cm
RX

Channel	Ch. 49	Ch. 50	Ch. 66
Center freq.	257.04GHz	265.68GHz	265.68GHz
Modulation	16QAM	16QAM	16QAM
Data rate	28.16Gb/s	28.16Gb/s	80Gb/s
Constellation (Equalized)		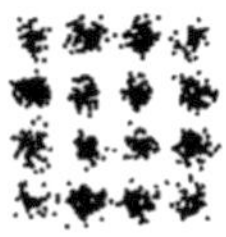	
Spectrum	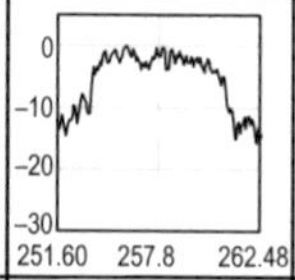 0, −10, −20, −30; 251.60 257.8 262.48	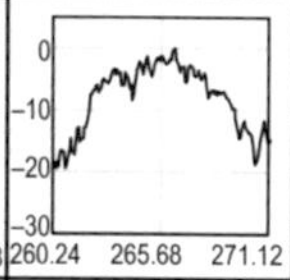 0, −10, −20, −30; 260.24 265.68 271.12	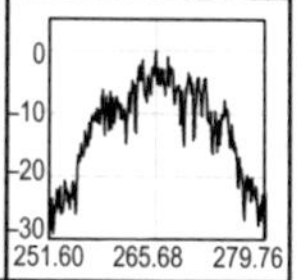 0, −10, −20, −30; 251.60 265.68 279.76
EVM	10.9%rms	11.3%rms	12.0%rms

Fig. 21.20 Measurement system for communication experiment using 300 GHz band CMOS transceiver and measurement result of constellation and output spectrum

$4.92 \times 2.25\,\text{mm}^2$, and a chip micrograph of the transceiver is shown in Fig. 21.19. Power consumption in transmitter mode and receiver mode was 890 and 897 mW, respectively. Figure 21.20 shows a photograph of the wireless communication measurement. A pair of 24 dBi horn antennas was used for wireless communication experiments. The output baseband signal from the receiver is analyzed using a real-time oscilloscope with vector signal demodulation and channel equalization. Figure 21.20 shows the frequency and bandwidth constellation and power spectrum for IEEE 802.15.3d channels 49, 50, and 66. It achieves a data transfer rate of 80 Gbit/s at 3 cm with 16QAM.

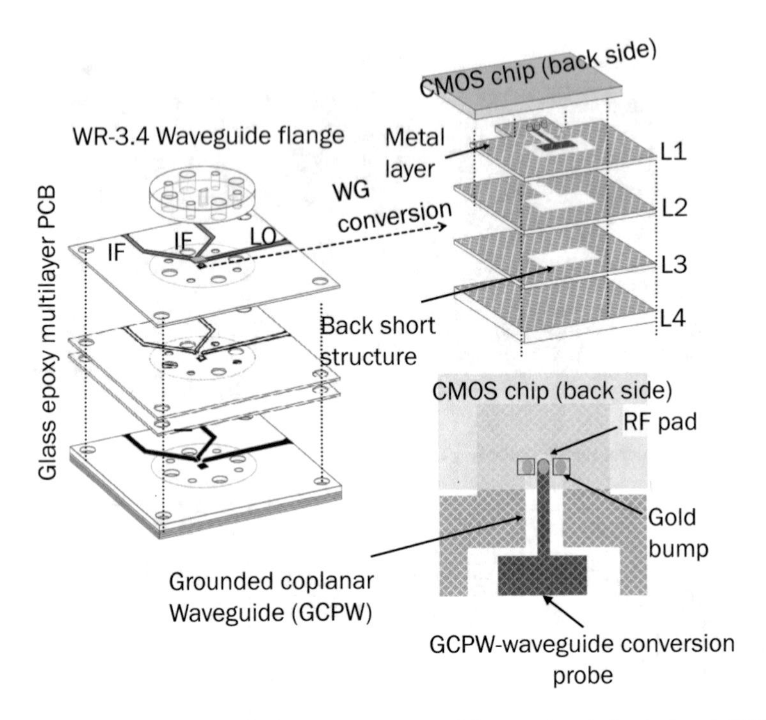

Fig. 21.21 CMOS-to-waveguide interface using a multilayer printed circuit board (PCB). A waveguide back short structure is made in the multilayer substrate

21.2.5 Module for 300 GHz Band Transceiver

A waveguide probe mounted on a wafer prober was used to measure the CMOS integrated circuit that is the core of the 300 GHz band transceiver. Although it is possible to attach an antenna to the waveguide probe, it is not possible to put the transceiver into practical use only by communication experiments with a wafer prober. The connection method between the CMOS chip and the antenna that converts the electric wave of the space and the electric signal of the circuit is also important. On-chip antennas may be used in millimeter wave band CMOS transceivers [37]. However, the on-chip antenna has a small antenna gain, and its application is limited to short-range wireless communication of a few centimeters. On the other hand, as described above, it is necessary to use an antenna with a high gain to extend the communication distance. Since the interface of a high gain antenna is generally a waveguide, an interface connecting the CMOS chip and the waveguide is required. Therefore, considering practical use, we prototyped a CMOS chip and a waveguide interface using an ordinary multilayer glass epoxy printed circuit board as shown in Fig. 21.21 [38]. The CMOS chip is flip-chip mounted on a multilayer printed circuit board. The 300 GHz band signal generated by the CMOS chip is transmitted using a short microstrip line of about 200μm on the printed circuit board. At the tip of the microstrip line, a waveguide conversion probe having a shape similar to a patch antenna is formed. The 1/4 wavelength back short structure required for waveguide conversion was formed in the multilayer substrate, and a metal waveguide flange was attached to the surface of the printed circuit board. This enables CMOS/waveguide conversion.

References

1. Heinemann, B., Rücker, H., Barth, R., Bärwolf, F., Drews, J., Fischer, G. G., Fox, A., Fursenko, O., Grabolla, T., Herzel, F., Katzer, J., Korn, J. Krüger, A., Kulse, P., Lenke, T., Lisker, M., Marschmeyer, S., Scheit, A., Schmidt, D. J., . . . , Wolansky, D. (2016). SiGe HBT with fx/fmax of 505 GHz/720 GHz. In *Proc. IEEE Int. Electron Devices Meeting (IEDM)* (pp. 3.1.1–3.1.4).
2. Deal, W. R. (2014). InP HEMT for sub-millimeter wave space applications: Status and challenges. In *Proc. and Terahertz waves (IRMMW-THz) 2014 39th Int. Conf. Infrared, Millimeter* (pp. 1–3).
3. Bameri, H., & Momeni, O. (2017). A high-gain mm-wave amplifier design: An analytical approach to power gain boosting. *IEEE Journal of Solid-State Circuits, 52*(2), 357–370.
4. Tokgoz, K. K., Abdo, I., Fujimura, T., Pang, J., Kawano, Y., Iwai, T., et al. (2019). A 273–301-GHz amplifier with 21-dB peak gain in 65-nm standard bulk CMOS. *IEEE Microwave and Wireless Components Letters, 29*(5), 342–344.
5. Han, R., & Afshari, E. (2013). A CMOS high-power broadband 260-GHz radiator array for spectroscopy," *IEEE Journal of Solid-State Circuits, 48*(12), 3090–3104.
6. Momeni, O., & Afshari, E. (2011). High power terahertz and millimeter-wave oscillator design: A systematic approach. *IEEE Journal of Solid-State Circuits, 46*(3), 583–597.
7. Wanner, R., Lachner, R., Olbrich, G., & Russer, P. (2007). A SiGe monolithically integrated 278 GHz push-push oscillator. In *IEEE MTT-S International Microwave Symposium* (pp. 333–336).
8. Sengupta, K., & Hajimiri, A. (2012). A 0.28 THz power-generation and beam-steering array in CMOS based on distributed active radiators. *IEEE Journal of Solid-State Circuits, 47*(12), 3013–3031.
9. Tousi, Y., Momeni, O., & Afshari, E. (2012). A novel CMOS high-power terahertz VCO based on coupled oscillators: Theory and implementation. *IEEE Journal of Solid-State Circuits, 47*(12), 3032–3042.
10. Seok, E., Cao, C., Shim, D., Arenas, D. J., Tanner, D. B., Hung, C. M., et al. (2008). A 410 GHz CMOS push-push oscillator with an on-chip patch antenna. In *IEEE International Solid-State Circuits Conference* (pp. 472–473).
11. Steyaert, W., & Reynaert, P. (2014). A 0.54 THz signal generator in 40 nm bulk CMOS with 22 GHz tuning range and integrated planar antenna. *IEEE Journal of Solid-State Circuits, 49*(7), 1617–1626.
12. Adnan, M., & Afshari, E. (2014). 14.8 A 247-to-263.5GHz VCO with 2.6mW peak output power and 1.14% DC-to-RF efficiency in 65nm bulk CMOS. In *2014 IEEE International Solid-State Circuits Conference Digest of Technical Papers (ISSCC)* (pp. 262–263).
13. Seo, M., Urteaga, M., Hacker, J., Young, A., Griffith, Z., Jain, V., et al. (2011). InP HBT IC technology for terahertz frequencies: Fundamental oscillators up to 0.57 THz. *IEEE Journal of Solid-State Circuits, 46*(10), 2203–2214.
14. Jameson, S., Halpern, E., & Socher, E. (2016). A 300GHz wirelessly locked 2 × 3 array radiating 5.4dBm with 5.1% DC-to-RF efficiency in 65nm CMOS. In *Proc. IEEE Int. Solid-State Circuits Conf. (ISSCC)* (pp. 348–349).
15. Jalili, H., & Momeni, O. (2017). A 318-to-370GHz standing-wave 2D phased array in 0.13 μm BiCMOS. In *Proc. IEEE Int. Solid-State Circuits Conf. (ISSCC)* (pp. 310–311).

16. Zhao, Y., Lu, H., Chen, H., Chang, Y., Huang, R., Chen, H., et al. (2015). A 0.54–0.55 THz 2×4 coherent source array with eirp of 24.4 dBm in 65nm CMOS technology. In *Proc. IEEE MTT-S Int. Microwave Symp* (pp. 1–3).
17. Kananizadeh, R., & Momeni, O. (2017). High-power and high-efficiency millimeter-wave harmonic oscillator design, exploiting harmonic positive feedback in CMOS. *IEEE Transactions on Microwave Theory and Techniques, 65*(10), 3922–3936.
18. Kananizadeh, R., & Momeni, O. (2018). Second-harmonic power generation limits in harmonic oscillators. *IEEE Journal of Solid-State Circuits, 53*(11), 3217–3231.
19. Lin, H., & Rebeiz, G. M. (2014). A 135–160 GHz balanced frequency doubler in 45 nm CMOS with 3.5 dBm peak power. In *Proc. IEEE MTT-S Int. Microwave Symp. (IMS2014)* (pp. 1–4).
20. Jalili, H., & Momeni, O. (2018). A standing-wave architecture for scalable and wideband millimeter-wave and terahertz coherent radiator arrays. *IEEE Transactions on Microwave Theory and Techniques, 66*(3), 1597–1609 (2018)
21. Jalili, H., & Momeni, O. (2019). A 0.34-THz wideband wide-angle 2-D steering phased array in 0.13-μ m SiGe BiCMOS. *IEEE Journal of Solid-State Circuits, 54*(9), 2449–2461.
22. Jalili, H., & Momeni, O. (2020). A 0.46-THz 25-element scalable and wideband radiator array with optimized lens integration in 65-nm CMOS. *IEEE Journal of Solid-State Circuits, 55*(9), 2387–2400.
23. Sharing and compatibility studies between land-mobile, fixed and passive services in the frequency range 275–450 GHz. *Report ITU-R SM.2450-0*, June 2019.
24. Deal, W.R., Leong, K., Yoshida, W., Zamora, A., & Mei, X.B. (2016). InP HEMT integrated circuits operating above 1,000 GHz. In *2016 IEEE International Electron Devices Meeting (IEDM)*, San Francisco (pp. 29.1.1–29.1.4). https://doi.org/10.1109/IEDM.2016.7838502
25. Heinemann, B., Rücker, H., Barth, R., Bärwolf, F., Drews, J., Fischer, G. G., et al. (2016). SiGe HBT with f_T/f_{max} of 505 GHz/720 GHz," *2016 IEEE International Electron Devices Meeting (IEDM)*, San Francisco (pp. 3.1.1–3.1.4). https://doi.org/10.1109/IEDM.2016.7838335
26. Diverse RF Semiconductor Technologies Are Driving the 5G Rollout. *Global Foundries*, July 2019.
27. Kim, S., Yun, J., Yoon, D., Kim, M., Rieh, J.-S., Urteaga, M., et al. (2015). 300 GHz integrated heterodyne receiver and transmitter with on-chip fundamental local oscillator and mixers. *IEEE Transactions on Terahertz Science and Technology, 5*(1), 92–101. https://doi.org/10.1109/TTHZ.2014.2364454
28. Sarmah, N., Grzyb, J., Statnikov, K., Malz, S., Vazquez, P. R., Föerster, W., et al. (2016). A fully integrated 240-GHz direct-conversion quadrature transmitter and receiver chipset in SiGe technology. *IEEE Transactions on Microwave Theory and Techniques, 64*(2), 562–574. https://doi.org/10.1109/TMTT.2015.2504930
29. Kang, S., Thyagarajan, S. V., & Niknejad, A. M. (2015). A 240 GHz fully integrated wideband qpsk transmitter in 65 nm CMOS. *IEEE Journal of Solid-State Circuits, 50*(10), 2256–2267. https://doi.org/10.1109/JSSC.2015.2467179
30. Katayama, K., Takano, K., Amakawa, S., Hara, S., Kasamatsu, A., Mizuno, K., et al. (2016). A 300 GHz CMOS transmitter with 32-QAM 17.5 Gb/s/ch capability over six channels. *IEEE Journal of Solid-State Circuits, 51*(12), 3037–3048. https://doi.org/10.1109/JSSC.2016.2602223
31. Takano, S., Amakawa, K., Katayama, S., Hara, R., Dong, A., Kasamatsu, I., et al. (2017). 17.9 A 105Gb/s 300 GHz CMOS transmitter. In *2017 IEEE International Solid-State Circuits Conference (ISSCC)*, San Francisco (pp. 308–309). https://doi.org/10.1109/ISSCC.2017.7870384
32. IEEE Standard for High Data Rate Wireless Multi-Media Networks, Amendment 2: 100 Gb/sWireless Switched Point-to-Point Physical Layer, IEEE Computer Society sponsored by the LAN/MAN Standards Committee. https://standards.ieee.org/standard/802_15_3d-2017.html
33. Vazquez, P. R., Grzyb, J., Sarmah, N., Heinemann, B., & Pfeiffer, U. R. (2017). A 219–266 GHz fully-integrated direct-conversion IQ receiver module in a SiGe HBT technology. *2017 12th European Microwave Integrated Circuits Conference (EuMIC)*, Nuremberg (pp. 261–264). https://doi.org/10.23919/EuMIC.2017.8230709

34. Hara, S., Katayama, K., Takano, K., Dong, R., Watanabe, I., Sekine, N., et al. (2017). A 416-mW 32-Gbit/s 300-GHz CMOS receiver. In *2017 IEEE International Symposium on Radio-Frequency Integration Technology (RFIT)*, Seoul (pp. 65–67). https://doi.org/10.1109/RFIT.2017.8048291
35. Hara, S., Katayama, K., Takano, K., Dong, R., Watanabe, I., Sekine, N., et al. (2017). A 32 Gbit/s 16QAM CMOS receiver in 300 GHz band. In *2017 IEEE MTT-S International Microwave Symposium (IMS)*, Honololu (pp. 1703–1706). https://doi.org/10.1109/MWSYM.2017.8058969
36. Lee, S., Dong, R., Yoshida, T., Amakawa, S., Hara, S., Kasamatsu, A., et al. (2019). 9.5 An 80 Gb/s 300 GHz-band single-chip CMOS transceiver. In *2019 IEEE International Solid- State Circuits Conference—(ISSCC)*, San Francisco (pp. 170–172). https://doi.org/10.1109/ISSCC.2019.8662314
37. Park, J., Kang, S., Thyagarajan, S., Alon, E., & Niknejad, A. (2012). A 260 GHz fully integrated CMOS transceiver for wireless chip-to-chip communication. In *2012 Symposium on VLSI Circuits (VLSIC)*, Honolulu, (pp. 48–49). https://doi.org/10.1109/VLSIC.2012.6243783
38. Takano, K., Katayama, K., Hara, S., Dong, R., Mizuno, K., Takahashi, K., Kasamatsu, A., Yoshida, T., Amakawa, S., & Fujishima, M. (2018). 300-GHz CMOS transmitter module with built-in waveguide transition on a multilayered glass epoxy PCB. In *2018 IEEE Radio and Wireless Symposium (RWS)*, Anaheim (pp. 154–156). https://doi.org/10.1109/RWS.2018.8304972

Part VI
Transceiver Technologies 2: III-V based Electronics

Chapter 22
III-V HBT

Ho-Jin Song

Abstract In this chapter, the recent state-of-the art III–V THz electronics for THz communications will be reviewed. In the early 2000s, abrupt advances in HBT technologies enabled us to reach THz frequencies and implement various fundamental functional blocks including amplifiers, mixers, oscillators, and phase-locked loop circuits. In addition, several full transceiver integrated circuits have been demonstrated for simple THz data transmission experiments.

22.1 Introduction

The idea of the heterojunction bipolar transistor (HBT) concept was first explained by Herbert Kroemer in 1957, who was the 2000 Nobel Prize winner in physics. The outstanding high-frequency performance of HBTs was realized by:

- hole injection suppression from the base into the emitter to increase the DC current gain, β and
- increased transport speed across the base by reducing the base transit time, τ_B

Since the first experimental demonstration in AlGaAs/GaAs system in 1980, various technical improvements have been made globally, in particular by means of changing material systems and aggressive scaling of transistor geometries. In general, HBTs provide a higher current density, a higher breakdown voltage at a given current gain, and a lower flicker noise than high electron mobility transistors (HEMTs), which is another strong candidate for terahertz applications, and thus are very advantageous for high-power amplifiers, mixers, and low-phase noise signal sources [1, 2]. In addition, excellent threshold uniformity and a high intrinsic transconductance enable highly integrated analog circuits, such as wireless communication transceivers at the lower end of the THz frequency. Recently, highly

H.-J. Song (✉)
POSTECH, Pohang, Republic of Korea
e-mail: hojin.song@postech.ac.kr

T. Kürner et al. (eds.), *THz Communications*, Springer Series in Optical Sciences 234, https://doi.org/10.1007/978-3-030-73738-2_22

scaled indium phosphide (InP) HBT technologies have been demonstrated with maximum oscillation frequencies f_{max} of 1 THz or greater [3].

22.2 Functional THz Integrated Circuits

22.2.1 *Power Amplifiers*

One of the biggest obstacles to using THz frequencies has been generating a high-power signal to maintain a sufficiently high signal-to-noise ratio or make the THz radiation travel substantial distances. From this perspective, power amplifiers (PAs) have been the most challenging and attractive devices.

Owing to the high radio-frequency (RF) figures of merit (f_t and f_{max}) with good breakdown voltage, state-of-the-art InP HBTs can achieve outstanding power densities. Because the bias currents of the InP HBT PA will be much higher, careful consideration of the DC bias network is required. Through novel design and layout, common emitter (CE) and cascode PA cells have been developed to overcome the high-current distribution challenges. In general, a CE topology provides good gain and stable operation at high current densities. However, as the frequency increases, the maximum available gain or maximum stable gain (MAG/MSG) drops significantly. In the case of state-of-the-art 250-nm InP HBTs [2] (f_t/f_{max} = 350/700 GHz), the CE topologies are expected to provide power densities of 0.97, 0.73, 0.45, and 0.13 W/mm at 90, 140, 220, and 330 GHz, respectively, which implies little usefulness of the CE topologies for PA at 330 GHz or higher. Conversely, a common base (CB) amplifier exhibits significantly more gain and achievable power density above 220 GHz of approximately $>$12 dB and 0.71 W/mm, respectively. More than 5-dB gain and 0.18-W/mm power density are expected even at 450 GHz.

To increase the output power from PAs, 4-, 8-, and 16-way combined HBTs have been demonstrated at the circuit or module level at THz frequencies. For 2^n corporate power combination, the losses from the on-chip combiner and path length between PA cells degrade combining efficiency. Once these overall losses reach 1.5 dB or more, spatial power combination must be considered for higher output power. In general, up to a 16-way PA-cell on-chip combination is likely feasible and has been demonstrated [4], as shown in Fig. 22.1a.

An increase in output power from a single PA MMIC is important because of the challenges associated with the power combination of multiple MMICs in a waveguide assembly even at low THz frequencies. To ensure a linear power increase associated with the number of MMICs in the module, PAs need to have similar gain and phase characteristics to maintain high RF isolation between PA chips and waveguide blocks. Build-to-build variations generate chip-to-chip phase mismatch. In addition, as the number of PAs increases, the DC bias complexity increases, as shown in Fig. 22.1b [5].

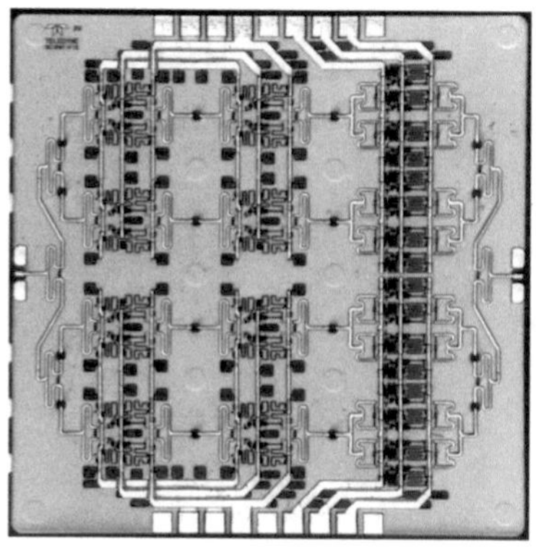

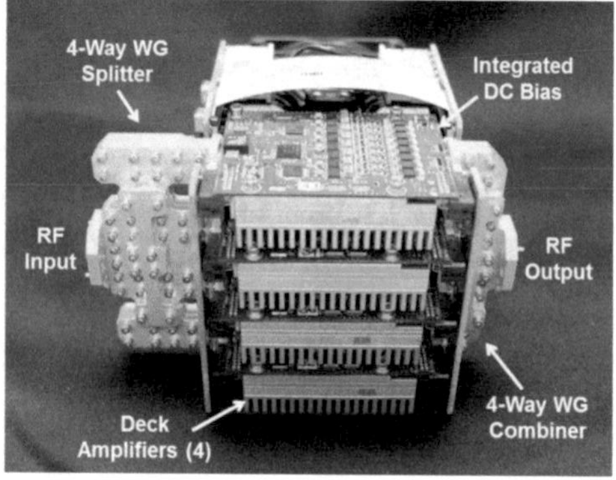

(a) (b)

Fig. 22.1 (**a**) 16-cell InP HBT SSPA [5], (**b**) 32-MMIC SSPA module [9]

Table 22.1 Summary of state-of-the-art InP HBT PAs above 150 GHz

Ref.	Circuit/module	Frequency	P_{out} (mW)	Gain (dB)	PAE (%)
[6]	3-stage	191–244	80–50	22–20	4.5–2.8
[7]	3-stage	191–244	139–75	20–14.5	–
[4]	3-stage	200–235	221–125	15.6–13.2	4–2
[8]	2-stage	205–235	112–62	6–4	5.1–2.2
[9]	32-MMIC module	230	710	>25	–
[5]	16-MMIC module	200–230	820–380	–	–

A few selected state-of-the-art HBTs PAs covering the G- and H-bands are summarized in Table 22.1. At the MMIC level, InP HBT-based PAs have demonstrated the highest output power level of 200 mW or more at approximately 200 GHz. At the module level, a 32-PA combined module has exhibited 710 mW at 230 GHz based on three-stage 4-PA cell solid-state power amplifiers (SSPAs). In [9], 823 mW at 216 GHz was reported with a 16-PA module using two-stage, 8-PA cell SSPAs. These results have radically improved upon state-of-the-art output power by a factor of 3–4 for a single PA module. As the InP HBT devices are optimized for further high power densities, the power from the HBT PAs will exceed 1 W in the near future.

22.2.2 *Signal Sources*

Signals for THz radio systems can be generated using on-chip oscillators or multiplier chains. Oscillator circuits can be designed for increased signal power at harmonics, enabling RF power generation at close to or beyond f_{max}, but fundamental operation would be advantageous for simpler and power efficient

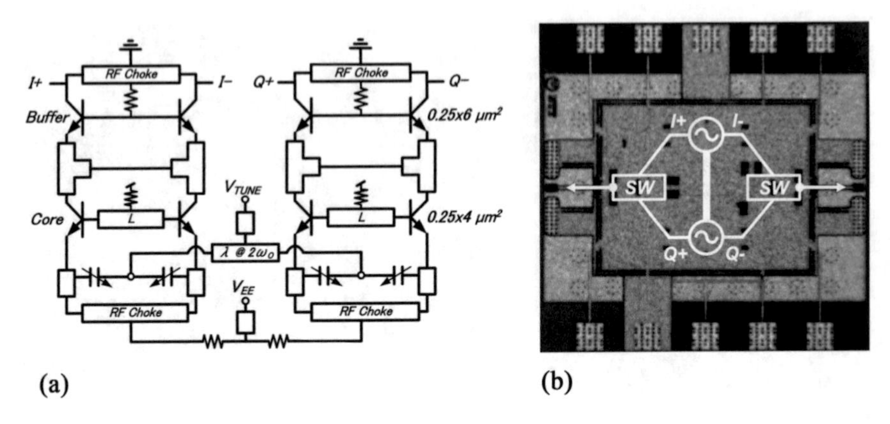

Fig. 22.2 (**a**) Schematic of quadrature output voltage controlled oscillator and (**b**) photo of fabricated MMIC [10]

with no need for filtering circuits to suppress subharmonics. Figure 22.2 shows a circuit schematic and chip photograph of a quadrature oscillator design. The differential topology is advantageous for integration in fully differential transceiver architectures, improving common-mode noise rejection and LO leakage evolve from handling only the critical signaling cancellation. In addition, quadrature outputs are essential in communication applications for generating or recovering quadrature phase modulation signals such as quadrature phase shift keying (QPSK) or quadrature amplification modulation. In general, HBT-based voltage-controlled oscillators (VCOs) utilize the HBT base-collector junction as a varactor diode. Although the varactors enable tuning of the oscillating frequency and phase, the tuning range is very limited because of the large fixed parasitics associated with the core HBTs at THz frequencies. From this point of view, multiplier chains would be more practical for various THz applications requiring a large frequency tuning range. The quadrature output voltage controlled oscillator (QVCO) in 250-nm InP HBTs provided an excellent phase balance of less than $\pm 12°$ at 325 GHz with an output power and power consumption of approximately -5 dBm and 90 mW, respectively, but the quadrature operation range was limited to approximately 1% fractional bandwidth [10].

Phase-locked loop (PLL) circuits have been demonstrated at above 200 GHz along with the VCOs with state-of-the-art HBT technologies. Static or dynamic frequency dividers, which are the key PLL components, have been implemented based on either conventional flip-flop-based digital circuits [11] or the inductive-loaded regenerative topology at up to 529 GHz in 130-nm HBTs [12]. Considering that the maximum operating frequency of the frequency divider is commonly limited by half of ft, phase-sensitive radio systems such as radar, sensing, and high-throughput communications operating at low THz frequencies will be well supported by the InP HBT technologies.

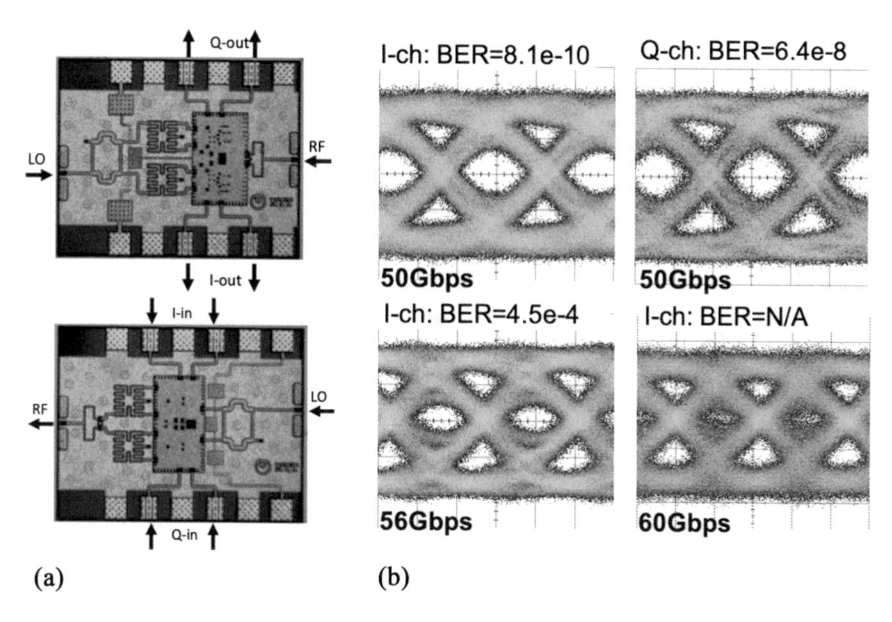

Fig. 22.3 (**a**) Photos of fabricated QPSK modulator and demodulator MMICs and (**b**) measured eye diagram at various data rate from back-to-back on-chip test [13]

22.2.3 Modulator and Demodulator [13]

Utilizing 700-GHz f_{max} InP HBT technologies, direct quadrature modulator and demodulator MMICs for future terahertz communications at 300 GHz have been experimentally demonstrated (see Fig. 22.3a for the fabricated MMICs). To modulate and demodulate signals, half-Gilbert cell mixers, which provide balanced signaling and moderate performance in conversion efficiency with a simple circuit configuration, were employed. To maintain the modulator and demodulator balance performance, passive baluns and couplers were implemented with thin-film microstrip lines, which exhibit less insertion loss than inverted microstrip lines (IMSLs), while the active mixers are based on IMSLs for short interconnections and less inductive parasitics. The half-Gilbert-cell mixers have a sufficiently wide operational bandwidth for high-throughput communications of more than 10% at 300 GHz. For QPSK operation, quadrature signal generation is necessary, which commonly relies on passive 90° couplers. Although the wavelength of THz waves is as short as 1 mm or less, it is too long to make integrated 90° couplers in MMICs. According to the static modulator constellation, an imbalance is expected to be less than approximately $\angle$ 0.6-dB $\pm$ 4°. An on-chip back-to-back experiment was conducted at up to 60 Gbps, and the 50-Gbps operation was verified with a low bit error rate on the order of 10^{-8} or less. The results demonstrate that the QPSK modulation scheme can be applied to double the data rate at terahertz frequencies.

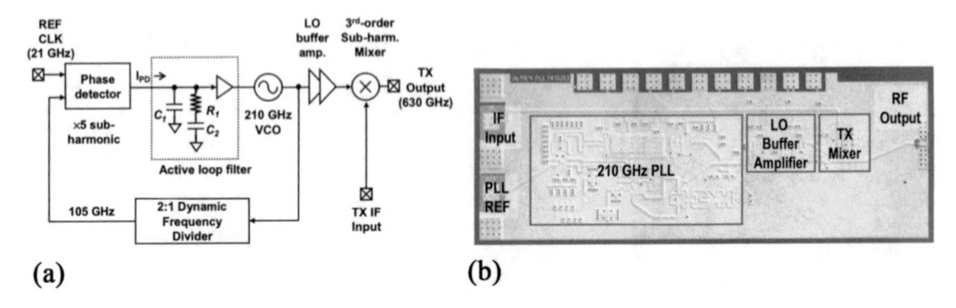

Fig. 22.4 (**a**) Schematic block diagram of 630-GHz Tx and (**b**) photo of fabricated MMIC [14]

22.3 THz Transceiver ICs

With the building block circuits such as PAs, amplifiers, mixers, and signal sources, integrated THz transmitter and receiver circuits have been realized. In this section, a few typical Tx and Rx are reviewed.

22.3.1 *Transmitter MMICs [14]*

A 630 GHz transmitter (Tx) consisting of a 210-GHz PLL, LO buffer amplifier, and up-converting mixer has been experimentally demonstrated in state-of-the-art 130 nm InP HBT technology [14]. Because of the differential topology, the Tx circuit is fairly insensitive to common-mode impedance associated with bias circuits and via inductances and tolerant to modeling uncertainties and errors. Figure 22.4 shows the schematic block diagram and photo of the fabricated Tx. To ensure the phase-locked operation, a 210-GHz PLL was integrated in the Tx. The VCO output is followed by a regenerative 2:1 dynamic frequency divider. Instead of traditional resistive or transimpedance-stage loading, inductive loading helps the PLL to operate at higher frequencies. Despite the dynamic divider architecture, a wide operating bandwidth of 40 GHz or more at 210 GHz has been obtained. The power consumption is also quite reasonable, that is, approximately 90 mW for the PLL block. A simple 18-dB resistive coupler was used to tap the VCO output for the PLL loop. The phase comparison was performed at a reasonably low frequency, a fifth-order subharmonic frequency. The phase detector was estimated to provide sufficiently high detection sensitivity even with a −20-dBm loop signal at 105 GHz.

The third-order subharmonic up-converting mixer was driven by the VCO with the two-stage cascode LO buffer amplifier. The double-balanced mixer is followed by a CB output stage that exhibits higher achievable gain than a CS configuration at THz frequencies. Capacitances at the intermediate nodes in the Gilbert cell were neutralized with transmission lines that improved the mixer conversion gain and stability. Approximately 200 mW of DC power was consumed by the mixer for a

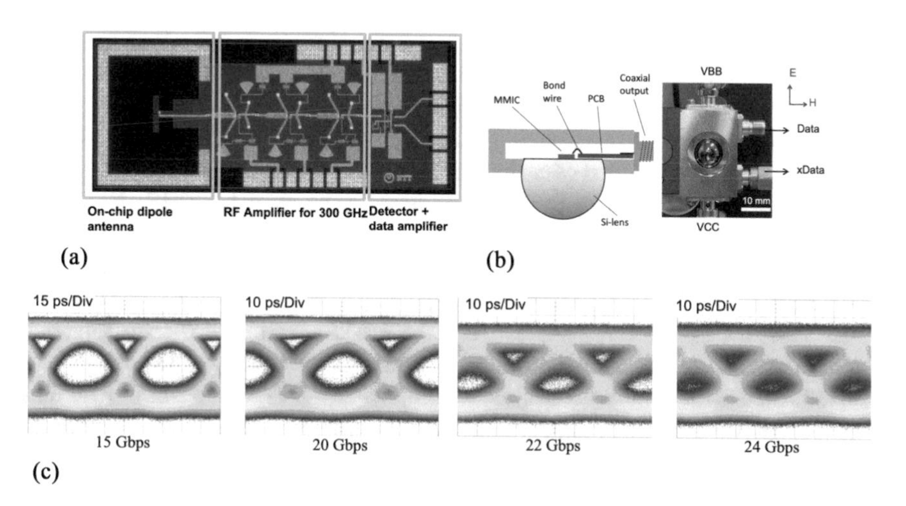

Fig. 22.5 (**a**) Photo of the fabricated receiver IC, (**b**) quasi-optical package with a silicon lens, and (**c**) measured eye diagram at 30-cm link [15]

conversion gain of approximately −20 to −25 dB at 600–640 GHz at 0 dBm of LO power, with a wide IF bandwidth of 30 GHz or more.

Because of the lack of an integrated PA, the measured transmitter saturated output was between −30 and −33 dBm after de-embedding losses from the measurement setups. It is worth noting that the energy efficiency of the very high-power PA presented in previous sections is quite poor compared to those operating in mm-wave frequencies. Thus, a large amount of DC power is converted into heat, causing thermal issues for large-scale integrated circuits such as a transmitter.

22.3.2 Receiver MMICs [15]

A fully integrated amplitude-shift keying (ASK) receiver MMIC operating at 300 GHz for future terahertz communications was experimentally demonstrated in a 250-nm InP HBT technology. Figure 22.5a shows the fabricated receiver IC. As shown, all necessary components such as a receiving dipole antenna, high-gain RF amplifier, and envelope detector for demodulating the ASK signal and output differential data amplifier were monolithically integrated in a $1.0 \times 2.5\,\text{mm}^2$ area. In this work, the on-chip antenna was designed to cooperate with a silicon lens on the back side of the receiver MMIC to compensate for the small on-chip antenna gain (see Fig. 22.5b). In addition, to ensure reliable and stable operation, a thin-film microstrip line (MSL) was employed as an on-chip transmission line and distributed circuit components, which is expected to suppress crosstalk between the on-chip antenna and the RF amplifier through the substrate and silicon lens. However, one drawback of using thick-film MSLs in high-frequency ICs is that the signal lines and transistors are on different metal layers; therefore, it is difficult to avoid the use of inductive interlayer vias. In other words, the parasitic inductance due to the interlayer vias should be carefully considered during the design. In this work, the parasitic inductance for a single via was extracted to be approximately 3–7 pH, depending on the layout and geometry of the via.

Because the peak $(f)_{max}$ of the HBTs used in the receiver MMIC is approximately 700 GHz, it is difficult to ensure a reasonable power gain at 300 GHz. Otherwise, the cascode configuration supplied by a single 3.3-V supply provides more than 10-dB power gain. A single-stage amplifier provided an approximately 8-dB gain, an 11-dB noise figure over a 15% bandwidth, and 0-dBm saturated output power at 300 GHz. An envelope detector for recovering ASK signals was implemented with a single series diode utilizing the HBT emitter-base junctions and a shunt MMI capacitor, followed by a differential transimpedance amplifier to convert the extracted data current to the final output voltage. The packaged receiver module with the silicon lens provided approximately 24-dBi beam directivity. The measured RF and baseband bandwidths were approximately 30 and 15 GHz, respectively, when a single bias of 3.3 V and total current of approximately 86 mA were applied. With the receiver module, simple wireless data transmission was conducted for up to 24 Gbps in the 300-GHz band. At 12.5 Gbps, error-free data transmission (bit error rate $<10^9$) over 0.3 m was achieved with a transmission power of −16 dBm and a 25-dBi transmitting antenna. With −10-dBm transmission power, the measured Q-factors of the received eye patterns were larger than 6 for up to 20 Gbps, which implies that the bit error rate will be less than 10^9.

It is worth noting that the antenna or radiators can be integrated into the MMICs or RFICs owing to the short wavelength of THz waves, which allows us to employ a quasi-optical THz package [16]. Figure 22.6 shows one example of a compact THz lens package for the receiver MMIC above [15]. Because there is no interconnection for high-frequency signals in package bodies, the assembly is not very sensitive to manufacturing tolerances or circuit parasitics. The 300-GHz receiver MMIC for wireless communications was mounted in the LTCC package with a flip-chip bonding technique for DC and data signal readouts, and the silicon lens was placed on the backside of the MMIC and fixed with a non-conductive adhesive epoxy. The performance of the compact module was evaluated in a data transmission experiment, and no degradation was observed in the bit error rate and the Q-factor of the recovered eye diagram.

22.4 Summary

The recent progress in highly scaled InP HBTs demonstrating maximum oscillation frequencies (f_{max}) of >1 THz clearly shows a path for achieving various THz applications with numerous functional circuits and integrated transceivers operating at up to 600 GHz. The circuit operation was extended to the lower end of the terahertz (THz) frequency band. The use of InP HBTs offers high RF output power

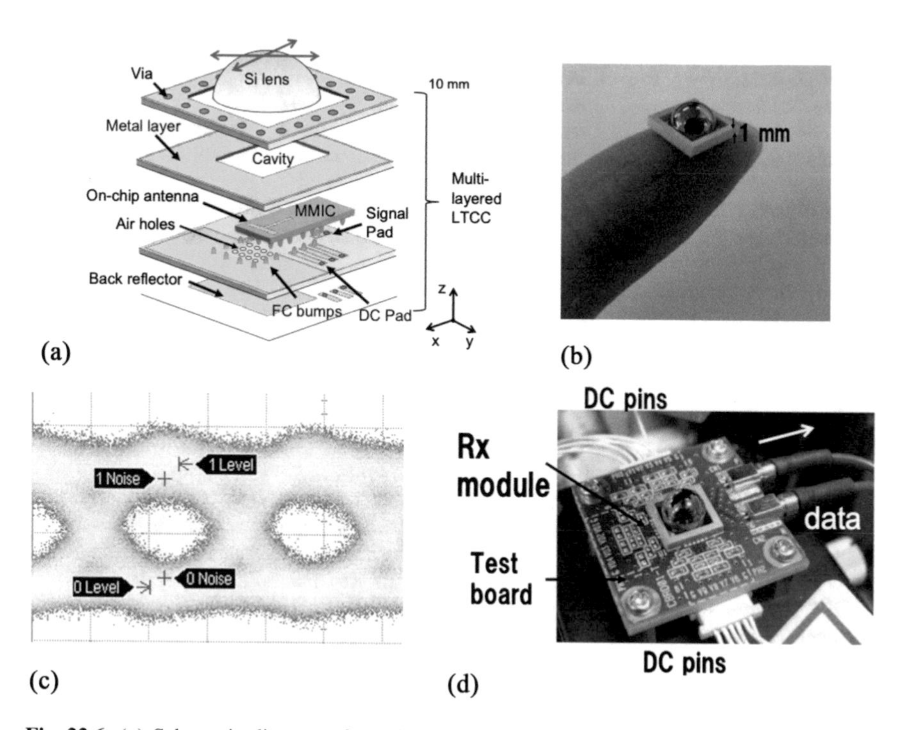

Fig. 22.6 (**a**) Schematic diagram of quasi-optical package, (**b**) photo of compact quasi-optical package with a silicon lens, (**c**) measured eye diagram at 30-cm link, and (**d**) photo of compact package on test PCB board

density, excellent threshold uniformity, and high levels of integration. Various circuit results reported from state-of-the-art InP HBT technologies include 200-mW PAs at 210 GHz, 670-GHz amplifiers, and QVCOs. In addition, fully integrated 600-GHz Tx and 300-GHz receiver circuits have been successfully demonstrated for specific THz applications.

References

1. Urteaga, M., Griffith, Z., Seo, M., Hacker, J., & Rodwell, M. J. W. (2017). InP HBT technologies for THz integrated circuits. *Proceedings of the IEEE, 105*(6), 1051–1067.
2. Rodwell, M. J. W., Urteaga, M., Mathew, T., Scott, D., Mensa, D., Lee, Q., Guthrie, J., Betser, Y., Martin, S. C., Smith, R. P., Jaganathan, S., Krishnan, S., Long, S. I., Pullela, R., Agarwal, B., Bhattacharya, U., Samoska, L., & Dahlstrom, M. (2001). Submicron scaling of hbts. *IEEE Transactions on Electron Devices, 48*(11), 2606–2624.
3. Urteaga, M., Pierson, R., Rowell, P., Jain, V., Lobisser, E., & Rodwell, M. J. W. (2011). 130 nm InP DHBTs with $ft > 0.52$ THz and $f_{\max} > 1.1$ THz. In *69th Device Research Conference* (pp. 281–282).

4. Z. Griffith, M. Urteaga, P. Rowell, & R. Pierson (2014). A 23.2 dBm at 210 GHz to 21.0 dBm at 235 GHz 16-way PA-cell combined InP HBT SSPA MMIC. In *2014 IEEE Compound Semiconductor Integrated Circuit Symposium (CSICS)* (pp. 1–4).
5. Rollin, J. M., Miller, D., Urteaga, M., Griffith, Z., & Kazemi, H. (2015). DC-240 GHz polystrata combining architecture for high efficiency and compact SSPAs. In *Proc. IEEE Compound Semiconductor IC Symposium* (pp. 11–14).
6. Griffith, Z., Urteaga, M., Rowell, P., & Pierson, R. (2014). A 50–80 mW SSPA from 190.8–244 GHz at 0.5 mW pin. In *2014 IEEE MTT-S International Microwave Symposium (IMS2014)* (pp. 1–4).
7. Griffith, Z., Urteaga, M., Rowell, P., & Pierson, R. (2015). 190–260 GHz high-power, broadband PA's in 250 nm InP HBT. In *2015 IEEE Compound Semiconductor Integrated Circuit Symposium (CSICS)* (pp. 1–4).
8. Radisic, V., Scott, D. W., Cavus, A., & Monier, C. (2014). 220-GHz high-efficiency InP HBT power amplifiers. *IEEE Transactions on Microwave Theory and Techniques, 62*(12), 3001–3005.
9. Gritters, D., Brown, K., & Ko, E. (2015). 200–260 GHz solid state amplifier with 700 mW of output power. In *2015 IEEE MTT-S International Microwave Symposium* (pp. 1–3).
10. Kim, J., Song, H., Ajito, K., Yaita, M., & Kukutsu, N. (2013). A 325 GHz quadrature voltage controlled oscillator with superharmonic-coupling. *IEEE Microwave and Wireless Components Letters, 23*(8), 430–432.
11. Griffith, Z., Urteaga, M., Pierson, R., Rowell, P., Rodwell, M., & Brar, B. (2010). A 204.8 GHz static divide-by-8 frequency divider in 250 nm InP HBT. In *2010 IEEE Compound Semiconductor Integrated Circuit Symposium (CSICS)* (pp. 1–4).
12. Seo, M., Hacker, J., Urteaga, M., Skalare, A., & Rodwell, M. (2015). A 529 GHz dynamic frequency divider in 130 nm InP HBT process. *IEICE Electron. Express, 12*, 20141118.
13. Song, H.-J., Kim, J., Ajito, K., Kukutsu, N., & Yaita, M. (2014). 50-Gb/s direct conversion QPSK modulator and demodulator MMICs for terahertz communications at 300 GHz. *IEEE Transactions on Microwave Theory and Techniques, 62*(3), 600–609.
14. Seo, M., Urteaga, M., Young, A., Hacker, J., Skalare, A., Lin, R., & Rodwell, M. (2012). A single-chip 630 GHz transmitter with 210 GHz sub-harmonic PLL local oscillator in 130 nm InP HBT. In *2012 IEEE/MTT-S International Microwave Symposium Digest* (pp. 1–3).
15. Song, H.-J., Kim, J. Y., Ajito, K., Yaita, M., & Kukutsu, N. (2013). Fully integrated ask receiver MMIC for terahertz communications at 300 GHz. *IEEE Transactions on Terahertz Science and Technology, 3*(4), 445–452.
16. Song, H.-J. (2017). Packages for terahertz electronics. *Proceedings of the IEEE, 105*(6), 1121–1138.

Chapter 23
III–V HEMT

Ingmar Kallfass

Abstract This chapter is concerned with the design of terahertz transmit and receive analog frontends on the basis of high electron mobility transistor technology. While HEMTs offer the highest cutoff frequencies and, hence, allow for the best gain-bandwidth tradeoffs at THz frequencies, the design paradigms of the implementation of HEMT-based transceiver MMICs for THz communication are very different from their counterparts in HBT and Si MOSFET technologies. This chapter highlights and discusses those design paradigms and reviews the state of the art in HEMT-based transceivers dedicated to THz communication.

23.1 Introduction

High electron mobility transistors (HEMT), also referred to as modulation-doped field effect transistors (MODFET), are today the fastest active devices. When scaled for small gate lengths of 35 nm or lower, they reach cutoff frequencies in excess of 1 THz. When combined with a full layer stack of an integrated circuit process, complete with passive components such as metal-insulator-metal (MIM) plate capacitors, thin-film resistors, and planar transmission lines, they allow for the implementation of millimeter-wave monolithic integrated circuits (MMIC) covering the entire millimeter-wave range and entering the sub-millimeter-wave range beyond 300 GHz. The still unsurpassed record operating frequency of 1.03 THz in a solid-state amplifier was achieved in 2014 by the team at Northrop Grumman Corp., using their in-house 25 nm InP HEMT technology [33]. Even if this amplifier had to use a cascade of ten stages to obtain a small-signal gain of merely 9 dB, it was an impressive demonstration of a true THz MMIC made possible by the HEMT's unique high-frequency performance.

I. Kallfass (✉)
University of Stuttgart, Stuttgart, Germany
e-mail: ingmar.kallfass@ilh.uni-stuttgart.de

T. Kürner et al. (eds.), *THz Communications*, Springer Series in Optical Sciences 234, https://doi.org/10.1007/978-3-030-73738-2_23

While its use as linear amplifying device in wideband low-noise amplifiers (LNA) and power amplifiers (PAs) remains unquestionably the HEMT's most spectacular purpose, the transistors can also be deliberately matched and biased for nonlinear operation, thus allowing for the implementation of frequency-converting circuits such as mixers and frequency multipliers, or to form voltage-controlled oscillators (VCO) and frequency dividers. The HEMT is therefore pre-destined to form the basis of high performance analog transmit and receive frontends in THz electronics.[1]

23.2 HEMT Technology for THz Circuits

The predominant HEMT technologies today make use of the high carrier mobilities in two-dimensional electron gases (2DEG) forming in the quantum well of an InAlAs/InGaAs hetero-junction. Indium-rich 2DEG channels can be realized as hetero-structures based on the III–V compound semiconductors InP or in the form of metamorphic HEMTs based on GaAs substrates (mHEMT). When combined with low access resistance in the drain and source contacts, low gate resistance achieved through the typical mushroom-shaped gate modules, and maintaining low parasitic fringing capacitances between the gate head and the source and drain contacts, a maximum frequency of oscillation f_{max}[2] up to and exceeding 1 THz can be achieved [31, 33, 41], equaled only by InP-based HBT technology [46]. A maximum transit frequency f_T[3] of 660 GHz has been reported in a 20 nm gate-length mHEMT [30]. Such speed-optimized HEMT technologies, in order to achieve the small gate footprints of down to currently about 20 nm, use electron-beam writing in the gate module. Here lies the first fundamental difference between III–V HEMT and silicon-based CMOS technologies with important implications. As opposed to the fully lithographic techniques used even in aggressively scaled Si-CMOS technology, the conventional III–V HEMT-based IC technologies with their sequential e-beam writing in the gate process result in time- and cost-intensive mass production. However, with their comparatively low number of lithographic mask layers and significantly lower mask set cost, the III–V HEMT technologies have a clear advantage over mass-market oriented Si-CMOS by providing the possibility of fast and cheap prototyping wafer runs and of addressing professional niche market applications which, at least up to today, are the predominant form of markets for THz electronics.

[1] The term "THz electronics" is used here in the commonly adopted and accepted lose definition of circuits operating beyond 0.1 THz center frequency.

[2] f_{max} is defined as the frequency where the combined maximum stable gain (MSG) and maximum available gain (MAG) becomes unity.

[3] f_T is defined as the unity short-circuit current gain.

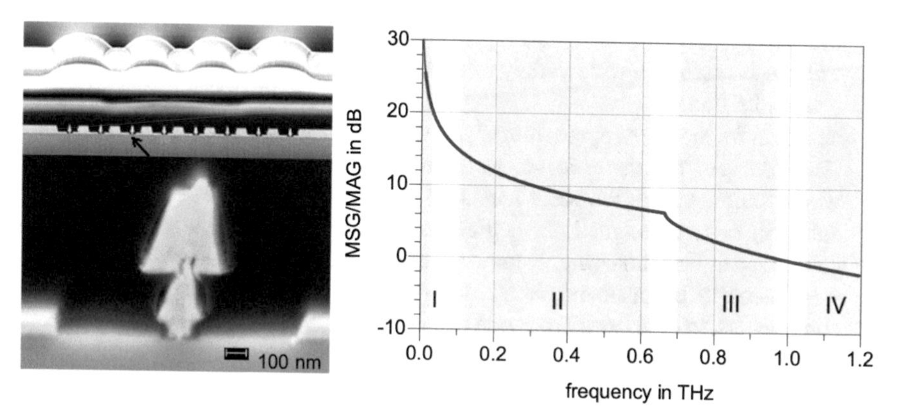

Fig. 23.1 Three-layer metal interconnect process of a GaAs-based high-speed IC process (top left) using mHEMTs with 35 nm footprint as active devices (bottom left). The corresponding MSG/MAG power gain characteristics (right) are used to highlight different operating ranges of the HEMT. SEM pictures are courtesy of Fraunhofer IAF

Figure 23.1 shows the cross-sectional images of a 35 nm gate-length HEMT with its associated IC process. The characteristic mushroom gate head is optimized for low gate resistance while maintaining low gate-to-source and gate-to-drain fringing capacitance. The three-layer metal interconnects can be used for the implementation of MIM plate capacitors, thin-film resistors, spiral inductors, and planar integrated waveguides. Transmission lines can be realized both as conventional microstrip or coplanar waveguides using the semi-insulating low-loss substrate as medium and as microstrip or coplanar waveguides suspended between the bottom metal layer used as ground plane and the signal conductors in the top metal layers, comparable to the backend-of-line processes used in Si-technologies. Figure 23.1 also shows the HEMTs' corresponding power gain characteristics. The maximum stable gain (MSG) and maximum available gain (MAG) are achieved when the HEMT is rendered unconditionally stable and is impedance-matched by ideal, lossless matching networks for maximum power transfer from the source to the load. Such unified MSG/MAG characteristics, which are also the most widely used technique to extrapolate f_{max}, can highlight different frequency operating ranges of the HEMT:

- area 1: at low microwave frequencies, the HEMT provides enormous power gain in excess of 20 dB per stage. Since, as motivated below, reactive impedance matching is usually employed in HEMT-based circuit design, the HEMT tends to produce unwanted low-frequency oscillations. Special care in the design of the monolithic integrated circuits and its mounting and packaging environment, and a comparatively high effort in stability analysis, especially at low frequencies, needs to be adopted to ensure unconditional stability of HEMT-based circuits.
- area 2: at frequencies up to the point where the Rollet stability factor k becomes unity, i.e., in the lower frequency range of only conditional stability where only

the MSG is defined, the HEMT provides a comfortable power gain per stage and can be used in a wide variety of circuit topologies, also making use of gain-bandwidth tradeoffs.

- area 3: at frequencies approaching $f_{\max}$, with an MAG dropping at approximately 20 dB/decade,[4] the rising losses in the reactive impedance-matching networks, which subtract from the ideal MAG of the HEMT, make power gain a scarce entity in circuit design. Less degrees of freedom exist in the design of gain cells, such as common-gate, common-source, or cascode topologies, and special care needs to be taken in circuit design to minimize the loss of the passive components, in order to still provide meaningful gain in what are typically carried out as multi-stage amplifiers.
- area 4: at frequencies beyond $f_{\max}$ any transistor acts only as a passive attenuating device but can still be employed in the design of frequency-converting circuits such as mixers or frequency multipliers, albeit at the cost of enormous conversion loss and RF noise, which cannot be compensated anymore by integrated RF amplification stages.

In an integrated circuit, the behavior of a HEMT is essentially governed by two factors: the biasing conditions and its source and load impedance conditions for a given input power level. For instance, to implement an LNA pre-amplifying stage in a receiver, the HEMT should be biased and matched for a minimum noise measure M [18], with M combining the noise-suppressing effect of the LNA's associated gain G_{ass} on subsequent stages with the noise factor F. The latter will take on its minimum value $F_{\min}$ if an optimum source impedance is presented to the HEMT:

$$M = \frac{F - 1}{1 - \frac{1}{G_{\text{ass.}}}} \tag{23.1}$$

The HEMT's unequalled low-noise properties are certainly the single most important motivation for its use in THz electronic circuits. Noise figures of uncooled HEMT-based LNAs of down to approx. 6 dB and 13 dB have been reported at 300 GHz [31] and 670 GHz [6] operating frequency, respectively. While the maximum linear gain of the HEMT is achieved under class-A conditions for current densities[5] of typically between 400 and 800 mA/mm, different biasing conditions are frequently adopted to implement other circuit functionalities, such as class-B operation for harmonic mixers or frequency multipliers, zero drain bias at close to threshold voltage for resistive mixers with minimum LO power requirements, and operation in the drain knee voltage for active mixers. By far the most frequently adopted gain cells at THz frequencies remain the common-source and the cascode

[4]Note that the typical drop of 10 dB/decade in MSG and 20 dB/decade in MAG becomes visible when the MSG/MAG characteristics are plotted on a logarithmic frequency axis

[5]Drain current per gate width

topologies. However, common-gate stages with resonant gain peaking have also been proven effective [37].

Reported output power of III–V HEMT-based solid-state power amplifiers lies at around 17 dBm at 220 GHz [36], 13 dBm at 300 GHz [24], and 2 dBm at 640 GHz [35] albeit at relative bandwidths of less than 10% and after mounting of the MMICs into metallic waveguide modules. Also, these values represent the maximum achievable output power, when the PA is saturated and in a strong gain compression regime. While the saturated output power is a relevant entity in radar applications, it is of very limited use in a transmitter dedicated to the transmission of modulated signals. Only on-off-keying (OOK) direct modulated signals and low-complexity phase-shift keying such as binary phase shift keying (BPSK) and quadrature phase shift keying (QPSK) will tolerate a high level of gain compression, and even they are better not operated beyond the 1 dB compression point of the transmitter. The most relevant output power-related figures of merit (FoM) for communication-oriented transmitters are describing the intermodulation behavior under multi-tone excitation, such as the third-order and fifth-order intermodulation products IM3 and IM5 at a given output power, and the third-order intercept points derived from it. The input- and output-related third-order intercept point, IIP3, respectively, OIP3, denotes the input, respectively, output power, at which the linearly extrapolated IM3 intercepts the linearly extrapolated fundamental tone power:

$$\mathrm{IIP3/OIP3} = P_{\mathrm{in/out}}\big|_{\mathrm{IM3}=0\,\mathrm{dBc}} \tag{23.2}$$

A similar definition can be adopted for the fifth-order intercept points. Using a two-tone measurement setup employing photonic mixing of three laser lines in a uni-traveling carrier photodiode (UTC), Ghanem and Schoch et al. measured the OIP3 and the OIP5 of a GaAs mHEMT-based 300 GHz power amplifier module to 13 and 11 dBm, respectively [11, 38]. The same amplifier reaches a single-tone saturated output power and output-related 1 dB compression point of 6 dBm and 1.6 dBm, respectively.

Some final words of caution are appropriate when stating these FoMs, especially in view of the comparison with other technologies: First, with noise and output power performance being in a general tradeoff with bandwidth, all FoMs should consider the associated bandwidth. Second, the loss incurred by the chip packaging generally rises with frequency. One therefore has to distinguish between the values measured on chip and on module level. Finally, while the best-performing packaging technique at THz frequencies remains the metallic waveguide, Si-based circuit design typically moves to on-chip RF radiation elements, for which typically the equivalent isotropic radiated power (EIRP) containing the gain of the radiating element is measured and stated.

23.3 Design Paradigms of III–V HEMT-Based THz Electronics

When comparing MMICs implemented in III–V HEMT-based technologies to their counterparts in III–V HBT and silicon-based high-speed CMOS and SiGe BiCOMS technologies, some striking differences in the paradigms underlying the respective circuit designs are evident. These can be understood, however, by considering the HEMT's technological particularities. Figure 23.2 highlights some of the pertinent specifics when designing HEMT-based integrated circuits at THz frequencies.

Most of the HEMT-specific design constraints can be understood when considering their poorly matched performance from device to device. Especially the strong variation of the threshold voltage of individual transistor devices on a single die, and the resulting strong variation of the gate voltage for maximum transconductance, has strong implications on the circuit design level. The threshold voltage is defined mostly by the thickness of the barrier layer between the HEMT's Schottky gate contact and the hetero-interface forming the 2DEG quantum well. Aggressively scaled HEMTs require thin barriers of only a few atomic layers, typically defined by a gate recess process, which is very hard to maintain constant across the wafer. Moreover, the semiconductor processing steps of III–V HEMT technology are not being optimized for robustness; as opposed to mainstream Si CMOS, the yield of III–V HEMT-based integrated circuits drops significantly with the transistor count. Limited yield and unmatched transistors are the main reasons why differential circuit topologies are rarely adopted in HEMT-based circuits, but single-ended approaches are preferred instead. This applies not only to amplifying stages in LNAs, PAs, and VGAs but also to the frequency-multiplying and frequency-converting stages. While the versatile Gilbert-cell topology [12] can in principle be implemented in HEMT technology, it would suffer from poor balancing characteristics due to a low even-mode suppression and, hence, poor isolation characteristics and risk of instability. Therefore, circuit functionalities like frequency multiplication, division, and conversion, which is likely to be implemented in some variation of the fully differential Gilbert cell in HBT and Si

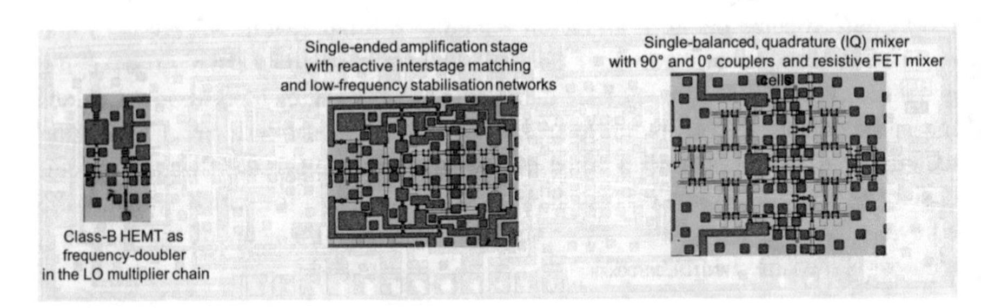

Fig. 23.2 Design paradigms in HEMT-based THz integrated circuits. The highlighted breakout circuits are taken from the 300 GHz receiver MMIC reported in [4]

MOSFET-based technologies, are usually implemented in single-ended circuits in HEMT-based technology.

To generate the local oscillator (LO) of heterodyne receivers and transmitters, a popular choice is to combine highly stable off-chip voltage-controlled oscillators (VCO) at low frequencies which HEMT-based on-chip frequency multiplier chains. The inherent degradation of the phase noise by a factor of $20 \cdot \log n$, n being the multiplication factor, is tolerated in view of the comparatively worse phase noise characteristics of HEMT-based fundamental frequency oscillators. While the Gilbert cell with its inherent rejection of unwanted harmonics is by far the most frequently encountered choice for the implementation of frequency multiplier stages in HBT- and MOSFET-based design, HEMT-based designs more typically make use of single-ended or single-balanced push-push stages biased under close to class B conditions for frequency doublers, or single-ended transistors in compressed class-A conditions. To reject the unwanted harmonics in single-ended frequency multipliers, special care needs to be applied to the harmonic terminations offered by the impedance matching networks.

The most popular mixer approach in HEMTs at THz frequencies therefore also remains the resistive mixer, where the HEMT is operated as a (gate-)voltage-controlled resistor. The popularity of the resistive mixer cell stems from its comparatively high linearity, the ease of biasing (only a gate supply voltage needs to be provided), the extremely high achievable bandwidth due to the ease of wideband impedance matching, and its bi-directionality for use, both in up- and down-conversion. Moreover, the relatively high conversion loss of a resistive mixer stage can easily be compensated by subsequent or preceding amplifier stages of transmitter and receivers, respectively.

When balanced or quadrature IQ topologies are desired in HEMT-based designs, the preferred option is to make use of 90° couplers, such as Lange or Tandem-X coupler, which are easily implemented on the high-quality semi-insulating GaAs or InP substrates. However, with the couplers usually relying on sections of $\lambda/4$ lines, their size is determined by the frequency of operation.

The rigorous adoption of conjugate-complex interstage impedance matching is another characteristic feature of HEMT-based circuit design. In view of avoiding loss of power gain, the use of integrated resistors is typically limited to stabilization purposes. Biasing circuits, impedance transformation networks, and load impedances are preferably realized as purely reactive networks, making extensive use of the impedance-transforming properties of distributed elements such as transmission lines. The reactive impedance matching, however, presents the additional challenge of low-frequency instability. In order to decouple the RF path from the off-chip DC supply networks, large on-chip shunting capacitors are required. These may create low-frequency resonances in conjunction with the inductive effects of the chip-interconnect and packaging environment, which, due to the high gain at low frequencies, may lead to low-frequency oscillations.

When looking at typical HEMT-based MMICs, therefore, the overall chip size is usually determined by the size of the integrated couplers, the reactive interstage impedance matching networks, and the on-chip shunting capacitors. In combination with the comparatively low transistor count per MMIC, this results in probably the most striking difference between HEMT-based circuits and their technology alternatives: the low integration density per MMIC.

23.4 III–V HEMT-Based Transmit-Receive Frontends for THz Communication

The pioneering work on wireless communication beyond 100 GHz based on active monolithic integrated circuits, rather than the previously used photonic or passive Schottky diode-based technologies, started in Japan shortly after 2000. Since then, several groups in Japan, Europe, and the United States have produced chip sets dedicated to THz communication on the basis of InP HEMT or GaAs mHEMT technologies. Most of these chip sets were further used in analog transmit and receive frontends for the transmission of wideband modulated signals with several tens of Gbit/s data rates, with maximum data rates reaching 100 Gbit/s, in point-to-point transmission experiments covering distances from the centimeter to the kilometer range (e.g., [17, 21]). While the initial focus was on direct modulation in transmitters and envelope detection in receivers using amplitude shift keying modulation (ASK), in continuation of the established photonic approaches, soon heterodyne architectures for the coherent detection of complex phase and amplitude modulated (APSK) signals were adopted. Figure 23.3 shows a summary of the pertinent HEMT-based chip sets for THz communication, aligned in terms of their publication date and their center frequency of operation.

23.4.1 ASK and Direct-Detection Receivers

For the transmitter, the direct modulation of a carrier tone by some ASK signal, usually an OOK bit sequence, can be adopted, effectively resulting in a BPSK transmission across the air interface. A direct-detection receiver may then recover the bit stream by envelope detection in a rectifying device such as a diode. Such an incoherent communication eliminates the need for carrier synchronization at the receiver. The main motivation for a HEMT-based transmit and receive frontend in this case is the increase of transmit power from a PA stage and the reduction of noise figure from an LNA stage in the receiver.

In 2003–2004, Kosugi et al. published a chip set consisting of a transmitter and receiver MMIC [28, 29], realized in 100 nm gate-length InP HEMT technology developed at NTT [45]. These MMICs, which saw the monolithic integration of RF amplifying stages with modulating and de-modulation stages, were at the heart of the analog frontend employed in the point-to-point transmission of 10 Gbit/s over

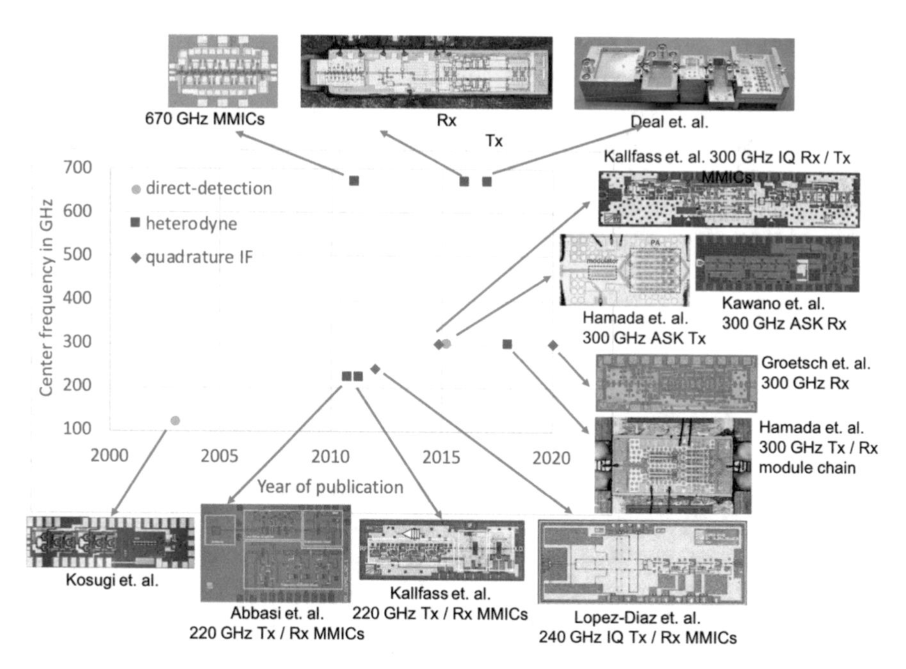

Fig. 23.3 Gallery of III–V HEMT-based transmit and receive MMICs for THz communication

a distance of 5.8 km at a carrier frequency of 120 GHz [19, 47]. A review of this inspiring work can be found, e.g., in [20].

In 2015–2016, Hamada et al. [15] at NTT and Kawano et al. [27] at Fujitsu presented InP HEMT-based direct-modulated ASK transmit and direct-detection receive MMICs operating at 300 GHz, which were later used by Song et al. at [39] for 20 Gbit/s ASK transmission in a media kiosk application scenario.

Direct-detection receivers have also been proposed for wireless communication in view of low power consumption and, hence, high energy-per-bit efficiency [44]. While CMOS-based technology will certainly be the technology of choice in this respect, the use of HEMT technology may still be argued in view of its superior noise performance.

23.4.2 APSK and Coherent Receivers

The appealing benefits of increased spectral efficiency and signal-to-noise ratio of coherent communication systems come at the cost of increased linearity requirements and the need for receiver synchronization. A mixer stage, driven by a local oscillator (LO), is used in the transmitter to up-convert a modulated signal presented to the intermediate frequency (IF) input port of the mixer. The up-conversion results in a double-sideband modulated RF signal centered around the LO frequency. Using its own LO, the receiver down-converts the captured RF signal to an IF frequency, from where the signal is typically sampled and digitized through fast analog-to-digital converters (ADC). With the availability of fast digital signal processing (DSP), the synchronization, channel selection, equalization, and de-modulation are typically carried out in the digital domain. In coherent communication it becomes possible to add phase modulation in addition to amplitude modulation (APSK), achieved through quadrature modulation in the transmitter. Thus, modern digital modulation schemes such as quadrature amplitude modulation (QAM) may be employed.

23.4.2.1 Double-Sideband Transmission and Heterodyne Detection

Initial work on coherent THz communication started out by adopting double-sideband transmission and heterodyne detection with chip sets, which are essentially an adaption of previously realized transmitters and receivers for high-resolution radar applications, modified by the addition of an up-converting mixer stage in the transmitter. Early chip sets operating at a center frequency of 220 GHz were developed in 2011 by Abbasi et al. [1] and Kallfass et al. [23] using GaAs mHEMT technology with 50 nm gate length [42]. Building on their in-house 30 nm InP HEMT technology with excellent high-frequency performance and cutoff frequencies f_T and f_{max} of 600 and 1200 GHz, respectively, the group around William Deal at Northrop Grumman Corp. developed MMIC components for amplifiers and frequency converters at 670 GHz, as early as 2011 [6]. These MMICs were later combined in a chain of metallic waveguide modules to form a heterodyne receiver [8] and transmitter [7] operating at 670 GHz, the highest operating frequency of active electronic transmit and receive frontends reported to date. More recently, in 2018–2020, Hamada et al. used a heterodyne transmit and receive frontend consisting of a waveguide module chain of individually packaged MMICs in 80 nm InP HEMT technology to demonstrate chip sets capable of supporting up to 100 Gbit/s [14] and even 120 Gbit/s [16] with 16QAM modulation [17].

23.4.2.2 Direct-Modulated Quadrature Channel Transmitters and Zero-IF Quadrature Receivers

In the heterodyne transmission scheme of single-ended up- and down-converters, the achievable symbol rate is mainly limited by the IF bandwidth of the mixer stages. By the direct conversion of two ASK-modulated analog baseband signals in a quadrature (IQ) modulator driven with the THz signal as local oscillator, i.e., by the direct up-conversion of the analog baseband signal into the RF domain, much higher modulation bandwidths can be adopted. Similarly, in the receiver, the intradyne down-conversion to near zero-IF using two quadrature down-conversion paths eliminates the IF bandwidth limitation to the achievable symbol rate. Consequently, the highest symbol rates to date have been achieved in THz communication by using the direct conversion from analog baseband to and from RF in quadrature-channel modulator and de-modulator stages. It is important to mention, however, that this approach is strongly limited by transceiver impairments such as the DC offset in the receiver, the local oscillator phase noise, and the magnitude and phase imbalances of the wideband quadrature channels [2]. The first chip set adopting this scheme was developed in the year 2012 in the frame of the MILLILINK project [ref.]. The analog frontend MMICs operating at a carrier frequency of 240 GHz are implemented in 35 nm GaAs mHEMT technology [31] and integrate sub-harmonically driven, quadrature modulating and demodulating stages with RF post- and pre-amplification stages in the transmitter and receiver, respectively [32]. Based on this chip set, wireless transmission of up to 96 Gbit/s data rate using 8PSK modulation over 40 m distance [1] and 64 Gbit/s data rate with QPSK modulation over 850 m distance were achieved [21]. Song et al. in 2014 reported on a chip set implemented in InP DHBT technology operating at 300 GHz center frequency and achieving 50 Gbit/s QPSK transmission with direct modulation and de-modulation [40]. Later, this principle was adopted in the TERAPAN chip set operating at 300 GHz and implemented in 35 nm mHEMT technology [22, 26], which were used in the realization of a four-channel linear phased array using LO phase-shifting [34] for electronic beam-steering [25].

23.4.2.3 Super-Heterodyne Architectures

In many reported transmission experiments, the bottleneck to the maximum achievable data rate is the bandwidth of the DSP rather than the available RF bandwidth at THz frequencies. While the DSP bandwidth effectively sets a limit to the achievable symbol rate in the modulation and de-modulation of complex-modulated signals, the exploitable RF bandwidth is in practice mostly limited by the transmission bandwidth of the analog transmit receive frontend. In [5], for instance, a 3dB transmission bandwidth of 38 GHz was measured using vectorial network analysis of the complex transfer function of a 300 GHz HEMT-based frontend. Moreover, most reported transmission experiments at $>$100 GHz carrier frequencies, rather than transmitting user data in real time, employ the generation of pseudo-random bit sequences (PRBS) at the transmitter and the offline receive signal analysis in terms of, e.g., error vector magnitude (EVM), bit error rate (BER), and signal-to-noise ratio (SNR). In an attempt to fill the entire available RF transmission bandwidth with real-time DSP and user data, frequency domain multiplexing (FDM) and a super-heterodyne receiver architecture is advantageous. A super-heterodyne receiver employs two down-conversion stages in two different IF ranges. The super-heterodyne approach, which is common in radios operating at microwave and millimeter-wave carrier frequencies, is a very convenient way to pre-select from many channels which are aggregated by FDM in the RF domain. Such FDM channel aggregation is also the governing principle of the recently established frequency standard IEEE802.15.3d for THz communication [10]. To link up with this frequency standard, and to enable the transmission of real-time modulated and de-modulated user data, the super-heterodyne approach is first adopted in THz communication by a Japanese-European consortium THoR [ref]. Proof-of-concept transmission experiments including channel aggregation reported in [3] have been carried out using the 300 GHz link components reported in [22]. The full bi-directional link will see the combination of Gigabit Modems operating in E-band (71–76 and 81–86 GHz) and in V-band (57–66 GHz, according to the IEEE802.11ad standard) with a 300 GHz transmit-receive frontend using the Modems' RF output as comparatively high IF input (57–86 GHz) to a 300 GHz up-converter, while in the receiver, the 300 GHz signal will be down-converted to a first IF frequency in V/E-band from where it will be taken up by the Modems' RF input. While no actual transmission experiments have been reported so far, the super-heterodyne receiver MMIC has recently been reported [13].

23.5 Conclusion and Outlook

With III–V HEMT-based integrated circuit technologies being historically the first to be adopted for the implementation of active electronic transmitters and receivers for THz communication, they are today still unrivaled by alternative technologies in terms of receiver noise and achievable gain-bandwidth products of their transmission characteristics. In combination with beam-focusing antennas, they allow for the highest combinations of data rate and transmission distance. However, the HEMT approach, while offering highest cutoff frequencies, suffers from comparatively poor transistor matching and yield, which strongly limits the reasonable transistor count per MMIC and renders more complex architectures such as multi-channel phased arrays, seen in modern Si-based millimeter-wave transceivers, impracticable on a single chip. What's more, the stand-alone HEMT technology, while offering the possibility to implement all required circuit functions of analog transmit-receive frontends with highest performances, lacks a reasonable possibility of mixed-signal integration of microwave, analog and digital functionalities. True system-on-chip approaches are therefore only practicable in CMOS and BiCMOS technologies.

To bring the full potential of III–V HEMT technology to the system level and allow for further advances in THz communication transceivers, ongoing research is focusing on several fronts.

Improvements on the device level, such as the introduction of MOS gates to control the HEMT channel for reduced gate leakage in aggressively scaled transistors [43], promise improvements in terms of yield, uniformity, and output power.

On the circuit level, as illustrated by the difference between the achievable output power of stand-alone PAs and that of the integrated transmitters reported so far, there is still significant headroom for optimized dynamic range by improved gain and linearity partitioning.

On the chip assembly level, the move from the high-performance but bulky and expensive metallic waveguide modules to a multi-chip system-in-package with integrated on-chip antennas mounted on PCB carrier substrates will allow for compact, versatile transceiver configurations [9].

References

1. Abbasi, M., Gunnarsson, S. E., Wadefalk, N., Kozhuharov, R., Svedin, J., Cherednichenko, S., Angelov, I., Kallfass, I., Leuther, A., & Zirath, H. (2010). Single-chip 220 GHz active heterodyne receiver and transmitter MMICs with on-chip integrated antenna. *IEEE Trans. Microwave Theory and Techniques, 59*(2), 466–478.
2. Antes, J., & Kallfass, I. (2015). Performance estimation for broadband multi-gigabit millimeter and sub-millimeter wave wireless communication links. *IEEE Transactions on Microwave Theory and Techniques, 63*(10), 3288–3299.
3. Dan, I., Ducournau, G., Hisatake, S., Szriftgiser, P., Braun, R. P., & Kallfass, I. (2019). A terahertz wireless communication link using a superheterodyne approach. *IEEE Transactions on Terahertz Science and Technology, 10*(1), 32–43.
4. Dan, I., Grötsch, C. M., Schoch, B., Wagner, S., John, L., Tessmann, A., & Kallfass, I. (2019). A 300 GHz quadrature down-converter s-MMIC for future terahertz communication. In *2019 IEEE International Conference on Microwaves, Antennas, Communications and Electronic Systems (COMCAS)* (pp. 1–6).
5. Dan, I., Rosello, E., Harati, P., Dilek, S., Kallfass, I., & Shiba, S. (2017). Measurement of complex transfer function of analog transmit-receive frontends for terahertz wireless communications. In *2017 47th European Microwave Conference (EuMC)* (pp. 1009–1012).
6. Deal, W., Mei, X. B., Leong, K.M.K.H., Radisic, V., Sarkozy, S., & Lai, R. (2011). THz monolithic integrated circuits using InP high electron mobility transistors. *IEEE Transactions on Terahertz Science and Technology, 1*(1), 25–32, 2011.
7. Deal, W. R., Leong, K., Zamora, A., Gorospe, B., Nguyen, K., & Mei, X. B. (2017). A 660 GHz up-converter for THz communications. In *2017 IEEE Compound Semiconductor Integrated Circuit Symposium (CSICS)* (pp. 1–4).
8. Deal, W. R., Leong, K., Zamora, A., Yoshida, W., Lange, M., Gorospe, B., Nguyen, K., & Mei, G. X. B. (2016). A low-power 670-GHz InP HEMT receiver. *IEEE Transactions on Terahertz Science and Technology, 6*(6), 862–864.
9. Dyck, A., Rösch, M., Tessmann, A., Leuther, A., Kuri, M., Wagner, S., Gashi, B., Schäfer, J., & Ambacher, O. (2019). A transmitter system-in-package at 300 GHz with an off-chip antenna and GaAs-based MMICs. *IEEE Transactions on Terahertz Science and Technology, 9*(3), 335–344.
10. IEEE standard for high data rate wireless multi-media networks—amendment 2: 100 Gb/s wireless switched point-to-point physical layer. In *IEEE Std 802.15.3d-2017 (Amendment to IEEE Std 802.15.3-2016 as amended by IEEE Std 802.15.3e-2017)* (pp. 1–55), Oct 2017.
11. Ghanem, H., Schoch, B., Kallfass, I., Szriftgiser, P., Zegaoui, M., Zaknoune, M., Danneville, F., & Ducournau, G. (2020). Non-linear analysis of a broadband power amplifier at 300 GHz. In *Proc. European Microwave Int. Circuits Conf. EuMIC, Utrecht* (pp. 1–4).
12. Gilbert, B. (1968). A precise four-quadrant multiplier with subnanosecond response. *IEEE Journal of Solid-State Circuits, 3*(4), 365–373.
13. Groetsch, C., Dan, I., John, L., Wagner, S., & Kallfass, I. (2021). A compact 281–319 GHz low-power downconverter MMIC for superheterodyne communication receivers. *IEEE Transactions on Terahertz Science and Technology, 11*(2), 231–239.
14. Hamada, H., Fujimura, T., Abdo, I., Okada, K., Song, H. J., Sugiyama, H., Matsuzaki, H., & Nosaka, H. (2018). 300-GHz. 100-Gb/s InP-HEMT wireless transceiver using a 300-GHz

fundamental mixer. In *2018 IEEE/MTT-S International Microwave Symposium—IMS* (pp. 1480–1483).
15. Hamada, H., Kosugi, T., Song, H. J., Yaita, M., El Moutaouakil, A., Matsuzaki, H., & Hirata, A. (2015). 300-GHz band 20-Gbps ask transmitter module based on InP-HEMT MMICs. In *2015 IEEE Compound Semiconductor Integrated Circuit Symposium (CSICS)* (pp. 1–4).
16. Hamada, H., Tsutsumi, T., Itami, G., Sugiyama, H., Matsuzaki, H., Okada, K., & Nosaka, H. (2019). 300-GHz 120-b/s wireless transceiver with high-output-power and high-gain power amplifier based on 80-nm InP-HEMT technology. In *2019 IEEE BiCMOS and Compound semiconductor Integrated Circuits and Technology Symposium (BCICTS)* (pp. 1–4).
17. Hamada, H., Tsutsumi, T., Matsuzaki, H., Fujimura, T., Abdo, I., Shirane, A., Okada, K., Itami, G., Song, H. J., Sugiyama, H., & Nosaka, H. (2020). 300-GHz-band 120-Gb/s wireless front-end based on InP-HEMT PAS and mixers. *IEEE Journal of Solid-State Circuits, 55*(9), 2316–2335.
18. Haus, H. A., & Adler, R. B. (1958). Optimum noise performance of linear amplifiers. *Proceedings of the IRE, 46*(8), 1517–1533.
19. Hirata, A., Kosugi, T., Takahashi, H., Takeuchi, J., Murata, K., Kukutsu, N., Kado, Y., Okabe, S., Ikeda, T., Suginosita, F., Shogen, K., Nishikawa, H., Irino, A., Nakayama, T., & Sudo, N (2010). 5.8-km 10-Gbps data transmission over a 120-GHz-band wireless link. In *2010 IEEE International Conference on Wireless Information Technology and Systems* (pp. 1–4).
20. Hirata, A., Kosugi, T., Takahashi, H., Takeuchi, J., Togo, H., Yaita, M., Kukutsu, N., Aihara, K., Murata, K., Sato, Y., Nagatsuma, T., & Kado, Y. (2012). 120-GHz-band wireless link technologies for outdoor 10-Gbit/s data transmission. *IEEE Transactions on Microwave Theory and Techniques, 60*(3), 881–895.
21. Kallfass, I., Boes, F., Messinger, T., Antes, J., Inam, A., Lewark, U., Tessmann, A., & Henneberger, R. (2015). 64 Gbit/s transmission over 850 m fixed wireless link at 240 GHz carrier frequency. *Journal of Infrared Millimeter and Terahertz Waves, 36*(2), 221–233.
22. Kallfass, I., Dan, I., Rey, S., Harati, P., Antes, J., Tessmann, A., Wagner, S., Kuri, M., Weber, R., Massler, H., Leuther, A., Merkle, T., & Kürner, T. (2015). Towards MMIC-based 300 GHz indoor wireless communication systems. *Trans. Institute of Electronics, Information and Communication Engineers IEICE, E98-C*(12), 1081–1090.
23. Kallfass, I., Antes, J., Schneider, T., Kurz, F., Lopez-Diaz, D., Diebold, S., Massler, H., Leuther, A., & Tessmann, A. (2011). All active MMIC based wireless communication at 220 GHz. *IEEE Trans. on Terahertz Science and Technology, 1*(2), 477–487.
24. John, L., Tessmann, A., Leuther, A., Neininger, P., Merkle, T., & Zwick, T. (2020). Broadband 300-GHz power amplifier MMICs in InGaAs mHEMT technology. *IEEE Transactions on Terahertz Science and Technology, 10*(3), 309–320.
25. Kallfass, I., Dan, I., Rey, S., Tessmann, A., Leuther, A., & Kuerner, T. (2018). MMIC electronic steerable 4-channel phased-array transmit-receive frontend for 300 GHz wireless personal area networks. In *Proc. 8th International Workshop on Terahertz Technology and Applications, Kaiserslautern*.
26. Kallfass, I., Harati, P., Dan, I., Antes, J., Boes, F., Rey, S., Merkle, T., Wagner, S., Massler, H., Tessmann, A., & Leuther, A. (2015). MMIC chipset for 300 GHz indoor wireless communication. In *2015 IEEE International Conference on Microwaves, Communications, Antennas and Electronic Systems (COMCAS)* (pp. 1–4).
27. Kawano, Y., Matsumura, H., Shiba, S., Sato, M., Suzuki, T., Nakasha, Y., Takahashi, T., Makiyama, K., Iwai, T., & Hara, N. (2015). A 20 Gbit/s, 280 GHz wireless transmission in InPHEMT based receiver module using flip-chip assembly. In *2015 European Microwave Conference (EuMC)* (pp. 562–565).
28. Kosugi, T., Shibata, T., Enoki, T., Muraguchi, M., Hirata, A., Nagatsuma, T., & Kyuragi, H. (2003). A 120-GHz millimeter-wave MMIC chipset for future broadband wireless access applications. In *IEEE MTT-S International Microwave Symposium Digest, 2003*, (Vol. 1, pp. 129–132).
29. Kosugi, T., Tokumitsu, M., Enoki, T., Muraguchi, M., Hirata, A., & Nagatsuma, T. (2004). 120-GHz Tx/Rx chipset for 10-Gbit/s wireless applications using 0.1/spl mu/m-gate InP HEMTs.

In *IEEE Compound Semiconductor Integrated Circuit Symposium, 2004.* (pp. 171–174).
30. Leuther, A., Koch, S., Tessmann, A., Kallfass, I., Merkle, T., Massler, H., Loesch, R., Schlechtweg, M., Saito, S., & Ambacher, O. (2011). 20 nm metamorphic HEMT with 660 GHz ft. In *IPRM 2011—23rd International Conference on Indium Phosphide and Related Materials* (pp. 1–4).
31. Leuther, A., Tessmann, A., Dammann, M., Massler, H., Schlechtweg, M., & Ambacher, O. (2013). 35 nm mHEMT Technology for THz and ultra low noise applications. In *2013 International Conference on Indium Phosphide and Related Materials (IPRM)* (pp. 1–2).
32. Lopez-Diaz, D., Kallfass, I., Tessmann, A., Leuther, A., Wagner, S., Schlechtweg, M., & Ambacher, O. (2012). A subharmonic chipset for gigabit communication around 240 GHz. In *2012 IEEE/MTT-S International Microwave Symposium Digest* (pp. 1–3).
33. Mei, X., Yoshida, W., Lange, M., Lee, J., Zhou, J., Liu, P. H., Leong, K., Zamora, A., Padilla, J., Sarkozy, S., Lai, R., & Deal, W. R. (2015). First demonstration of amplification at 1 THz using 25-nm InP high electron mobility transistor process. *IEEE Electron Device Letters, 36*(4), 327–329.
34. Merkle, T., Tessmann, A., Kuri, M., Wagner, S., Leuther, A., Rey, S., Zink, M., Stulz, H., Riessle, M., Kallfass, I., & Kürner, T. (2017). Testbed for phased array communications from 275 to 325 GHz. In *2017 IEEE Compound Semiconductor Integrated Circuit Symposium (CSICS)* (pp. 1–4).
35. Radisic, V., Leong, K. M. K. H., Mei, X., Sarkozy, S., Yoshida, W., & Deal, W. R. (2011). Power amplification at 0.65 THz using InP HEMTs. *IEEE Transactions on Microwave Theory and Techniques, 60*(3), 724–729.
36. Radisic, V., Leong, K. M. K. H., Mei, X., Sarkozy, S., Yoshida, W., Liu, P. H., Uyeda, J., Lai, R., & Deal, W. R. (2010). A 50 mW 220 GHz power amplifier module. In *2010 IEEE MTT-S International Microwave Symposium* (pp. 45–48).
37. Sato, M., Takahashi, T., & Hirose, T. (2010). 68–110-GHz-band low-noise amplifier using current reuse topology. *IEEE Transactions on Microwave Theory and Techniques, 58*(7), 1910–1916.
38. Schoch, B., Tessmann, A., Leuther, A., Szriftgiser, P., Ducournau, G., & Kallfass, I. (2021). Two-tone intermodulation performance of a 300 GHz power amplifier MMIC. In *Proc. IEEE Int. Microwave Symposium, Los Angeles*.
39. Song, H. J., Kosugi, T., Hamada, H., Tajima, T., El Moutaouakil, A., Matsuzaki, H., Kawano, Y., Takahashi, T., Nakasha, Y., Hara, N., Fujii, K., Watanabe, I., Kasamatsu, A., & Yaita, M (2016). Demonstration of 20-Gbps wireless data transmission at 300 GHz for KIOSK instant data downloading applications with InP MMICs. In *2016 IEEE MTT-S International Microwave Symposium (IMS)* (pp. 1–4).
40. Song, H.-J., Kim, J.-Y., Ajito, K., Kukutsu, N., & Yaita, M. (2014). 50-Gb/s Direct Conversion QPSK Modulator and Demodulator MMICs for Terahertz Communications at 300 GHz. *IEEE Transactions on Microwave Theory and Techniques, 62*(3), 600–609.
41. Takahashi, T., Kawano, Y., Makiyama, K., Shiba, S., Sato, M., Nakasha, Y., & Hara, N. (2016). Enhancement of f_{max} to 910 GHz by adopting asymmetric gate recess and double-side-doped structure in 75-nm-gate inalas/ingaas HEMTs. *IEEE Transactions on Electron Devices, 64*(1), 89–95.
42. Tessmann, A. (2005). 220-GHz metamorphic HEMT amplifier MMICs for high-resolution imaging applications. *IEEE Journal of Solid-State Circuits, 40*(10), 2070–2076.
43. Tessmann, A., Leuther, A., Heinz, F., Bernhardt, F., John, L., Massler, H., Czornomaz, L., & Merkle, T. (2019). 20-nm $in_{0.8}ga_{0.2}$ as MOSHEMT MMIC technology on silicon. *IEEE Journal of Solid-State Circuits, 54*(9), 2411–2418.
44. Thome, F., Maroldt, S., & Ambacher, O. (2015). Novel destructive-interference-envelope detector for high data rate ask demodulation in wireless communication receivers. In *2015 IEEE MTT-S International Microwave Symposium* (pp. 1–4).
45. Umeda, Y., Enoki, T., Osafune, K., Ito, H., & Ishii, Y. (1996). High-yield design technologies for InAlAs/InGaAs/InP-HEMT analog-digital ICs. *IEEE Transactions on Microwave Theory and Techniques, 44*(12), 2361–2368.

46. Urteaga, M., Pierson, R., Rowell, P., Jain, V., Lobisser, E., & Rodwell, M. J. W. (2011). 130 nm InP DHBTs with $ft > 0.52$ THz and $f_{max} > 1.1$ THz. In *69th Device Research Conference* (pp. 281–282).
47. Yamaguchi, R., Hirata, A., Kosugi, T., Takahashi, H., Kukutsu, N., Nagatsuma, T., Kado, Y., Ikegawa, H., Nishikawa, H., & Nakayama, T. (2008). 10-Gbit/s MMIC wireless link exceeding 800 meters. In *2008 IEEE Radio and Wireless Symposium* (pp. 695–698).

Chapter 24
Resonant Tunneling Diode

Masahiro Asada and Safumi Suzuki

Abstract Terahertz (THz) sources using resonant tunneling diodes (RTDs) are described in this chapter. Room-temperature fundamental oscillation up to 1.98 THz has been obtained by reducing the electron delay time in RTDs and conduction loss in antennas. Output power of 0.7 mW was obtained at 1 THz by a large-scale array. For spectral and polarization properties, electrical frequency tuning with the integration of varactor diode, spectral narrowing by the phase-locked loop, and radiation of circular polarized and vortex waves are briefly introduced. Applications of RTDs to wireless communication and radar system using direct intensity modulation are also described.

24.1 Introduction

Compact and coherent solid-state sources are important components for various applications of the terahertz (THz) frequency range, in particular for wireless communications. Electron devices are being studied from the millimeter-wave side. In particular, transistors including HBTs, HEMTs, and CMOS transistors are recently making significant progress in operation frequency, as described in other chapters. Resonant tunneling diodes (RTDs) have also been considered as one of the candidates for THz oscillators at room temperature [1–5]. In this chapter, THz sources using resonant tunneling diodes (RTDs) are described. Although RTDs are also used as detectors [6, 7], only a very brief description will be given in 24.5 for RTD detectors.

M. Asada (✉)
Institute of Innovative Research, Tokyo Institute of Technology, Tokyo, Japan
e-mail: asada@pe.titech.ac.jp

S. Suzuki
Department of Electrical and Electronic Engineering, Tokyo Institute of Technology, Tokyo, Japan

T. Kürner et al. (eds.), *THz Communications*, Springer Series in Optical Sciences 234, https://doi.org/10.1007/978-3-030-73738-2_24

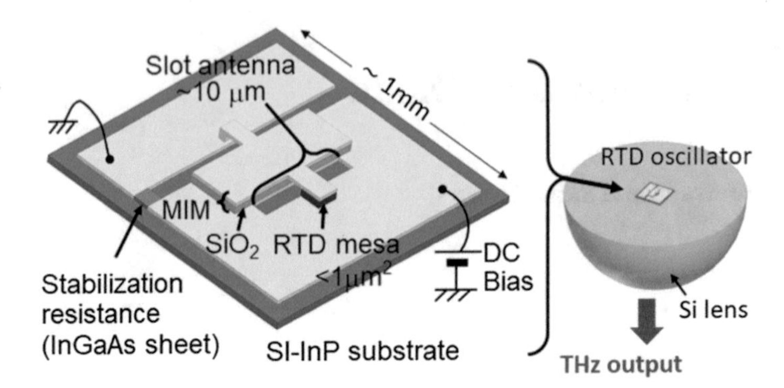

Fig. 24.1 Structure of RTD THz oscillator with slot antenna (reproduced from [3]. Copyright (2016) Springer Nature)

24.2 Device Structure

The structure of our RTD oscillator is shown in Fig. 24.1 [3]. The RTD is located at the center of a slot antenna which works as a resonator and a radiator. The RTD is composed of AlAs/GaInAs double-barrier structure on semi-insulating InP substrate. The peak current density and the peak-valley current ratio are ~30 mA/μm^2 and 2, respectively. The mesa bottom of the RTD is connected to one side of the slot, and the upper electrode on the RTD is connected to the other side of the slot via a metal bridge and a large capacitance of the metal-insulator-metal (MIM) layer. The oscillation occurs when the negative differential conductance (NDC) of RTD compensates for the conductance of the antenna including conduction and radiation losses. The oscillation frequency is determined by the resonance circuit constructed with the capacitance of RTD and slot antenna. The output power is extracted from the substrate side through a hemispherical Si lens, as shown in Fig. 24.1.

24.3 Toward High-Frequency and High-Power Operation

For high-frequency oscillation, a short electron delay time in RTD is required to maintain the NDC at high frequency, and the conduction loss in the antenna must be reduced. The electron delay time is composed of the dwell time in the resonant tunneling region and the transit time in the collector depletion region [3]. The dwell time was reduced by a narrow quantum well and barriers. For the transit time, the thickness of the collector spacer was optimized to simultaneously reduce the capacitance. The conduction loss was reduced by optimizing the antenna length and improving the structure and fabrication process of the metal bridge [4]. To further reduce the conduction loss, the thickness of the antenna electrode was increased,

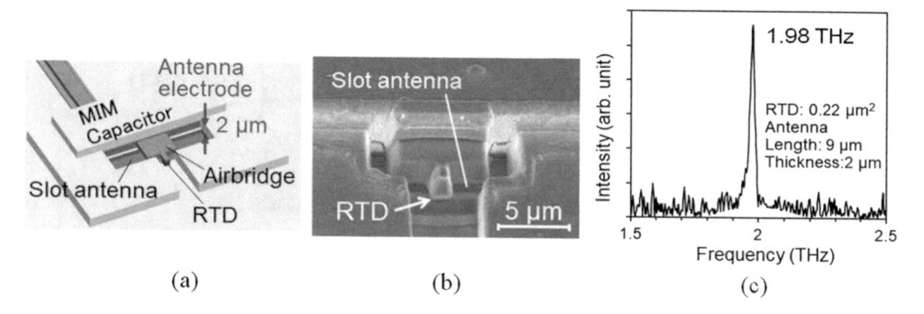

Fig. 24.2 Oscillator with thick antenna electrode. (**a**) Structure, (**b**) scanning electron microscope image, and (**c**) oscillation spectrum. (Reproduced, with permission, from [5]. Copyright (2017) IEEE)

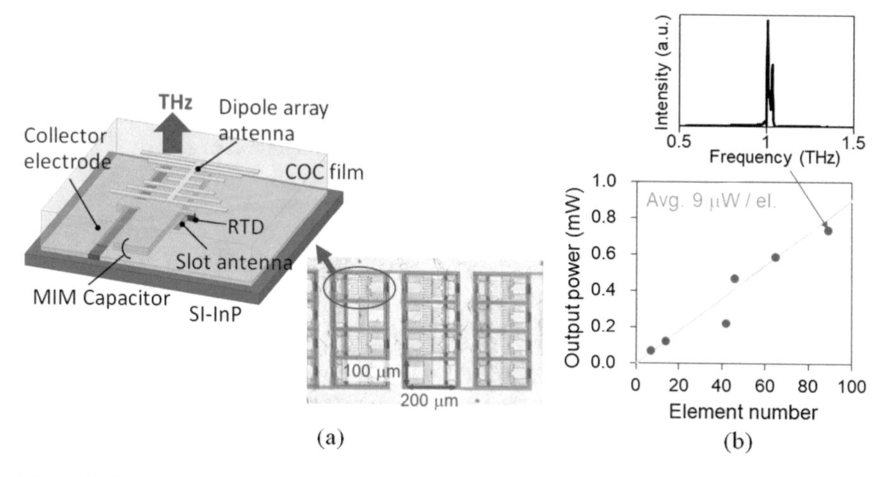

Fig. 24.3 Large-scale array of RTD oscillators with dipole array antennas. (**a**) Structure and (**b**) output power and oscillation spectrum. (Reproduced, with permission, from [9]. Copyright (2019) AIP Publishing)

and oscillation at 1.98 THz was obtained [5], as shown in Fig. 24.2. This is the highest frequency of room-temperature electronic single oscillators to date.

For high-output power, we fabricated an offset slot antenna with high radiation conductance and achieved a high-output power of 0.61 mW at 620 GHz with a synchronized two-element array [8]. For a large-scale array, 0.73 mW was obtained at 1 THz with 89 elements [9], as shown in Fig. 24.3. The output coherence of this array was imperfect, and a few peaks were observed in the oscillation spectra, because the coupling was not intentionally made between the elements in this device. This property may be useful for imaging application, because the interference fringe is suppressed.

New structures for high-frequency and high-output power were proposed [10, 11], in which a cavity and a bow-tie antenna separately work as the resonator and radiator, respectively. An output power of ~2 mW at 1 THz in a single oscillator is expected for a rectangular cavity resonator [10], and also, oscillation frequency of 2–3 THz is theoretically possible with a cylindrical cavity resonator [11].

24.4 Spectral and Polarization Characteristics

To achieve electrical frequency tuning in the RTD oscillators, a varactor diode was integrated in the slot antenna. A wide frequency tuning from 420 to 970 GHz was obtained in an array of the frequency-tunable oscillators covering different frequency ranges [12]. Spectral narrowing of the frequency-tunable oscillator by the phase-locked loop system was also demonstrated [13]. Spectral width as narrow as 1 Hz was achieved in the whole tuning range.

A circular polarized THz wave was generated by an RTD oscillator with radial line slot antenna (RLSA) [14]. In a preliminary experiment, the axial ratio of the polarization was 2.2 dB, and the directivity was 15 dBi at 500 GHz. This structure was also applied for vortex wave by changing the arrangement of RLSA [15].

24.5 Applications to Wireless Communication and Radar System

Researches toward applications of RTDs to imaging [16], spectroscopy [12], wireless communication [6, 7, 18–20], and radar system [21–23] have been reported.

For wireless communication, the output power of RTD oscillator can be modulated directly by superposition of signals on the bias voltage. This property is convenient for compact transmitter. The cutoff frequency of direct modulation up to 30GHz was reported [17]. Using the direct modulation, wireless data transmission with the data rate up to 44 Gbps with the bit error rate of 5×10^{-4} (< FEC limit) and 25 Gbps without error was demonstrated at 650 GHz [18]. Multichannel data transmission was also demonstrated [19]. 56 Gbps was obtained by the frequency division multiplexing with 500 and 800 GHz. The same data rate was also obtained in the polarization division multiplexing. Data transmission with a circular polarized wave, which was almost insensitive to the rotation of the transmitter, was demonstrated at 500 GHz with 1 Gbps using an RTD oscillator with RLSA [14]. RTD can also be used for a detector because it has nonlinear current-voltage characteristics. THz wireless communications with RTD detectors have been demonstrated [6, 7].

THz radar system, which is useful for 3D imaging, was also demonstrated [21–23]. The amplitude-modulated (AM) THz output of an RTD oscillator is irradiated to an object, and the reflected wave is demodulated. By the comparison of the phase of the demodulated signal with that of the reference signal, the distance between the

RTD and the detector with reflection at the object is obtained. Using two frequencies for the modulation, absolute distance was measured with an accuracy of 0.063 mm [22]. Radar system with the frequency-modulated subcarrier was also reported [23]. Multiple objects can be observed in this system.

24.6 Conclusion

THz sources using RTD oscillators were described in this chapter, including oscillation characteristics and applications to wireless communication and radar system. Attempts are being made to achieve higher performance, such as higher frequencies and higher-output power. Researches aimed at various applications are also being conducted. An important issue for future applications will be high-power operation. In addition, various advanced functions, such as output beam forming, are important for applications such as wireless communication, and further research is desired. In any case, the advantage of an RTD is its compactness, and it is expected that a variety of applications will be expanded utilizing this feature.

References

1. Brown, E. R., Sonderstrom, J. R., Parker, C. D., Mahoney, L. J., Molvar, K. M., & McGill, T. C. (1991). Oscillations up to 712 GHz in InAs/AISb resonant-tunneling diodes. *Applied Physics Letters, 58*, 2291–2293.
2. Reddy, M., Martin, S. C., Molnar, A. C., Muller, R. E., Smith, R. P., Siegel, P. H., Mondry, M. J., Rodwell, M. J. W., Kroemer, H., & Allen, S. J. (1997). Monolithic Schottky-collector resonant tunnel diode oscillator arrays to 650 GHz. *IEEE Electron Device Letters, 18*, 218–221.
3. Asada, M., & Suzuki, S. (2016). Room-temperature oscillation of resonant tunneling diodes close to 2 THz and their functions for various applications. *Journal of Infrared, Millimeter, and Terahertz Waves, 37*, 1185–1198.
4. Maekawa, T., Kanaya, H., Suzuki, S., & Asada, M. (2016). Oscillation up to 1.92 THz in resonant tunneling diode by reduced conduction loss. *Applied Physics Express, 9*, 024101.
5. Izumi, R., Suzuki, S., & Asada, M. (2017). *1.98 THz resonant-tunneling-diode oscillator with reduced conduction loss by thick antenna electrode*. The International Conference on Infrared, Millimeter and THz Waves (IRMMW-THz), MA3.1, Cancun.
6. Shiode, T., Mukai, T., Kawamura, M., & Nagatsuma, T. (2011). Giga-bit wireless communication at 300 GHz using resonant tunneling diode detector. *Proceedings of Asia Pacific Microwave Conference. (APMC), 1122-1125.*
7. Nishida, Y., Nishigami, N., Diebold, S., Kim, J., Fujita, M., & Nagatsuma, T. (2019). Terahertz coherent receiver using a single resonant tunneling diode. *Scientific Reports, 9*, 18125.
8. Suzuki, S., Shiraishi, M., Shibayama, H., & Asada, M. (2013). High-power operation of terahertz oscillators with resonant tunneling diode using impedance-matched antennas and array configuration. *IEEE Journal of Selected Topics in Quantum Electronics, 19*, 8500108.
9. Kasagi, K., Suzuki, S., & Asada, M. (2019). Large-scale array of resonant-tunneling-diode terahertz oscillator for high output power at 1 THz. *Journal of Applied Physics, 125*, 151601.

10. Kobayashi, K., Suzuki, S., Han, F., Tanaka, H., Fujikata, H., & Asada, M. (2020). Analysis of a high-power resonant-tunneling-diode terahertz oscillator integrated with a rectangular cavity resonator. *Japanese Journal of Applied Physics, 59*, 050907.
11. Izumi, R., Sato, T., Suzuki, S., & Asada, M. (2019). Resonant-tunneling-diode terahertz oscillator with a cylindrical cavity for high-frequency oscillation. *AIP Advance, 9*, 085020.
12. Kitagawa, S., Mizuno, M., Saito, S., Ogino, K., Suzuki, S., & Asada, M. (2017). Frequency-tunable resonant-tunneling-diode oscillators applied to absorbance measurement. *Japanese Journal of Applied Physics, 56*, 058002.
13. Ogino, K., Suzuki, S., & Asada, M. (2018). Phase locking and frequency tuning of resonant-tunneling-diode terahertz oscillators. *IEICE Transactions on Electronics, 101-C*, 183–185.
14. Horikawa, D., Chen, Y., Koike, T., Suzuki, S., & Asada, M. (2018). Resonant-tunneling-diode terahertz oscillator integrated with a radial line slot antenna for circularly polarized wave radiation. *Semiconductor Science and Technology, 33*, 114005.
15. Chen, Y., Suzuki, S., & Asada, M. (2018). *Generation of terahertz vortex waves in resonant-tunneling-diode oscillators by integrated radial line slot antenna*. The International Conference on Infrared, Millimeter and THz Waves (IRMMW-THz), Tu-POS-25, Nagoya.
16. Miyamoto, T., Yamaguchi, A., & Mukai, T. (2016). Terahertz imaging system with resonant tunneling diodes. *Japanese Journal of Applied Physics, 55*, 032201.
17. Ikeda, Y., Kitagawa, S., Okada, K., Suzuki, S., & Asada, M. (2015). Direct intensity modulation of resonant-tunneling-diode terahertz oscillator up to ~30GHz. *IEICE Electronics Express, 12*, 20141161.
18. Asada, M. & Suzuki, S. (2017). *THz oscillators using resonant tunneling diodes and their functions for various applications*. Workshop in European Microwave Week (EuMW), Nuernberg, WTu-01.
19. Oshima, N., Hashimoto, K., Suzuki, S., & Asada, M. (2017). Terahertz wireless data transmission with frequency and polarization division multiplexing using resonant-tunneling-diode oscillators. *IEEE Transactions on Terahertz Science and Technology, 7*, 593–598.
20. Wasige, E. (2018). Over 10 Gbps mm-wave and THz wireless links. *European Microwave Week, WTh04-03*.
21. Dobroiu, A., Wakasugi, R., Shirakawa, Y., Suzuki, S., & Asada, M. (2018). Absolute and precise terahertz-wave radar based on an amplitude-modulated resonant-tunneling-diode oscillator. *Photonics, 5*, 52.
22. Dobroiu, A., Wakasugi, R., Shirakawa, Y., Suzuki, S., & Asada, M. (2020). Amplitude-modulated continuous-wave radar in the terahertz range using lock-in phase measurement. *Measurement Science Technology, 31*, 105001.
23. Dobroiu, A., Shirakawa, Y., Suzuki, S., Asada, M., & Ito, H. (2020). Subcarrier frequency-modulated continuous-wave radar in the terahertz range based on a resonant-tunneling-diode oscillator. *Sensors, 20*, 6848.

Chapter 25
Plasma-Wave Devices

Taiichi Otsuji and Akira Satou

Abstract This chapter describes the recent advances in plasma-wave devices for terahertz applications. The first topic deals with nonlinear hydrodynamic properties of massive two-dimensional plasmons in compound semiconductor heterostructure material systems to demonstrate ultrahigh sensitive detection of terahertz radiation and intense broadband emission. The second topic focuses on graphene, a monolayer carbon-atomic honeycomb lattice crystal, exhibiting peculiar carrier transport and optical properties owing to massless and gapless energy spectrum. Theoretical and experimental studies on active Dirac plasmonics in graphene toward the creation of graphene terahertz injection lasers are described.

25.1 Massive 2D Plasmonic Devices

More than 40 years ago, a new direction in physics opened up with the arrival of plasma-wave electronics [1, 2]. The possibility that plasma waves could propagate faster than electrons fascinated all, suggesting that the so-called "plasmonic" devices could work at terahertz (THz) frequencies, too high for standard electronic devices. The original investigation in plasmonic THz devices was given to the inversion layer in Si MOSFETs [1] and two-dimensional (2D) electron systems in quantum wells of semiconductor heterostructures [2]. All those materials serve massive electrons and holes. The 2D electrons' fluidic and hydrodynamic nonlinear natures reflect the 2D plasmonic functionalities with highly sensitive rectification, instability-driven emission, and frequency conversion of THz radiation. We call such plasmons as massive 2D plasmons. Due to the excellent electron transport properties, III–V compound semiconductor heterostructures like InGaAs/InAlAs/InP, GaAs/AlGaAs, as well as GaN/AlGaN have been well investigated for high-end highly sensitive THz detectors [3]. On the other hand, due to very

T. Otsuji (✉) · A. Satou (✉)
Research Institute of Electrical Communication, Tohoku University, Sendai, Japan
e-mail: otsuji@riec.tohoku.ac.jp; a-satou@riec.tohoku.ac.jp

T. Kürner et al. (eds.), *THz Communications*, Springer Series in Optical Sciences 234, https://doi.org/10.1007/978-3-030-73738-2_25

large integration capability with highest manufacturing cost-effective reproducibility, Si-CMOS detectors have been well developed to make focal plane THz detector array cameras for various industrial/military applications [3].

When THz photons are coupled with 2D plasmons, 2D surface plasmon polaritons (SPPs) are excited, and the original THz photon fields are extremely confined by two to three orders of magnitude in general due to the slow-wave nature of the plasma waves. The wavelength of THz electromagnetic waves is situated in 10s to 100s of micrometers, whereas the THz plasma waves are situated in 10s to 1000s nanometers. This could dramatically increase the interaction of THz photons with 2D electrons of the substance and then dramatically increase the quantum efficiency of the device operation based on conversion of the THz photons to electrons. Also, standard semiconductor planar integrated device processes nanotechnology fits to design THz plasmonic cavities in sub-micrometer-scaled transistor device structures.

The detectors can work for nonresonant, broadband operation with rather high detection responsivity (due to excellent 2D plasmon nonlinear natures) that does not need high-quality factors of the 2D plasmonic structures, which is easy to perform in comparison with resonant, narrowband operation that needs high-quality factors for the 2D plasmonic structures in the devices [4]. On the other hand, in terms of THz emitter and oscillator applications, it is difficult to develop coherent plasmonic THz sources that are the primary demand for various applications. So far, room-temperature coherent plasmonic THz emitters/oscillators have been yet to be realized, and only broadband incoherent or resonant quasi-coherent plasmonic THz emitters have been demonstrated [5].

Highly efficient coupling of THz photons and 2D plasmons is one of the important figures of merits in the 2D plasmonic devices. Metallic grating structures have been utilized for a long time as broadband, efficient THz antennae to efficiently couple THz photons with 2D plasmons. An advanced grating antenna structure called an asymmetric dual-grating gate (A-DGG) was proposed and implemented in HEMT structures [6], resulting in record responsivity ranging from 200 GHz to 2 THz [3, 5]. Two grating gates are interdigitated with an asymmetric spacing between adjacent gate fingers. Under complementary (Hi/Lo)-biased DGG conditions, high electron density region under the high-biased DG works for a plasmonic rectifier, whereas depleted region under the low-biased DG works for a resistive load region. In such a unit section of one A-DGG finger pair, the rectified plasmonic dc current is converted to a photovoltaic signal. The photovoltaic signal in such a unit section is superposed along with the entire A-DGG sections in a cascading manner, resulting in a high photovoltaic output [5].

One drawback for A-DGG HEMTs is the polarization sensitivity that is insensitive to THz radiation whose polarization is parallel to the DGG fingers. To cope with this issue, a 2D diffraction-grating gate ((2D-DGG) structure was proposed and polarization-insensitive THz detection demonstrated (Fig. 25.1a and b) [7]. Very recently, a novel gate readout scheme was also proposed (Fig. 25.1a). Due to high-output impedance of the drain terminal, the drain readout cannot match the 50-Ω impedance of high-speed interconnection, resulting in multi-reflection-caused

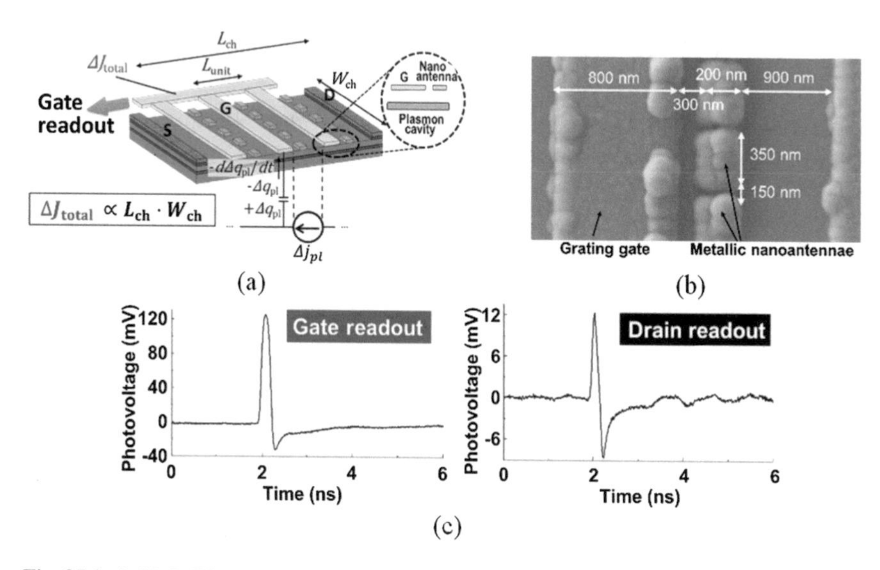

Fig. 25.1 A 2D-DGG HEMT plasmon THz detector [8]. (**a**) Device bird's view, (**b**) scanning electron microscopic (SEM) image of a magnified device top view, (**c**) measured photovoltaic signals by gate readout (left), and drain readout (right) in response to pulsed CW THz radiation incidence

severe waveform distortion in response to a high-speed on/off coded THz radiation incidence. Also, the high-impedance output could not drive a 50-Ω matching load at the far end, resulting in severely lowering the effective responsivity under 50-Ω loaded conditions. On the other hand, the gate terminal serves a low-output impedance that is easy to match the 50-Ω interface. This could successfully resolve the aforementioned critical two problems simultaneously, providing an excellent fidelity photoresponse signal integrity with a record high responsivity (Fig. 25.1c) [8].

In terms of 2D plasmonic THz emitters, various experimental demonstrations have been reported, including broadband emission due to thermally excited incoherent plasmons in single-gate and DGG HEMTs and frequency tunable quasi-coherent resonant THz emission due to resonant plasmon excitations in a single-gate AlGaN/GaN HEMT. A major interest for use in THz emission from the 2D plasmon emitter devices is the promotion of 2D plasmonic instability. A dc channel current driven by a dc drain-source bias voltage may promote the plasmon instability, leading to self-oscillatory coherent plasmon emissions. There are several different mechanisms to promote the instability such as Dyakonov-Shur Doppler-shifting reflection-amplification type [9], Ryzhii-Satou-Shur electron-transit time-modulation type [5], and Cherenkov plasmonic-boom type [10]. However, the electron transport properties of compound semiconductors limit the quality factors of the 100-nanometer-scaled 2D plasmon cavities up to a few even at low temperatures. This substantially makes it difficult to realize coherent intense THz oscillatory emission in any THz emitters driven by massive 2D plasmons [5].

25.2 Massless 2D Plasmonic Devices

Graphene, a monolayer carbon-atomic honeycomb lattice sheet, originates from gapless and linear dispersive energy band structure, providing relativistic charged quanta of massless Dirac fermions in electrons and holes. Graphene massless Dirac fermions of electrons and holes serve extremely higher carrier mobilities by more than two orders of magnitude than those of massive electrons and holes in any conventional semiconductors. The optical properties of graphene are exotic; interband transition-based optical absorption is independent of photon energy with absorption coefficient of 2.3% per layer. The graphene Dirac plasmons also take extremely low damping and high nonlinearities. These unique features have opened a new paradigm to explore any unexplored science and engineering fields including THz active plasmonic devices [11].

The most popular demonstrations on graphene plasmonic THz devices were plasmonic detection of THz radiation in graphene field effect transistors (GFETs) [12]. There exist several different rectification mechanisms including plasmonic, photo-thermoelectric, as well as bolometric detections. The former two mechanisms are primarily concerned in the GFET implementations, whereas the last one is associated with resistive two-terminal device implementations. Depending on the applied gate bias condition, the plasmonic and/or photo-thermoelectric detection takes place [3]. The plasmonic detection serves fastest photoresponse with rather high responsivity. Due to immaturity of the graphene synthesis and GFET process technology, the obtained detection performance stays at rather unexpected low levels that do not reflect idealistic graphene carrier transport properties [3]. One of the hot topics in this field is clear observation of plasmonic higher harmonic resonant photoresponses in a GFET under cryogenic temperatures, which was never ever observed in any massive semiconductor materials/devices [13].

Linear and gapless energy spectrum of graphene carriers enables population inversion under optical and electrical pumping, giving rise to gain in a wide THz frequency range. We first theoretically discovered this phenomenon [14] and resultant THz gain and recently demonstrated experimental observation of amplified spontaneous THz emission and single-mode THz lasing with rather weak intensity at 100 K in current-injection pumped GFETs featured by a distributed feedback dual-gate structure [15]. A high quality of epitaxial graphene synthesized by the thermal decomposition of a C-face SiC substrate was utilized [15].

Present issues of poor gain overlapping and poor quantum efficiency (limited by the interband absorption coefficient of 2.3%) can be resolved by introducing the graphene SPPs [16]. Graphene plasmon instability could help substantially boost the THz gain as a new physical mechanism. We've recently succeeded in experimental observation of giant amplification of stimulated emission of THz radiation at 300 K driven by graphene-plasmon instability in A-DGG GFETs [17]. A highest quality of exfoliated graphene was utilized with h-BN encapsulated layers (Fig. 25.2a). The ADGG-GFETs introduce periodically modulated carrier-density profiles by applying a high bias to one GG and a charge neutral point bias to the other GG.

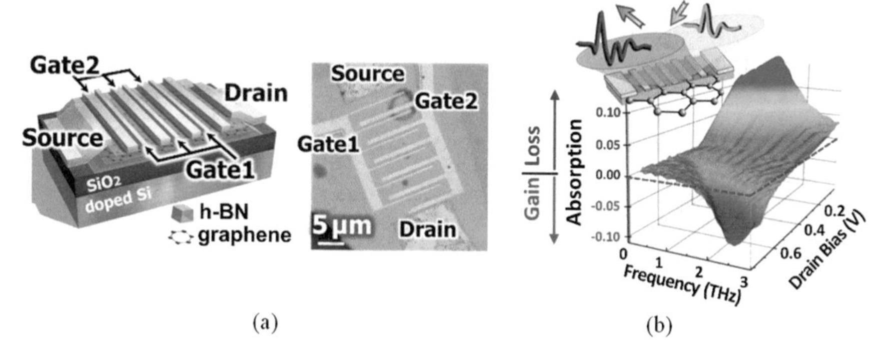

Fig. 25.2 Room temperature amplification of THz radiation by A-DGG GFETs [17]. (**a**) A device's bird's view and SEM image, (**b**) measured absorption spectra for different drain biases. The red-colored portions turn to amplification (negative absorption) with gain of up to 9%

The finger width of the highly biased GG defines the SPP cavity size so that an SPP resonant absorption is obtained to the THz radiation when drain is unbiased. Depending on the carrier density of the SPP cavity, the absorption peak frequency shifted. With the increasing drain dc bias level, the absorption peak exhibited a red shift and weakened. When the drain bias approaches a low-end threshold level, the sample became perfectly transparent over the measured frequency range. When the drain bias exceeds a high-end threshold level, the sample exhibited a resonant amplification with a maximum gain of 9%. The gain spectra showed a blue shift with the increasing drain bias (Fig. 25.2b) [17].

Such an overall response from absorption to amplification with respect to the drain bias can be qualitatively interpreted by the graphene SPP instability theory [10, 17]. Integrating the graphene SPP amplifier into a current-injection graphene THz laser transistor will be a promising solution toward room-temperature intense THz lasing. A more sophisticated approach based on a gated double-graphene-layered van der Waals nanocapacitor heterostructure has been proposed as a new type of quantum cascade THz laser structures [18]. Its proof of concept has been experimentally demonstrated [19]. Further study will make it a reality.

References

1. Allen, S. J., Jr., Tsui, D. C., & Logan, R. A. (1977). Observation of the two-dimensional Plasmon in silicon inversion layers. *Physical Review Letters, 38*, 980–983.
2. Hopfel, R., Lindemann, G., Gornik, E., Stangl, G., Gossard, A. C., & Wiegmann, W. (1982). Cyclotron and plasmon emission from two-dimensional electrons in GaAs. *Surface Science, 113*, 118–123.
3. Otsuji, T. (2015). Trends in the research of modern terahertz detectors: Plasmon detectors. *IEEE Transactions on Terahertz Science and Technology, 5*, 1110–1120.

4. Dyakonov, M., & Shur, M. (1640-1645). Detection, mixing, and frequency multiplication of terahertz radiation by two-dimensional electronic fluid. *IEEE Transactions on Electron Devices, 43*, 1996.
5. Otsuji, T., Watanabe, T., Boubanga Tombet, S., Satou, A., Knap, W., Popov, V., Ryzhii, M., & Ryzhii, V. (2013). Emission and detection of terahertz radiation using two-dimensional electrons in III-V semiconductors and graphene. *IEEE Transactions on Terahertz Science and Technology, 3*, 63–71.
6. Popov, V. V., Fateev, D. V., Otsuji, T., Meziani, Y. M., Coquillat, D., & Knap, W. (2011). Plasmonic terahertz detection by a double-grating-gate field-effect transistor structure with an asymmetric unit cell. *Applied Physics Letters, 99*, 243504.
7. Suzuki, M., Hosotani, T., Otsuji, T., Suemitsu, T., Takida, Y., Ito, H., Minamide, H., & Satou, A. (2018). *Coupling of 2D plasmons in grating-gate plasmonic THz detector to THz wave with lateral polarization*. the 43rd IRMMW-THz Dig., Tu-A2-1a-2, Nagoya, Aichi, Japan, Sept. 11, 2018.
8. Negoro, T., Saito, T., Hosotani, T., Otsuji, T., Takida, Y., Ito, H., Minamide, H., & Satou, A. (2020). Gate-readout of photovoltage from a grating-gate plasmonic THz detector. the 45th IRMMW-THz Dig. Buffalo, NY, USA (Online WEB), Nov. 12, 2020.
9. Dyakonov, M., & Shur, M. (1993). Shallow water analogy for a ballistic field effect transistor: New mechanism of plasma wave generation by dc current. *Physical Review Letters, 71*, 2465–2468.
10. Aizin, G. R., Mikalopas, J., & Shur, M. (2016). Current driven “plasmonic boom” instability in gated periodic ballistic nanostructures. *Physical Review B,**93***, 195315. S.A. Mikhailov, “Plasma instability and amplification of electromagnetic waves in low-dimensional electron systems,” Physical Review B, 58, 1517–1532, 1998.
11. Geim, A. K., & Novoselov, K. S. (2007). The rise of graphene. *Nature Materials, 26*, 183–191.
12. Vicarelli, L., Vitiello, M. S., Coquillat, D., Lombardo, A., Ferrari, A. C., Knap, W., Polini, M., Pellegrini, V., & Tredicucci, A. (2012). Graphene field-effect transistors as room-temperature terahertz detectors. *Nature Materials, 11*, 865–871.
13. Bandurin, D. A., Svintsov, D., Gayduchenko, I., Shuigang, G., Xu, G., Principi, A., Moskotin, M., Tretyakov, I., Yagodkin, D., Zhukov, S., Taniguchi, T., Watanabe, K., Grigorieva, I. V., Polini, M., Goltsman, G. N., Geim, A. K., & Fedorov, G. (2018). Resonant terahertz detection using graphene plasmons. *Nature Communications,**9***, 5392.
14. Ryzhii, V., Ryzhii, M., & Otsuji, T. (2007). Negative dynamic conductivity of graphene with optical pumping. *Journal of Applied Physics, 101*, 083114.
15. Yadav, D., Tamamushi, G., Watanabe, T., Mitsushio, J., Tobah, Y., Sugawara, K., Dubinov, A. A., Satou, A., Ryzhii, M., Ryzhii, V., & Otsuji, T. (2018). Terahertz light-emitting graphene-channel transistor toward single-mode lasing. *Nanophotonics, 7*, 741–752.
16. Watanabe, T., Fukushima, T., Yabe, Y., Boubanga-Tombet, S. A., Satou, A., Dubinov, A. A., Aleshkin, V. Y., Mitin, V., Ryzhii, V., & Otsuji, T. (2013). The gain enhancement effect of surface plasmon polaritons on terahertz stimulated emission in optically pumped monolayer graphene. *New Journal of Physics, 15*, 075003.
17. Boubanga-Tombet, S., Knap, W., Yadav, D., Satou, A., But, D. B., Popov, V. V., Gorbenko, I. V., Kachorovskii, V., & Otsuji, T. (2020). Room temperature amplification of terahertz radiation by grating-gate graphene structures. *Physical Review X, 10*, 031004.
18. Dubinov, A. A., Bylinkin, A., Ya Aleshkin, V., Ryzhii, V., Otsuji, T., & Svintsov, D. (2016). Ultra-compact injection terahertz laser using the resonant inter-layer radiative transitions in multi-graphene-layer structure. *Optics Express, 24*, 29603–29612.
19. Yadav, D., Boubanga-Tombet, S., Watanabe, T., Arnold, S., Ryzhii, V., & Otsuji, T. (2016). Terahertz wave generation and detection in double-graphene layered van der Waals heterostructures. *2D Mater,**3***, 045009.

Part VII
Transceiver Technologies 3: Photonics

Chapter 26
Photonics-Based Transmitters and Receivers

Tadao Nagatsuma and Guillaume Ducournau

Abstract This chapter describes latest advances in THz communication research based on photonic technologies. Photonics-based transmitters and receivers are explained focusing on system configurations and enabling technologies.

26.1 Research Trend

There have been numerous publications since early 2000, which report wireless transmission experiments using carrier frequencies beyond 100 GHz [1–9]. Figure 26.1 summarizes reported data rates of over 2 Gbit/s, experimentally achieved by transmission experiments at carrier frequencies from 100 GHz to 700 GHz. Data points with triangles are demonstrations using electronics-based transmitter, while data points with circles are achieved by photonics-based ones. Filled and open marks denote data measured with a real-time transmission experiment and with an off-line signal processing, respectively.

Carrier frequencies below 450 GHz are the most actively studied bands due to the availability of devices and components for transmitters with enough output power. Si-electronics-based transceivers now enable 100-Gbit/s wireless links at frequencies from 200 GHz to 300 GHz [10–13], and GaAs/InP-electronic technologies can cover frequency bands up to 300 GHz band with a data rate of over 64 Gbit/s at a transmission distance of 850 m [13] and 120 Gbit/s at 10 m [14].

The photonics-based approach, particularly employed in transmitters, is not only a technology driver to show unprecedented performance of THz communications but also a technology demonstrator to get early users or customers. In addition,

T. Nagatsuma (✉)
Osaka University, Toyonaka, Japan
e-mail: nagatuma@ee.es.osaka-u.ac.jp

G. Ducournau
Université de Lille, Lille, France

T. Kürner et al. (eds.), *THz Communications*, Springer Series in Optical Sciences 234, https://doi.org/10.1007/978-3-030-73738-2_26

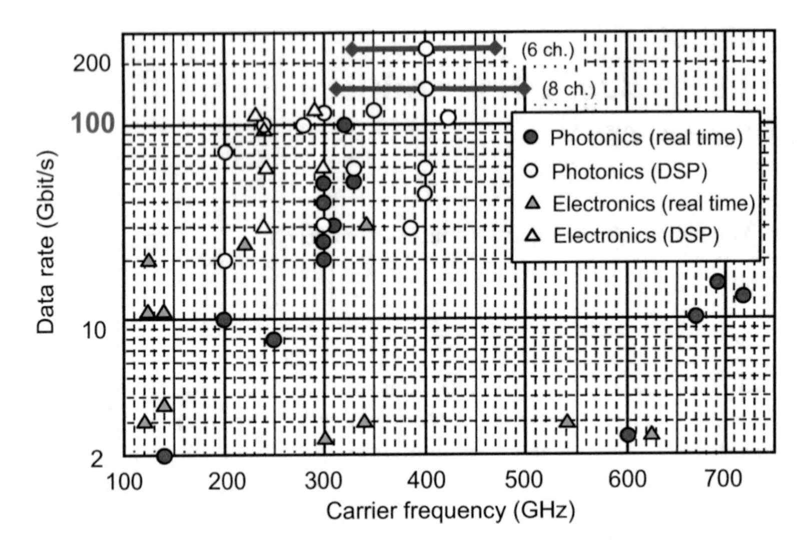

Fig. 26.1 Carrier frequency vs. data rate obtained for single-channel THz wireless communication systems with real-time transmission (solid circle/triangle) and with off-line digital signal processing (open circle/triangle)

photonics-based systems might be deployed in the future convergence of fiber-optic and wireless communication networks.

26.2 System Configurations

Figures 26.2a and b show schematic diagrams of a transmitter and a receiver, respectively. THz communication research was initiated with the use of photonic techniques for signal generation and modulation as shown in building blocks of the photonics-based transmitter of Fig. 26.2a. Compared to the all-electronic transmitter, it has proven to be effective to achieve spurious-free carrier signals [15] and higher data rates of over 50 Gbit/s. This could be realized thanks to the availability of telecom-based high-frequency components such as lasers, modulators, and photodiodes (O–E converters). The use of optical fiber cables enables us to distribute high-frequency RF signals over long distances and makes the size of transmitter front-ends compact and light. Moreover, ultimate advantage of the photonics-based approach is that wired (fiber-optic) and wireless communication networks could be connected seamlessly in terms of data rates and modulation formats.

In the transmitter, first, intensity-modulated optical signals, whose envelope is sinusoidal at a designated THz frequency, are generated with the use of dual-wavelength light sources generating different wavelengths, λ_1 and λ_2. Then, these two wavelengths of lights are injected to the photodiode, which leads to the generation of RF signals or THz waves at a frequency given by $f_{RF} = c\Delta\lambda/\lambda_1\lambda_2$,

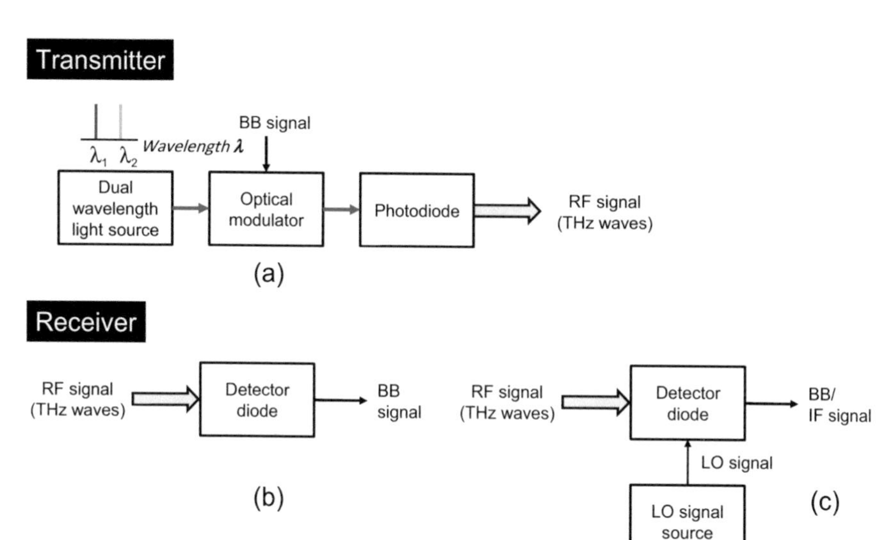

Fig. 26.2 Block diagram of the wireless link using a photonics-based transmitter (**a**) and receivers for direct (envelope) detection (**b**) and coherent (heterodyne and homodyne) detection (**c**). BB: base band (i.e., unmodulated digital signal at input or recovered digital signal at output)

where $\Delta\lambda$ is a difference in wavelength of lights and c is the vacuum velocity of light. The converted signals are finally radiated into free space by an antenna connected at the photodiode current output.

In the receiver, the RF signal is demodulated into the baseband data signal using detector diodes with direct detection or envelope detection scheme (Fig. 26.2b) and coherent homodyne or heterodyne detection scheme (Fig. 26.2c). Usually, the demodulated signal is amplified with a baseband amplifier and reshaped with a limiting amplifier. Transmission characteristics are evaluated with baseband testing instruments such as sampling oscilloscopes and bit error rate testers (BERTs). There are several options for detector diodes such as Schottky barrier diodes (SBDs) [16], plasma-wave FET detectors [17], resonant tunneling diodes (RTDs) [18, 19], and Fermi-level managed barrier diodes (FMBDs) [20–22]. SBDs are the most commonly used detectors for THz communications, as a kind of workhorse. One of the practical features of plasma-wave FET detectors is a uniformity of device performance over the wafer and might be applicable to detector arrays. RTDs have recently proven to be used as sensitive receivers when they are operated as a self-injection mixer [19]. FMDBs are verified to exhibit their superior performance with respect to sensitivity and required LO power [22].

Figure 26.3 shows different approaches to implement the dual-wavelength light source. The most straightforward source involves combining the light from two different single-frequency semiconductor lasers (Fig. 26.3a). While its main advantage is the broad tuning range of frequency, the weakness is that the frequency stability is generally poor, requiring locking techniques for the two optical wavelengths [23].

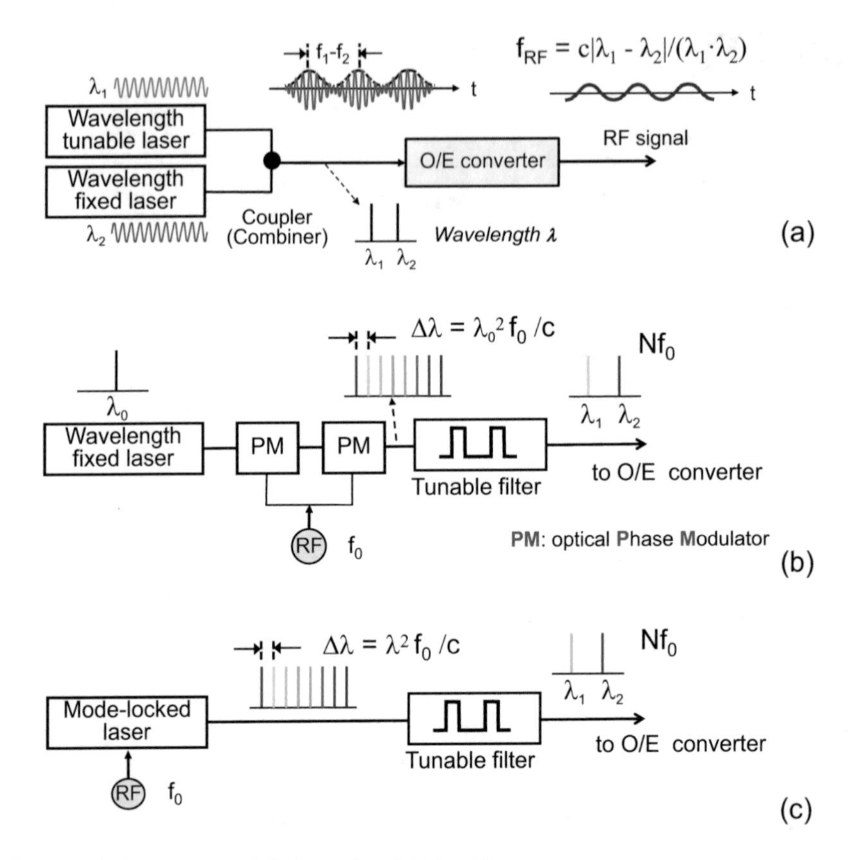

Fig. 26.3 Block diagrams of dual-wavelength light sources

One way to achieve stable signals is to use an external optical modulator driven by an electronic RF source. When an optical intensity or phase modulator is used after a CW single-frequency semiconductor laser, sidebands around the optical wavelength are generated, being spaced by the modulation frequency applied to the modulator. Higher-order harmonics of over 10th can be achieved cascading modulators [24]. In this case, optical filters such as arrayed waveguide gratings (AWGs) and other tunable filters should be used to select only two of harmonic components showing optical frequency comb signals (Fig. 26.3b). Another optical CW signal generator with THz frequencies is based on optical pulsed sources. The output of the semiconductor laser is a continuous stream of pulses, spaced in time by the inverse of the repetition frequency, $f_0 = c/2n_gL$, where n_g is a refractive index of the laser active layer and L is a cavity length of the laser. When the external RF signal at f_0 or f_0 /m (m is an integer) is applied to the laser, the optical spectrum becomes a comb of modes, spaced by this frequency. The main characteristic is that each of the modes is locked in phase to the adjacent ones (Fig. 26.3c).

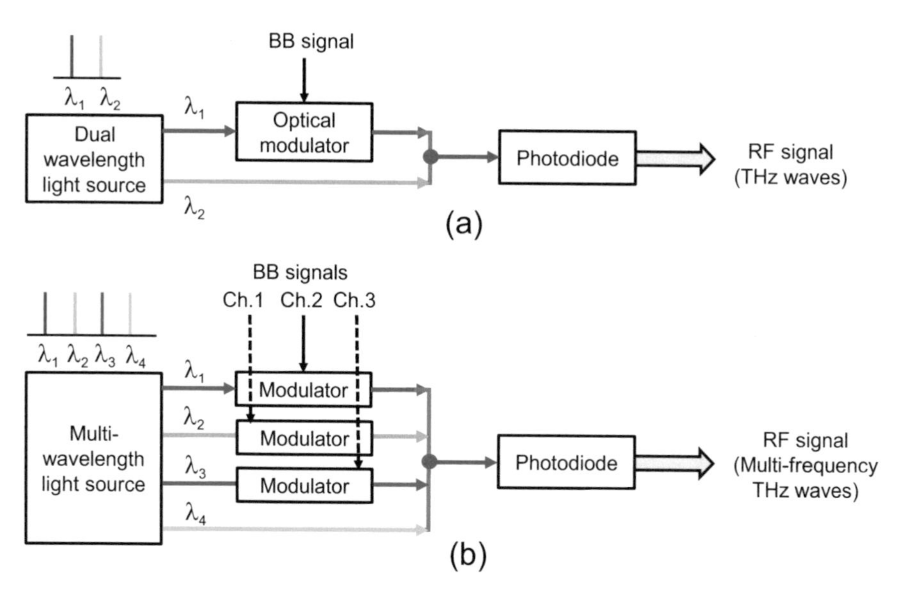

Fig. 26.4 Modulation schemes for light sources

One of the practical issues in the comb-based RF signal generation is a phase noise, since the signal generation scheme is essentially a frequency multiplication, in which the phase noise increases quadratically with the harmonic number. Low-phase noise and narrow-linewidth photonic sources based on the spectral narrowing effect of the stimulated Brillouin scattering (SBS) phenomenon in an optical cavity such as fibers have been studied; the phase noise of generated RF signals does not scale up as the frequency increases [25].

In photonics-based transmitters for communications, data signals or baseband signals can be easily applied by using optical modulators. Figure 26.4a shows the most common way to modulate the THz carrier with a designated modulation format, which includes OOK (ASK), BPSK, QPSK, QAM, etc., where one of the two light waves is modulated in the optical domain. In the case of OOK (ASK) modulation, both light waves can be simultaneously modulated to double the efficiency. Multichannel transmitter can be realized by using multi-wavelength light sources like the optical frequency comb as shown in Fig. 26.4b [26].

26.3 O–E Converters

In photonics-based transmitters, the most critical component is the O–E converter which offers broad bandwidth, high-output power (high saturation level), and high linearity. To meet these requirements, there have been quite a lot of research and development on THz photodiodes and photoconductors operating at telecom

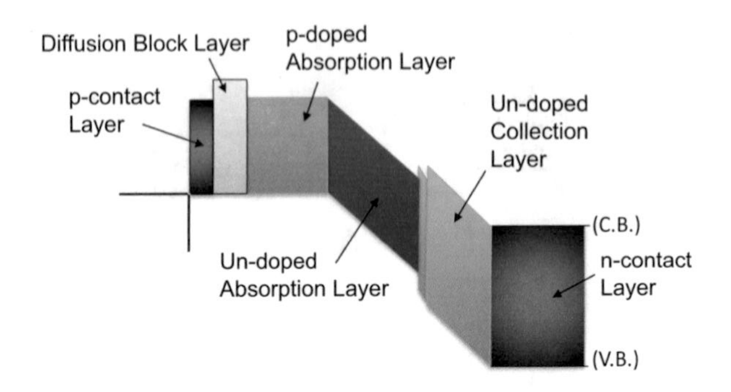

Fig. 26.5 Band diagram of modified UTC-PD.

wavelength (1.55 μm). Uni-travelling-carrier photodiodes (UTC-PDs) and their modified structures have been widely used for THz transmitters, since UTC-PDs exceed the performance of conventional PIN photodiodes with respect to bandwidth and output power [27–29]. In UTC-PDs, only electrons act as active carriers to avoid the transport of holes, whose velocity is much slower than that of electrons. Performance improvement in PIN photodiodes as well as photoconductors has also been continuing [30, 31].

Figure 26.5 shows a structure of modified UTC-PD which optimizes a trade-off between the bandwidth and the responsivity [26]. It looks like a combination of UTC-PD with p-doped absorption layer and un-doped carrier collection layer and PIN-PD with un-doped absorption layer.

Figure 26.6 shows a photo of resonant cavity-enhanced broadband UTC-PDs (RCEUTC-PDs) optimized for 300-GHz-band signal generation. It uses a novel semitransparent top contact utilizing subwavelength apertures for an enhancement of optical transmission. The responsivity of the device is improved by introducing a metallic mirror below the diode mesa through wafer bonding, producing an optical resonant cavity [27]. This photodiode has been successfully applied to 300-GHz-band wireless communications at 100 Gbit/s [32].

Figure 26.7 shows a waveguide-output UTC-PD module, which operates in 600-GHz band [33]. A broadband coupler, which connects the UTC-PD and the hollow waveguide, is monolithically integrated with the UTC-PD chip. As a result, 3-dB bandwidth of the module exceeds 200 GHz, and the module has been successfully applied to the 600-GHz-band wireless link.

Further continuous research for THz photodiodes should be focused on an increase of the output power. THz photodiodes should be inherently operated at as higher current as possible to increase the conversion efficiency from optical to RF power. One of the critical problems is a thermal management of photodiodes, since the failure of THz photodiodes is often caused by the thermal effect due to horrific increase in the current density, as the optical input power increases. Transferring the substrate from InP to other substrates, which have better thermal conductivity, such as SiC or heatsink materials has proven to be effective even

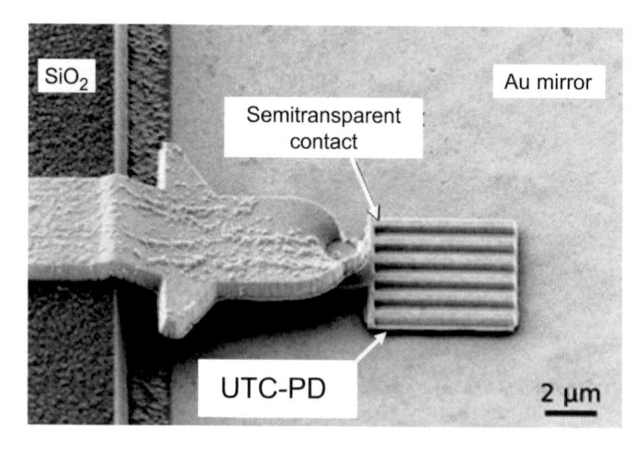

Fig. 26.6 SEM image of RCEUTC-PD, integrated with conductor-backed CPW lines on SiO_2

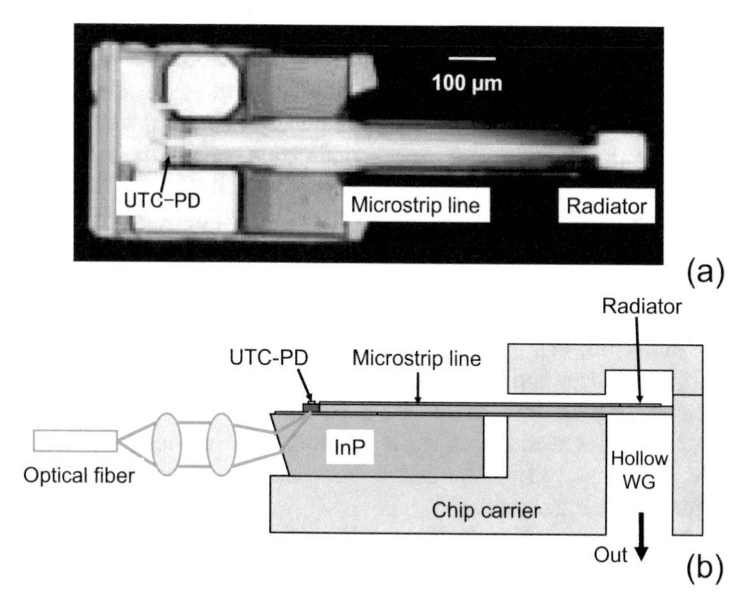

Fig. 26.7 (**a**) 600-GHz-band UTC-PD chip with an integrated coupler to be packaged on the hollow waveguide (WR-1.5). (**b**) Packaging structure of the UTC-PD module with optical fiber input and waveguide (WG) output

at microwave and millimeter-wave frequencies [34, 35]. Another way is a power combining by collecting the output power from multiple photodiodes in the circuit level via waveguides or the spatial level via antennas [36–38].

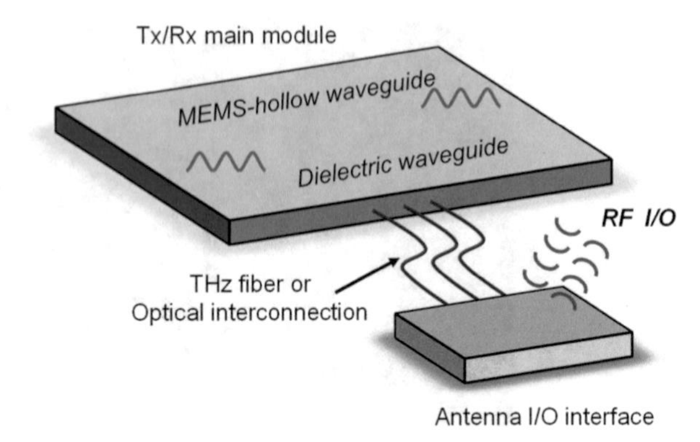

Fig. 26.8 Conceptual illustration of future THz transceiver module

26.4 Integration Technologies

For photonics-based transmitters to compete with all electronic ones in terms of size and cost, integration technologies are of greatest importance with monolithic as well as hybrid manners. A fully monolithically integrated transmitter on the InP substrate, which includes dual lasers, modulators, amplifiers, and photodiodes, has been demonstrated at 90 GHz [39]. Before reaching such a full-component integration, integrated photonic sources would be the first priority, since the cost and size of most THz communication systems are governed by dual-wavelength sources, which satisfy required stability, controllability, and output power. Integrated laser sources on InP, Si, and polymeric materials have successfully applied to THz communications [40, 41].

In the hybrid integration scheme particularly for much higher-frequency operation, the choice of platform technologies is still under consideration among several promising candidates such as MEMS-based hollow waveguide [42] and dielectric waveguide structures [43] to minimize loss and dispersion in the THz signal transmission and/or interconnects.

Figure 26.8 shows a concept of future THz systems, where dielectric waveguides or hollow waveguides act as a broadband interconnect platform for active functional devices, such as signal generators, modulators, and detectors. In some cases, optical interconnection can be used for loss- or dispersion-sensitive parts such as antenna remoting. THz cables, or THz fibers, are also considered as alternative media to the optical interconnect, when the conversion efficiency from baseband to THz signals becomes competitive with that from baseband to optical signals.

26.5 Experimental Systems Using Photonic Transmitters

Many examples of system demonstrations are described in Part X. Here, some representative wireless links using photonics-based transmitters at carrier frequencies from 200 GHz to 500 GHz are summarized in Table 26.1. The highest single-channel data rate achieved to date with an error-free condition (bit error rate (BER) of less than 10^{-11}) is 50 Gbit/s, where the on-off keying (OOK) modulation in the photonics-based broadband transmitter and the direct detection in the receiver are employed at 320 GHz [8]. In order to increase data rates, coherent transmission systems with multilevel modulation formats such as quadrature phase shift keying (QPSK), 16 quadrature amplitude modulation (QAM), and 64QAM have been examined with a single carrier and multi-carriers [32, 44–50]. A record single-channel data rate has reached 100 Gbit/s with BER of 3.4×10^{-3} at 240 GHz [44] and with BER of 2.0×10^{-2} at 425 GHz [49], which are measured by the digital signal processor (DSP) installed in commercially available oscilloscopes.

Table 26.1. Summary of recent THz communication research using photonic transmitters

Figure 26.9a shows a block diagram of a single-channel 100-Gbit/s transmission system, using the photonics-based transmitter with 16 QAM modulation format in 300-GHz band [32]. The achievable BER was lower than forward error correction (FEC) limit value of $\sim 4 \times 10^{-3}$ with clear constellation as shown in Fig. 26.9b.

In contrast, Fig. 26.10 shows a block diagram of real-time 100-m transmission system using the photonics-based transmitter with QPSK modulation format and an "analog" coherent receiver without DSPs [50–52]. This is in order not only to exam-

Table 26.1 Summary of recent THz communications research using photonic transmitters

Authors	Frequency	Modulation	Data rate	Signal processing
S. Koenig et al. [44] (KIT, 2013)	~240 GHz	16QAM (25 Gbaud)	100 Gbit/s	**DSP** (off-line w/ real-time scope)
V. K. Chinni et al. [32] (IEMN, 2018)	280 GHz	16QAM (25 Gbaud)	100 Gbit/s	**DSP** (off-line w/ real-time scope)
H. Shams et al. [45] (UCL, 2015)	~200 GHz	QPSK (12.5 Gbaud) ***4 carriers***	100 Gbit/s **(4 ch)**	**DSP** (off-line w/ real-time scope)
X. Yu et al. [46] (DTU, 2016)	300 GHz ~500 GHz	QPSK (10 Gbaud) ***8 carriers***	160 Gbit/s **(8 ch)**	**DSP** (real-time scope)
X. Pang et al. [47] (NETLAB, 2016)	300 GHz ~500 GHz	16QAM (12.5 Gbaud) ***6 carriers***	260 Gbit/s **(6 ch)**	**DSP** (real-time scope)
A. Stöhr et al. [48] (Duisburg U., 2017)	328 GHz	64QAM (7.4 Gbaud)	59 Gbit/s	**DSP** (real-time scope)
X. Pang et al. [49] (NETLAB, 2017)	425 GHz	16QAM (32 Gbaud)	106 Gbit/s	**DSP** (real-time scope)
T. Nagatsuma et al. [50] (Osaka U., 2016)	320 GHz	QPSK (50 Gbaud)	100 Gbit/s	**Analog Real time**

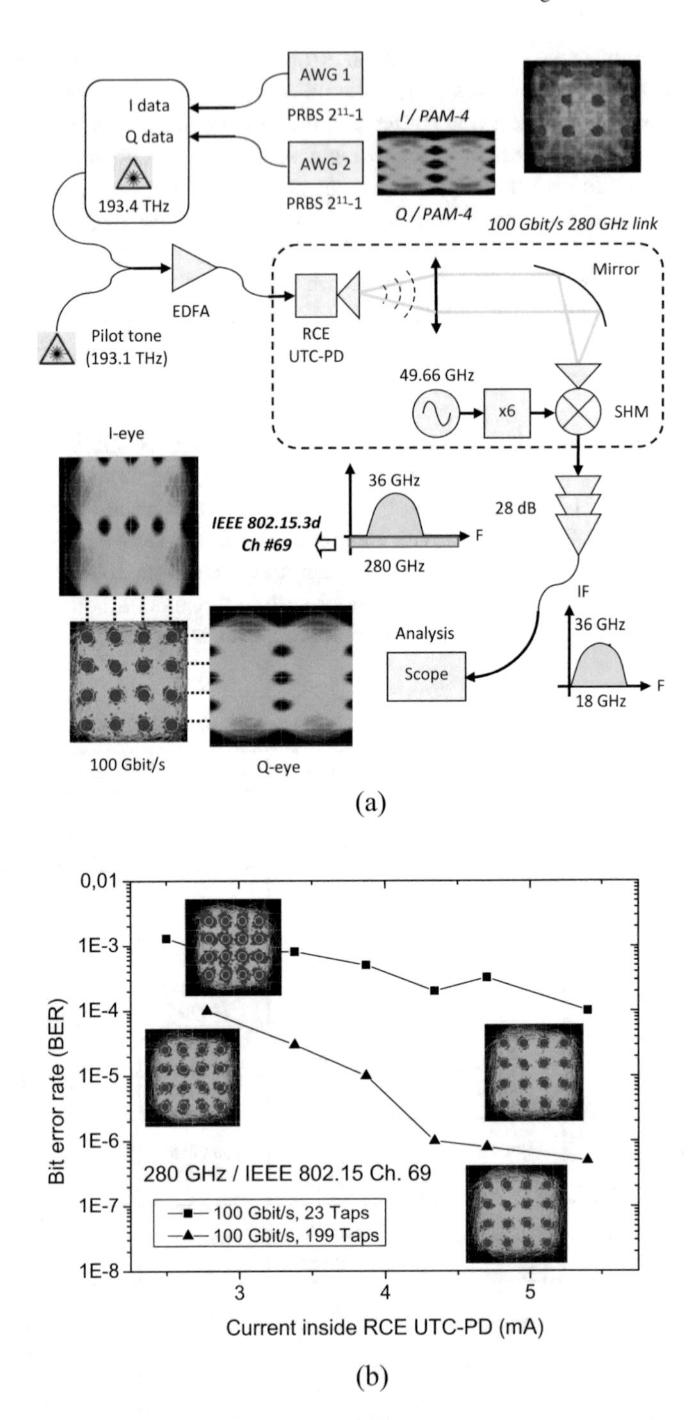

Fig. 26.9 (**a**) 100 Gbit/s wireless link. Carrier frequency 280 GHz with 16QAM modulation scheme. (**b**) BER characteristics at 25 GBaud using 23 (squares) and 199 taps (triangles) for equalization

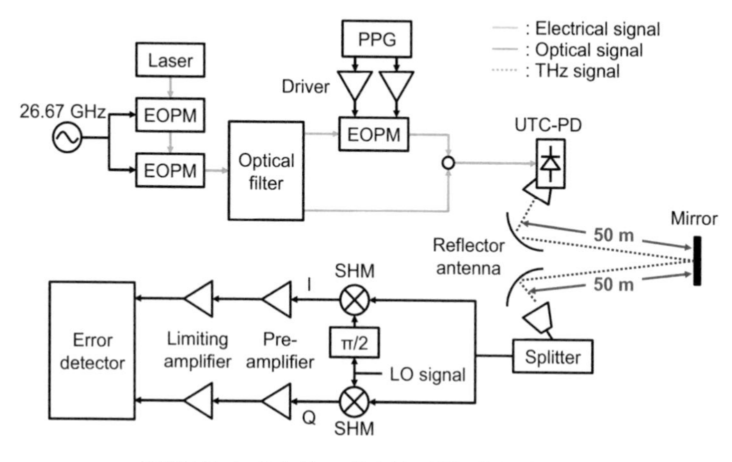

(a)

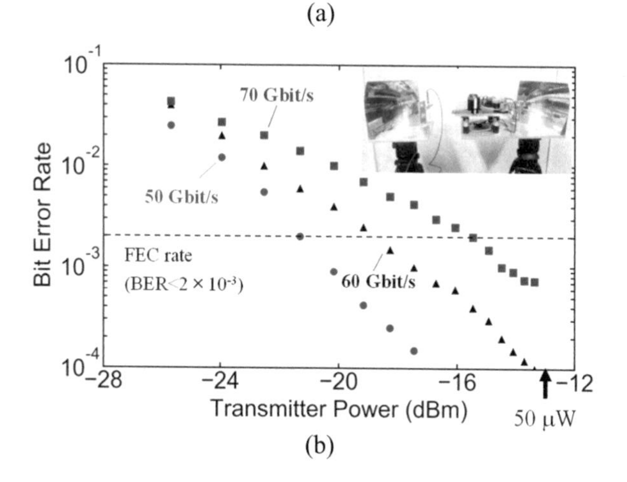

(b)

Fig. 26.10 (**a**) Experimental setup for 100-m wireless link using a photonic QPSK transmitter at 320 GHz. (**b**) BER characteristics of 100-m QPSK transmission system. Photo: transmitter and receiver mounted on high-gain reflector antenna

ine a long-time stability of the transmission as well as an error-free transmission but also to conduct a real-time transmission of uncompressed ultrahigh-definition TV (4 K/8 K) signals without the use of application-specific DSPs and/or high-speed DA/AD converters.

26.6 Photonics-Based Receivers

Ultimate motivation of photonics-enabled THz communications is a convergence of fiber-optic and wireless radio networks. Thus, in the receiver, the output signals should be converted to optical signals with simpler and more efficient ways. One of practical ways is IF signal transmission and demodulation as schematically shown in Fig. 26.11. The operation point of the electro-optic intensity modulator is usually set to the null point to suppress the optical subcarrier signal. This method has proven to be effective not only to overcome an IF signal leakage problem, which is common with electronic devices, but also to shape the optical spectrum for efficient demodulation in communication experiments in 300-GHz and 600-GHz bands [53, 54].

Figure 26.12 shows other types of receiver configurations, where photonic devices and techniques are employed. In Fig. 26.12a, photonically generated signals are applied to the diode mixer as a local oscillator (LO) signal, and it is often referred to as a photonic LO. In case of mixers using Schottky barrier diodes (SBDs), they usually require the LO power of a few mW, and thus the amplifier is used after the photonic LO source, since the output power of THz photodiodes is less than 1 mW at 300 GHz. Fermi-level managed barrier diodes (FMBDs) require the LO power of less than 0.1 mW, because of their low barrier height (<0.1 eV) compared to SBDs (0.6 ~ 0.9 eV). FMBD mixers directly pumped by the UTC-PD have successfully been applied to the receiver in the 300-GHz band [22, 55].

Receivers of Fig. 26.12b are directly pumped by optical signals and are often called as photonic mixers, and photodiodes and photodetectors have been demonstrated as photonic receivers for millimeter-wave and THz communications [56, 57]. The most crucial issue in photonic mixers is high conversion loss, which is as high as 30–50 dB. Even in such a situation, photoconductive receivers have been applied

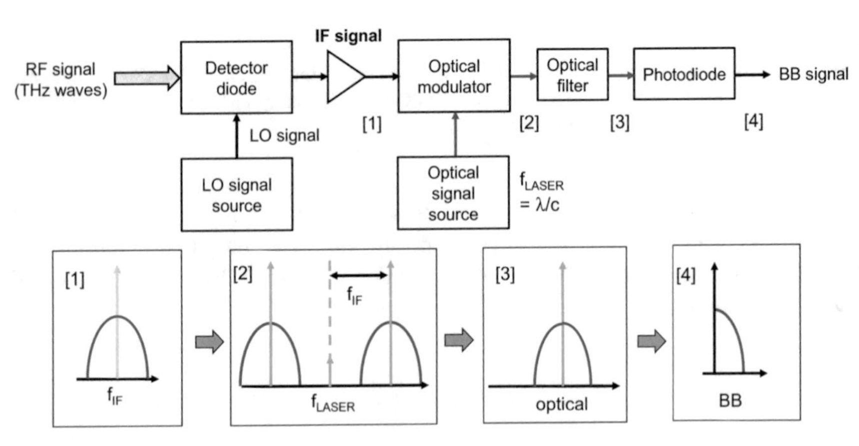

Fig. 26.11 Block diagram of the optical fiber output-type receiver, in which photonic techniques are efficiently used in the demodulation process

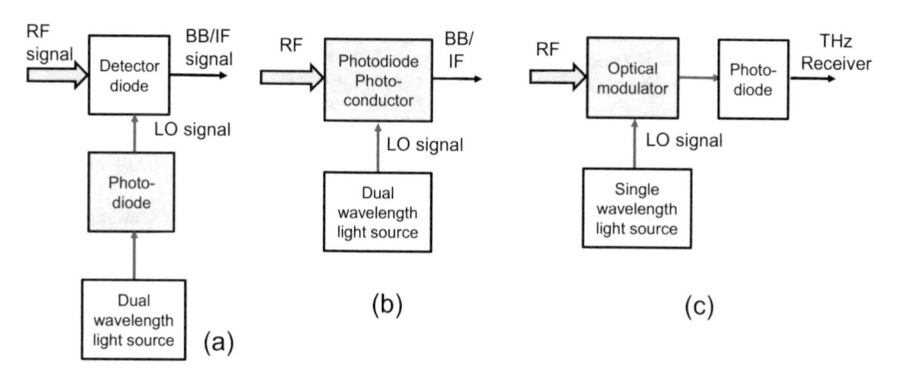

Fig. 26.12 Photonics-enabled receiver configurations. (**a**) Photonic local oscillator (LO) type. (**b**) Photonic receiver type. (**c**) Photonic transducer type

to the receiver for 10-Gbit/s QPSK data link at 300 GHz by using a THz preamplifier in the front-end [58].

Receivers of Fig. 26.12c are based on the optical modulator which can operate at THz frequencies and are more simplified optical fiber output-type compared to the configuration of Fig. 26.10. Recent progress in optical modulator technologies using highly nonlinear polymeric EO materials and plasmonic enhancement in interaction between THz waves and optical waves in materials have enabled a 300-GHz-band operation for communications [59].

References

1. Federici, J., & Moeller, L. (2010). Review of terahertz and subterahertz wireless communications. *Journal of Applied Physics, 107*, 111101.
2. Kleine-Ostmann, T., & Nagatsuma, T. (2011). A review on terahertz communications research. *Journal of Infrared, Millimeter, and Terahertz Waves, 32*, 143–171.
3. Song, H.-J., & Nagatsuma, T. (2011). Present and future of terahertz communications. *IEEE Transactions on Terahertz Science and Technology, 1*, 256–263.
4. Nagatsuma, T., et al. (2013). Terahertz wireless communications based on photonics technologies. *Optics Express, 21*, 23736–23747.
5. Kürner, T., & Priebe, S. (2014). Towards THz communications - status in research, standardization and regulation. *Journal of Infrared, Millimeter, and Terahertz Waves, 35*, 53–62.
6. Seeds, A., et al. (2015). Terahertz photonics for wireless communications. *IEEE Journal of Lightwave Technology, 33*, 579–589.
7. Ducournau, G., et al. (2015). THz communications using photonics and electronic devices: The race to data-rate. *Journal of Infrared, Millimeter, and Terahertz Waves, 36*, 198–218.
8. Nagatsuma, T., & Carpintero, G. (2015). Recent progress and future prospect of photonics-enabled terahertz communications research. *IEICE Transactions on Electronics, E98–C*, 1060–1070.
9. Nagatsuma, T., Ducournau, G., & Renaud, C. C. (2016). Advances in terahertz communications accelerated by photonics. *Nature Photonics, 10*, 371–379.

10. Fujishima, M., et al. (2015). Terahertz CMOS design for low-power and high-speed wireless communication. *IEICE Transactions on Electronics, E98–C*, 1091–1104.
11. Lee, S., et al. (2019). An 80-Gb/s 300-GHz-band single-chip CMOS transceiver. *IEEE Journal of Solid-State Circuits, 54*(10), 3577–3588.
12. Rodriguez-Vazquez, P., et al. (2020). A QPSK 110-Gb/s polarization-diversity MIMO wireless link with a 220–255 GHz tunable LO in a SiGe HBT technology. *IEEE Transactions on Microwave Theory and Techniques, 68*, 3834–3851.
13. Kallfass, I., et al. (2015). Towards MMIC-based 300GHz indoor wireless communication systems. *IEICE Transactions on Electronics, E98-C*, 1081–1090.
14. Hamada, H., & a. (2020). 300-GHz-band 120-Gb/s wireless front-end based on InP-HEMT PAs and mixers. *IEEE Journal of Solid-State Circuits, 55*(9), 2316–2335.
15. Takahashi, H., et al. (2014). 10-Gbit/s close-proximity wireless system meeting the regulation for extremely low-power radio stations. *IEICE Electronics Express, 11*(3), 20130989.
16. Hesler, J. L. (2015). Chapter 5: Terahertz Schottky diode technology. In H.-J. Song & T. Nagatsuma (Eds.), *Handbook of terahertz technologies: Devices and applications* (pp. 104–131). Pan Stanford Publishing.
17. Blin, S., et al. (2013). Wireless communication at 310 GHz using GaAs high-electron-mobility transistors for detection. *Journal of Communications and Networks, 15*(6), 559–568.
18. Diebold, S., et al. (2016). Modeling and simulation of terahertz resonant tunneling diode-based circuits. *IEEE Transactions on Terahertz Science and Technology, 6*(5), 716–723.
19. Nishida, Y., et al. (2019). Terahertz coherent receiver using a single resonant tunnelling diode. *Scientific Reports, 9*, 18125.
20. Ito, H., & Ishibashi, T. (2017). InP/InGaAs Fermi-level managed barrier diode for broadband and low-noise terahertz-wave detection. *Japanese Journal of Applied Physics, 56*(1), 014101-1-014101-7.
21. Ito, H., & Ishibashi, T. (2019). Broadband heterodyne detection of terahertz-waves using rectangular-waveguide-input Fermi-level managed barrier diode module. *Electronics Letters, 55*(16), 905–907.
22. Nagatsuma, T. et al.. (2019). *300-GHz-band wireless communication using Fermi-level managed barrier diode receiver*. Proceedings of 2019 IEEE MTT-S International Microwave Symposium (IMS2019), June 2019.
23. Friederich, F., et al. (2010). Phase-locking of the beat signal of two distributed-feedback diode lasers to oscillators working in the MHz to THz range. *Optics Express, 18*(8), 8621–8629.
24. Gao, Y., et al. (2014). Microwave generation with photonic frequency sextupling based on cascaded modulators. *IEEE Photonics Technology Letters, 2*(12), 1199–1202.
25. Li, Y., et al. (2020). 300-GHz-band wireless communication using a low phase noise photonic source. *International Journal of Microwave and Wireless Technologies, 12*, 551–558.
26. Yu, X., et al. (2016). 160 Gbit/s photonics wireless transmission in the 300-500 GHz band. *APL Photonics, 1*(8), 081301.
27. Nagatsuma, T. (2009). Generating millimeter and terahertz waves. *IEEE Microwave Magazine, 10*(4), 64–74.
28. Latzel, P., et al. (2017). Generation of mW level in the 300-GHz band using resonant-cavity-enhanced unitraveling carrier photodiodes. *IEEE Transactions on Terahertz Science and Technology, 7*(6), 800–807.
29. Renaud, C. C., et al. (2018). Antenna integrated THz uni-traveling carrier photodiodes. *IEEE Journal of Selected Topics in Quantum Electronics, 24*(2), 1–11.
30. Peytavit, E., et al. (2013). CW source based on photomixing with output power reaching 1.8 mW at 250 GHz. *IEEE Electron Device Letters, 34*(10), 1277–1279.
31. Nellen, S., et al. (2020). Experimental comparison of UTC- and PIN-photodiodes for continuous-wave terahertz generation. *Journal of Infrared, Millimeter, and Terahertz Waves, 41*, 343–354.
32. Chinni, V. K., et al. (2018). Single-channel 100 Gbit/s transmission using III–V UTC-PDs for future IEEE 802.15. 3d wireless links in the 300 GHz band. *Electronics Letters, 54*, 638–640.

33. Nagatsuma, T. et al.. (2018). *600-GHz-band waveguide-output uni-traveling-carrier photodiodes and their applications to wireless communication.* Proceedings of 2018 IEEE/MTT-S International Microwave Symposium (IMS2018), pp. 1180–1183.
34. Li, N., et al. (2006). High power photodiode wafer bonded to Si using Au with improved responsivity and output power. *IEEE Photonics Technology Letters, 18*(23), 2526–2528.
35. Beling, A., Xie, X., & Campbell, J. C. (2016). High-power, high-linearity photodiodes. *Optica, 3*(3), 328–338.
36. Shimizu, N., & Nagatsuma, T. (2006). Photodiode-integrated microstrip antenna array for subterahertz radiation. *IEEE Photonics Technology Letters, 18*(6), 743–745.
37. Song, H. J., et al. (2012). Uni-travelling-carrier photodiode module generating 300 GHz power greater than 1 mW. *IEEE Microwave and Wireless Components Letters, 22*(7), 363–365.
38. Che, M., et al. (2020). Optoelectronic THz-wave beam steering by arrayed photomixers with integrated antennas. *IEEE Photonics Technology Letters, 32*(16), 979–982.
39. Van Dijk, F., et al. (2014). Integrated InP heterodyne millimeter wave transmitter. *IEEE Photonics Technology Letters, 26*(10), 965–968.
40. Carpintero, G., et al. (2014). Microwave photonic integrated circuits for millimeter-wave wireless communications. *Journal of Lightwave Technology, 32*(20), 3495–3501.
41. Carpintero, G., et al. (2018). Wireless data transmission at terahertz carrier waves generated from a hybrid InP-polymer dual tunable DBR laser photonic integrated circuit. *Scientific Reports, 8*(1), 1–7.
42. Campion, J., et al. (2019). Toward industrial exploitation of THz frequencies: Integration of SiGe MMICs in silicon-micromachined waveguide systems. *IEEE Transactions on Terahertz Science and Technology, 9*(6), 624–636.
43. Nagatsuma, T. (2016). Millimeter-wave and terahertz-wave applications enabled by photonics. *IEEE Journal of Quantum Electronics, 52*(1), 1–12.
44. Koenig, S., et al. (2013). Wireless sub-THz communication system with high data rate. *Nature Photonics, 7*, 977–981.
45. Shams, H., et al. (2015). 100 Gb/s multicarrier THz wireless transmission system with high frequency stability based on a gain-switched laser comb source. *IEEE Photonics Journal, 7*, 1–11.
46. Yu, X., et al. (2016). 160 Gbit/s photonics wireless transmission in the 300-500 GHz band. *APL Photonics, 1*, 081301.
47. Pang, X. et al. (2016). *260 Gbit/s photonic-wireless link in the THz band.* Proceedings of IEEE Photonics Conference., Postdeadline paper.
48. Stöhr, A. et al. (2017). *Coherent radio-over-fiber THz communication link for high data-rate 59 Gbit/s 64-QAM-OFDM and real-time HDTV transmission.* Tech. Dig. Optical Fiber Communications Conf. (OFC 2017), Tu.3B2.
49. Pang, X. et al. (2017). *Single channel 106 Gbit/s 16QAM wireless transmission in the 0.4 THz band.* Tech. Dig. Optical Fiber Communications Conf. (OFC 2017), Tu.3B5.
50. Nagatsuma, T. et al. (2016). *Real-time 100-Gbit/s QPSK transmission using photonics-based 300-GHz-band wireless link.* Tech. Dig. IEEE International Topical Meeting on Microwave Photonics (MWP 2016), TuM1.1.
51. Nagatsuma, T. et al. (2016). *300-GHz-band wireless transmission at 50 Gbit/s over 100 meters.* Tech. Dig. International Conference on Infrared, Millimeter and Terahertz Waves (IRMMW-THz 2016), F2D.2.
52. Iwamoto, K. et al. (2017). *100-meter wireless transmission at 70 Gbit/s in 300-GHz-band.* Conference of the 4th Microwave/THz Science and Applications (MTSA2017), S2-5.
53. Takiguchi, K. (2019). Method for converting high-speed and spectrally efficient terahertz-wave signal into optical signal. *Optics Express, 29*(5), 6598–6606.
54. Y. Uemura et al. (2020). *600-GHz-band heterodyne receiver system using photonic techniques.* Tech. Dig. IEEE International Topical Meeting on Microwave Photonics (MWP2020), P. 35.
55. Nagatsuma, T. et al.(2020). *Wireless communication using Fermi-level-managed barrier diode receiver with J-band waveguide-input port,*" Proceedings of 2020 IEEE/MTT-S International Microwave Symposium (IMS2020), pp. 631–634.

56. Mohammad, A. W., et al. (2018). 60-GHz transmission link using uni-traveling carrier photodiodes at the transmitter and the receiver. *Journal of Lightwave Technology, 36*(19), 4507–4513.
57. Harter, T. et al. (2017). *Wireless multi-subcarrier THz communications using mixing in a photoconductor for coherent reception.* Tech. Dig. 2017 IEEE Photonics Conference (IPC), pp. 147–148.
58. Harter, T., et al. (2019). Wireless THz link with optoelectronic transmitter and receiver. *Optica, 6*(8), 1063–1070.
59. Ummethala, S. (2019). THz-to-optical conversion in wireless communications using an ultra-broadband plasmonic modulator. *Nature Photonics, 13*(8), 519–524.

Part VIII
Transceiver Technologies 4: Vacuum Electronic Devices

Chapter 27
Vacuum Electronic Devices

Claudio Paoloni

Abstract The exploitation of the spectrum beyond 100 GHz is the solution for the full implementation of 5G and the development of 6G concepts. Low-power electronics is already available, but technology advancements are needed to overcome the increasing atmosphere and rain attenuation above 100 GHz, which presently limit the transmission distance.

New solid-state power amplifiers (SSPA) based on GaN or InP processes have been produced in the recent years, but the best output power achieved is far below the watt level needed for transmission by high modulation schemes over long range.

Vacuum electronic devices, namely, traveling-wave tubes (TWTs), offer more than one order of magnitude and more power than SSPA, over wide bandwidth, representing a promising solution for ultrahigh-capacity long links.

However, the short wavelength above 100 GHz poses substantial fabrication challenges, which require new technology approaches for high-volume production to bring TWTs into the wireless communication market. This chapter will explore the working mechanism, the potentiality, the features, the state of the art, and the possible deployment scenarios for millimeter-wave and sub-THz TWTs.

27.1 Introduction

The advancements in solid-state electronics have permitted outstanding performance with multi-Gb/s data rate in the millimeter-wave and terahertz portion of the spectrum. Unfortunately, the simultaneous decrease of output power of solid-state power amplifiers (SSPA) [1, 2] and the increase of link attenuation at the increase of frequency prevent long links above 100 GHz, even with the use of very high-gain antennas.

C. Paoloni (✉)
Engineering Department, Lancaster University, Lancaster, UK
e-mail: c.paoloni@lancaster.ac.uk

T. Kürner et al. (eds.), *THz Communications*, Springer Series in Optical Sciences 234, https://doi.org/10.1007/978-3-030-73738-2_27

Simple link budget calculations highlight that the minimum transmission power needed to support high-modulation schemes and range longer than 100 meters above 100 GHz has to be in the watt range.

Recently, the use of traveling wave tubes (TWTs) as wideband high-power amplifier in new generation wireless networks is investigated [3, 4]. Potentially, a TWT can provide a minimum of 40 W at 100 GHz and 1 W at 300 GHz. This output power is more than one order of magnitude higher than the power available from any available SSPAs at the same frequencies.

The availability of this level of transmission power is opening new perspectives in wireless transport of unprecedented data rate over long distance, enabling new paradigms for backhaul and fixed wireless access.

The main challenge is the design and fabrication of affordable TWTs, due to the short wavelength at sub-THz range (e.g., 3 mm at 100 GHz), which determines their dimensions. In the following, properties, fabrication processes, and applications of TWTs for sub-THz wireless communications will be discussed.

27.1.1 How a TWT Works

Electrons can be easily accelerated at a fraction of the velocity of light c (e.g., $v_e = 0.2\ c$ where $c = 3\ 10^8$ m/s) by an electric field established by a voltage V applied between the source of electrons (cathode) and an electrode (anode) placed at a given distance. The electron velocity v_e is a function of the accelerating voltage V ($v_e = \sqrt{\frac{2e\ V}{m}}$ where m and e are the mass and the charge of the electron, respectively).

An electron beam is a flow of electrons, namely, a current I_b, confined in space with a defined cross section, usually circular, that travels at a speed given by the accelerating voltage V. The DC energy of the beam is given by the product of the accelerating voltage V by the current I_b.

Typically, millimeter-wave TWTs have the beam voltage V in the order of 5–20 kV corresponding to accelerated electrons above 0.1 c and the current I_b in the order of tens to a few hundreds of mA. The current depends on the area cross section of the electron beam and electron emission parameters of the cathode. As an example, an electron beam with $V = 15$ kV voltage and $I_b = 50$ mA current has 750 W power.

The property of a TWT is to convert part of the high DC energy of the electron beam into RF energy. The TWT working mechanism is extensively described in literature [5]; however, a short summary is given to provide the reader some background. The phase velocity of an RF wave propagating in space is defined as the velocity of its phase fronts. The transfer of energy in a TWT happens if the electron velocity v_e is slightly higher than the phase velocity of the RF wave.

The main components for amplification in a TWT are the beam optics and the interaction structure (Fig. 27.1).The beam optics is the subassembly that permits the

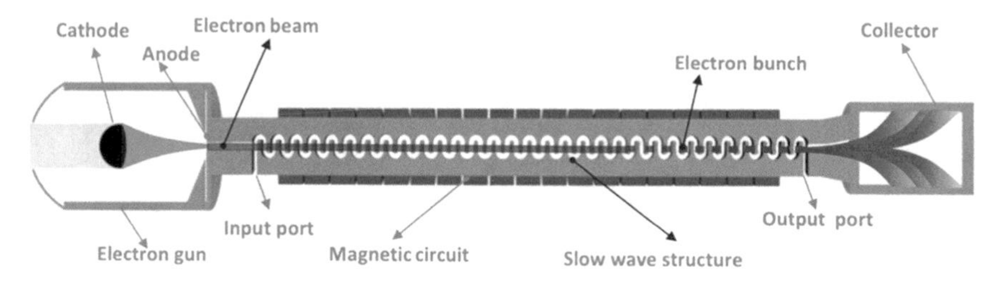

Fig. 27.1 Schematic of TWT

generation and the confinement of the electron beam with a given electron velocity and current. It consists of an electron gun that produces the electron beam, the magnetic focusing system that keeps the electron beam confined in a well-defined shape along its path, and the collector that receives the electrons at the end of their travel and partially recovers their energy. The electron beam needs high vacuum (10^{-8} torr or higher) to flow without degradation, therefore the definition of vacuum tubes.

The interaction structure is typically a waveguide, modified to slow down the phase velocity of the RF wave, also called slow wave structure (SWS). The SWS is characterized by the dispersion curve that represents the variation of the phase velocity as a function of the frequency. Given a beam voltage and consequently the speed of the electrons (independent from the frequency), the SWS is designed to reduce the phase velocity of the RF input signal slightly lower than the electron velocity. The wider is the frequency region where the defined phase velocity is constant and consequently in synchronism with the electron velocity, the wider is the TWT bandwidth.

The electrons from the electron gun enter in the SWS from the electron gun through a connecting tiny tunnel. When in the SWS, a portion of electrons interacts with the decelerating fronts of the RF field, the other portion with an accelerating front. Accelerated electrons reach the decelerated electrons forming, along the SWS path, high-density aggregations of electrons (bunches) alternate to regions of low density of electrons. Bunches have a high mass given by the high density of electrons. Due to the phase velocity slightly lower than the electron velocity, the formation of the bunches is mostly in the decelerating region of the wave field, where they are slowed down traveling toward the end of the SWS. The kinetic energy lost by the electrons, for the principle of conservation of energy, is transferred to the wave in the form of electromagnetic energy producing its amplification. The bunches contain the full frequency content of the RF signal.

This apparently simple process permits the amplification at any frequency and over a wideband (up to two octaves) depending on the SWS characteristics and feasibility. The limits for the highest achievable frequency f are set by the fabrication technology and the product $Pxf^2 = const$ where P is the average output power.

A TWT includes other important parts that will not be described for brevity, such as the RF windows, to maintain the vacuum and permit the transmission of the signal

Fig. 27.2 W-band TWT built by Thales (France) in the frame of H2020 TWEETHER project [4]

at the ports, the magnetic focusing systems, the sever to avoid oscillations for high-gain TWT, and the vacuum flanges [5]. A TWT is usually enclosed in a compact case. The TWT in 27.2 produced by THALES (France) has dismensions: Length 340 mm, width 72 mm, and height 54 mm.

27.1.2 Why TWTs Generate More Power Than Solid-State Amplifiers

Solid-state devices (BJT, HEMT, HBT, etc.) work by modulating a current flowing between two electrodes. The electrons flow in a short region of semiconductor experiencing continuous collisions with the semiconductor reticule that limit the electron speed. The collisions convert the kinetic energy of electrons in heat concentrated in that very small device region, which has to be dissipated, making challenging the heat sinking in case of relatively high currents.

The electrons in the TWT flow in high vacuum, which, different from any other material, allows in principle unlimited velocity due to the lack of scattering and collision with a reticule and no heat generation. The heat dissipation is mostly in the collector that is separated from the section where the RF signal gains energy from the electron beam. The collector is designed both for thermal dissipation and to recover the electron current for high efficiency. Finally, thermal issues are practically absent in the RF interaction region different from solid-state devices. This different thermal behavior of TWTs makes them suitable for handling very high RF power.

Other effect of the internal collisions in the semiconductor is the reduction of electron mobility in comparison to electrons flowing in vacuum, affecting the high-frequency performance.

Solid-state devices are limited to low voltages due to the low electric field breakdown, function of the short electrode distance, and materials. TWTs, due to the high vacuum and size, support higher critical electric field for breakdown, which permits very high-voltage and consequently high RF power at the ports. Recently, GaN solid-state power amplifiers, due to the GaN wide bandgap, start to replace TWTs at microwave, but the power performance is still not comparable at millimeter waves and above.

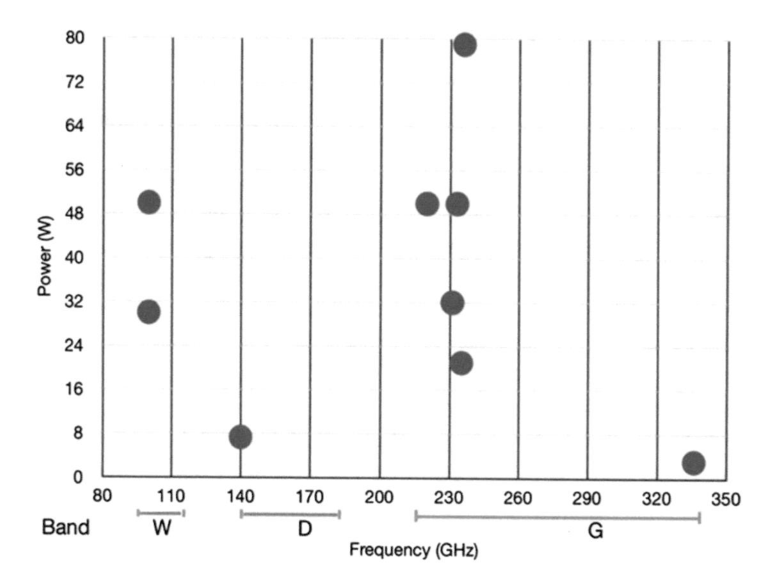

Fig. 27.3 Map of state of the art of millimeter-wave TWTs (more data in [8])

27.1.3 *State of the Art*

The production of millimeter-wave and sub-THz TWTs is still at the state of the art (Fig. 27.3). A large number of studies on different configurations, topologies, and fabrication methods for TWTs above 100 GHz are available in literature [6]. On the contrary, only a few working prototypes were reported. Intrinsic fabrication challenges, limited market, and the lack of facilities for large production are among the reasons of low availability of millimeter-wave TWTs.

Among the most relevant examples, it is remarkable a 231.5–235-GHz TWT providing a peak output power of 32 W. It was built with space qualification for enabling the VISAR [7]. A TWT with 60 W of output power at 214.5 GHz using a 12.1 kV, 118 mA electron beam is reported in [8]. A 320 GHz TWT using a folded waveguide as SWS was fabricated and tested. This TWT achieves 19.6 dB gain at 318.24 GHz with 16.9 kV beam voltage, but in a very narrow band [9].

The given examples and some more not cited here demonstrate the potential of TWTs above 100 GHz. As rough estimation, TWTs in the range between 100 GHz and 300 GHz could provide 30–40 dB gain and 2–50 W output power. Above 300 GHz, a few watts could be achievable.

27.2 Fabrication Technology for Sub-terahertz TWTs

A TWT is a micromechanical device. Different from the mass production of solid-state amplifiers based on patterning the surface of a semiconductor substrate, a TWT is produced by assembling three-dimensional metal parts built by different fabrication processes, most of them manually. Electron gun, collector, RF window, and SWS need specific technologies. The most challenging part to fabricate above 100 GHz, where wavelengths are shorter than 3 millimeters, is the slow wave structure. An SWS is a metal (usually high conductivity oxygen-free copper) waveguide modified to slow down the phase velocity, connected to an input and output couplers to the waveguide flanges. Dimensions of the smaller features could be below 100 microns.

In addition, the skin depth of copper above 100 GHz is in the range of 100 or 200 of nanometers depending on the frequency. This requires a high-surface finishing to ensure a surface roughness better than 50% of the skin depth to avoid high ohmic losses. High-precision micromechanical fabrication processes are needed based on computer numerical control (CNC) milling or photolithographic process.

CNC milling is normally used for producing mechanical parts, but when the features are below 100 microns, CNC machines with high accuracy and the availability of small tooling are crucial for quality of the parts (Fig. 27.4(a)).

Probably, the most advanced CNC milling is the DMG Mori Seiki NN1000 nano-CNC mill used at the University of California Davis [10]. The nano-CNC mill has five axes and a movement position accuracy of 1 nm/100 mm and a repeatability of 5 nm/100 mm. The nano-CNC milling permitted to build SWSs up to 0.346 THz. So far, 350 GHz seems a maximum limit for fabrication of SWS by CNC milling both for accuracy and micro-tooling size (minimum diameter 76 μm). In general, depending on the features and the level of accuracy, different quality of CNC milling machines can be used.

MEMS technology has been recently considered to build SWSs above 100 GHz and up to 1 THz [11, 12]. The LIGA microfabrication process (from German Lithographe, Galvonoformung, und Abfornung) is based on photolithography and electroplating. A layer of photoresist specific for thick layer (the most common is

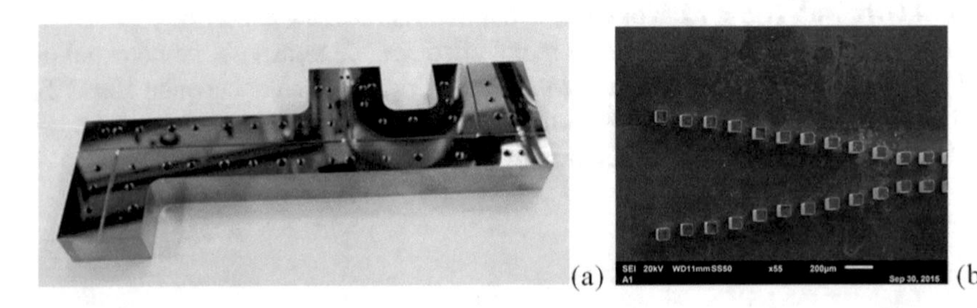

Fig. 27.4 (**a**) D-band (141–148.5 GHz) SWS fabricated by CNC milling [10]. (**b**) LIGA 300 GHz double corrugated waveguide [13]

SU-8) is deposited on copper or silicon, depending on the approached used. Then a mask with the pattern of the SWS is placed on the baked resist to be exposed to ultraviolet (UV) light. After the development, the SU-8 is removed where the metal has to be grown by electroforming. Finally, after the electroforming has produced the desired metal structure, the SU-8 is fully removed, and the SWS is obtained. Further, CNC milling could be needed for adding specific features.

Figure 27.4(b) shows a 300 GHz SWS fabricated by LIGA (pillar size is 70 x 70 x 150) [13].

The LIGA process is affordable and can be used for large-scale production of SWS. The fabrication of beam optics for sub-THz TWTs is not different from microwave TWTs. The assembly is more challenging due to the alignment with the precision at a fraction of degree to assure the correct flow of the electron beam that could have less than 100 micron radius.

27.3 Sub-terahertz TWTs for Wireless Communications

The availability of transmission power at watt level at the millimeter-wave and sub-THz frequencies is very attractive to extend link range, reduce antenna size, and enable point-to-multipoint coverage or long point-to-point links.

It is not new that TWTs could be considered promising enabling devices for millimeter-wave links with high capacity, wide bandwidth, and long range. The major obstacle has been so far the fabrication process, high production cost, low production capacity, and lack of a robust use case.

The advent of 5G first and 6G recently, the need of high-capacity backhaul for densification of small cells, and the digital divide in rural and suburban areas where the deployment of the fiber is not affordable pose new challenges to wireless networks. The question is how to bring wireless Internet everywhere with almost unlimited capacity.

A first response was given by the European Commission Horizon 2020 TWEETHER project [14]. The project proposed to distribute wireless Internet in point to multipoint at W-band (92–95 GHz) with about 3.5 Gb/s/km^2 area capacity over circular sector 1 km long and with up to 90-degree aperture angle, providing 99.99% availability up to ITU zone K. The link budget, due to the low-gain (16 dBi) antenna to provide sectors with 90-degree aperture angle, requires 40 W saturated of transmission power. GaN SSPAs can provide not more than 1 or 2 W at this frequency. TWEETHER project introduced for the first time a TWTs as enabling amplifier for point to multipoint at W-band. TWTs compensate the low- gain antenna and provide long links unattainable by SSPAs. A TWT with 40 W output power and 40 dB gain was designed and build [4].

The European Commission H2020 ULTRAWAVE project [15, 16] moved forward the use of TWTs proposing a point-to-multipoint distribution at D-band (141–148.5 GHz) with point-to-point transport between D-band hubs at G-band (275–305 GHz). At the time of the writing of the chapter, the D-band TWT, with about 12 W output power, is in advanced fabrication stage. The simulations of the G-band TWT provide about 1 W output power.

Then other projects using TWTs for sub-THz links followed. The European Commission H2020 ThoR project is developing a 300 GHz TWT for PtP links [17].

The EPSRC (UK Engineering and Physics Science Research Council) DLINK project is developing two TWTs for a D-band PtP link in frequency domain duplex (FDD) with 45 Gb/s for 1 km range. One TWT is for the lower link (151.5–161.5 GHz) and one for the upper link (161.5–174.8 GHz) [18].

27.3.1 Main TWT Parameters for High-Capacity and Long-Range Wireless Links

27.3.1.1 Linearity and Intermodulation

Sub-THz long links could support multi-Gb/s data rate exploiting multi-GHz frequency bands, if signal-to-noise ratio (SNR) and linearity allow it.

TWTs exhibit high power and efficiency when operated at saturation, where high distortion and intermodulation products are present. The P_{out}/P_{in} (output power versus input power) curve of a TWT shows a linear region and a saturation region marked by 1 dB gain compression point where the gain decreases of 1 dB from the linear gain [19].

The third-order intermodulation (IM3) distortion measures the generation of unwanted frequency occurring in any amplifiers when operate in nonlinear regime.

To support high-modulation schemes (e.g., 16QAM or 64QAM), a TWT has to operate in the linear region, well below the saturation region. This operation regime is called back-off. The level of back-off depends on the linearity needed for the used modulation scheme. Typically, 6–8 dB back-off is needed to operate at 64 QAM. The typical value of carrier to third-order intermodulation ratio (C/IM3) at 6 dB back-off is about 30 dBc.

To note that the back-off operation reduces the efficiency because the electron beam power does not change, the RF power is reduced well below saturation. However, a relevant portion of the electron beam energy can be recovered by the collector. Different from solid-state devices, recovering energy from the spent beam permits a much higher overall efficiency, e.g., with 6 dB back-off, the overall efficiency can be better than 40%.

But, in case of a D-band TWT with 40 dBm (10 W) saturated output power, after 6 dB back-off, the output power reduces to 34 dBm (2.5 W). A D-band InP DHBT SSPA with 17 dBm (50 mW) output power after 6 dB back-off reduces to 11 dBm (12.5 mW). 200 times less power!

27.3.1.2 Noise Figure

The noise figure of a TWT is typically high (more than 10 dB). The use of a low-noise solid-state amplifier as driver reduces the overall noise figure and helps to increase the overall gain.

27.3.1.3 Efficiency

The overall efficiency is linked to linearity and back-off level of operation. It is mainly a function of interaction efficiency, defined as the ratio of the RF power over the beam power, and of the collector efficiency in recovering the electron beam. With a proper collector design, the collector efficiency could reach 90%. On the contrary, the interaction efficiency at millimeter wave could be lower than 10%. Finally, the overall efficiency could be in the range of 30–50%.

27.3.1.4 Size and Deployment

The compact size of a TWT permits to allocate it in a cabinet with small size and footprint. As an example, the first TWT [4] (Fig. 27.2) for wireless W-band, PmP links was designed to be enclosed in a cabinet (44 x 35 cm footprint) with the motherboard including the low-power electronics and the horn antennas. In Fig. 27.5a, it is shown how the TWT is assembled in the W-band transmission hub realized in the TWEETHER project [10]. The photo in Fig. 27.5b shows the transmission hub on the mast. An optimization of the assembly and the smaller size of TWT above 100 GHz would bring a further reduction of size and footprint.

27.4 Conclusions

Traveling wave tubes are a promising solution to provide high transmission power at millimeter-wave and sub-THz frequency. Their availability would enable long and wide links with high data rate and high-modulation scheme exploiting the wide frequency band above 100 GHz. The development of the first millimeter-wave TWTs above 100 GHz designed for wireless links is in progress.

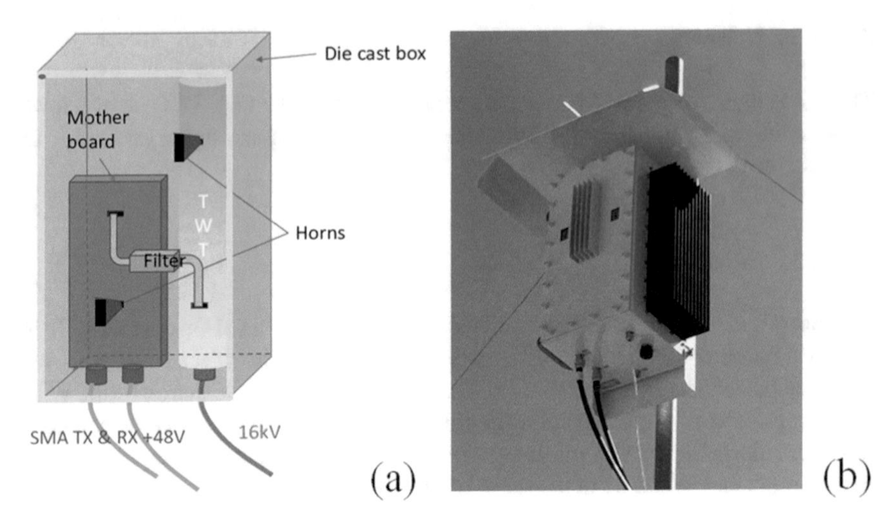

Fig. 27.5 (**a**) Schematic of W-band transmission hub with the TWT. (**b**) Realized TH (footprint 44 × 35 cm)

References

1. Rollin, J.-M., et al. (2015). A polystrata 820 mW G-band solid state power amplifier. In *Proceedings of IEEE compound semiconductor Integrated Circuit Symposium (CSICS)*.
2. M. Ćwikliński *et al.*, "D-band and G-band high-performance GaN power amplifier MMICs," IEEE Transactions on Microwave Theory and Techniques, vol. 67, no. 12, pp. 5080–5089, Dec. 2019.
3. Basu, R., et al. (2018). Design of sub-THz traveling wave tubes for high data rate long range wireless links. *Semiconductor Science and Technology, 33*, 124009.
4. F. André et al., "Technology, assembly, and test of a W-band traveling wave tube for new 5G high-capacity networks," IEEE Transactions on Electron Devices, vol. 67, no. 7, pp. 2919–2924, July 2020, doi: https://doi.org/10.1109/TED.2020.2993243.
5. Barker, R. J., Booske, J. H., Luhmann, N. C., & Nusinovich, G. S. (2005). *Modern Microwave and Millimeter-Wave Power Electron*. Piscat-away: IEEE.
6. Paoloni, C., Gamzina, D., Letizia, R., Zheng, Y., & Luhmann, N. C., Jr. (2020). Millimeter wave traveling wave tubes for the 21st century. *Journal of Infrared, Millimeter, and Terahertz Waves*. https://doi.org/10.1080/09205071.2020.1848643.
7. Armstrong, C. M., et al. (2018). A compact extremely high frequency MPM power amplifier. *IEEE Transactions on Electron Devices, 65*(6), 2183–2188.
8. Field, M., et al. (2018). Development of a 100-W 200-GHz high bandwidth mm-wave amplifier. *IEEE Transactions on Electron Devices, 65*(6), 2122–2128. https://doi.org/10.1109/TED.2018.2790411.
9. Hu, P., et al. (2018). Development of a 0.32-THz folded waveguide traveling wave tube. *IEEE Transactions on Electron Devices, 65*(6), 2164–2169.
10. Gamzina, D., et al. (2016). Nano-CNC machining of sub-THz vacuum Electron devices. *IEEE Transactions on Electron Devices, 63*(10), 4067–4073. https://doi.org/10.1109/TED.2016.2594027.
11. Paoloni, C., et al. (2013). Design and realization aspects of 1-THz Cascade backward wave amplifier based on double corrugated waveguide. *IEEE Transactions on Electron Devices, 60*(3), 1236–1243.

12. C. D. Joye *et al.*, "Demonstration of a high power, wideband 220-GHz traveling wave amplifier fabricated by UV-LIGA," IEEE Transactions on Electron Devices, vol. 61, no. 6, pp. 1672–1678, Jun. 2014.
13. Malek Abadi, S. A., & Paoloni, C. (2016). UV-LIGA microfabrication process for sub-terahertz waveguides utilizing multiple layered SU-8 photoresist. *Journal of Micromechanics and Microengineering, 26, 9, 8*, 095010.
14. C. Paoloni et al. (2018) Transmission Hub and Terminals for Point to Multipoint W-Band TWEETHER System. *2018 European Conference on Networks and Communications (EuCNC)*, Ljubljana, Slovenia.
15. http://www.ultrawave2020.eu
16. C. Paoloni *et al.*, "Technology for D-band/G-band ultra capacity layer," *2019 European Conference on Networks and Communications (EuCNC)*, Valencia, Spain, 2019, pp. 209–213.
17. https://thorproject.eu
18. Paoloni, C, *et al.*, R 2020, Long-range millimetre wave wireless links enabled by travelling wave tubes and resonant tunnelling diodes, IET Microwaves Antennas and Propagation, vol. 14, no. 15, pp. 2110–2114. https://doi.org/10.1049/iet-map.2020.00.
19. D. M. Goebel, R. R. Liou, W. L. Menninger, Xiaoling Zhai and E. A. Adler, "Development of linear traveling wave tubes for telecommunications applications," in IEEE Transactions on Electron Devices, vol. 48, no. 1, pp. 74–81, Jan. 2001.

Part IX
Baseband Processing and Networking Interface

Chapter 28
High-Bandwidth, Analogue-to-Digital Conversion for THz Communication Systems

Thomas Schneider and Mladen Berekovic

Abstract Terahertz wireless communication with its immanent wide bandwidth is seen as one of the potential solutions for ultra-high-speed resilient communication systems. One of the major challenges in THz systems, however, is high-bandwidth signal processing and especially analogue-to-digital conversion. Although electronic converters have recently reached very high bandwidths, the decrease in Moore's law future scaling advances makes it mandatory to look for alternatives. Electronic and photonic combination for high-speed signal processing on a single integrated platform might enable ultra-high bandwidth and cost-effective processing units for next-generation THz wireless systems. In this chapter, we will demonstrate all-optical sampling by Nyquist pulses. These pulses are fully reconfigurable and simple to control by tuning external RF signals without using any optical tuning. Therefore, the pulses can precisely be adapted to the sampling requirements. The method offers sampling rates with three to four times the radio frequency of the used modulators. Even higher sampling rates are obtained with existing low-bandwidth electronics by spectral or time-interleaving.

28.1 Introduction

The increase in data traffic demands ultra-high-speed resilient communication systems with several hundred Exabytes per month [1]. While the strain is felt in all parts of the network, it is particularly significant in the wireless channels going from mobile to backhaul and vice versa. Forecasts indicate that the expected data rate in the wireless channel can exceed 100 Gbit/s within the next few years [2]. Terahertz wireless communication (typically 0.3 THz–3 THz), with its inherent

T. Schneider (✉)
THz-Photonics Group, Technische Universität Braunschweig, Braunschweig, Germany
e-mail: thomas.schneider@ihf.tu-bs.de

M. Berekovic
Universität Lübeck, Lübeck, Germany

T. Kürner et al. (eds.), *THz Communications*, Springer Series in Optical Sciences 234, https://doi.org/10.1007/978-3-030-73738-2_28

high bandwidth, is seen as one of the promising ways to furnish such high data rate requirements. One of the major challenges in THz systems, however, is high-bandwidth signal processing and especially analogue-to-digital conversion. The combination of electronic and photonic signal processing on single integrated, CMOS-compatible silicon chips might enable cost-effective, ultra-high bandwidth signal processing units for future wireless THz systems.

For the transformation and processing of information carried by electromagnetic signals into their binary correspondent a precise analogue-to-digital conversion (ADC) is essential. For the computer technology, the powerful software improvement has kept pace with the more complex hardware development. However since the sampling rate of signals is a limiting factor for the performance, this was not the case for electronic ADC. Electronic ADC schemes are limited by the jitter of the electronic clock and by several noise sources such as quantisation noise thermal noise, and nonlinearities [3, 4].

Typically, electronic ADC are based on sample and hold circuits. In the last few years, the CMOS and SiGe technology has seen a tremendous progress. Thus, data converters with sampling rates of tens of GSa/s are available. Following the Nyquist sampling theorem, an ADC requires a sampling rate of at least twice the bandwidth of the signal to sample. If a spectral efficiency of 3 bit/s/Hz is assumed, the transmission of 120 Gbit/s signals with THz wireless systems requires baseband widths of 40 GHz and thus sampling rates of 80 GSa/s. An ADC with 72 GSa/s and 36 GHz input bandwidth in 14 nm CMOS technology has been shown recently [5], for instance.

The current limits and technology front in electronic converters and their trend over the last 20 years can be seen in Fig. 28.1. For frequencies higher than a corner frequency, the efficiency of the design deteriorates by 10 dB per decade. However, this corner frequency has shifted by a factor of 60 over the last two decades, largely due to technological improvements. As can be seen from the latest data, 100 GSa/s sampling rates are becoming within reach. Although the technological advancement has slowed down significantly due to the demise of silicon scaling to Moore's law [6], there still is improvement. Therefore, it can be expected that the achievable sampling rate will saturate around 100 GSa/s in the foreseeable future, creating the need for optical conversion if higher sampling rates are required.

Figure 28.2 shows the aperture of published ADCs over the last 20 years. The influence of jitter is highlighted by two asymptotic limitation curves. This data shows that 100 GSa/s are nearly within reach, but further significant improvements are unlikely for the near future.

Much higher input bandwidths can be achieved by spectral interleaving [9]. However, state-of-the-art electronic ADC show a trade-off between bandwidth and resolution, i.e., the higher the bandwidth, the lower the resolution. This is due to the phase noise or jitter of the clock, i.e., the inability to sample with a precise repetition rate. Corresponding to the jitter of ultralow phase noise quartz oscillators, electronic ADC exhibits jitter levels of 50–80 fs [10]. The accompanied sampling error results in a decreasing resolution for high-bandwidth ADC and especially spectral interleaving.

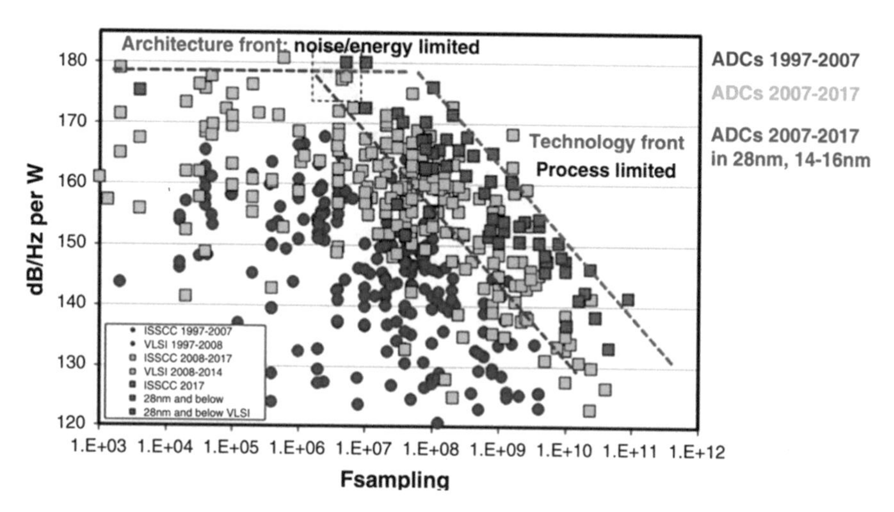

Fig. 28.1 Schreier FoM (figure of merit) taken from [7] with data from [8]. The numbers are taken near the Nyquist frequency. The limits for lower frequencies are mainly defined by architectural parameters while the asymptotic behavior at the higher frequencies is mostly determined by process technology and F_T in particular. It degrades by 10 dB/decade. Over the last 10 years, it has moved by a factor of 60x to the right

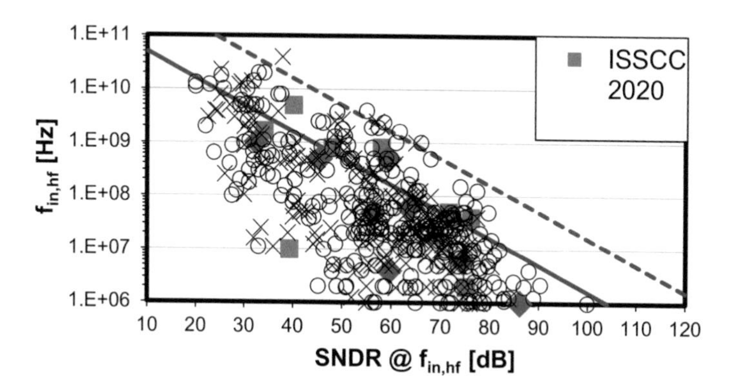

Fig. 28.2 Aperture for converter designs published over the last decade. Data taken from [8]

During the last 25 years, there has been a great interest in reconciling the ADC requirement with optical components. This was mainly driven by the rapid growth of photonic technologies, which might enable high-bandwidth, precise optical sampling. The use of photonic components to make or improve an analogue-to-digital converter (ADC) has attracted interest since the early 1970s, and today one can easily find many textbooks and scientific papers related to "photonic ADCs." Photonic-assisted ADCs are electronic ADCs that use photonics to improve the limited performance of the electronics like the resolution or effective number of bit (ENOB), jitter, quantization noise, etc. Generally, the sampling is performed in the optical domain, while the quantization is carried out in the electrical domain. For better performances of such ADCs, the sampling becomes the most important issue. High-resolution optical sampling can be achieved by several methods. The simplest one is to utilize a fast photodetector in combination with electronic sample-and-hold circuits. However, the method is restricted by the bandwidth of the electronic components, since the signal is completely sampled in the electrical domain. Much higher sampling rates can be achieved if the signal to sample is multiplied by a train of short laser pulses. For the multiplication, either a modulator, driven by the electrical signal [3], or a nonlinear element can be used [11, 12]. For the first method, the sampling rate is restricted by the bandwidth of the modulator, and the second requires a stable, short-period optical pulse source. Optical pulse sources like a mode-locked laser (MLL) can show very low jitter values in the attosecond range [13, 14]. However, the integration of MLL on a chip is still quite challenging. Additionally, for the multiplication of the pulse sequence with the signal, a nonlinear device like a crystal, a nonlinear effect in a fiber, or a Mach-Zehnder modulator (MZM) is required [13, 15]. These nonlinearities might lead to additional distortions of the signal. Furthermore, the optical path length of the MLL cavity defines the repetition rate of the pulses and therefore the sampling frequency. Thus, a tuning of the sampling frequency is only possible in a limited range and rather complicated. Last but not least, tunable delay lines for each branch and an exact calibration are required for a parallelization of the sampling in the time domain, and for a parallelization in the frequency domain, interleaved pulses of different wavelengths are necessary [16].

In this chapter, we will discuss about the basics of sampling and the optical sampling with Nyquist pulses [17–25]. An all-optical sampling technique for full-field sampling (amplitude and phase) followed by some examples [26–32] will be introduced.

28.2 Sampling Basics

Sampling can be described in the frequency and the time domain, both related by the well-known Fourier theorems. In time domain, ideal sampling can be perceived as the multiplication between a bandlimited signal $s(t)$ and a Dirac-delta sequence $\delta(t)$ as:

$$s_a(t) = s(t) \times \sum_{n=-\infty}^{\infty} \delta\left(t - nt_s\right) \tag{28.1}$$

where $t_s = 1/\Delta f$ is the sampling rate. The multiplication in the time domain corresponds to a linear convolution in the frequency domain (*) between the signal spectrum $S(f)$ and the Fourier transform of the Dirac-delta sequence, which is again a Dirac-delta sequence. This can be mathematically expressed as:

$$S_a(f) = S(f) * \Upsilon\left(\Delta f\right) \tag{28.2}$$

Here, Υ(f) is an unlimited frequency comb with Δf as the spacing between the frequency lines. Thus, Eq. (28.2) can be rewritten as:

$$S_a(f) = \sum_{n=-\infty}^{\infty} S\left(f - n.\Delta f\right) \tag{28.3}$$

In the frequency domain, ideal sampling produces an unlimited number of equal copies of the signal spectrum separated by Δf. In electronics, sampling is usually carried out with sample-and-hold (S/H) circuits. Since this functionality can be approximated as rectangular pulses, weighted by the sampling values, this results in a multiplication of the spectral copies with a sinc-shaped function in the frequency domain. Thus, the copies of the signal spectrum have a sinc-shaped envelope, which leads to a distortion of the sampled signal. Digital signal processing can be used to reconstruct the signal. However, the power requirement and processing time limit the overall sampling rate of the system. Photonic sampling, on the contrary, can offer ultra-wideband sampling. The short laser pulses that are used for photonic sampling can be approximated by a Gaussian function. Thus, the spectral copies of the signal have a Gaussian envelope, again leading to a distortion of the sampled signal.

The only possibility to avoid such a distortion is the multiplication of the spectral copies with a rectangular function in the frequency domain. Thus, all spectral copies are equal – as for ideal sampling – but in order to make this kind of sampling practically feasible, the number of copies is limited. Therefore, in the time domain, the signal to sample has to be multiplied with sinc-shaped Nyquist pulses.

28.3 Sinc-Shaped Nyquist Pulses

The single sinc-shaped Nyquist pulse with a perfect rectangular optical spectrum is unlimited in time. Thus, the generation of single pulses is practically as impossible as ideal sampling. A single sinc pulse can be represented by a rectangular spectrum. If a number of single equidistant frequencies are carved out of this rectangular spectrum, this results in an unlimited train of sinc pulses in the equivalent time domain. Thus, a flat frequency comb with phase-locked, equally spaced components within the bandwidth defined by the single-pulse spectrum corresponds to a train of sinc pulses. Such a flat frequency comb can practically be generated with one or two coupled intensity modulators [17, 18], driven with one or several radio frequencies. If the first modulator is driven with n and the second with m frequencies, the resulting flat, phase-locked optical comb consists of $N = (2n + 1)(2\,m + 1)$ optical frequencies. The corresponding Nyquist pulse sequence is completely characterized by the spacing between the adjacent frequency lines Δf and the total bandwidth

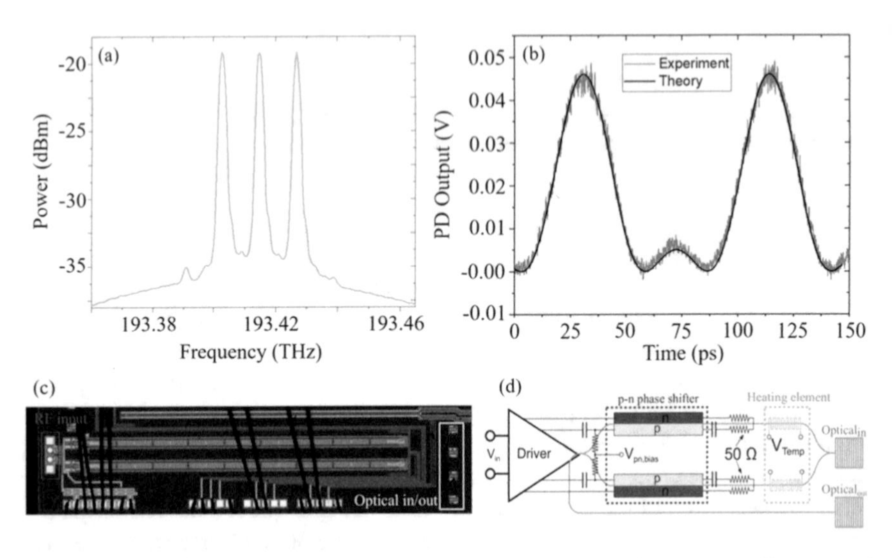

Fig. 28.3 Three-line frequency comb (**a**) and corresponding sinc-pulse sequence (**b**) generated with an integrated Mach-Zehnder modulator (**c**) and [32] grating couplers (GC) (**d**) for the optical input and output, p-n junction biased with the DC voltage $V_{pn,bias}$ for an optical phase shift in the two arms of the Mach-Zehnder interferometer. The temperature controllers are driven with the voltage V_{Temp}, and the sinusoidal RF signal V_{in} is fed to the input driver

$N\Delta f$. The pulse repetition period is $T = 1/\Delta f$, and the rectangular bandwidth $B_s = N\Delta f$ defines the pulse duration (from the maximum to the first zero crossing) to $\tau_p = 1/(B_s)$. Thus, pulse width and repetition rate can be varied by simply tuning the frequency comb parameters Δf and the number of frequency comb lines N. A time shift of the sinc-pulse sequences can be achieved by a phase change of the electrical signal driving the modulators.

A frequency comb with just three lines (one modulator driven with $n = 1$) and the corresponding sinc-pulse sequence generated by a Mach-Zehnder modulator integrated in a silicon chip are shown in Fig. 28.3a and b, respectively [32]. The layout and setup of the modulator are depicted in Fig. 28.3c and d. As can be seen, the generated comb is with 0.04 dB very flat, and the out-of-band 16 dB suppression of higher harmonics is sufficient for high-quality optical sampling with an integrated device.

28.4 Photonic Sampling with Sinc-Shaped Nyquist Pulses

Consider any bandlimited analogue signal $s(t)$ with a baseband bandwidth of B. According to the sampling theory (Nyquist criteria), the signal is fully recoverable if the sampling rate (number of samples per second) is at least twice the maximum frequency present in the signal spectrum $f_s = 1/t_s > 2B$. Here, t_s is the time duration between two samples, and f_s is the sampling frequency. The signal, ideally sampled with an unlimited Dirac-delta sequence, in the frequency domain is defined by Eq. 28.3. If the signal is sampled with a sinc-pulse sequence, instead, it can be written as [30]:

$$S_a(f) = \sum_{n=-\infty}^{\infty} S\left(f - n\Delta f\right) \times \sqcap_N \tag{28.4}$$

with $\sqcap_N$ as a rectangular function with the bandwidth of the rectangular frequency comb $B_s = N\Delta f$, representing the sinc-pulse sequence in the frequency domain. Since the sampling is carried out in the optical domain, the comb bandwidth is defined around the frequency of the optical carrier. Thus, the only difference to ideal sampling in the frequency domain is that the number of spectral copies is restricted to the number of frequency lines in the frequency comb N.

The maximum sampling rate for real-time signals is $f_{sR} = \Delta f$. For periodical signals or if the sampling is parallelized, it corresponds to the bandwidth of the frequency comb $f_{sP} = B_s$. For one single modulator, the bandwidth of the comb can correspond to three times its RF bandwidth, and for two cascaded modulators, it can be increased to four times. Thus, with integrated modulators with an RF bandwidth of 100 GHz [33] or even 160 GHz [34], sampling rates of up to 640 GSa/s would be possible, corresponding to baseband bandwidth of the signals to sample of up to 320 GHz. This is already much more than the baseband width of THz communication signals. However, a further increase of the sampling rate can be achieved by spectrum slicing [9] or the time lens concept [35].

Figure 28.4 shows some experimental examples for the optical sampling based on time-frequency coherence [26–32]. In Fig. 28.4a and b, a nine-line frequency comb, generated with two coupled modulators and the corresponding sinc-pulse sequence, is shown. One of the two coupled modulators has an RF bandwidth of 40 and the other of 20 GHz. Since the frequency difference between the comb lines is 10 GHz, the comb bandwidth is 90 GHz, resulting in pulses with a duration of $\tau_p \approx 11$ ps. The red trace in Fig. 28.4b shows the ideal theoretical and the black trace the measured sinc-pulse sequence. As can be seen, the measured curve comes very close to the ideal one. In Fig. 28.4c, a 40 GHz sinusoidal signal is sampled with such a 90-GHz sinc-pulse sequence. The colored pulses are the time shifted sinc pulses, measured with an electrical sampling oscilloscope with a bandwidth of just 40 GHz. Since the bandwidth of the electrical device is lower than the bandwidth of the pulses, they seem to be distorted. However, the sampling points are the red squares, achieved by an integration over the repetition rate of the pulses. These sampling points follow the signal to sample very well. Thus, they are not affected by the distortions due to the limited bandwidth of the electrical measurement devices.

The sampling of a sinusoidal signal with sinc-pulse sequences with 32 zero crossings between two pulses (the corresponding rectangular frequency comb consists of 33 spectral lines) is shown in Fig. 28.4d. This was again achieved with two coupled modulators, where the first one was driven with five and the second with

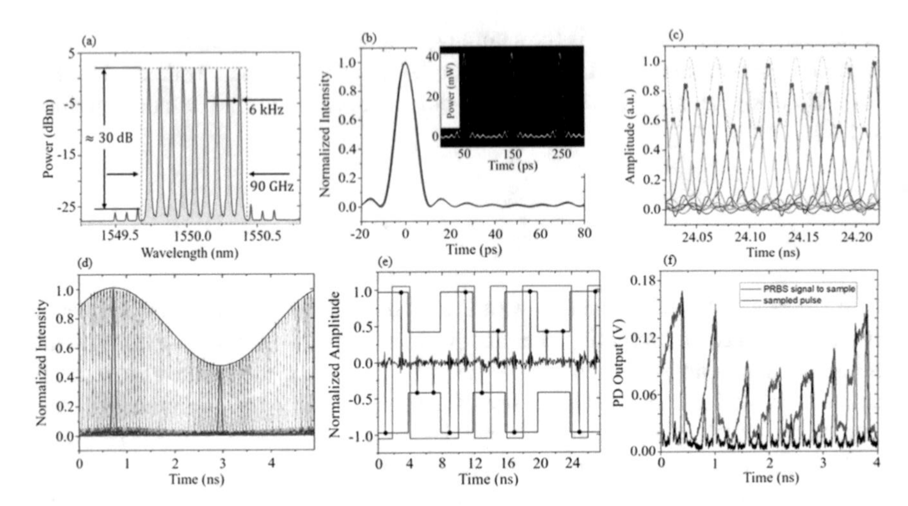

Fig. 28.4 (**a**) Optical frequency comb generated with two coupled intensity modulators and (**b**) corresponding sinc-pulse sequence with the theoretical (red) and experimental (black) curve [17]. (**c**) Sampled 40 GHz sinusoidal signal. The red squares are the sampling points [17]. (**d**) Sinusoidal signal sampled with a sinc-pulse sequence with 32 zero crossings between two pulses. The colors show the different time-shifted sinc pulses [31]. (**e**) Full-field sampling of amplitude (blue) and phase (red) [30] and (**f**) sampling with an integrated Mach-Zehnder modulator driven with two RF input frequencies, resulting in an optical comb with five lines [32]

one RF input frequencies. The different colors represent different pulse sequences, time shifted by a phase change of the electrical signal. That the method can be used to sample the full field (amplitude and phase) is shown in Fig. 28.4e. This is a binary phase shift keying (BPSK) signal with two amplitude levels (blue); the two different phases are depicted in red. The sampling of an arbitrary signal with an integrated device is shown in Fig. 28.4f. Here, an integrated MZM was driven with two RF frequencies, resulting in an optical frequency comb with five lines.

Thus, simple integrated photonic-electronic complementary metal oxide (CMOS) compatible chips could lead to small-footprint and cheap sampling devices with a very high quality for the sampling of THz signals.

References

1. Cisco; and/or its Affiliates, The Zettabyte Era: Trends and Analysis. Cisco Vis. Netw. Index, 1–24 (2014).
2. Song, H., & Nagatsuma, T. (2011). Present and future of terahertz communications. *IEEE Transactions on Terahertz Science and Technology, 1*(1), 256–263.
3. Valley, G. C., Li, L., & Yang, Z. (2007). Photonic analog-to-digital converters. *Optics Express, 15*, 1955–1982.
4. P.W. Juodawlkis, *et. al.* Optically sampled analog-to-digital converters. IEEE Trans. Microw. Theory Techn. 49, (2001).

5. Kull, L., et al. (2018). A 24-to-72GS/s 8b time interleaved SAR ADC with 2.0-to-3.3pJ/conversion and >30dB SNDR at nyquist in 14nm CMOS FinFET," in *Proceedings of 2018 IEEE ISSCC* (pp. 358–360), San Francisco, CA, USA.
6. Collaert, N. (2020) *Future Scaling: Where Systems and Technology Meet*. IEEE International Solid-State Circuits Conference, ISSCC.
7. Doris, K. (2017). *Hybrid data converters*. In Hybrid ADCs, Smart Sensors for the IoT, and Sub-1V & Advanced Node Analog Circuit Design, from P. Harpe, K. Makinwa and A. Baschirotto [Editors, Springer series on] Advances in Analog Circuit Design Series, pp. 3–15.
8. Murmann, B.: ADC performance survey. http://www.stanford.edu/~murmann/adcsurvey.html.
9. Fontaine, N. K., Scott, R. P., Zhou, L., Soares, F. M., Heritage, J. P., & Yoo, S. J. B. (2010). Real-time full-field arbitrary optical waveform measurement. *Nature Photonics, 4*, 248–254.
10. Vectron International (2016). *Ultra low phase noise oven-controlled crystal oscillator OX-305 at 100 MHz*. Datasheet.
11. Westlund, M., et al. (2005). High-performance optical-fiber-nonlinearity- based optical waveform monitoring. *Journal of Lightw. Technology, 23*, 2012–2022.
12. Li, J., et al. (2004). 0.5-Tb/s eye-diagram measurement by optical sampling using XPM-induced wavelength shifting in highly nonlinear fiber. *IEEE Photonics Technology Letters, 16*, 566–568.
13. Khilo, A., Spector, S., Grein, M., Nejadmalayeri, A., Holzwarth, C., Sander, M., et al. (2012). Photonic ADC: Overcoming the bottleneck of electronic jitter. *Optics Express, 20*(4), 4454–4469.
14. Nejadmalayeri, A. H., Grein, M. E., Spector, S. J., et. al (2012). Attosecond Photonics for Optical Communications. In *Proceedigs of OFC*, Los Angeles, pp. 1–3.
15. Salem, R., Foster, A., Turner-Foster, A. C., Geraghty, D. F., Lipson, M., & Gaeta, A. L. (2009). High-speed optical sampling using a silicon-chip temporal magnifier. *Optics Express, 17*(6), 4324–4329.
16. Bhushan, A. S., Coppinger, F., Jalali, B., Wang, S., & Fetterman, H. F. (1998). 150 Gsample/s wavelength division sampler with time-stretched output. *Electronics Letters, 34*, 474–475.
17. Soto, M. A., Alem, M., Shoaie, M. A., Vedadi, A., Brès, C. S., Thévenaz, L., & Schneider, T. (2013). Optical sinc-shaped Nyquist pulses of exceptional quality. *Nature Communications, 4*, 2898. https://doi.org/10.1038/ncomms3898.
18. Soto, M. A., Alem, M., Shoaie, M. A., Vedadi, A., Brès, C.-S., Thévenaz, L., & Schneider, T. (2013). Generation of Nyquist sinc pulses using intensity modulators. *CLEO*.
19. Schneider, T. et al. (2013). *Optical Nyquist-pulse generation with a power difference to the ideal sinc-shape sequence of < 1%*. 2013 *ITG* -Photonic Networks, Leipzig, pp. 1–5.
20. Soto, M. A., Alem, M., Shoaie, M. A., Vedadi, A., Brès, C., Thévenaz, L., & Schneider, T. (2013). Highly tunable method to generate sinc-shaped Nyquist pulses from a rectangular frequency comb. In *Advanced Photonics 2013*, paper SPT2D.6.
21. Soto, M. A., Alem, M., Shoaie, M. A., Vedadi, A., Brès, C., Schneider, T., & Thévenaz, L. (2013). Optical sinc-shaped Nyquist pulses with very low roll-off generated from a rectangular frequency comb. *ACPC*. paper AF2E.6.
22. Preußler, S., Wenzel, N., & Schneider, T. (2014). Flexible Nyquist pulse sequence generation with variable bandwidth and repetition rate. *IEEE Photonics Journal, 6*(4), 1–8. Art no. 7901608.
23. Schneider, T., Wenzel, N., & Preußler, S. (2014). Generation of almost-ideally, sinc-shaped Nyquist pulse sequences with arbitrary bandwidth and repetition rate. *Photonic Networks; 15. ITG Symposium*, Leipzig, Germany, pp. 1–6
24. Preußler, S., Wenzel, N., & Schneider, T. (2014). *Generation of flat, rectangular frequency combs with tunable bandwidth and frequency spacing* (pp. 1–2). San Jose: 2014 CLEO.
25. Preußler, S., Wenzel, N., & Schneider, T. (2014). Flat, rectangular frequency comb generation with tunable bandwidth and frequency spacing. *Optics Letters, 39*, 1637–1640.
26. Preußler, S., Raoof Mehrpoor, G., & Schneider, T. (2016). Frequency-time coherence for all-optical sampling without optical pulse source. *Scientific Reports, 6*, 34500. https://doi.org/10.1038/srep34500.

27. da Silva, E. P., et al. (2016). Combined optical and electrical Spectrum shaping for high-baud-rate Nyquist-WDM transceivers. *IEEE Photonics Journal, 8*(1), 1–11. Art no. 7801411.
28. Preußler, S., & Schneider, T. (2016). *Frequency-time coherence for all-optical sampling.* In ACPC paper AS2I.2.
29. Schneider, T., & Preussler, S. (2017). All-optical sampling without optical source. *Proceedings of. SPIE 10119*, Slow Light, Fast Light, and Opto-Atomic Precision Metrology X, 101191Q.
30. Meier, J., Misra, A., Preußler, S., & Schneider, T. (2019). Orthogonal full-field optical sampling. *IEEE Photonics Journal, 11*(2), 1–9. Art no. 7800609.
31. J. Meier, A. Misra, S. Preußler, and T. Schneider, "Optical convolution with a rectangular frequency comb for almost ideal sampling," in Proceedings of SPIE, vol. 10947 (SPIE, 2019).
32. Misra, A., Kress, C., Singh, K., Preußler, S., Scheytt, J. C., & Schneider, T. (2019). Integrated source-free all optical sampling with a sampling rate of up to three times the RF bandwidth of silicon photonic MZM. *Optics Express, 27.*
33. Wang, C., Zhang, M., Chen, X., Bertrand, M., Shams-Ansari, A., Chandrasekhar, S., Winzer, P., & Lončar, M. (2018). Integrated lithium niobate electro-optic modulators operating at CMOS-compatible voltages. *Nature, 562*, 101–104.
34. Hoessbacher, C., Josten, A., Baeuerle, B., et al. (2017). Plasmonic modulator with >170 GHz bandwidth demonstrated at 100 GBd NRZ. *Optics Express, 25*, 1762–1768.
35. Misra, A., Preußler, S., Zhou, L., & Schneider, T. (2019). Nonlinearity- and dispersion-less integrated optical time magnifier based on a high-Q SiN microring resonator. *Scientific Reports, 6*, 34500.

Chapter 29
Modulation Formats

David A. Humphreys

Abstract Modulation is essential to convey information through a communication system. This chapter gives an overview of the modulation schemes and their use to date in THz systems. The choice of a modulation scheme is a trade-off of factors such as noise, RF channel, cost/complexity, and component technologies. THz systems fall at a technology cusp between optical and RF systems. Spectral efficiency is important at lower frequencies where the spectrum is congested, but THz frequency bands are currently underused, and technical challenges like power, high-channel losses, and oscillator-phase stability dominate the design choices.

29.1 Overview of Digital Modulation

Information is carried through modulation of the amplitude, phase, frequency, or polarization of the RF carrier. The modulation bandwidth is typically a few percent of the carrier frequency, and so operating at a higher frequency is an attractive solution to meet increased capacity demand [1] and escape the congestion in the lower-frequency spectrum [2]. THz carrier frequencies offer the potential for modulation bandwidth of many GHz. Factors such as target bit efficiency, RF power, linearity, RF channel, coexistence, and cost will influence the choice of modulation scheme.

The bandwidth and data rates used for coherent optical communications and THz communication are comparable. Also, 5G mmWave is aimed at a global market. For THz communications to be an attractive proposition, the build and material costs will be important when selecting technologies.

D. A. Humphreys (✉)
NPL, Teddington, UK
e-mail: david.humphreys@npl.co.uk

T. Kürner et al. (eds.), *THz Communications*, Springer Series in Optical Sciences 234, https://doi.org/10.1007/978-3-030-73738-2_29

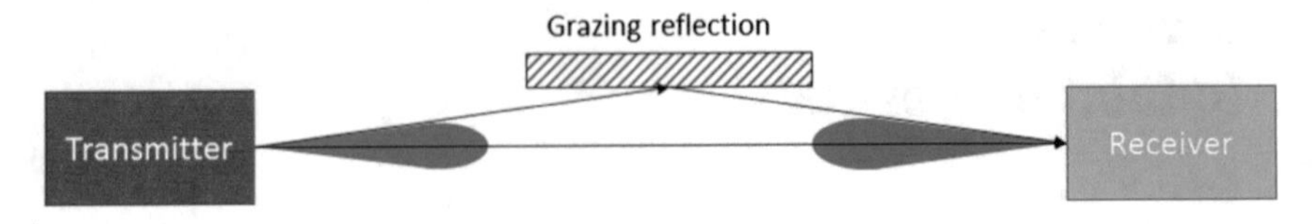

Fig. 29.1 Grazing incidence reflection from an object close to the beam transmission path

29.1.1 RF Channel

As with any frequency band, the propagation channel influences the modulation choices. At THz frequencies, molecular absorptions and scattering by fog and rain are significant [3] and add to the *Friis* transmission losses [4].

Analysis of point-to-point links with high-gain antennas with low side-lobe power for fronthaul/backhaul suggested that multipath reflection is minimal [5]. The main path is line of sight, but there may be an additional delayed path due to grazing reflections (see Fig. 29.1). Using gain (15–40 dBi) and angular pattern data from commercial antennas and recalculating for 300 GHz showed that the additional path delay is mainly dependent on the link power. The delay varied from 49 ps (183 m) to 110 ps (0.64 m) at −5 dBm RF power.

29.1.2 Scalar and Vector Modulation

The clearest separation of modulation formats is between scalar (power) and vector (magnitude and phase) modulations. Vector modulation can be further subdivided into single-carrier and multicarrier schemes. Polarization is normally reserved as a diversity scheme to increase the number of available channels.

The simplest scalar modulation is the presence or absence of the RF carrier, On-off keying (OOK), corresponding to a 1-bit representation of the states (RZ (return to zero)) or 1-bit representation of the transition timing (NRZ (nonreturn to zero)). Similar intensity modulation schemes have formed the backbone of optical fiber communication for data rates of up to 10 Gb/s.

Amplitude shift keying (ASK) is a variant of OOK where the RF power at the lower level is not zero. This removes the ambiguity present in OOK where "0" and loss of signal are identical. The coding should be chosen to avoid long sequences of "0" or "1" to ensure good timing recovery.

Vector modulation adds an additional degree of freedom as both I and Q (or phase and amplitude) components are modulated simultaneously, allowing a much wider range of modulation schemes. These techniques are extensively used at lower RF frequencies and in coherent optical communications where the data rate is comparable to that available at THz frequencies.

Multicarrier modulation schemes encode the data in terms of an M-ary constellation on each frequency component of the signal. The carrier spacing is set by the bit-length T through the Fourier relationship. The commonest of these modulation schemes is orthogonal frequency-domain multiple access (OFDMA) schemes which have been used for IEEE 802.11 series standards and for the fourth- and fifth-generation (4G and 5G) mobile communications [6, 7] because of their bit efficiency and immunity to multipath effects in the RF channel. The number of bits that can be assigned to any frequency component will depend on the channel quality. The symbol time incorporates a "cyclic prefix," typically about 7% of the bit duration T, created by appending the copy of the first few points to the end of the time-domain waveform to allow for the different multipath delays. OFDMA is the subject of many books [8, 9]. At sub-6 GHz frequencies, the symbol duration is typically a few milliseconds because cell sizes are large giving long multipath delays. A positive advantage for multicarrier schemes is their immunity to dispersion and multipath, but the slow adjacent carrier power roll-off, caused by the rectangular time window, makes the scheme vulnerable to frequency offsets that affect many carriers and produces significant out-of-band power, masking adjacent signals.

There are other multicarrier schemes that use windowing to improve their out-of-band performance. A popular candidate, though not adopted for 5G, is filter bank multicarrier, which separates the real and imaginary components and maps these to alternate half-symbols [10].

29.1.3 Constellation Diagram and Data Coding

The simplest representation of data symbols is through a constellation diagram that shows the decoded and recovered and normalized result at each symbol time. The separation of the constellation points (data) indicates the sensitivity to noise of the communication scheme by calculating the energy in the signal transmitted and comparing this with the noise power required to create a specific error rate.

Filtering, phase, amplitude, and sample-timing optimizations are applied before the data is mapped onto the constellation diagram. The in-phase (I) and quadrature (Q) components are often represented in a normalized Cartesian or polar format, depending on the modulation represented. 16-QAM, shown in Fig. 29.2, uses 2 bits on each of I and Q axes to represent 4 bits per symbol. In the example shown here, the bits are selected as blocks b0123 as b0, b1 (I) and b2, b3 (Q).

Noise is normally the limiting factor for any communication system [11]. The noise statistics are normally assumed to be Gaussian and uncorrelated, AWGN, and the separation of constellation points. The constellation coding can minimize the number of bit errors. Gray-coded adjacent constellation points give a single-bit error. This coding scheme can be applied to two-dimensional constellations, such as 16 QAM, by applying the coding strategy to each of the two orthogonal dimensions.

For a square, fully filled, M-QAM constellation, the constellation point separation decreases as the number of bits is increased, requiring a higher SNR to

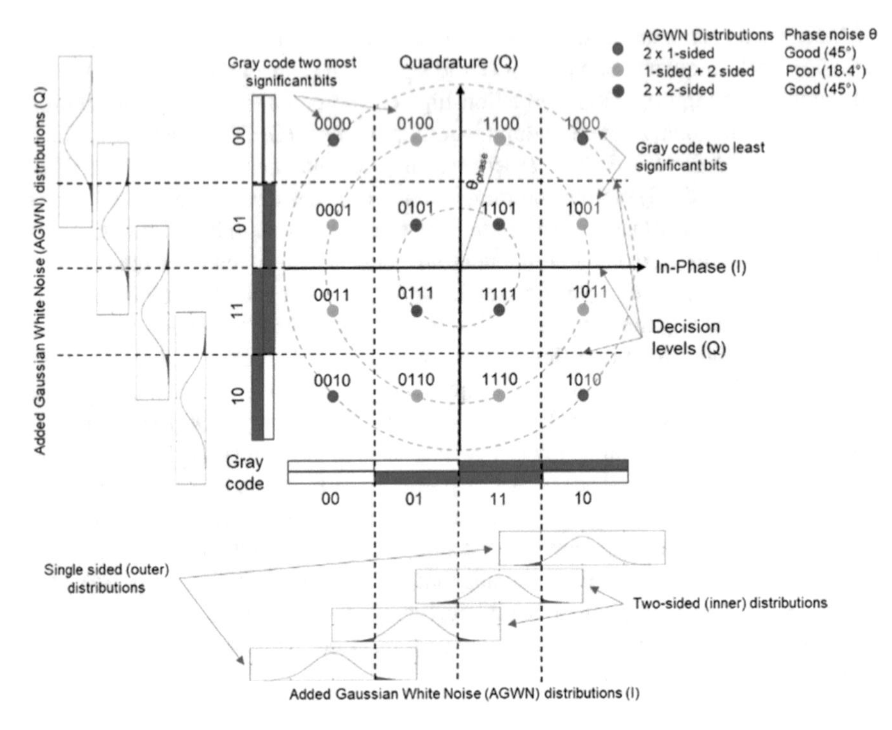

Fig. 29.2 A Gray-coded 16-QAM constellation (4 bits). The probability of error depends on the constellation point position. Points in the core of the constellation (red) overlap with more neighbors than the points of the edge or the corners. Phase noise, due to the THz source or imperfect correction of Doppler shift increase, will disproportionally affect the constellation points (green) with the smallest angular difference (e.g., $0100 - 1100$ or $1001 - 1011$ on the center circle)

achieve the same BER. Assuming evenly weighted constellation points, AWGN noise only n_k with no phase-noise component and considering only the nearest neighbor symbols, the BER is given by a received signal (x_k) [12]:

$$P_b^{\mathrm{QAM}}(M) \approx \frac{2}{\log_2(M)}\left(\frac{\sqrt{M}-1}{\sqrt{M}}\right) erfc\left(\sqrt{\frac{3\mathrm{snr}}{2\,(M-1)}}\right) \tag{29.1}$$

where

$$\mathrm{snr} = \frac{\left\langle |x_k|^2 \right\rangle}{\left\langle |n_k|^2 \right\rangle} \tag{29.2}$$

A signal-to-noise increase of about 7 dB is required to increase the bit efficiency by 2 bits per symbol.

29.1.4 Error Vector Magnitude

Error vector magnitude (EVM) represents the normalized rms value of the residual error vector at each constellation point. EVM has a reciprocal relationship with signal-to-noise ratio (SNR) [13, 14] directly related through:

$$\mathrm{EVM}_{\mathrm{rms}} = \sqrt{\frac{1}{\mathrm{SNR}} + 2 - \left(1 + g_t\right)\sqrt{\frac{1 + \cos\left(\phi_t\right)}{1 + g_t^2}}}, \tag{29.3}$$

where (g_t) is the gain imbalance ratio and (ϕ_t) is the quadrature error [15].

EVM has also been used in ETSI 3GPP and other standards [16], but the normalization and pre-correction variables differ between standards. This is an issue for higher-order QAM waveforms where normalization to the I and Q limiting or to the rms power scales the EVM results. A second issue is data symbol recovery. If the symbol values are known beforehand (directed), then error vectors that exceed the decision boundaries are correctly handled, but if the symbol value is taken from the recovered values, then the error vector is referred to the closest constellation points, which may not necessarily be the correct symbol point. Traceable assessment of the EVM contribution arising from the source and receiver, connected back to back, has been made at the frequencies used for 3G, assuming uncorrelated noise components [17], and at 44 GHz [18] using the NIST uncertainty framework that handles correlation terms [19]. The RF waveform can be directly measured with a digital sampling oscilloscope (DSO), but this approach is limited to carrier frequencies less than 100 GHz.

Traceable EVM measurements with a real-time digital oscilloscope (RTDO) at 0.9–2.56 GHz, with a bandwidth of 3.84 MHz, ignored any correlated uncertainty corrections as the back-to-back measurements of impedance match and instrument flatness corrections are negligible and effectively a constant [17]. Although a DSO can measure over millisecond epochs, this is impractical at low trigger rates.

Figure 29.3 shows the upper half of the 16-QAM constellation with AWGN. Although the received data corresponds to a “1100” symbol, it is incorrectly decoded as “1000.” This error lowers EVM calculated from recovered data as the SNR degrades.

EVM and its associated uncertainties are currently being studied in the IEEE P1765 Standards Working Group [20]. In this standard, all the calculations are “data-directed” rather than based on “recovered” data symbols, selected as the constellation point with the minimum individual error vector. At THz frequencies, the modulation bandwidth spans several GHz, and a mixer-based receiver is needed as direct acquisition is impractical. Impedance-match and flatness corrections will filter the noise components, giving correlated uncertainty components.

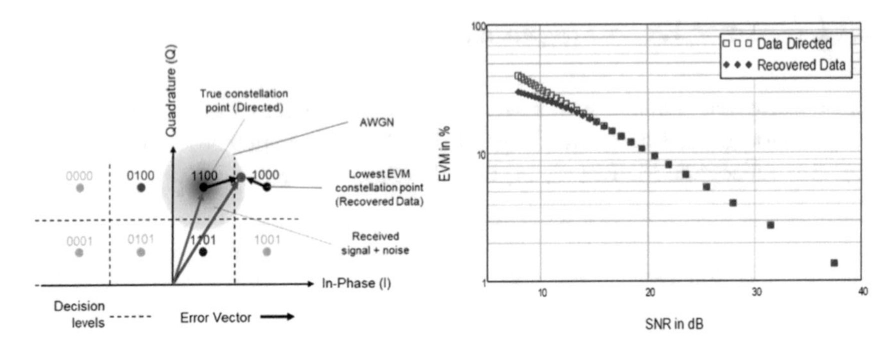

Fig. 29.3 (**a**) Different symbol recovery result for data-directed and data-recovered approaches. (**b**) shows the lowered EVM result caused by misclassifying the symbol

29.2 Scalar-Modulated THz Systems

Various scalar systems have been reported: 1–8 Gb/s at 250 GHz [21] with a BER of 10^{-10}, 5–12.5 Gb/s (ASK) at 300 Gb/s [22], 25 Gb/s [23] at 220 GHz, and 40 Gb/s at 300 GHz [24] with a BER of better than 10^{-10}. This system was extended to 48 Gb/s (SBD detector-limited) using polarization diversity. The same author reported 50 Gb/s over 100 m with a BER of 9.5×10^{-4} also using OOK modulation [25].

In the optical technology-based systems, the THz signal is generated within a uni-traveling carrier UTC photodiode by photo-mixing (heterodyne difference frequency mixing) within the photodiode. The radiated THz power is typically a few tens to a few hundreds of microwatts [26] for photocurrents of about 10 mA. Work is in progress to increase the achievable THz power from the UTC photodiodes [27]. As part of this work, two lasers to generate the optical heterodyne laser sources have been monolithically integrated with the UTC as a compact source [28].

Although incoherent modulation schemes have bandwidth and diversity limitations, they may find applications for short-range transmission, where cost is a critical factor.

29.3 Vector Modulation THz Systems

Vector-modulated formats provide an additional independent degree of freedom for modulation, giving a higher bit efficiency but require a phase-stable THz carrier to avoid bit errors. Scalar systems typically achieve a bit efficiency of about 0.5 bits/Hz. Vector modulation-based systems have achieved higher bit efficiencies and data rates: 2.86 bits/Hz at 340 GHz using 16-QAM modulation [29] and 105 Gb/s at 3.5 bits/Hz using 32 QAM at 300 GHz over 6 channels [30]. In addition, the reported EVM performance varied from 4.8% to 8.5% across the channels.

Stöhr and Hermello et al. have achieved a bit-efficiency of 5.9 bits/Hz with OFDMA modulation [31]. This work used optical signal generation with a UTC photodiode. The 84 OFDM subcarriers were each modulated with 64 QAM (6 bits) giving 8.54 ns/symbol.

Both optical heterodyne and RF harmonic multiplication have been used to achieve a phase-stable RF carrier. At optical frequencies, a phase-stabilized optical frequency comb [24] or an optical phase-locked loop. Coherent optical transmitters containing compound Mach-Zehnder integrated optic modulators are commercially available and can generate the required complex modulation for use with a UTC diode; however, in general, this is still a discrete-based process.

There has been considerable progress with the electronic systems using a variety of technologies and material systems. Electronic systems also offer a degree of flexibility about the ordering of components to compensate for technology limitations such as low gain or low power or linearity.

29.3.1 Modulation Comparison for Existing Published Systems

Both optical and electrical generations have been developed for THz demonstrator systems. Several authors have summarized the state of the art in terms of carrier frequency, data rate, link distance, or technology. Propagation path distance is affected by the signal-to-noise ratio required for modulation format, the available transmitter power, and the RF channel. Figure 29.4 shows the timeline of the published work in terms of the modulation format and data rate [21–25] and [29–46]. The scalar results are primarily based on optical technology with unstabilized optical frequencies so the THz carrier may have a linewidth of over 10 MHz. The later demonstrators show data rates above 100 Gb/s, and the highest data rates have been achieved with OFDMA and 16-QAM systems.

29.4 Choosing a Modulation Scheme

The selection of a modulation format needs to take account of both the objectives and the constraints set by technology and the environment. To date, the objectives have been to develop and explore what is possible with the preexisting technologies and the tools available for research. The technology demonstrations mainly have fixed point-to-point links, and in many cases, the signal processing is off-line.

The closest analogues to THz communications are the development of mm-Wave 5G-NR systems and coherent optical communications. 5G-NR specifies the same cyclic prefix OFDMA modulation for both the lower and mm-Wave frequencies. Coherent optical communication achieves a comparable bandwidth and data throughput to THz systems [47]. At these high data-rates, it is important to

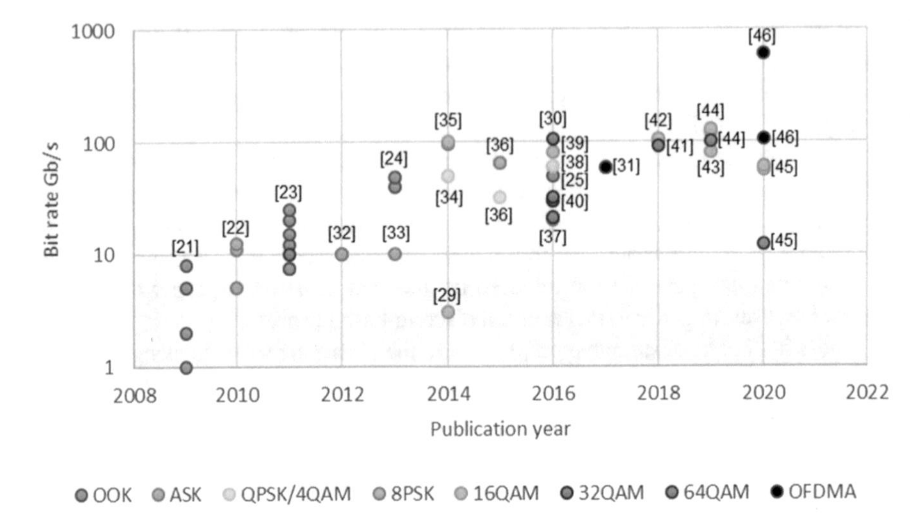

Fig. 29.4 Scalar and vector THz demonstrators, classified by their modulation format

minimize the signal processing overheads so that the fast ADC will have a low ENOB and this constraint will also have implications for constellation complexity.

Lucky and Foschini proposed optimal constellation designs [47–50]. Foschini's optimum layout assumes that for any three constellation points, the separation should be equal, leading to an equilateral triangle primitive giving symmetric constellations of 4, 7, and 19 points. A partially filled 19-point constellation would carry 4 bits and have an asymmetry (no training required). AWGN added to the equilateral triangle primitive will always result in both 1-bit and 2-bit errors. By contrast, a square M-QAM has a lower probability of a 2-bit error and requires a lower ADC resolution.

29.4.1 Noise in Measured Systems

The AWGN assumption applied to many communication systems may not be correct at THz frequencies. Data from experimental results published by Dan et al. [51] for a 35 nm mHEMT-based system at 300 GHz have been reanalyzed to investigate the phase-noise behavior of the 2 Gbaud for the 8-APSK measurements. In Fig. 29.5a, the back-to-back AWG to oscilloscope results (green) show random noise only, but the measurements over a 1 m path at symbol rates show both AWGN and phase noise around the constellation points. The evolution of the phase error over time shows random and systematic components, with a periodic ripple of about 1 MHz. In THz systems, phase noise will be an issue because of the high multiplication factor and greater Doppler shift offset.

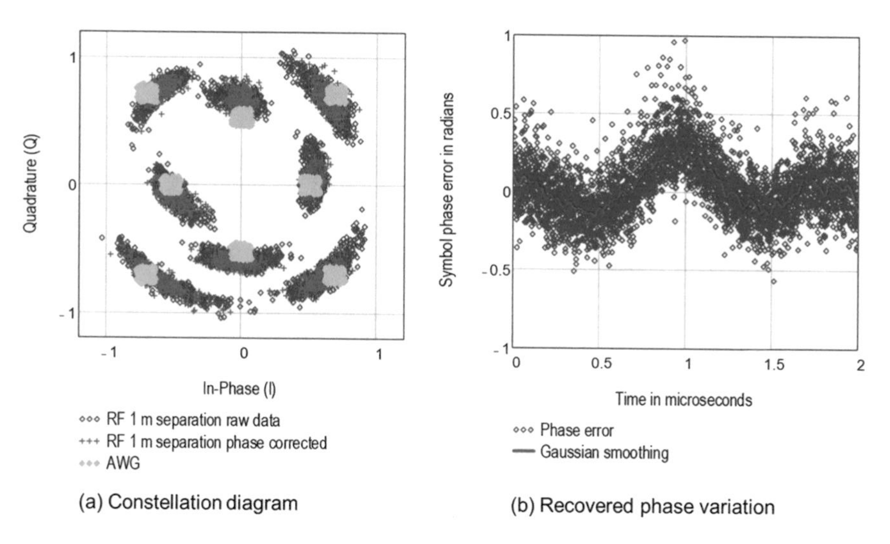

Fig. 29.5 Analysis of experimental results at 300 GHz showing AWGN and phase noise. The phase noise can be corrected, improving the BER by a factor of 5.5. (Data from the TERAPAN project was provided by I. Dan and is published in [51])

The BER calculated from the results improved from 1.9×10^{-2} to 3.4×10^{-3} after correction.

Other measurements made at comparable baud rates have demonstrated higher-density constellation without phase-noise impairments.

29.4.2 Alternatives and Choices

The application, capacity, and cost are important guides for choosing a modulation scheme. For a short-range system where the capacity and range requirements are not excessive, a UTC-based scalar system may be suitable.

For single-carrier and multicarrier systems where phase noise is a problem, the angular separation of constellation points becomes a critical issue. How the constellation is decoded may also be important as the assumptions made for AWGN may not apply. An 8-APSK modulation will allow up to 90° phase rotation, but the two constellation rings may need to be separated in terms of magnitude. This will provide a lower protection against AWGN. The nearest neighbor points form a triangle, and Gray coding will not provide single-bit errors for all transitions from the inner to the outer ring, increasing the BER by 20%, as shown in Fig. 29.6a. Altering the angular relationship between the two rings reduces the probability of the 2-bit transition but also increases the probability of the single-bit errors.

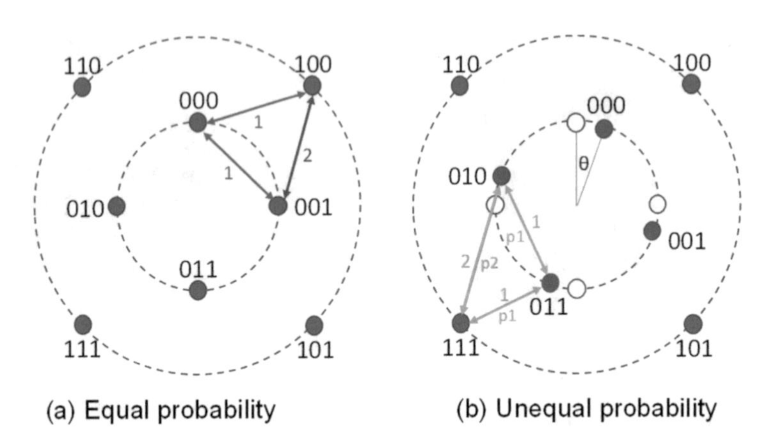

Fig. 29.6 Original 8-APSK constellation and modified constellation to reduce the probability of the 2-bit errors

Constellations using multiple amplitude rings provide advantages for phase noise, at the expense of AWGN and the need for higher-resolution ADCs. Figure 29.2 showed that certain symbol errors will have a higher probability of error with phase noise. It should be possible to extract this information from the FEC correction process. The quality of service reporting in the current standards is quite simplified, with the aim of adjusting the modulation to maximize the transmission rate [6]. It may be possible in future systems to feedback additional information to improve the transmission quality.

29.4.3 Multicarrier

The estimated residual multipath delay discussed in 29.1.1 and 29.1.3 suggested multipath delays of (49–110) ps; inferring OFDM subcarrier spacings of 640 MHz to 1400 MHz would be quite feasible. Experimental results by Katayama et al. [30] show multiple conventional wideband time-domain transmission, each with a bandwidth of 4.7 GHz and a subcarrier separation of 5 GHz. No over-the-air measurements are reported. Hermelo et al. [31] use a 64-QAM modulation with 84 OFDM subcarriers and 7% overhead and 10 GHz bandwidth to achieve a baud rate of 59.06 Gb/s. The RF transmission link is 5 cm in this case, and the data was generated off-line.

OFDM modulation may offer the highest spectral efficiency, but PAPR and computational overhead may be an issue. The short multipath time delay will allow a high symbol rate, with the additional benefit of providing some protection against phase noise and Doppler shift, and the use of pilot tones will aid signal recovery.

29.4.4 Summary of Key Points

1. The choice of a modulation scheme should not be carried out in isolation as it links to almost all other aspects of the system, e.g., data capacity, coding, FEC, power, power efficiency, SNR, range. and cost.
2. Scalar modulation is quite crude, but based on optical/UTC technology, it is potentially low cost with a reasonable bandwidth. This may be a cost-effective short-range OTA link for indoor, data center, and kiosk applications.
3. Impressive single-carrier and classical multicarrier research results (106 Gbit/s) have been obtained. These systems are more complex and at present may offer benefits as line-of-sight links where there is a demand for the highest bandwidth. As this technology matures, it will benefit from the mm-Wave research for New Radio (5G) systems.
4. Because of the high-carrier frequency, oscillator stability and Doppler shifts will add phase noise to the received signal, affecting signal-recovery and constellation design choices.

Glossary

5G	Fifth-generation cellular radio system
AWGN	Added white Gaussian noise
ASK	Amplitude shift keying
APSK	Amplitude and phase-shift keying
EVM	Error vector magnitude
FEC	Forward error correction
LTE	Long-term evolution (fourth-generation cellular radio system)
OFDMA	Orthogonal frequency-domain multiple access
OOK	On-off keying
OTA	Over the air (testing)
M-QAM	Quadrature amplitude modulation for an arbitrary number of constellation points (e.g., 16 QAM)
SNR	Signal-to-noise (power) ratio, normally expressed in decibels

References

1. Cherry, S. (July 2004). Edholm's law of bandwidth. *IEEE Spectrum, 41*(7), 58–60.
2. Mokole, E. L., et al. Spectrum use, congestion, issues, and research areas at radio-frequencies (Radar, Sonar & Navigation, 2018). In *Radar and communication spectrum sharing* (Chap. 5, pp. 135–173). https://digital-library.theiet.org/content/books/10.1049/sbra515e_ch5

3. Siles, G. A., & Riera, J. M. (2009, January). An introduction to THz atmospheric propagation and passive remote sensing applications. In *Conference: XXIV Simposium Nacional URSI, Santander, Espana*.
4. Friis, H. T. (1946, May). A note on a simple transmission formula. *Proceedings of the IRE, 34*(5), 254–256.
5. Rey, S., & Kuerner, T. (2015, September). *Why/when is AWGN a suitable channel model for wireless front-/backhaul?*. IEEE 802.15-15-15-15-0681-00-003d.
6. Rumney, M. (2013, July). *LTE and the evolution to 4G wireless – Design and measurement challenges*, 2nd ed., CH2. Wiley, ISBN:9781119962571.
7. Kim, Y., et al. (2019, June). New Radio (NR) and its evolution toward 5G-advanced. *IEEE Wireless Communications, 26*(3), 2–7.
8. Li, Y. G., & Stuber, G. L. (Eds.). (2006). *Orthogonal frequency division multiplexing for wireless communications*. Springer. ISBN 978-0-387-29095-9.
9. Cho, Y. S., et al. (2010). Introduction to OFDM. In *MIMO-OFDM wireless communications with MATLAB®*. IEEE, pp. 111–151.
10. Arthur, J. K., et al. (2019). Comparative analysis of orthogonal frequency division modulation and filter bank-based multicarrier modulation. In *2019 International Conference. on Communications, Signal Processing and Networks (ICCSPN), Accra, Ghana* (pp. 1–10).
11. Shannon, C. E. (1948, July, October). A mathematical theory of communication. *The Bell System Technical Journal, 27*, 379–423, 623–656.
12. Proakis, J. G. (2001). *Digital communications* (4th ed.). New York: McGraw-Hill.
13. Shafik, R. A., et al. (2006, November). On the error vector magnitude as a performance metric and comparative analysis. In *2nd International Conference on Emerging Technologies, Peshawar, Pakistan* (pp. 27–31).
14. Mahmoud, H. A., & Arslan, H. (2009). Error vector magnitude to SNR conversion for nondata-aided receivers. *IEEE Transactions on Wireless Communications, 8*(5), 2694–2704.
15. Hudlička, M., et al. (2016). BER estimation from EVM for QPSK and 16-QAM coherent optical systems. In *2016 IEEE 6th International Conference on Photonics (ICP), Kuching* (pp. 1–3).
16. 6.5.2 Error Vector Magnitude, ETSI TS 136 104 V8.3.0 (2008–2011). LTE; Evolved Universal Terrestrial Radio Access (E-UTRA); Base Station (BS) radio transmission and reception (3GPP TS 36.104 version 8.3.0 Release 8), p. 17.
17. Humphreys, D. A., & Miall, J. (2013, June). Traceable measurement of source and receiver EVM using a real-time oscilloscope. *IEEE Transactions on Instrumentation and Measurements, 62*(6), 1413–1416.
18. Remley, K. A., et al. (2015, May). Millimeter-wave modulated-signal and error-vector-magnitude measurement with uncertainty. *IEEE Transactions on Microwave Theory and Techniques, 63*(5), 1710–1720.
19. Jargon, J. A., et al. (2012). Establishing traceability of an electronic calibration unit using the NIST microwave uncertainty framework. In *79th ARFTG Microwave Measurement Conference, Montreal, QC* (pp. 1–5).
20. P1765 – trial-use recommended practice for estimating the uncertainty in error vector magnitude of measured digitally modulated signals for wireless communications. https://standards.ieee.org/project/1765.html
21. Song, H. J., et al. (2009, October). 8 Gbit/s wireless data transmission at 250 GHz. *Electronics Letters, 45*(22), 1121–1122.
22. Song, H. J., et al. (2010, October). Terahertz wireless communication link at 300 GHz. In *IEEE topical meeting Microwave Photonics (MWP)* (pp. 42–45).
23. Kallfass, I., et al. (2011, November). All active MMIC-based wireless communication at 220 GHz. *IEEE Transactions on Terahertz Science and Technology, 1*(2), 477–487.
24. Nagatsuma, T., et al. (2013). Terahertz wireless communications based on photonics technologies. *Optics Express, 21*(20), 23736–23747.

25. Nagatsuma, T., et al. (2016). 300-GHz-band wireless transmission at 50 Gbit/s over 100 meters. In *2016 41st International Conference on Infrared, Millimeter, and Terahertz waves (IRMMW-THz), Copenhagen* (pp. 1–2).
26. Ito, H., et al. (2004). Continuous THz-wave generation using uni-travelling-carrier photodiode. In *15th International Symposium on Space Terahertz Technology, April 27–29, Northampton, MA, USA* (pp. 143–150).
27. *High power uni-travelling carrier photodiodes for THz wireless communications*. UK Research and Innovation project 2259136, Oct 2015 – Feb. 2021. https://gtr.ukri.org/projects?ref=studentship-2259136#/tabOverview
28. Mohammad, A. W., et al. (2018). Optically pumped mixing in photonically integrated uni-travelling carrier photodiode. In *2018 43rd International Conference on Infrared, Millimeter, and Terahertz Waves (IRMMW-THz), Nagoya* (pp. 1–2).
29. Wang, C., et al. (2014, January). 0.34-THz wireless link based on high-order modulation for future wireless local area network applications. *IEEE Transactions on Terahertz Science and Technology, 4*(1), 75–85.
30. Katayama, K., et al. (2016, December). A 300 GHz CMOS transmitter with 32-QAM 17.5 Gb/s/ch capability over six channels. *IEEE Journal of Solid-State Circuits, 51*(12), 3037–3048.
31. Hermelo, M. F., et al. (2017). Spectral efficient 64-QAM-OFDM terahertz communication link. *Optics Express, 25*, 19360–19370.
32. Hu, S., et al. (2012, November). A SiGe BiCMOS transmitter/receiver chipset with on-chip SIW antennas for terahertz applications. *IEEE Journal of Solid-State Circuits, 47*(11), 2654–2664.
33. Chung, T. J., & Lee, W.-H. (2013, June). 10-Gbit/s wireless communication system at 300 GHz. *ETRI Journal, 35*(3), 386–396.
34. Song, H., et al. (2014, March). 50-Gb/s direct conversion QPSK modulator and demodulator MMICs for terahertz communications at 300 GHz. *IEEE Transactions on Microwave Theory and Techniques, 62*(3), 600–609.
35. Koenig, S., et al. (2014, October). Wireless sub-THz communication system with high data rate enabled by RF photonics and active MMICtechnology. In *Proceedings of the IEEE Photonics Conference* (pp. 414–415).
36. Kallfass, I., et al. (Feb. 2015). 64 Gbit/s transmission over 850 m fixed wireless link at 240 GHz carrier frequency. *Journal of Infrared, Millimeter, and Terahertz Waves, 36*(2), 221–233.
37. Katayama, K., et al. (2016, May). CMOS 300-GHz 64-QAM transmitter. In *IEEE MTT-S International Microwave Symposium Digest* (pp. 1–4).
38. Yu, X., et al. (2016, November). 400-GHz wireless transmission of 60-Gb/s Nyquist-QPSK signals using UTC-PD and heterodyne mixer. *IEEE Transactions on Terahertz Science and Technology, 6*(6), 765–770.
39. Jia, S., et al. (2016). THz photonic wireless links with 16-QAM modulation in the 375-450 GHz band. *Optics Express, 24*, 23777–23783.
40. Song, H., et al. (2016). Demonstration of 20-Gbps wireless data transmission at 300 GHz for KIOSK instant data downloading applications with InP MMICs. In *2016 IEEE MTT-S International Microwave Symposium (IMS), San Francisco, CA* (pp. 1–4).
41. Chinni, V. K., et al. (2018). Single-channel 100 Gbit/s transmission using III–V UTC-PDs for future IEEE 802.15.3 d wireless links in the 300 GHz band. *Electronics Letters, 54*(10), 638–640.
42. Jia, S., et al. (2018, January). 0.4 THz photonic-wireless link with 106 Gbps single channel bitrate. *Journal of Lightwave Technology, 36*(2), 610–616.
43. Lee, S., et al. (2019, February). 9.5 an 80 Gb/s 300GHz-band single-chip CMOS transceiver. In *Proceedings of the IEEE international solid-state circuits conference*.
44. Hamada, H., et al. (2019). 300-GHz 120-Gb/s wireless transceiver with high-output-power and high-gain power amplifier based on 80-nm InP-HEMT technology. In *2019 IEEE BiCMOS and Compound semiconductor Integrated Circuits and Technology Symposium (BCICTS), Nashville, TN, USA* (pp. 1–4).

45. Dan, I., et al. (2020, January). A terahertz wireless communication link using a superheterodyne approach. *IEEE Transactions on Terahertz Science and Technology, 10*(1), 32–43.
46. Jia, S., et al. (2020, September 1). 2 × 300 Gbit/s Line Rate PS-64QAM-OFDM THz Photonic-Wireless Transmission. *Journal of Lightwave Technology, 38*(17), 4715–4721.
47. Goodwins, R. 5G New Radio: The technical background," ZDnet, February 1, 2019. https://www.zdnet.com/article/5g-new-radio-the-technical-background/
48. Lucky, R., & Hancock, J. (1962, June). On the optimum performance of m-ary systems having two degrees of freedom. *IRE Transactions on Communications, CS-10*, 185–192.
49. Foschini, G., et al. (1974, January). Optimization of two-dimensional signal constellations in the presence of Gaussian noise. *IEEE Transactions on Communications., 22*(1), 28–38.
50. Hanzo, L., et al. (2004). *Quadrature amplitude modulation: From basics to adaptive Trellis-Coded, Turbo-Equalised and Space-Time Coded OFDM, CDMA and MC-CDMA Systems* (2nd edn.). Wiley/IEEE Press. ISBN 0-470-09468.
51. Dan, I., et al. (2017). Impact of modulation type and baud rate on a 300GHz fixed wireless link. In *2017 IEEE Radio and Wireless Symposium (RWS), Phoenix, AZ* (pp. 86–89).

Chapter 30
Forward Error Correction: A Bottleneck for THz Systems

Onur Sahin and Norbert Wehn

Abstract A truly practical and widely deployed THz technology, and ultra-high data rate enabled, i.e. Tb/s, link-level solutions will rely on significant progress on multiple technical design fronts, including the transceiver and digital baseband components. The digital baseband is responsible for processing and corresponding storage/memory procedures of the bitstreams and their computations in order to enable efficient, reliable, and high-speed transmit and receive operations. In THz and primarily Tb/s data rates that will be enabled in the THz spectrum, these computation and memory operations reach out to the levels that are orders of magnitude higher than state-of-the-art baseband solutions. Forward error control (FEC) on the other hand stands out as the most complex and highest power-consuming component in the end-to-end digital baseband chain for THz and Tb/s link-level technologies. In this chapter, we provide a detailed FEC implementation and design requirement analysis targeting Tb/s data rates while considering the underlying practical deployment constraints including power and device physical area. The chapter also presents an FEC design framework that marries implementation and code design domain. Finally, a comprehensive summary of the most promising FEC encode, decode, and implementation solutions is provided along with the remaining challenges toward the Tb/s data rate goal.

30.1 Introduction

Ultra-high throughput, or data-rate, communications with target data rates approaching Tb/s, are arguably one of the most critical capabilities that is promised by THz systems. Existing 5G standards, with foundations laid in 3GPP Release

---

O. Sahin (✉)
InterDigital, London, UK
e-mail: Onur.Sahin@InterDigital.com

N. Wehn
Technische Universität Kaiserslautern, Kaiserslautern, Germany

T. Kürner et al. (eds.), *THz Communications*, Springer Series in Optical Sciences 234, https://doi.org/10.1007/978-3-030-73738-2_30

15 and 16, are built on throughputs in the access of 20 Gbps peak data rates, whereas several technology evolution observations, from data traffic trends to network infrastructure models, put practical wireless Tb/s technology as the next major milestone [1]. However, there are many technological challenges to develop a practical and feasible Tb/s wireless technology, which relies on the frequency bands above 100 GHz bands, and broadly named as THz spectrum.

A key challenge in THz systems, particularly the ones aiming ultrahigh throughputs, requires highly efficient radio frequency (RF) front-ends. Recent advances in the RF front-ends, e.g., electronic, photonic, and hybrid solutions, have put THz and Tb/s systems one step closer to practical use [2]. Yet, another key challenge is related to the implementation of digital baseband processing algorithms at data rates approaching Tb/s. Forward error correction (FEC), a.k.a. channel coding, is a basic component of any wireless link technology, constituting the most complex and computationally intense component in the digital baseband chain. In a wireless system, FEC provides detection and correction of errors in the information bitstream. These errors could occur during transmission over a noisy channel that induces errors in the bitstream due to physical effects such as interference, thermal noise, and device imperfections. Considering its instrumental benefits, FEC has been adopted in almost all wireless communication products and generations and will be even more critical for THz systems considering the underlying substantially higher throughputs. There is no question that the design of communication systems is no longer just a matter of spectral efficiency or bit/frame error rate. Rather, it is mandatory to jointly consider the implementation efficiency and implementation cost. In the past, progress in microelectronic silicon technology driven by Moore's law was an enabler of large leaps in throughput, lower latency, and lower power. However, we have reached a point where microelectronics can no more keep pace with the increased requirements from communication systems. In addition, advanced semiconductor technology nodes imply new challenges such as power density. Thus, channel coding for beyond 5G THz and ultrahigh data rate systems requires a cross-layer approach, covering information theory, decoding algorithm development, parallel hardware architectures, and semiconductor technology.

In this chapter, we provide FEC requirements for THz systems from a practical implementation perspective. Then we summarize state-of-the-art FEC solutions that target these requirements.

30.2 FEC Performance Requirements for Ultrahigh Throughput THz Systems

Wireless link-level or digital baseband design takes various key performance indicators (KPIs) into account, all of which collectively determine the underlying system requirements. Some of the key KPIs for wireless system design include error correction capability, operating frequency band, transmission range, operating

Table 30.1 FEC KPI definitions

KPI	Unit	Explanation
Throughput	Gb/s	The net information throughput
Area	mm^2	Silicon area of the FEC IP
Power	Watt	Total power dissipation of the FEC IP
Area efficiency	$Gb/s/mm^2$	Throughput per unit area
Power density	W/mm^2	Power dissipation per unit area
Energy efficiency	pJ/bit	Energy required for decoding one information bit
Latency	ns	Time for decoding one code word
Frequency	MHz	Clock frequency of the FEC IP

conditions (in terms of channel characteristics), latency constraints, transmission direction, hybrid automatic repeat request (HARQ) admissibility, code flexibility (in terms of block length and code rate), and cost. FEC design requirements depend on these communication-related KPIs but also in the same way on the hardware implementation KPIs.

The most important implementation KPIs are information bit throughput, area, and power consumption, which are directly related to cost (silicon area, package), area efficiency and energy efficiency, power density, and latency. These KPIs are summarized in Table 30.1. For an FEC intellectual property (IP) block, the relation between the KPI parameters can be given by

$$\text{Area efficiency} = \frac{\text{Throughpout}}{\text{FEC}_{\text{area}}} \left(\text{bit/s/mm}^2\right) \tag{30.1}$$

$$\text{Energy efficiency} = \frac{\text{FEC}_{\text{power}}}{\text{Throughput}} \left(\text{J/bit}\right) \tag{30.2}$$

$$\text{Power density} = \text{Area efficiency} * \text{Energy efficiency} = \frac{\text{FEC}_{\text{power}}}{\text{FEC}_{\text{area}}} \left(\text{W/mm}^2\right) \tag{30.3}$$

where $\text{FEC}_{\text{power}}$ and FEC_{area} correspond to power consumption at the FEC IP block and area of the FEC IP block, respectively. A large value for area efficiency and a low value for energy efficiency indicate high efficiency of the FEC implementation solution. The affordable chip area of the FEC IP block depends on various factors such as technology node (IC design cost in very advanced technology nodes can be more than \$100 million in today's terms), production/market volume, system prizes, etc. We assume that the total area of a baseband SoC is limited to 100 mm^2 due to cost issues. A maximum of 10% of this area should be reserved for the FEC

IP. Hence, a feasible FEC_{area}, as widely deployable solution for THz use cases, is assumed to be *less* than 10 mm^2 [3].

In the following, we discuss the FEC design challenges and define a broad performance requirement framework for THz systems and use cases.

BER: The BER requirements for THz use cases are bound to vary significantly depending on the categorization of these under either infrastructure or end-user domains. For instance, IEEE 802.15.3d standard sets very stringent BER requirements of 10^{-12} for infrastructure-type use cases such as wireless back-haul/fronthaul, and data centers, whereas relatively relaxed BER requirement of 10^{-6} for close-proximity communications with applications in personal area networks.

Throughput: Beyond 5G era is expected to be dominated by new definition of multimedia transportation and use cases, including 360-degree ultra-HD videos and augmented and virtual reality applications some of which might be realized by novel AR/VR terminals at the end users. Various detailed analyses and predictions place the throughput requirements in the order of 100 Gb/s-1 Tb/s, up to 50× order of magnitude improvement over 5G peak throughputs [4].

Power and Power Density: Thermal design power (TDP), which is the maximum amount of heat the package/cooling system of an SoC chip is designed for, sets an upper limit on the power and corresponding power density. Because of cost reasons (package/cooling cost), the power of the FEC IP is typically limited to around 1 Watt, which yields a power density in the order of 0.1 W/mm^2 as a feasible value for practical FEC IP implementation.

Energy Efficiency: A total power budget of 1 Watt for FEC IP and 1 Tb/s throughput requirement results in 1 pJ/bit energy efficiency constraint. Power density and energy efficiency belong to the biggest implementation challenges. For comparison, 1 pJ is in the same order of magnitude as performing a 32-bit multiplication in 28 nm technology or one order of magnitude lower than accessing a single bit from a state-of-the-art external dynamic random-access memory (DRAM). If the power is constrained, increasing the throughput requires decreasing the energy efficiency per decoded bit by the same order.

Area Efficiency: A 100 Gb/s/mm^2 requirement can be directly obtained from corresponding throughput (1 Tb/s) and chip area (10 mm^2) values.

Frequency: A maximum feasible frequency limit for an FEC IP is 1 GHz. This limit exists mainly for two reasons. First, the dynamic power consumption scales linear with the frequency. Second, FEC IPs are typically implemented as synthesizable IP cores. Managing routing complexity and clock skew beyond 1 GHz in synthesis-based design methodologies becomes extremely challenging and becomes even more difficult with future technology nodes. Achieving 1 Tb/s throughput under a frequency constraint of 1 GHz implies that 1000 information bits must be decoded in a single clock cycle. Hence, extremely parallel signal processing and corresponding parallel architectures become mandatory.

Code Flexibility: This feature includes flexible adaptation of code length and coding rate during the transmission. In some of the use cases, such as intra-

Table 30.2 FEC implementation requirements projected onto 7 nm technology node

Area limit	10 mm^2
Area efficiency limit	100 Gb/s/mm^2
Energy efficiency limit	~1 pJ/bit
Power density limit	0.1 W/mm^2

device communications where relatively constant channels are observed, the code flexibility can be low. On the other hand, for use cases with dynamically varying wireless links, e.g., VR/AR connectivity, the FEC needs to support high flexibility.

Latency: FEC latency corresponds to the duration of one full-code word decoding at the decoder. The latency constraint directly relates to system-level latency or latency in the air interface transmission. It should be noted that high throughput does not necessarily imply low latency. Achieving short latency is often even more challenging than high throughput. For THz applications with stringent application or service-level latency requirements, the target FEC latency is expected to be around 25 ns, whereas for latency-tolerant applications, FEC latency can be in the order of 250 ns [3].

The following table summarizes the implementation targets for B5G THz systems in achieving 1 Tb/s throughput as described above and assuming an advanced technology node, i.e., 7 nm technology node (Table 30.2).

30.3 FEC Candidates for Ultrahigh Throughput THz Systems

FEC solutions for throughput beyond 100 Gbit/s already exist for some applications, e.g., optical communications. However, these solutions are mainly restricted to algebraic hard decision decoding techniques, whereas the most powerful FEC techniques exploit soft decision information. Hence, in this chapter, we put the focus on soft decision techniques, i.e., Turbo codes, LDPC codes, and Polar codes, that are FEC techniques with outstanding error correction capabilities. These codes are adopted by many wireless communication standards, e.g., Turbo codes are adopted by 3G UMTS and 4G LTE; LDPC codes by Wi-Fi, WiMAX, WiGig, DVB-S2, 10 GBase-T, and 5G eMBB data channels; and more recent Polar codes by 5G eMBB control channels.

As already stated, achieving throughput toward 1 Tb/s under 1 GHz frequency constraints requires the decoding of up to 1000 information bits in a single clock cycle. Hence, a very high degree of parallelism of decoding is a must [5]. An achievable parallelism strongly depends on the code itself, the complexity and properties of the decoding algorithms, and the underlying architecture. For example, in Turbo codes, the structure of the interleaver has a strong impact on the achievable parallelism. In general, sub-functions of a decoding algorithm that have no mutual data dependencies can easily be parallelized by spatial parallelism. This is, for instance, the case for the belief propagation (BP) algorithm that is frequently used to decode LDPC codes. In the BP algorithm, all check nodes can be processed independently from each other. The same applies for the variable nodes. The situation is different for the MAP algorithm used in Turbo decoding where the calculation of a specific trellis step depends recursively on other trellis steps. This results in a sequential behavior, and therefore different trellis steps cannot be calculated in parallel. Such data dependencies exist also in iterative decoding algorithms between the various iterations.

Other important features for efficient high-throughput implementations are locality and regularity to minimize power consumption and interconnection [5]. Interconnection can largely contribute to area, delay, and energy consumption. Let us assume a FEC IP block of size 10 mm^2. A signal must travel at least 7 mm if it is to be transmitted from one corner to the other in the IP block. This will take in the order of 3–4 ns in a 14 nm technology node. If we run the IP block with 1 GHz, several pipeline stages should be inserted into the wiring, or alternatively the frequency must be largely decreased. This example shows that interconnect delay can largely decrease the throughput and/or increase power and latency. Thus, data transfers can be as important as calculations. This is especially the case for BP-based decoding of LDPC codes in which data transfers dominate.

Although channel decoding algorithms for advanced codes are mainly data flow-dominated, they imply irregularity and restricted locality, since efficient channel coding for Turbo and LDPC codes is grounded on some randomness, that is, the interleaver for Turbo codes and the Tanner graph for LDPC codes, respectively. Any regularity and locality in these structures improve the implementation efficiency but have negative impact on the communication performance. Therefore, there is some fundamental discrepancy between information theory and efficient implementations. Table 30.3 summarizes important implementation properties for the three code classes.

30.4 High-Throughput Decoders

The throughput of an FEC decoder depends on various communication and implementation parameters. Let N be the code block size and R the rate of a channel code. Let I denote the number of iterations that a corresponding iterative decoder requires to decode a code block ($I = 1$ in the case of a non-iterative decoding algorithm). Let π denote the degree of achievable parallelism in an FEC architecture. π is defined as the ratio of the operations that are performed in one clock cycle to the total number of operations necessary to perform one decoding iteration for a complete code block of length N. Note that "operation" in this context can be a computation but also a data transfer. Let f be the clock frequency.

The throughput T_{inf} (information bits per second) of an FEC architecture can be estimated by

Table 30.3 Implementation properties of different code classes

Code	Dec. algorithms	Parallel vs. serial	Locality	Compute kernels	Transfers vs. compute
Turbo	MAP	Serial/iterative	Low (interleaver)	Add-compare-select	Compute-dominated
LDPC	Belief propagation	Parallel/iterative	Low (Tanner graph)	Min-Sim/add	Transfer-dominated
Polar	Successive cancelation/list	Serial	High	Min-sum/add/sorting	Balanced

$$T_{\text{inf}} = N * R * \frac{1}{I} * \pi * f * (1 - \omega) \tag{30.4}$$

where ω is a normalized value between 0 and 1 that represents the timing overhead due to, for example, data distribution, interconnection, memory access conflicts, etc. f is determined by the critical path in the compute kernels of the corresponding decoding algorithms and/or delay due to interconnection and is upper limited to 1 GHz as already stated. The overhead ω increases with increasing N and π and is larger for decoding algorithms that have limited locality and need more data transfers. The impact of ω on the throughput can be considered as an effective reduction of f or decrease in π, if additional clock cycles are mandatory, such as memory conflicts, which cannot be hidden. To achieve a large T_{inf}, a high code rate R is beneficial, whereas π is to be maximized and ω minimized. The following dependencies exist between error correction parameters and implementation KPIs:

- The dynamic power consumption P_{dyn} depends first-order linear on f, N, and π, while it is independent of I and R.
- The energy efficiency is first-order independent of f, N, and π but inversely proportional to R and linear dependent on I.
- The area is first-order proportional to N and π.
- The area efficiency depends first-order linear on R and f and is inversely proportional to I.
- The error correction capability for a given code depends on R, N, and I, i.e., small R, large N, and large I in general improve the error correction capability.

From these dependencies, it can be seen that there is no optimal parameter setup that meets all KPI requirements. Rather, there are conflicting requirements with regard to the KPIs and error correction capability. Thus, many trade-offs on code design, decoding algorithms, architectures, and implementation in advanced technology nodes exist.

Mathematically speaking, this suggests a Pareto optimization problem. Even though channel coding itself is based on a solid theoretical mathematical framework of information theory, no theory exists that describes the interrelation between communication performance, i.e., error correction capability, and implementation KPIs. To tackle this challenge, it is necessary to design space for the different code classes that contains all relevant parameters. This design space should be intensively explored to understand the interrelation between all the parameters and to find the best trade-off under given constraints. Figure 30.1 shows an exemplary design space for LDPC codes. It consists of four branches, representing the relevant code design, decoding algorithm, architecture parameters, and constraints, respectively. To limit the complexity of exploration, the design space has to be pruned, i.e., cut-off branches that are not promising to achieve the requested KPIs. Similar design spaces exist also for Turbo codes and Polar codes. A detailed discussion of these design spaces is out of scope of this chapter.

Although Table 30.3 shows that there are significant differences for Turbo, LDPC, and Polar codes, there is a commonality for all code classes from an

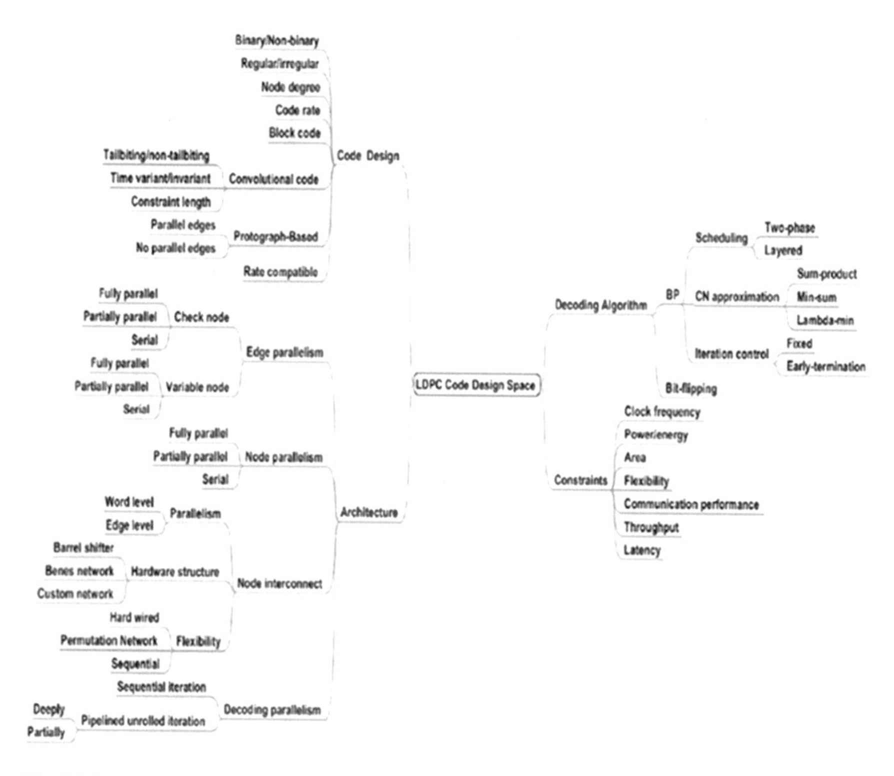

Fig. 30.1 Exemplary design space exploration (DSE) diagram of LDPC codes

architecture and implementation perspective. To achieve throughput toward Tb/s, intense functional and spatial parallelisms have to be employed in all architectures for the different code classes. A straightforward solution is to use spatial parallelism by placing several low-throughput decoders in parallel in a multicore solution and connect them with a distribution network. Such an architecture is scalable and easy to design but suffers from a low implementation efficiency and a large latency for individual frames/blocks. Moreover, the distribution network can become an implementation bottleneck due to the huge amount of necessary wiring. A better solution is to increase the throughput of the core itself. Here, functional parallelism, also named pipelining, is a very efficient technique to speed up algorithms with data dependencies. Functional parallelism enables the "unrolling" of iterative loops. Buffers are inserted between the different iteration stages. In this way, the data dependencies are broken up, and the iterations can be executed in parallel on different data sets.

In the next section, we present FEC solutions for the three code families for ultrahigh throughput use cases. It is out of scope of this chapter to introduce the basics of the different coding techniques. We assume that the reader is familiar with the basics of Turbo codes, LDPC codes, and Polar codes and refers to [6, 7], and references therein for further information.

30.4.1 Turbo Decoder

Among the three code classes, decoders for Turbo codes are most challenging to implement for throughputs toward Tb/s. In its basic form, a Turbo decoder consists of two component decoders connected through an interleaver and a de-interleaver. It applies an iterative loop, exchanging extrinsic information between its two components, cooperatively improving the decoding result [8]. State-of-the-art hardware architectures for Turbo decoding use one hardware instance alternatingly performing operations of the two component decoders. Furthermore, the code block can be divided into smaller sub-blocks that can be processed in parallel. However, additional computations are necessary at the sub-block boundaries to mitigate a degradation in error correction performance due to the broken data dependencies. Various hardware architecture archetypes exist according to the dominant type of parallelization at architectural level:

- Parallel MAP (PMAP): These decoders spatially parallelize the decoding of different sub-blocks of the code trellis on multiple sub-decoder cores [9, 10]. However, for smaller sub-blocks and at high code rates, mitigation measures for avoiding BER performance loss are necessary that limits the maximum degree of parallelization [11].
- Pipelined MAP (XMAP): The XMAP decoder, named for its X-shaped pipeline structure, uses a functional parallelization approach where the state-metric recursions of the MAP algorithm are unrolled and pipelined [12–14]. It also operates on smaller sub-blocks, here called windows, and therefore suffers from the same limitations as the PMAP architecture with respect to parallelization and BER performance loss.

For both architectures, dependent on the structure of the interleaver, memory access conflicts can result which in turn decreases the throughput. To avoid such access conflicts, an architecture-aware interleaver design is mandatory. State-of-the-art implementations of PMAP decoders achieve a throughput of 1–2 Gb/s [15–17], and similarly 1–2 Gb/s have been demonstrated for XMAP decoders [14, 18] in recent technology nodes. The fully parallel MAP (FPMAP) decoder architecture is the extreme case of the PMAP with a sub-block size reduced to 1 trellis stage in combination with a shuffled decoding schedule [19]. The FMAP has been shown to achieve a throughput of 15 Gb/s, an order of magnitude more than previously published PMAP implementations [20], but it suffers from a reduced BER performance for high code rates [21].

The parallelism π of these decoders is $\pi \leq 1$. In order to enable a throughput beyond 100 Gb/s for Turbo decoder architectures, spatial or functional parallelization at the MAP component decoder level is not sufficient, and functional parallelism at the decoder iteration level becomes mandatory, i.e., $\pi > 1$. Pipelining the iterations leads to a further architecture archetype, named fully pipelined iteration unrolled XMAP, denoted as UXMAP in the following. In this decoder architecture, many blocks are processed in parallel in the pipeline. At each clock

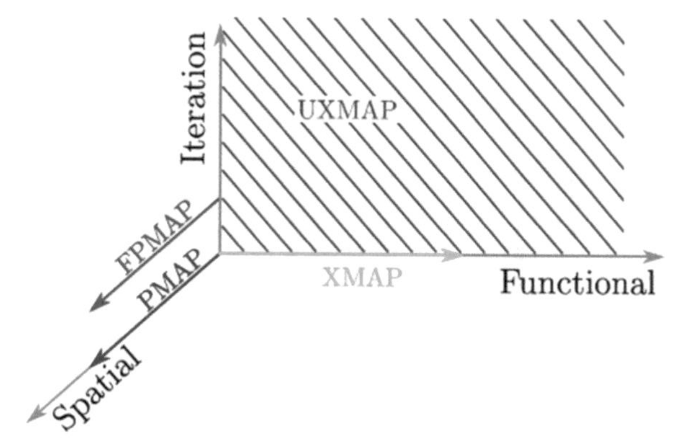

Fig. 30.2 Turbo decoder architecture and types of parallelization [23]. ((c) 2020 IEEE, reproduced with permission)

cycle, a new block is read and output [21, 22]. This allows for a very high throughput that is determined by the block size and the clock frequency only.

Figure 30.2 shows different decoder architectures dependent on parallelization type. The FMAP lies parallel to the spatial parallelism axis due to the shuffled decoding schedule which can be seen as an iteration parallelism of 2. The UXMAP is located in a plane spanned by the functional and iteration parallelism axes. It yields, however, a large decoder area. In a recent work, the first Turbo decoder which achieves 100 Gb/s and occupies almost 24 mm^2 for a frame size (in terms of information bits) of N = 128 in 28 nm FD-SOI technology has been reported [21].

To reduce the decoder area and enable block size flexibility while keeping the throughput, we can combine the UXMAP architecture with spatial parallelism. In relation to Fig. 30.2, this means that the UXMAP plane moves along the spatial axes. In this architecture, the iteration pipelines are split into smaller X-windows of, for example, sizes 32 and 4 of these placed in parallel when targeting a block size of 128. Since the decoder area first order quadratically increases with the window size, four parallel smaller windows consume less area than a single pipeline processing a large window of size 128. The corresponding area reduction is up to 40% when the 128 window size is changed into 4 windows of size 28 each. Moreover, the splitting enables frame size flexibility to support various block sizes: 128, 64, and 32. Interleaving is a further challenge in this architecture since (1) the structure of the interleaver impacts the error correction capability and (2) the interleaver structure has to support various block sizes. It has been shown that a modified, almost regular permutation interleaver can provide scalability as well as very good error correction capability [23].

The drawback of unrolled architectures is the static number of iterations, i.e., pipeline stages, which have to be fixed at design time. To enable good error correction capability over the SNR range, worst-case assumptions on the iteration numbers should be made. Serial architectures leverage stopping criterions to exploit the dynamics in the channel and stop iterations as early as possible. This dynamic behavior can also be exploited in the UXMAP architecture by setting the iteration

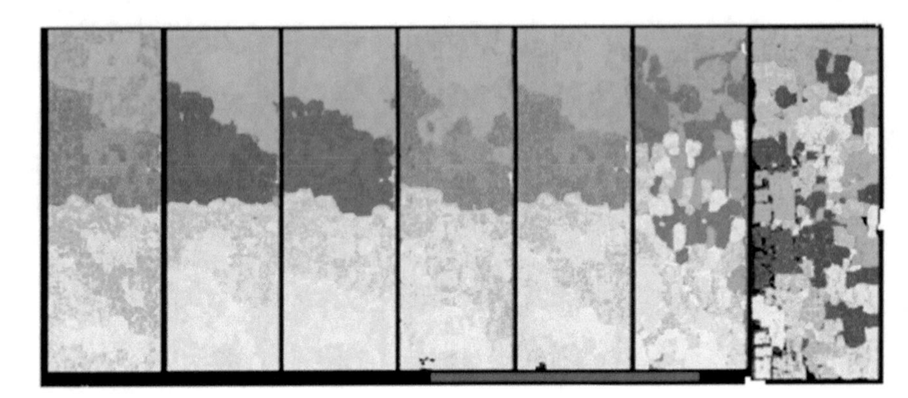

Fig. 30.3 Turbo decoder architecture and layout in 28 nm FD-SOI technology [23]. ((c) 2020 IEEE, reproduced with permission)

to the average number to achieve a given target error correction rate. In the case that this number is not sufficient when decoding a code block, this block is fed to a so-called afterburner. This afterburner performs further iterations in a sequential way. Since the afterburner is also pipelined, it can process several blocks in parallel. This approach reduces not only the area but also the power consumption and energy efficiency. To avoid a loss in error correction performance, a reliable, low-complexity stopping criterion is mandatory to detect blocks that should be fed into the afterburner. Since hard decisions on the decoded frames are any way available in the pipeline with no overhead, the hard-decision-aided (HDA) stopping criterion [24] is selected as stopping criterion.

Figure 30.3 shows the layout of the described decoder architecture that is implemented in 28 nm FD-SOI technology under worst-case assumption. It consumes 14.32 mm^2 area and achieves 800 MHz. The achieved throughput is 102.4 Gbit/s that results in an area efficiency of 7.15 Gbit/s/mm^2. The decoder can process block sizes up to 128 bits, split into 4 windows, with up to 7 iterations of which 3 are processed in the unrolled UXMAP and up to 4 in the afterburner stage. The afterburner can decode up to 32 blocks in parallel. A full decoding iteration consists of two half-iteration to process each component decoder, respectively. The left six boxes in Fig. 30.3 represent the six half-iterations. The four different colors in a half-iteration highlight the four windows, each of size 32. The rightmost box is the afterburner stage.

Recent investigations have shown that new low-complexity reduction techniques based on the local soft-output Viterbi algorithm enable throughput up to 500 Gbit/s in 28 nm technology with marginal loss in BER performance in comparison with the max-Log-MAP algorithm. In advanced technology nodes, the processing of larger block sizes becomes possible. Hence, throughputs approaching 1 Tb/s become feasible for Turbo codes.

30.4.2 LDPC Decoder

Unlike Turbo decoders, which suffer from sequential behavior, LDPC codes have an inherent parallelism. State-of-the-art LDPC decoders apply BP algorithm for decoding that iteratively exchanges messages between variable and check nodes on the Tanner graph. The structure of the Tanner graph is defined by the parity-check matrix H. In BP, all check nodes can be processed independently from each other. In the same way, all variable nodes can be processed in parallel. The result of each check node (variable node) calculation is then spread via the edges of the Tanner graph to all connected variable nodes (check nodes). Although the parity check matrix is sparse, LDPC decoding is dominated by data transfers and not by the node computations. This is in contrast to Turbo codes where the computations dominate the overall decoding (see also Table 30.3). Moreover, the Tanner graph has very limited locality to provide good error correction capability, which in turn challenges an efficient implementation of the data transfers in highly parallel LDPC decoder architectures.

The throughput of a BP-based LDPC decoder is determined by the number of edges that are processed in parallel. Let us consider an LDPC block code with a parity check matrix H that has $\Omega(H)$ 1 entry (the number of 1s in H equals the number of edges in the Tanner graph). Let $\Psi(A)$ denote the number of edges that can be processed in one clock cycle by an LDPC decoder architecture. Then the corresponding parallelism of an architecture can be evaluated as $\pi = \Psi(sA)/\Omega(H)$. Three archetypes exist:

- Partially parallel architectures: Only a subset of edges and nodes are processed in parallel, i.e., $\pi < 1$. These architectures are very common for large block sizes that use quasi-cyclic (QC) block codes. An example is the DVB-S2 standard. The resulting throughput is in the order of magnitude of 10 Gbit/s in 28 nm technology.
- Fully parallel architectures at iteration level: All edges are processed in parallel, i.e., $\pi \approx 1$. All check nodes and variable nodes are instantiated one to one as hardware units and the corresponding edges are hardwired. Because of the low locality in the Tanner graph, routing congestion is a big challenge in these architectures. The throughput for this architecture is in the order of magnitude of 100 Gbit/s in 28 nm technology.
- Unrolled fully parallel architectures: These architectures are similar to the fully parallel architectures, but, in addition, the iterations are unrolled and pipelined analogous to the iteration unrolling in Turbo decoding. With $\pi = I$, a new block is processed in every clock cycle. Only this architectural approach is feasible to achieve throughput toward 1 Tb/s.

The unrolled LDPC decoder architecture has the same disadvantage as the unrolled Turbo decoder with regard to the number of iterations that has to be fixed at design time. To mitigate the problem of worst-case iterations, the aforementioned afterburner approach can be exploited as in the Turbo decoder. A further option to reduce the number of iterations, and moreover the routing complexity, is to apply layered decoding. In unrolled layered decoding, the parity matrix H is horizontally divided into several sub-matrices, e.g., two sub-matrices H_1 and H_2, as shown below:

$$H = \begin{pmatrix} H_1 \\ H_2 \end{pmatrix} \tag{30.5}$$

In this case, the processing of the sub-matrices H_1 and H_2 is performed in a two-stage pipeline. When processing sub-matrix H_2, its variable node values are already updated with the check node values from the processing of the previous sub-matrix H_1. This results in a faster convergence. The layered unrolled architecture processes as in the original unrolled architecture, i.e., one block per clock cycle, and has two advantages compared to the standard unrolled architecture.

- First, layered decoding reduces the number of iterations for a given error correction performance and, hence, improves the area efficiency and energy efficiency.
- Second, the routing complexity between two subsequent pipeline stages is reduced that further improves area and energy efficiency.

Figure 30.4 shows the layout of an unrolled layered LDPC decoder performing four iterations. The decoder is implemented in 22 nm FD-SOI technology for an (1032,860) LDPC code. The decoder consumes 3.24 mm^2 and achieves 459 MHz that results in a coded throughput of 474 Gbit/s. The area efficiency is 146 Gbit/s/mm^2. The power consumption is 2.67 W and the energy efficiency thus 5.64 pJ/bit.

Block sizes of up to 2000 are feasible with the unrolled LDPC decoder approach in today's technology nodes [25]. For larger block sizes, spatially coupled LDPC (SC-LDPC) or LDPC convolutional codes are better suited. These codes were first introduced by Jimenez-Feltström and Zigangirov (1999) and enable continuous transmission and block transmission of arbitrary size with reasonable decoding complexity. Small sub-codes defined by parity check matrices H_i of size (c − b) × c (typical values for c are 100–2000 bits) are coupled in a diagonal dominant matrix H with code rate R = b/c. In this way, H forms a block code of arbitrary size as shown below:

Fig. 30.4 Layout of unrolled LDPC layer block decoder in 22 nm FD-SOI technology. (The decoding iterations are highlighted in different colors)

$$H = \begin{pmatrix} H_0 & & & & \\ H_1 & H_0 & & & \\ \vdots & H_1 & & & \\ H_{m_{cc}} & \vdots & & & \\ & H_{m_{cc}} & \ddots & H_0 & \\ & & & \vdots & H_0 \\ & & & H_{m_{cc}-1} & \vdots \\ & & & H_{m_{cc}} & H_{m_{cc}-1} \\ & & & & H_{m_{cc}} \end{pmatrix} \tag{30.6}$$

Similar to conventional LDPC block codes, SC-LDPC codes can be decoded with the BP algorithm. However, the structural characteristic of the SC-LDPC code enables the so-called window decoding. Here, a window W defines a sub-matrix of H and thus a subset of variable and check nodes in the corresponding Tanner graph. During the decoding process, the window moves from the upper left to the lower right of H. In every processing step t, the nodes corresponding to the current processing window $W(t)$ are updated, and the window is moved forward by one sub-block over the diagonal structure. The decoder only needs to store the sub-blocks and corresponding messages inside a window and can start the decoding as soon as the first sub-block is available at the input of the decoder. Since the size of the window is much smaller than the size of H, the initial latency and the memory requirements are much smaller compared to a conventional LDPC block code of similar length. Different window decoding schemes/architectures exist that primarily differ in the number of performed iterations and the scheduling inside the window. However, only two architectures are suited for very high-throughput decoding: (1) the layered window decoder [26] and (2) the iteration unrolled (fully parallel) window decoder [27]. For this type of decoders, the block size N should be replaced by c in Eq. (30.2) to calculate the corresponding throughput.

The layered window decoder performs the decoding layer by layer, similar to a layered block decoder. Here, a layer consists of the sub-matrices $H_{mcc} \dots H_0$, i.e., one row of H. Different layers in a window that are not overlapping w.r.t. the involved variable nodes are processed in parallel.

In contrast, the unrolled window decoder operates on multiple decoding windows simultaneously. The windows each have size W and are offset to each other by one sub-block. The processing of the decoding windows is performed in parallel on dedicated unrolled two-phase decoder cores (sub-decoders). The sub-decoders exchange extrinsic messages to resolve data dependencies of neighboring windows. The handling of these exchange messages can become very complex for large window sizes and due to routing limitations create a bottleneck for the throughput. Therefore, for high-throughput applications, a window size of $W = 1$ is advantageous. In this case, a sub-decoder operates only on one column of H, i.e., the sub-matrices $H_0 \dots H_{mcc}$, opposed to the layered window decoder that works on rows.

When comparing the layered window decoder and the unrolled window decoder (with $W = 1$), the unrolled decoder achieves a higher throughput due to the fact that the sub-decoders operate on smaller blocks compared to the layered window decoder (one sub-block vs. $(m_{cc} + 1)$ sub-blocks). These smaller blocks reduce the routing complexity, which has a large impact on the critical path and the energy consumption. The error correction performance is inversely proportional to the throughput increase, i.e., the layered window decoder exhibits a better error correction performance. This can be explained by the fact that the shortening of the critical path becomes only possible by breaking up data dependencies, i.e., splitting the check nodes. Every breaking of data dependency delays the updating of messages by at least one clock cycle, which negatively impacts the error correction performance. This degradation could be counterbalanced by an increase in the iteration numbers, i.e., increasing the number of decoding windows. This does not impact the throughput but has negative impact on the area and energy efficiency. Hence, various trade-offs exist between error correction performance and implementation efficiency. An unrolled window decoder with sub-block size $c = 640$, coupling width $m_{cc} = 1$, and code rate $R = 0.8$ achieves a coded throughput of 424 Gb/s and consumes 2.35 mm^2 area in 22 nm FD-SOI technology for 8 decoding iterations. This yields an area efficiency of 180 Gb/s/mm^2. The energy efficiency is 8 pJ/bit.

30.4.3 Polar Decoder

Polar codes, invented by Arikan [28], has recently received considerable attention in the context of 5G. Successive cancellation (SC) and successive cancellation list (SCL) [29] are the most prominent decoding algorithms for Polar codes. SC comes with a low algorithmic decoding complexity but a limited error correction performance for finite code lengths. SCL applies list decoding on the SC algorithm which significantly improves the error correction at the cost of higher algorithmic decoding complexity. SC and SCL decoding algorithms traverse the Polar Factor Tree (PFT) in a depth-first manner [30] resulting in a sequential decoding procedure. The sequential nature of these decoding algorithms poses a bottleneck for very high throughput operations, whereas the BP algorithm used in LDPC codes allows very high throughputs thanks it is inherent parallelism. Due to this behavior, BP was also adopted for Polar decoding [31]. However, BP applied to Polar decoding needs large number of decoding iterations to approach the error correction performance of SC [32]. This large number of iterations decreases the throughput and increases the latency. Moreover, even for a very high number of iterations, BP cannot compete with the error correction performance of SCL decoding [33].

To achieve a very high throughput, the PFT traversal can be unrolled and pipelined [34], in a way similar to the iteration unrolling in Turbo and LDPC decoding as described above. Whenever a node is visited during the tree traversal, a corresponding pipeline stage can be instantiated. In this way, for a block length of N (= number of leaves in the PFT), the maximum number of pipeline stages is 2*(2 N-2) + 1. This allows all N*logN operations to be executed in parallel and yielding $\pi = 1$. Obviously, the complexity of the decoding architecture is directly proportional to the size of the PFT. However, for a given code, the tree can be reduced by various transformations. For example, if a subtree represents a repetition code or a parity check code, the corresponding subtree can be replaced by a single node. Alike, one can merge rate 0 and rate 1 nodes into its parent nodes [35] or use majority logic decoding in subtrees. The achievable tree reduction strongly depends on the underlying Polar code. In a similar way, SCL decoder for very high throughput can be implemented. However, in contrast to SC, SCL considers at the leaves of the PFT both possible values, i.e., 0 and 1, for the estimation of an information bit. The number of code word candidates, named paths, therefore doubles for every bit estimation that results in an exponential path increase. Therefore, the number of paths is limited to a list size L in order to keep decoding complexity manageable. This implies that after each bit estimation, paths must be discarded if their number exceeds L. The reliability of each path is rated by an appropriate path metric that determines if a path is discarded or not. A short CRC can assist the selection of the best candidate at the end of the decoding process. This comes with low overhead but improves the decoding performance. The SCL decoding complexity increases with increasing list size since the lists must be managed and correspondingly sorted. On the other side, the error correction performance improves with increasing list site. Hence, there is a trade-off between

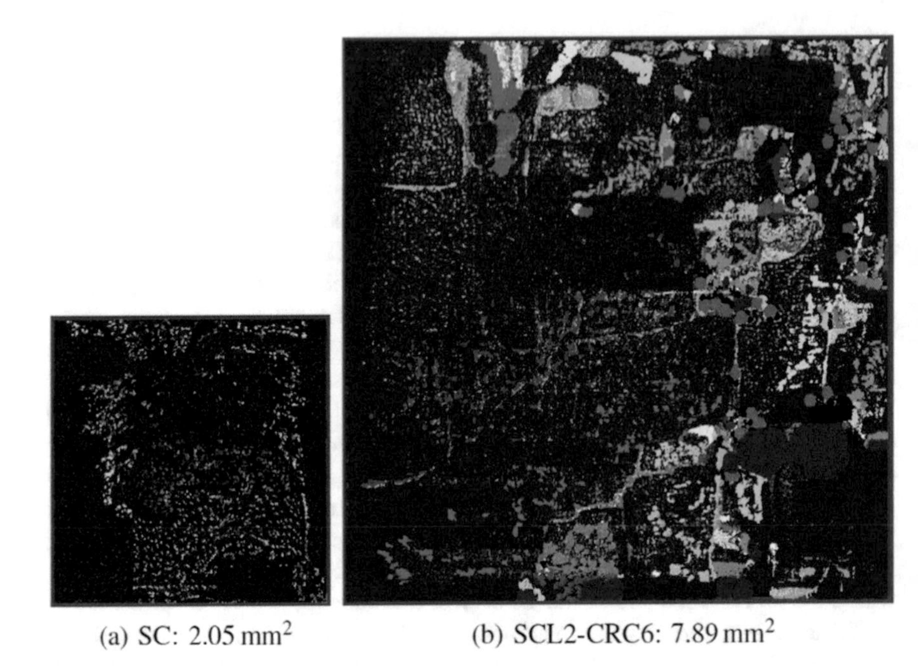

(a) SC: $2.05\,\text{mm}^2$ (b) SCL2-CRC6: $7.89\,\text{mm}^2$

Fig. 30.5 Layout of an SC and an SCL polar decoder in 28 nm FD-SOI technology [36]. ((c) 2020 IEEE, reproduced with permission)

error correction capability and implementation efficiency. Note that the SC decoding algorithms corresponds to an SCL with list size 1 [36]. To show this trade-off, we compare an SC decoder with an SCL decoder of list size 2 and a 6-bit CRC. Both decoders were implemented in 28 nm FD-SOI technology and use the same Polar code with code block size 1024 and code rate ½. At a FER of 10^{-5}, the SCL has a ~0.7 dB better error correction performance than the SC decoder. The SC decoder achieves a coded throughput of 517 Gbit/s, an area of 2.05 mm^2, and a power consumption of 1.65 W. The SCL decoder achieves a throughput of 506 Gbit/s, an area of 7.89 mm^2, and a power consumption of 4.47 W. Hence, the improvement in error correction capability of ~0.7 dB compared to the SC decoder comes at the cost of a 3.8$\times$ larger area and 2.9$\times$ in power. Figure 30.5 shows both decoders using the same scale.

In summary, throughput toward Tb/s becomes feasible for Polar codes even for the more complex SCL decoding.

30.5 Conclusion

In this chapter, we have shown that forward error correction for advanced channel coding schemes is very challenging when targeting Tb/s throughput under power density and energy efficiency constraints. A cross-layer approach becomes mandatory, since code and decoding algorithms have a large impact on the final implementation efficiency. Here, parallelism and large locality are the most important factors that strongly depend on the code structure and the decoding algorithms. Since the decoding algorithms largely differ for the three considered coding schemes, the challenges are different for each code family. Turbo codes require low-complexity decoding algorithms to further boost implementation efficiency, whereas throughput of LDPC decoding is strongly dominated by the interconnect complexity inherent to the Tanner graph structure. All levels of parallelism must be efficiently exploited to reach Tb/s throughput, and pipelining is mandatory to break up data dependencies. However, heavily pipelined architectures require a huge amount of registers. As a result, the clock tree becomes a major bottleneck in terms of clock skew and power consumption. This is especially a challenge for high-throughput Polar decoders. Quantization directly impacts all implementation KPIs. Hence, efficient quantization is another important optimization parameter. New approaches such as information bottleneck techniques are promising candidates especially for data transfer-dominated coding schemes like LDPC decoding to further boost implementation efficiency. Except for SC-LDPC codes, all other coding techniques are limited to small block sizes when targeting Tb/s throughput. Alike, iterative decoding algorithms are limited to a small number of iterations. However, small block sizes and low number of iterations have negative impact on the error correction performance. Hence, achieving very good error correction capabilities under such constraints is still challenging when high implementation efficiency has high priority.

References

1. Saad, W., Bennis, M., & Chen, M. (2019). A vision of 6G wireless systems: Applications, trends, technologies, and open research problems. *IEEE Network, 34*(3), 134–142.
2. Nagatsuma, T., Ducournau, G., & Renaud, C. C. (2016, June). Advances in terahertz communications accelerated by photonics. *Nature Photonics, 10*(6), 371–379.
3. H2020 EPIC Project Technical Report. (2018, March). *B5G wireless Tb/s FEC KPI requirement and technology gap analysis*. https://epic-h2020.eu/downloads/EPIC-D1.2-B5G-Wireless-Tbs-FEC-KPI-Requirement-and-Technology-Gap-Analysis-PU-M07.pdf
4. Latva-aho, M., & Leppänen, K. (2019, September). *Key drivers and research challenges for 6G ubiquitous wireless intelligence*. Oulu: University of Oulu.
5. Kestel C., Herrmann M., Wehn N. (2018, December). *International Symposium on Turbo Codes & Iterative Information Processing (ISTC)*. Hong Kong, China.
6. Shao, S., Hailes, P., Wang, T., Wu, J., Maunder, R. G., Al-Hashimi, B. M., & Hanzo, L. (2019, January). Survey of Turbo, LDPC, and Polar Decoder ASIC implementations. *IEEE Communications Surveys & Tutorials, 21*(3), 2309–2333.
7. Niu, K., Chen, K., Lin, J., & Zhang, Q. T. (2014, July). Polar codes: Primary concepts and practical decoding algorithms. *IEEE Communications Magazine, 52*(7), 192–203.
8. Berrou, C., Glavieux, A., & Thitimajshima, P. (1993, May). Near shannon limit error-correcting coding and decoding: Turbo-codes. In *IEEE International Conference on Communications (ICC)*, S. 1064–S. 1070.

9. Roth, C., Belfanti, S., Benkeser, C., & Huang, Q. (2014, June). Efficient parallel turbo-decoding for high-throughput wireless systems. *IEEE TCAS I: Regular Papers, 61*(6), S. 1824–S. 1835.
10. Sun, Y., & Cavallaro, J. (2010). Efficient hardware implementation of a highly-parallel 3GPP LTE/LTE-advance turbo decoder. *Integration VLSI Journal.*
11. Weithoffer, S., Kraft, K., & Wehn, N. (2017, September). Bit-level pipelining for highly parallel turbo-code decoders: A critical assessment. In *2017 IEEE AFRICON*, S. 121–S. 126.
12. Worm, A., Lamm, H., & Wehn, N. (March 2001). Design of low-power high-speed maximum a priori decoder architectures. In *Design, Automation, and Test in Europe Conference (DATE)*, S. 258–S. 265.
13. May, M., Ilnseher, T., Wehn, N., & Raab, W. (2010, March). A 150Mbit/s 3GPP LTE turbo code decoder. In *Design, Automation, and Test in Europe Conference (DATE)*, S. 1420–S. 1425.
14. Weithoffer, S., Pohl, F., & Wehn, N. (2016, September). On the applicability of trellis compression to turbo-code decoder hardware architectures. In *International Symposium on Turbo Codes and Iterative Information Processing. (ISTC)*, S. 61–S. 65.
15. Shrestha, R., & Paily, R. P. (2014, September). High-throughput turbo decoder with parallel architecture for lte wireless communication standards. *IEEE TCAS I: Regular Papers, 61*(9), S. 2699–S. 2710.
16. Weithoffer, S., & Wehn, N. (2018). Where to go from here? New cross layer techniques for LTE turbo-code decoding at high code rates. *Advances in Radio Science, 16*(C), S. 77–S. 87.
17. Ilnseher, T., Kienle, F., Weis, C., & Wehn, N. (2012, August). A 2.15gbit/s turbo code decoder for lte advanced base station applications. In *International Symposium on Turbo Codes and Iterative Information Processing (ISTC)*, S. 21–S. 25.
18. Wang, G., Shen, H., Sun, Y., Cavallaro, J. R., Vosoughi, A., & Guo, Y. (May 2014). Parallel interleaver design for a high throughput HSPA+/LTE multi-standard turbo decoder. *IEEE TCAS I: Regular Papers, 61*(5), S. 1376–S. 1389.
19. Zhang, J., & Fossorier, M. P. (February 2005). Shuffled iterative decoding. *IEEE Transactions on Communications, 53*(2), S. 209–S. 213.
20. Li, A., Xiang, L., Chen, T., Maunder, R. G., Al-Hashimi, B. M., & Hanzo, L. (2016). VLSI implementation of fully parallel lte turbo decoders. *IEEE Access, 4*, S. 323–S. 346.
21. Weithoffer, S., Nour, C. A., Wehn, N., Douillard, C., & Berrou, C. (2018, December). 25 years of turbo codes: From Mb/s to beyond 100 Gb/s. In *International Symposium on Turbo Codes and Iterative Information Processing (ISTC)*, S. 1–S. 6.
22. Weithoffer, S., Herrmann, M., Kestel, C., & Wehn, N. (2017, Octobre 1–6). Advanced wireless digital baseband signal processing beyond 100 Gbit/s. In *IEEE International Workshop on Signal Processing Systems (SiPS).*
23. Weithoffer, S., Griebel, O., Klaimi, R., Nour, C. A., & Wehn, N. (2020, May). Advanced hardware architectures for turbo code decoding beyond 100 Gb/s. In *2020 IEEE Wireless Communications and Networking Conference (WCNC)* (pp. 1–6). IEEE.
24. Shao, R. Y., Lin, S., & Fossorier, M. P. (1999, August). Two simple stopping criteria for turbo decoding. *IEEE Transactions on Communications, 47*(8), S. 1117–S. 1120.
25. Ghanaatian, R., Balatsoukas-Stimming, A., Müller, T. C., Meidlinger, M., Matz, G., Teman, A., & Burg, A. (2018). A 588-Gb/s LDPC decoder based on finite-alphabet message passing. *IEEE Transactions on Very Large Scale Integration (VLSI) Systems, 26*(2), S. 329–S. 340.
26. Jimenez-Feltström, A., & Zigangirov, K. S. (1999, September). Time-varying periodic convolutional codes with low-density parity-check matrix. *IEEE Transactions on Information Theory, IT-45*, S. 2181–S. 2191.
27. Hassan, N. U., Schlüter, M., & Fettweis, G. P. (2016). *Fully parallel window decoder architecture for spatially-coupled LDPC codes* (S. 1–S. 6). Kuala Lumpur.
28. Arikan, E. (2009). Channel polarization: A method for constructing capacity-achieving codes for symmetric binary-input memoryless channels. *IEEE Transactions on Information Theory, 55*(7), 3051–3073.
29. Tal, I., & Vardy, A. (2015). List decoding of polar codes. *IEEE Transactions on Information Theory, 61*(5), 2213–2226.

30. Alamdar-Yazdi, A., & Kschischang, F. R. (2011, December). A simplified successive-cancellation decoder for polar codes. *IEEE Communications Letters, 15*(12), S. 1378–S. 1380.
31. Arıkan, E. (2010, July). Polar codes: A pipelined implementation. In *Proceedings of the 4th International Symposium on Broad Communications ISBC 2010* (pp. 11–14).
32. Abbas, S. M., Fan, Y., Chen, J., & Tsui, C.-Y. (2017, March). High-throughput and energy-efficient belief propagation polar code decoder. *IEEE Transactions on Very Large Scale Integration (VLSI) Systems, 25*(3), 1098–1111.
33. Elkelesh, A., Ebada, M., Cammerer, S., & Brink, S. T. (2018). Belief propagation decoding of polar codes on permuted factor graphs. In *IEEE Wireless Communications and Networking Conference (WCNC)*, (S. 1–S. 6). Barcelona.
34. Giard, P., Sarkis, G., Thibeault, C., & Gross, W. J. (May 2015). 237 Gbit/s unrolled hardware polar decoder. *Electronics Letters, 51*(10), S. 762–S. 763.
35. Sarkis, G., Giard, P., Vardy, A., Thibeault, C., & Gross, W. J. (2014, May). Fast polar decoders: Algorithm and implementation. *IEEE Journal on Selected Areas in Communications, 32*(5), S. 946–S. 957.
36. Kestel, C., Johannsen, L., Griebel O., Jimenez, J., Vogt, T., Lehnigk-Emden, T., & Wehn N. (2020, September). A 506 Gbit/s Polar Successive Cancellation List Decoder with CRC. Accepted for publication, IEEE 31st PIMRC'20 – Workshop on Enabling Technologies for Terahertz Communications.

Chapter 31
MAC and Networking

Alexandros-Apostolos A. Boulogeorgos, Admela Jukan, and Angeliki Alexiou

Abstract This chapter presents medium access control (MAC) approaches that are tailored to the terahertz (THz) communications systems. In more detail, after presenting the THz wireless systems particularities that are expected to influence the MAC, an initial access (IA) scheme is presented that can be used to guarantee alignment between the base station and user equipment, and its performance is quantified under different wireless environments. Moreover, a hierarchical beam tracking approach is discussed that ensures alignment between basestation (BS) and mobile user equipement (UE) with low overhead. Furthermore, random access and scheduled access issues are addressed. Finally, low-complexity relaying approaches are presented as countermeasures to antenna misalignment and blockage.

31.1 MAC Layer Functionalities

MAC layer protocols are the first step toward feasible THz networking. To this end, the following specific features need to be considered:

- Deafness caused by narrow beams and directivity: Deafness is a misalignment between transmitter and receiver beams. This effect challenges the MAC design for directional millimeter-wave (mmW/THz) communications.
- Control channel (CC) selection: Control channel between received and transmitter plays a critical role to addressing the issue of throughput, transmission coverage, and deafness and ultimately decides the effectiveness of node discovery and coupling process. The latter needs to decide the transmission frequency and antenna directionality (omni, semi, or fully directional).

A.-A. A. Boulogeorgos · A. Alexiou (✉)
Digital Systems, University of Piraues, Piraeus, Greece
e-mail: al.boulogeorgos@ieee.org; alexiou@unipi.gr

A. Jukan
Chair for Communication Networks, Technische Universität Braunschweig, Braunschweig, Germany

T. Kürner et al. (eds.), *THz Communications*, Springer Series in Optical Sciences 234, https://doi.org/10.1007/978-3-030-73738-2_31

- Line-of-sight (LoS) blockage: As a result of ultra-high data rate, data loss is an important parameter that could affect the link robustness. Identification of the blocked channel and implementation of a suitable anti-blockage solutions are the main challenges in MAC design [1].
- Mobility management: Frequent interruptions and disconnections disrupt the connectivity; as a result the mobile endpoints require some methods to reestablish the link robustness. The use of narrow beams makes links susceptible to user mobility challenges [2], since mobility can change the quality of the established beam pairs and impact the feasibility of the established link.
- Spatial reuse: This effect refers to the simultaneous transmissions (spatial time reuse), which also challenges MAC design. To support spatial reuse, two mechanisms could be used based on (1) links that are away which requires large path loss to synchronize and (2) adjacent links.
- Relaying and multihop protocols design: Since the reachability between different nodes inside the THz applications could be increased, novel approaches to design and implementing relaying techniques are needed.
- Interference management: The MAC layer protocol needs to be designed in consideration of channel interference for dense indoor scenarios. To this end, new interference models need to be implemented with rapid scheduling and fast channel access and switching features based on channel interference information.
- Transceiver design: Researchers should focus on designing an efficient transceiver for the THz MAC layer, which is able to satisfy requirements on synchronization, framing, and scheduling.
- Scheduling algorithms: The use of radio resources for a given policy such as maximizing throughput, minimizing the total interference in the network, or reducing system delay could boost the overall quality of service. Therefore, the joint consideration of the scheduling module, medium access layer, and the physical layer protocols is critical to engineering the scheduler to meet the channel conditions and data traffic requirements.
- Cross-layer design: New THz cross-layer should be designed using network and physical layer aware algorithms at the MAC layer.
- Synchronization: The design and implementation of new algorithms for node synchronization is the main block at the MAC level due to its efficiency in increasing the frame collisions between nodes.

The existing MAC protocols in wireless networks cannot be directly applied, because they do not consider the unique features of THz band. Yet, designing an efficient MAC protocol for THz wireless communication is crucial for future high-speed networks. As a consequence, new MAC layer mechanisms need to be developed. Motivated by this, in what follows, we present such the building blocks of THz MAC protocols.

31.1.1 Initial Access

In wireless THz systems, there are two possible handshaking mechanisms to start the communication: receiver (RX)- and transmitter (TX)-initiated communication. The TX and RX play different roles here. While the RX is responsible for reducing the number of transmissions, the TX is responsible to making sure that quality of transmission and channel performance are guaranteed.

Most of the THz MAC protocols are following the TX-initiated communication due to its simplicity and distributed nature. This mechanism is responsible for data transmission and synchronization. For nanoscale networks, this mechanism is applied to allow the nodes to transmit when they have data to send. An example of TX-initiated communication scheme is proposed in [3] to take benefits of a low weight channel coding scheme, minimize the interference, and maximize the probability of efficient decoding of the receiving information. A node which wants to set up the communication will send transmission request (TR), including synchronization trailer, transmission identifier (ID), packet ID, transmitting data symbol rate, and error detecting code, to a receiving node that will generate an acknowledgment (ACK) with the agreement of those communication parameters and send back a transmission confirmation (TC) message. This provides benefits related to delay and throughput, but there exist a few limitations, such as handshake process overhead and limited computational power of nanodevices. In macroscale networks on the other hand, the steerable narrow beam with directional antennas in macroscale network can overcome the high path loss and increase transmission distance at THz band, and the goal of TX-initiated communication utilized in macroscale applications is to improve throughput. In [4–6], the authors use 2.4 GHz band for signaling and omnidirectional antennas with alignment for overcoming the antenna facing problem. IEEE 802.11 request to send (RTS)/clear-to-send (CTS) is used for the initial access (IA) and control information, while the directional antennas are used to transmit data among nodes, which can reach the transmission distance up to 1 m.

The RX-initiated MAC protocols are designed to saving energy and reducing excess message overhead, which is especially important in nanoscale networks. This was studied in [7, 8]. Paper [9] presents a receiver-initiated communication model for centralized networks in which a ready-to-receive (RTR) packet is transmitted to a nearby nodes by the server and then those ones that send back ACK in a random-access manner with probability p to set up the communications among nodes. The goal of a MAC protocol discussed in [8] is for optimal energy consumption and allocation problem to maximize data rate, where one packet can sometime need to take multiple time slots to transmit when the amount of energy harvested is not enough. In macroscale networks, papers [10, 11] propose to improve the channel utilization by using a sliding window flow control mechanism with a one-way handshake. In these schemes, a CTS message from a node, containing the information of RXs' sliding window side, is broadcast to other nodes by using a dynamically turning narrow beam, then that CTS will be checked, and the sender will point its directions for the required period toward the receiver. The challenges for these schemes are as follows: collisions can happen in the case of multiple transmitters, and the packet reception guarantee and the initial neighbor discovery of the neighbor nodes are not still considered.

Next, we present an IA process that was analyzed in [12] and is compatible with several directional MAC protocols, such as IEEE 802.11ad and IEEE 802.11ay. Also, we revisit closed-form expressions for the IA procedure performance in terms of IA success probability. Finally, numerical results are given that highlight the impact of blockage and antenna misalignment on the performance of IA procedure.

31.1.1.1 System Model

We consider a wireless THz system, which is composed of a base station (BS) and a UE. The BS and the UE are equipped with antenna arrays of N_b and N_u antenna elements, respectively. Both the BS and UE support analog beamforming. The system time is divided into non-overlapping IA cycles of T period. In each cycle, a CS is initiated by the BS. In the CS, the BS uses different codebooks in order to sweep the main lobe toward N_b non-overlapping directions. In each one of the N_b directions, synchronization signals are broadcasted. On the other side, the UE sweeps through N_u receive beamforming directions to detect the received signal. As a result, $L = N_b N_u$ spatial channels (i.e., BS-UE direction pairs) are forming. This procedure is followed by a RA phase, in which a connection request is performed to the BS by the UE and the BS replies with a RA response. Finally, the DT phase follows, which can support both random and scheduled access.

CS Phase By indexing the transmissions in each scan cycle with $l = 1, 2, \cdots, L$ and based on the analysis in [12], the baseband equivalent received signal at time t, in which the $l-$th beam-pair is used, at the UE can be expressed as $r_l = \delta_{l,l_o}\theta_{l_o}h_{l_o}(t)s_l(t) + n_l(t)$, where $n_l(t)$ is the additive white Gaussian noise (AWGN) with variance $N_o W$ and $s_l(t)$ is the deterministic synchronization signal transmitted by the BS. Likewise, W and N_o respectively denote the single-sided signal bandwidth and the single-sided noise power spectral density. Moreover, l_o denotes the perfectly aligned UE and BS beamspace, while $\theta_{l_o} \in \{1, 0\}$ stands for the two hypotheses, namely, LoS and non-LoS (nLoS) at the pair l_o. Likewise, $\delta_{l,l_o} = \begin{cases} 1, & \text{for } l = l_o \\ 0, & \text{otherwise} \end{cases}$. Finally, $h_{l_o}(t)$ represents the complex channel gain and can be obtained as $h_{l_o}(t) = h_p h_f(t)$, with h_p and $h_f(t)$ respectively being the deterministic path-gain and the small-scale fading. The deterministic path-gain can be further analyzed as $h_p = \frac{c\sqrt{G_b G_u}}{4\pi f d} \exp\left(-\frac{1}{2}\kappa(f)d\right)$, where c, f and d, respectively, stand for the speed of light, the operating frequency, and the transmission distance, while G_b and G_u represent the BS and UE antenna gains, respectively. Moreover, $\kappa(f)$ models the absorption coefficient and can be evaluated as in [13]. For the small-scale fading, the following cases are considered:

1. $|h_f(t)|$ follows Rician distribution with PDF $f_{|h_f|}(x) = x \exp\left(-\frac{x^2+v^2}{2}\right) \mathrm{I}_o(vx)$, where v is connected with the shape parameter K through $K = \frac{v^2}{2}$.
2. $|h_f(t)|$ follows Nakagami-m distribution with PDF $f_{|h_f|}(x) = \frac{2m^m}{\Gamma(m)} x^{2m-1} \exp\left(-mx^2\right)$.
3. there is no fading; thus, $|h_f(t)| = 1$.

The received signal is subject to filtering, squaring, and integration over time interval T, which can be expressed as $Y_l = \frac{2}{N_o} \int_0^T |r_l(t)|^2 \, \mathrm{dt}$. Note that Y_l is a measure of the energy of the received waveform in the $l-$th pair at the UE, which acts as a test statistic that determines whether the received energy in the $l-$th pair corresponds only to noise or to the energy of both the synchronization signal and noise.

RA Phase In the RA phase, the UE initiates a connection request by transmitting a RA preamble sequence. The UE sweeps through N_u^{RA} orthogonal directions, while the BS through N_b^{RA} directions, utilizing $L_{\mathrm{RA}} = N_u^{\mathrm{RA}} N_b^{\mathrm{RA}}$ beam-pairs. Assuming that during the CS phase the UE identified the correct BS-UE pair, the baseband equivalent received signal at the BS can be similarly obtained as

$$r_b = \theta_{l_o}^b h_l^b s_l^b + n_l^b, \tag{31.1}$$

where s_l^b and n_l^b, respectively, stand for the RA preamble with transmitted power E_s^u and the AWGN, whereas $\theta_{l_o}^b = \begin{cases} 1, \text{ LoS} \\ 0 \text{ nLoS} \end{cases}$ and $h_l^b = h_m^b h_f^b$, with h_f^b being the small-scale fading coefficients at the RA phase. Note that without loss of generality, we assume that the h_m^b and h_f^b follow the same distributions as h_m and h_f, respectively. Based on (31.1), the instantaneous SNR can be obtained as $\rho^b = \frac{\theta_{l_o}^b |h_l^b|^2 E_s^u}{N_o^b}$, where N_o^b is the single-side noise power at the BS.

31.1.1.2 Performance Analysis

CS Phase Theorems 1, 2, and 3 provide the detection and false-alarm probabilities for the cases in which $|h_f(t)|$ follows Rician and Nakagami-m distributions and is deterministic, respectively.

Theorem 31.1 *In the case of Rician fading, the detection and false-alarm probabilities can be respectively obtained as*

$$P_{\mathrm{d}}(\lambda) = \left(1 - \Pr\left(\theta_{l_o} = 0\right)\right) P_d\left(\lambda \left|\theta_{l_o} = 1\right.\right) + \Pr\left(\theta_{l_o} = 0\right) P_d\left(\lambda \left|\theta_{l_o} = 0\right.\right) \tag{31.2}$$

and

$$P_{\mathrm{fa}} = \frac{\Gamma\left(N, \frac{\lambda}{2}\right)}{\Gamma(N)}, \tag{31.3}$$

where $N = 2TW$ *and* λ *are the test statistics threshold, whereas* $\Pr\left(\theta_{l_o} = 0\right)$ *is the* l_o *beam-pair blockage probability. Moreover, in* (31.2)

$$\begin{aligned} P_d\left(\lambda \left|\theta_{l_o} = 1\right.\right) &= \frac{N_o}{2E_r} \exp\left(-\frac{\upsilon^2}{2}\right) \mathrm{F_A}\left(1; 1; \frac{\upsilon^2}{2 + 4\frac{E_r}{N_o}} \frac{1}{\frac{N_o}{2E_r} + 1}\right) \\ &\quad - \frac{N_o}{2E_r} \left(\frac{\lambda}{2}\right)^N \exp\left(-\frac{\upsilon^2}{2}\right) \frac{1}{N\Gamma(N)} \\ &\quad \times \mathrm{H_A}\left(N, 1; N, N+1, 1; \frac{1}{\frac{N_o}{2E_r} + 1} \frac{\lambda}{2}, -\frac{\lambda}{2}, \frac{\upsilon^2}{2 + 4\frac{E_r}{N_o}}\right) \end{aligned} \tag{31.4}$$

and

$$P_d\left(\lambda \left|\theta_{l_o} = 1\right.\right) = \frac{\Gamma\left(N, \frac{\lambda}{2}\right)}{\Gamma(N)}. \tag{31.5}$$

Proof The proof is presented in [12]. □

Theorem 31.2 *In the case of Nakagami-m fading, the detection and false-alarm probabilities can be respectively obtained as in* (31.2) *and* (31.3) *by replacing* $P_d\left(\lambda \left|\theta_{l_o} = 1\right.\right)$ *by*

$$\begin{aligned} P_d^N\left(\lambda \left|\theta_{l_o} = 1\right.\right) &= \frac{2m^m}{E_r^m \Gamma(m)} \exp\left(\frac{mN_o}{E_r}\right) \sum_{k=0}^{m-1} \sum_{n=0}^{N-1} \binom{m-1}{k} \frac{2^{k-n}}{n!} (-N_o)^{m-1-k} N \\ &\quad \times \left(\frac{E_r}{2m}\right)^{k-n+1} \Gamma\left(k-n-1, \frac{mN_o}{E_r}, \frac{mN\lambda}{E_r}, 1\right). \end{aligned} \tag{31.6}$$

Proof The proof is presented in [14]. □

Theorem 31.3 *In the absence of fading, the detection and false-alarm probabilities can be respectively evaluated as*

$$P_d^{wo} = \left(1 - \Pr\left(\theta_{l_o} = 0\right)\right) \mathrm{Q}\left(\frac{\lambda - N_o\left(1 + \frac{E_r}{N_o}\right)}{\frac{N_o}{\sqrt{N}}\left(1 + \frac{E_r}{N_o}\right)}\right) + \Pr\left(\theta_{l_o} = 0\right) \mathrm{Q}\left(\frac{\lambda - N_o}{\frac{N_o}{\sqrt{N}}}\right) \tag{31.7}$$

and

$$P_{\mathrm{fa}} = \mathrm{Q}\left(\frac{\lambda - N_o}{\frac{N_o}{\sqrt{N}}}\right). \tag{31.8}$$

Proof The proof can be easily performed by employing the central limit theorem. □

RA Phase Theorems 31.4 and 31.5 present the outage probability for the case in which the l_o−th link experiences Rice and Nakagami-m fading, respectively, while Theorem 31.6 returns the outage probability in the absence of fading.

Theorem 31.4 *Under the assumption of Rician fading, the outage probability can be evaluated as*

$$P_o\left(\gamma_{th}\right) = \Pr\left(\theta_{l_o}^b = 0\right) + \left(1 - \Pr\left(\theta_{l_o}^b = 0\right)\right)\left(1 - \mathrm{Q}_1\left(\upsilon, \sqrt{\frac{\gamma_{th} N_o^b}{|h_p|^2 E_s^u}}\right)\right), \tag{31.9}$$

where γ_{th} is the SNR threshold.

Proof The proof of this theorem is provided in [12]. □

Theorem 31.5 *Under the assumption of Nakagami-m fading, the outage probability can be computed as*

$$P_o^N\left(\gamma_{th}\right) = \Pr\left(\theta_{l_o}^b = 0\right) + \left(1 - \Pr\left(\theta_{l_o}^b = 0\right)\right)\left(1 - \frac{\Gamma\left(m, \frac{m\gamma_{th} N_o^b}{|h_p|^2 E_s^u}\right)}{\Gamma\left(m\right)}\right). \tag{31.10}$$

Proof The proof of this theorem is provided in [15]. □

In the absence of fading, the outage probability can be evaluated as

$$P_o^{wo} = \Pr\left(\theta_{l_o}^b = 0\right). \tag{31.11}$$

IA Probability of Success The IA procedure is considered successful when both the CS and RA phases are successful. Since the two phases are independent, the success probability can be obtained as

$$P_s = P_d\left(\lambda\right)\left(1 - P_o\left(\gamma_{th}\right)\right). \tag{31.12}$$

31.1.1.3 Numerical Results

In this section, we investigate the joint effect of blocking and small-scale fading in the CS and RA phase performance as well as the system's overall performance, by illustrating analytical and Monte Carlo simulation results. The following insightful scenario is investigated. Unless otherwise stated, the communication bandwidth is set to 1 MHz, the synchronization signal period is 1 μs, the relative humidity is 50%, while the atmospheric pressure is 101325 Pa. Moreover, the temperature is 296°K, and the transmission distance is equal to 0.4 m. The TX/RX antenna gains are set to 20.7 dBi, and the blocking probability is 1%. Finally, the BS-UE channel is assumed to experience Rician fading.

Figure 31.1 reveals the impact of the channel and transmission characteristics to the CS and RA phases performance. On the one hand, Fig. 31.1.a presents the joint effect of small-scale fading and blocking in the CS. Particularly, receiver operation curves (ROCs) are given for different values of K and probability of blockage, P_B. From this figure, it becomes apparent that for a fixed K, as P_B decreases, the system performance improves. Moreover, for a given P_B, as K increases, the CS efficiency improves. On the other hand, Fig. 31.1.b demonstrates the impact of blockage in the RA. In more detail, the outage probability is plotted as a function of P_B for different

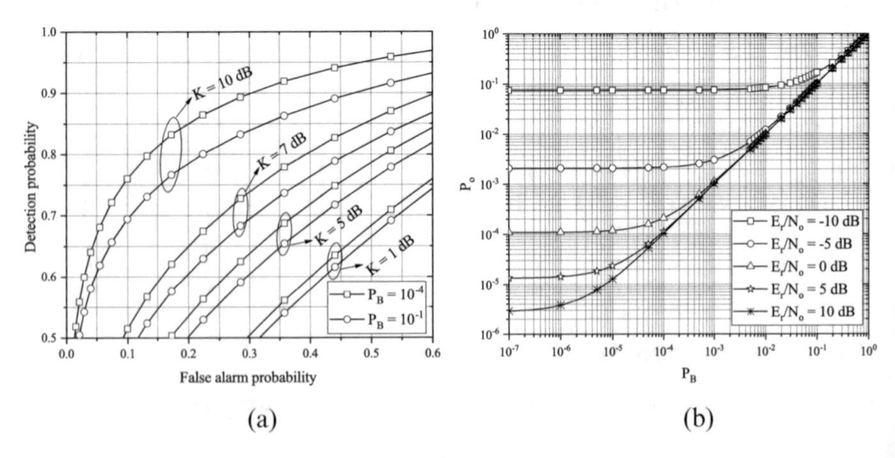

Fig. 31.1 (**a**) ROC for the CS phase, assuming different values of blocking probability and K. (**b**) Outage probability vs blocking probability for different values of E_r/N_o. ©2019, IEEE, reproduced with permission

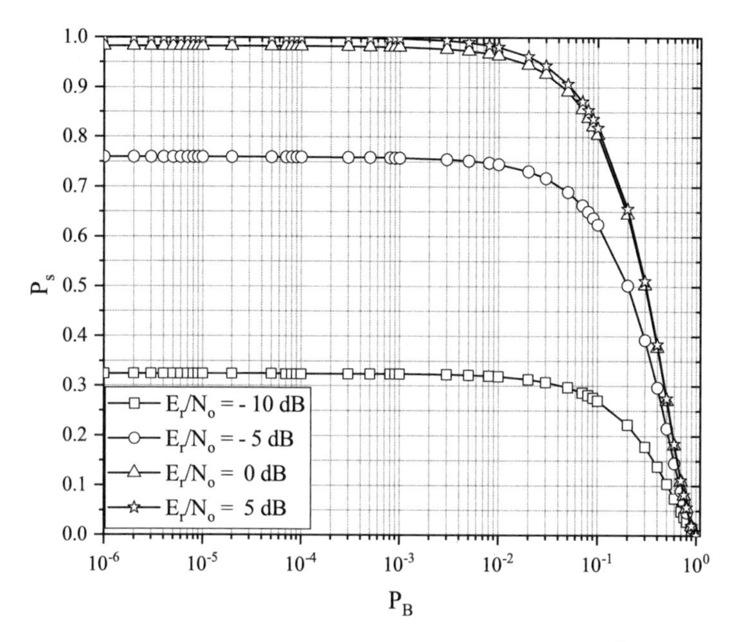

Fig. 31.2 Success vs blocking probabilities for different values of E_r/N_o. ©2019, IEEE, reproduced with permission

E_r/N_o values, assuming $K = 10$ dB. We observe that for a given E_r/N_o, as P_B increases, the outage performance degrades. Moreover, for a fixed P_B, as E_r/N_o increases, the outage performance improves. Finally, from this figure, it becomes evident that in the high P_B regime, the outage probability tends to P_b. This indicates the significance of taking into account blockage, when assessing the performance of the RA phase.

Figure 31.2 demonstrates the probability of successful IA, P_s, against the blocking probability, for different E_r/N_o values. From this figure, it becomes evident that for a given E_r/N_o, as P_b increases, P_s decreases. Moreover, in the low P_B regime, i.e., $P_B \leq 10^{-2}$, P_s is determined by the level of the E_r/N_o. On the other hand, in the high P_b regime, P_s is mainly influenced by P_B.

31.1.2 Channel Access Mechanisms

We consider here three channel access mechanisms: random, scheduled, and hybrid access channel.

Random channel access is used both in nanoscale networks (large number of nodes) and macroscale networks (comparably lower number of nodes). For macro networks, ALOHA and CSMA are two known techniques of random mechanisms, shown to increase throughput but also likely to increase delay and cause collisions.

Paper [10] is an example of scheme-based carrier-sense multiple access (CSMA) proposed to reduce the message overhead with one-way handshake. A node listens for messages from other nodes until one is received in its transmission mode. However the challenge is the antenna facing problem due to directional antennas used, which is addressed in [16] by using highly directional antennas to reduce high path loss. Papers [4, 6] are examples of schemes based on CSMA with multiple radios or hybrid systems to divide IA and data transmissions. The problems of these works are high message overhead and antenna switching delay and repeated discovery phase with mobility schemes which the channel is accessed by sending RTS/CTS packet including node positions. In [6] the transmitter gets TTS packet, instead of CTS packet, from the receiver based on estimation of angle of arrival and can switch as well as adjust its directional antenna toward the receiver antenna to transmit data. While this prior work has shown that it can reduce the message overhead, the issues of high path and absorption loss and uncertainty of packet loss during user association phase remain unsolved.

Random network access has also been used in the context of nanonetworks. Paper [17] presents a slotted CSMA/collision avoidance (CA)-based channel access mechanism and energy harvesting model, considers the super-frame duration and packet size for slotted CSMA protocols, and shows an efficient slots usage with slotted CSMA method. A beacon of these types of networks can synchronize the subsequent transmissions or request a data transmission, where collision can happen when directly sending beacon transmission for data transmission as two nodes can operate at the same time. An simple ALOHA-based channel access mechanism is presented in [18].

When random channel access is not an option, a scheduled channel access can be considered. In time division multiple access (TDMA) approach, each node occupies a time period to transmit its data [19]. An example for this approach is dynamic scheduling scheme based on TDMA presented in [20], where the length time slots assigned for each node depend on the amount of transmission data. For frequency and time division multiple access (FTDMA) approach, an example in [21] proposes a dynamic frequency selection algorithm (DYNAMIC-FH) which uses FTDMA, where multiple frequencies and time slots scheduling are proposed for a different number of users to avoid collisions. Another FTDMA approach is presented in [22] which divides the frequency slots and time slots in its approach. The frequency sequences must be orthogonal when many users are using them to transmit data at the same time.

To increase the transmission range in THz band, the directional narrow beams need to be used, which in turn leads to high delay for handovers, beam tracking, and establishment of IA. The papers [23, 24] address that a TDMA with multiple radios is designed for beam alignment scheduling, channel access, and synchronization among nodes. In [24], the authors use software define network (SDN)-based controller (SDNC) to switch between mmWave and THz band for vehicular communication for high bandwidth data transfer operation to maximize the bits exchange between the cell tower and vehicle.

Finally, random and scheduled channel access can be combined into a hybrid approach which has only been used in the context of macroscale networks. This mechanism is suitable in order to overcome the limitations of random access and the scheduled access mechanisms and improve the performance. Examples of hybrid channel access mechanisms can be found in [25–27].

31.1.2.1 Random Access

System Model We consider a THz wireless network and a homogeneous Poisson network of transmitters on the plane with density λ_t per unit area, each associated to a RX. In order to quantify the collision performance of the network, we consider a reference link between a reference RX and its intended TX, which has a geographical/spatial length of L. In what follows, we call the RX and the TX of the reference link as the reference RX and the tagged TX. Since all the nodes are distributed according to a Poisson process, according to Slivnyak's theorem, the conditional distribution of the potential interference given the typical RX at the origin is another homogeneous point Poisson precess (PPP) with the same density.

We consider a directional ALOHA protocol, without power control. In this scenario, the transmission power of all the links is P_s. Moreover, we assume that every interferer can be active with probability p_a. In other words, the probability that it transmits in a specific slot is p_a. In slotted ALOHA, the transmissions are regulated to start at the beginning of a time slot. As a result, it is a good model for random access during the data transmission phase (DTP) in IEEE802.11ay. By assuming that during the beamforming association training phase (BATP) the TX and RX have obtained the perfect beamforming vectors, we can assume that in DTP the TX of every link is spatially aligned with its intended RX. Based on these assumptions, in what follows, we will ignore the beam-training overhead. Likewise, due to the specular nature of the THz propagation, in this study, we assume that only LoS links can be established.

We consider the frequency, environment, and distance-dependent path loss presented in [15], and we use the protocol model of interference, i.e., for a given distance between a reference RX and its intended TX, a collision occurs if there is at least another interfering TX no farther than a certain distance for the reference RX. As the probability of having LoS condition on a link decreases exponentially with the distance, far away TXs will be most probably blocked and therefore cannot contribute in the interference the RX experiences. Therefore, we consider only the impact of spatially close TXs.

At the MAC, the beamforming is represented by using the ideal sector antenna pattern. Thus, the TX and RX antenna gains can be obtained as $G_{t/r} = \begin{cases} \frac{2\pi-(2\pi-\theta_{t/r})\epsilon_{t/r}}{\theta_{t/r}}, & \text{main lobe} \\ \epsilon_{t/r}, & \text{side lobe} \end{cases}$, where $\theta_{t/r}$ and $\epsilon_{t/r}$, respectively, stand for the half-power beamwidth and the strength of the side lobe.

Due to the THz link extreme sensitivity to obstacles, we need to introduce a proper blockage model that accommodates the following properties: (1) blockers can randomly appear in a THz link, and (2) one blocker can block multiple angularly close THz links. As a consequence, we assume that the centers of the blockers follow a PPP with density λ_o, which is independent of the wireless network. To capture the second property, we define a coherence angle, θ_c, over which the LoS conditions are statistically correlated. That means that inside a coherence angle, an obstacle blocks all the interference behind itself; hence, there is nLoS in distances $d \geq l$ with respect to the typical RX, if there is a blocker at distance l. Notice that θ_c increases with the size and density of the obstacles.

Interference Profile Let $d_{\max}$ be the interference range; β the minimum signal to interference plus noise ratio (SINR) threshold at the typical RX, below which the impact of interference is neglectable; and σ_N the noise power. Then, the SINR due to the transmission of the intended TX and an aligned LoS interferer located at distance d can be obtained as $\gamma = \frac{\mathrm{G} G_t G_r P_s}{\mathrm{G}_I G_t G_r P_s + \sigma_N}$, where G and G_I, respectively, stand for the path gains of the typical TX-RX and interferer-RX links, while P_s denotes the transmission power. As a consequence, $d_{\max}$ is the distance for which $\gamma = \beta$.

A TX at distance d from the typical RX can cause collision provided that the following conditions hold: (1) it is active; (2) the typical RX is inside its main lobe; (3) it is inside the main lobe of the typical Rx; (4) it is located inside the interference range $d \leq d_{\max}$; and (v) it is in the LoS condition with respect to the typical Rx. The following theorem returns the probability of having at least one LoS interferer.

Theorem 31.6 *The probability of having at least one LoS interferer can be evaluated as*

$$P_{\mathrm{LI}} = 1 - \exp\left(\lambda_I A_L\right) + \frac{\lambda_I}{\lambda_I + \lambda_o} \exp\left(\lambda_o A_L\right) \times \left(\exp\left(-\left(\lambda_o + \lambda_I\right) A_L\right) - \exp\left(-\left(\lambda_o + \lambda_I\right) A_{d_{\max}}\right)\right), \tag{31.13}$$

where $\lambda_I = \rho_a \lambda_t \frac{\theta_t}{2\pi}$, $A_L = \pi L^2$, $A_{d_{\max}} = \pi d_{\max}^2$, *with* λ_t *being the density of the number of TXs per unit area and* ρ_a *denoting the the average probability of a TX to be active.*

Proof For brevity, the proof is presented in [28]. □

Figure 31.3 illustrates that the interference probability as a function of λ_t for different values of λ_o, assuming $\theta_t = 0.1°$, is also depicted. In the considered scenario, the transmission frequency and power are, respectively, set to 275 GHz and 10 dBm, the transmission distance is 1 m, while the transmission probability equals 50%. Finally, standard environmental conditions are considered. From this figure, we observe that, for a given λ_o, as λ_t increases, the interference probability also increases. Moreover, for a fixed λ_t, as λ_o increases, the interference probability decreases. Likewise, for relatively low values of λ_o, we observe that the impact of blockage on the interference profile is negligible.

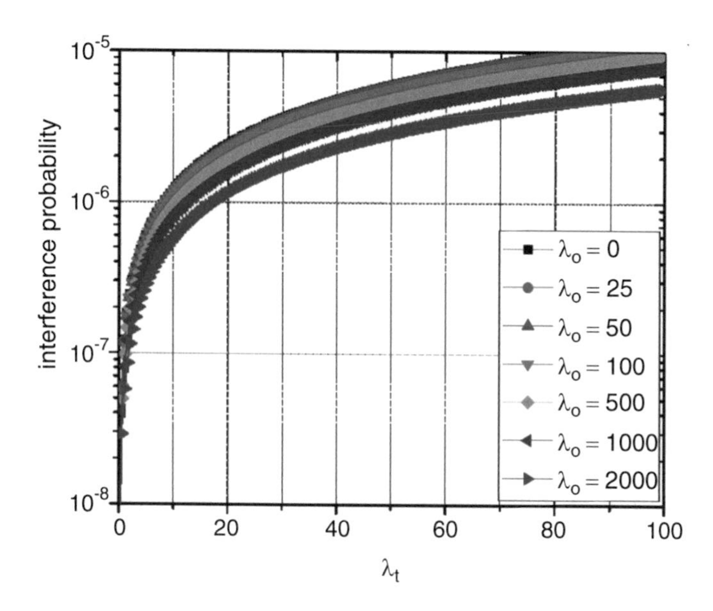

Fig. 31.3 Interference probability vs λ_t for different values of λ_o

31.1.2.2 Directional ALOHA

Since wireless THz systems are not interference limited, there is no need to implement a collision avoidance mechanism. Motivated by this, a directional ALOHA protocol for random access could be an option to consider. This section is devoted to present the performance of the directional ALOHA protocol, in terms of collision probability, successful transmission probability, throughput, and area spectral efficiency (ASE).

Collision Probability The collision probability can be defined as the probability of having at least one interferer or, equivalently, as the complementary probability of having no interference at all. Thus, it can be evaluated as $P_c = \int_0^{d_{\max}} P_{c|L}\, f_l(l)\, \mathrm{dl}$, where $P_{c|L} = 1 - (P_{\mathrm{LI}})^k$, $f_{l_i}(x) = \frac{2x}{d_{\max}^2}$, and $k = \left\lceil \frac{\theta}{\theta_r} \right\rceil$, with θ being the total angle of the RX observation area and l_i being the distance between the reference RX and the interferer.

Figure 31.4 demonstrates the collision probability as a function of the TX density, for different values of obstacle density. From this figure, it is evident that the impact of blockage is not important enough to be taken into account. Moreover, for a given λ_o, as λ_t increases, the collision probability also increases. For realistic values of λ_t, i.e., $\lambda_t < 10$, we observe that the collision probability is very low. This verifies that the THz wireless systems can be considered interference-free.

Throughput For a fixed L, the success transmission occurs when there is no blockage on the link and no collision. As a consequence, the conditional successful

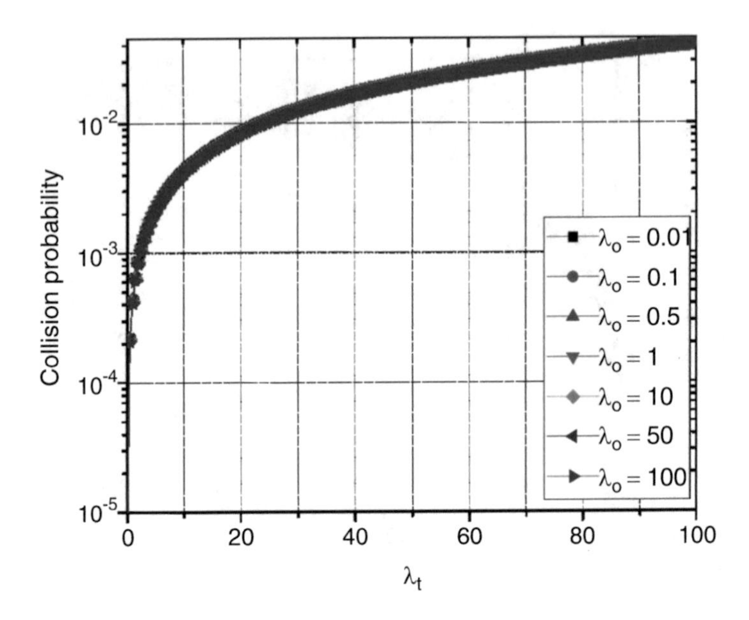

Fig. 31.4 Collision probability vs λ_t, for different values of λ_o

transmission probability can be expressed as $P_{s|L}(l) = \rho_a \exp(-\lambda_o A_l)(P_{\text{LI}})^k$. Hence, the throughput can be computed as $r_{\text{ALOHA}} = \int_0^{d_{\max}} P_{s|L}(x) f_{l_i}(x)\, \mathrm{d}x$.

Figure 31.5 plots the per-link throughput as a function of the transmission probability, for different values of λ_o, assuming $\lambda_t = 25$, $\theta_t = \theta_r = 0.01^o$ and transmission power equals 10 dBm. As expected, for a given obstacle density, as the transmission probability increases, the per-link throughput also increases. Moreover, for a given transmission probability, as λ_o increases, the throughput performance degrades. Interestingly, for realistic values of λ_o, i.e., $\lambda_o < 10$, the impact of blockage on the ALOHA per-link throughput performance is negligible.

Area Spectral Efficiency (ASE) In order to calculate the ASE, i.e., the network throughput normalized to the network size, of the directional ALOHA, we consider a large area. Let us assume that the number of TXs inside A is $n_t + 1$ with *nt* being a Poisson distributed random variable and that the number of retransmissions of each TX is set to infinity. As a consequence, the network throughput can be obtained as $\text{ASE}_{\text{ALOHA}} = \left(\frac{1}{A} + \lambda_t\right) r_{\text{ALOHA}}$.

Next, we provide numerical results that illustrate the ALOHA performance in terms of ASE. In particular, Fig. 31.6 demonstrates the ASE as a function of the transmission probability, for different values of TX density in a rectangular area of 4 m^2, assuming that the obstacle density is set to 0.0025 and the half-power beamwidth of both the TX and RX is 0.05°. As expected, for a given TX density, as the transmission probability increases, the ASE also increases. Likewise, for a fixed transmission probability, as the TX density increases, ASE also increases.

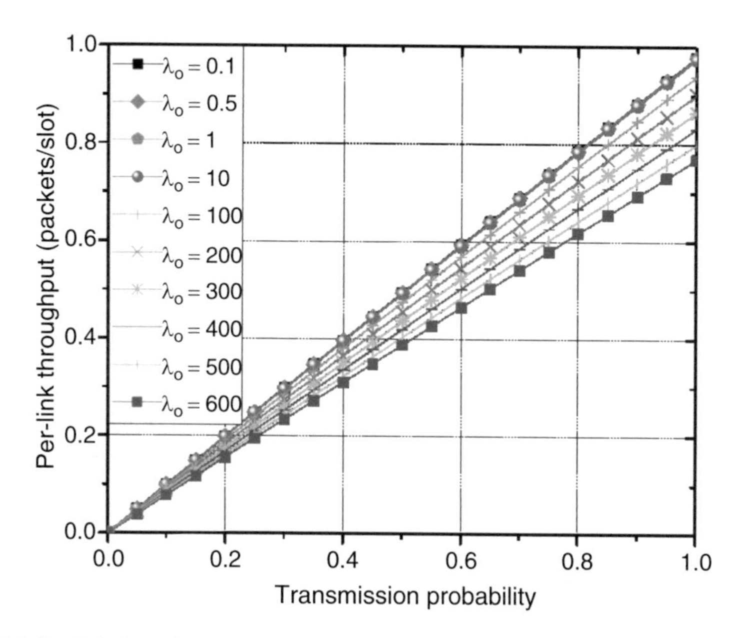

Fig. 31.5 Per-link throughput vs transmission probability for different values of λ_o

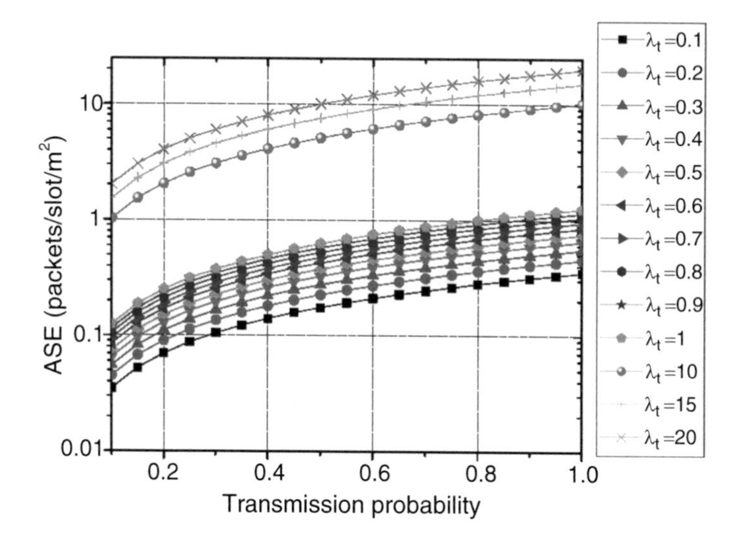

Fig. 31.6 ASE vs transmission probability for different values of λ_t

31.1.2.3 Scheduled Access

For scheduled access, we select the TDMA protocol, which is in line with IEEE 802.11ay. Unlike the directional ALOHA, TDMA activates only one link at a time. This ensures an interference-free communication between the UEs and the

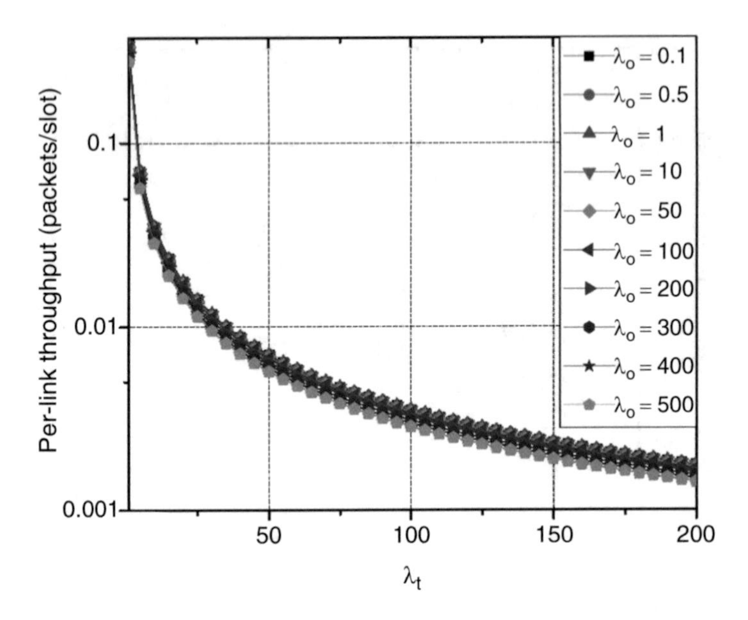

Fig. 31.7 Per-link throughput vs λ_t for different λ_o

BSs. In other words, in the scheduled access the collision probability equals zero. Therefore, the key performance indicators (KPIs) that we use in order to evaluate the performance of the directional TDMA protocol are the per-link throughput and the ASE. In what follows, we consider the same system model and assumption that were used for the RA scenario.

Per-link Throughput Since there are no concurrent transmissions in TDMA, the success probability can be analytically evaluated as $\Pr\left(\theta_{L_i} = 1 \,|L_i, n_t\right) = \exp\left(-\lambda_o A_{L_i}\right)$. The TDMA scheduler allocates only $\frac{1}{n_t+1}$ of the total resources to each link. Hence, by assuming transmission of one packet per slot, the MAC throughput of each link i in TDMA can be obtained as

$$r_{\text{TDMA}} = \sum_{n_t=0}^{\infty} \frac{(A\lambda_t)^{n_t} \exp\left(-A\lambda_t\right)}{n_t!\,(n_t+1)} \int_0^{d_{\max}} \exp\left(-\frac{\lambda_o \theta_c x^2}{2}\right) \mathrm{dx}, \tag{31.14}$$

which, after some algebraic manipulations can be simplified to

$$r_{\text{TDMA}} = \frac{2}{d_{\max}} \frac{1 - \exp\left(-A\lambda_t\right)}{A\lambda_t} \frac{1 - \exp\left(-A_{d_{\max}}\lambda_o\right)}{A_o\theta_c}. \tag{31.15}$$

Next, we present illustrative results that quantify the performance of the TDMA scheduler, assuming the scenario described in Sect. 31.1.2.1. In particular, Fig. 31.7 plots the per-link throughput as a function of the TX density for different values of obstacles density. As expected, for a given obstacle density, as the TX density increases, the resources are allocated to more users; hence, the per-link throughput decreases. Moreover, for a fixed TX density, as the obstacle density increases, the per-link throughput decreases. Finally, by comparing these results with the respective results of the ALOHA protocol, we observe that due to the low interference levels as well as the spatial resource reuse, random access outperforms scheduled access in terms of throughput.

ASE The following theorem returns a closed-form expression for the evaluation of ASE in TDMA systems.

Theorem 31.7 *The ASE of TDMA scheduler can be obtained as*

$$\mathrm{ASE}_{\mathrm{TDMA}} = \frac{1-\exp\left(-\lambda_o A_{d_{\max}}\right)}{\lambda_o A A_{d_{\max}}}. \tag{31.16}$$

Proof For brevity, the proof is presented in [28]. □

From (31.16), it becomes evident that the ASE depends only on the obstacle density, the network area, and the maximum transmission area and not from the TX density.

31.2 THz Networking

THz networking is today in its infancy. The reasons behind the slow evolution toward the full networking architecture are fundamental in nature, as previously described. To enable multihop communication and efficient routing and forwarding in THz network nodes, significant challenges in the physical layer communication need to be overcome, including path losses, challenges in MAC design, and the related challenges in estimating reliable transmission distance and multihop communication reach. We envision that the quality of transmission (QoT) will play the significant role in determining the quality of service (QoS), the basic standard for new services in carrier-grade networks.

To this end, a proper design of THz network data plane, control plane, and management plane is needed. In the THz data plane, we need network and node architecture that can enable multihop communication, switching, routing, and forwarding of the THz data flows. In multihop communication links, one of the key questions to answer relates to the optimal relaying distance ("THz reach") which is due to the distance-dependent behavior of the available bandwidth and the decreasing of the transmission distance, due to the deterioration of the transmission parameters. With distance also the SNR deteriorates, while increasing SNR will make it easier to cross more number of hops, which in turn increases the end-to-end delay. This illustrates that a number of tradeoffs need to be found with optimizations, to design the network nodes and topologies (Fig. 31.8).

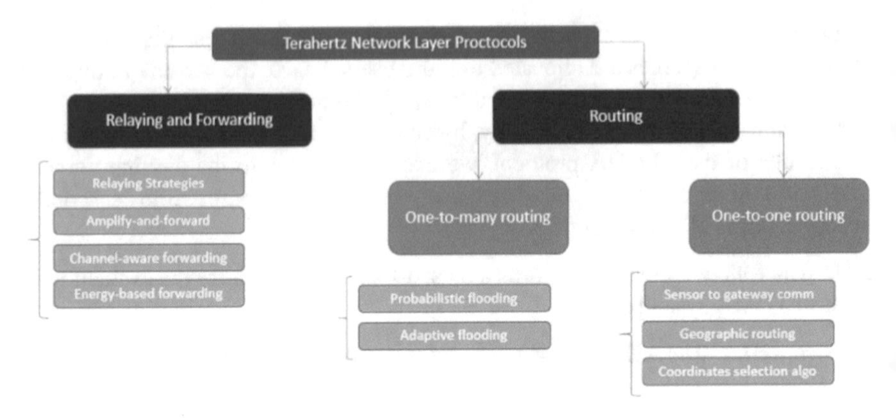

Fig. 31.8 Classification of network layer protocols, see [38] for more insights

In the THz control plane, we will need methods to accurately estimate the modulation format, transmission distance, and additional QoS parameters, e.g., coding-dependent packet error rate (PER) at a given SNR and bit error rate (BER) prior to set up multihop THz paths. This can be done by deploying either measurements in real time, applying the known channel models, or developing machine learning (ML) techniques to accurately estimate the modulation format, transmission distance, and additional QoS parameters, e.g., coding-dependent packet error rate (PER) at a given SNR and BER in the THz networked system. If we deal with measurements supplied by a system under real conditions, they may dynamically vary depending on the channel state, and there is a challenge of real-time monitoring and admission control into the THz network. To guarantee a given QoS, the control plane needs to make sure that the THz networked systems can adapt to the varying channel conditions. To this end, it may be appropriate to define a separate control plane channel that can allow for real-time communication between transmitter and receiver, to be able to communicate the related configuration parameters. This is similar to GHz systems today, where such as out-of-band channel (for instance, a WiFi channel) uses measurements and fast optimization algorithms of important channel and system parameters provided by the receiver itself to determine the adaptive system parameters as output. The influence and correlation of the system parameters, e.g., modulation, channels, and data coding, to the measurements available at the input and output of the control plane system is an open research topic in the research community.

31.2.1 An Overview of THz Network Layer Protocols

To maintain the communication between different THz nodes, under the assumption of dynamic link lengths and reach, THz networks also require new network layer protocols as current protocols are likely inappropriate. As it is typically the case in networks, two families of protocols are needed: forwarding and routing protocols. Forwarding is generally a function in a node responsible about choosing the way of sending the information to the neighboring nodes, including choice of antennas, transmission ports, or spatial interfaces. Routing on the other hand is a network wide function and is a control plane function responsible for identifying the paths and nodes to send/forward a message from a source to a specific destination, usually in a multihop communication.

31.2.1.1 Relaying and Forwarding

Due to the limited transmission power and the high path losses in THz band communication, the use of directional antennas is a must, which in turn affects the network layer and the relaying strategies. To this end paper, [29] investigates the said relaying strategies for THz-band communication networks by introducing a mathematical framework to study the optimal relaying distance that maximizes the network throughput. Amplify and forward is a technique that takes into consideration the spreading loss and molecular loss to investigate the bit error rate (BER) performance along the relaying path [30]. Channel aware forwarding refers to the ability to determine the performance of THz network before sending data based on relaying and forwarding [31]. Energy-based forwarding has also been studied to optimize the throughput and lifetime along with the energy [32].

31.2.1.2 Routing

The routing categories are divided between one-to-one routing and one-to-many routing. The former is still a major challenge in THz communications, while the latter has been addressed by flooding. To this end, literature reports adaptive and probabilistic flooding. Paper [33] investigates an adaptive flooding mechanism based on deactivation of wireless nanonodes using their perceived signal to interference and resource levels. Probabilistic flooding is studied in [34] whereby known flooding routing algorithms are analyzed on their applicability in THz networks.

For one-to-one routing, flooding is clearly inefficient and other methods have been proposed. The technique referred to as *sensor to gateway* was reported in [35] where data delivery is provided by relaying over nano-routers as an intermediate between the nanonodes and the gateway connected to large-scale network. The so-called geographic routing was studied in [36] which uses geolocation techniques by measuring distances between different nodes in terms of hopes. Finally, a coordinates selection algorithm was presented in [37] which introduces a stateless addressing and routing scheme for 3D nanonetworks.

31.3 Conclusions, Open Issues, and Challenges

This chapter discussed the impact of antenna misalignment and blockage in the performance of the fundamental mechanisms of the MAC in THz wireless systems. Toward this direction, we presented suitable models and performance metrics that accommodate and quantify the aforementioned phenomenon impact in the IA random and scheduled access procedures. The derived expressions were verified through Monte Carlo simulations that revealed the significance of taking into account the THz channel particularities in the design of the THz MAC.

Regarding THz networking, several open research and challenges are worth summarizing. The network architectures in THz spectrum are an open issue, as is its hybrid integration with the traditional telecom infrastructure. Mobility is a grand challenge in wireless networking in general and in THz networking in particular. Similarly, the important issues of privacy and security remain widely unaddressed, despite significant importance to wider adoption. It is furthermore clear that the scalability is one of the most area that will play a critical role especially for nanonodes but also for macroscale network, due to extreme data rates. Many of the applications supported by the THz band communications also require high reliability of the nodes.

References

1. Singh, S., Ziliotto, F., Madhow, U., Belding, E., & Rodwell, M. (2009). Blockage and directivity in 60 GHz wireless personal area networks: From cross-layer model to multihop MAC design. *IEEE Journal on Selected Areas in Communications, 27*(8), 1400–1413.
2. Han, C., Zhang, X., & Wang, X. (2019). On medium access control schemes for wireless networks in the millimeter-wave and terahertz bands. *Nano Communication Networks, 19*, 67–80.
3. Jornet, J. M., Pujol, J. C., & Pareta, J. S. (2012). Phlame: A physical layer aware MAC protocol for electromagnetic nanonetworks in the terahertz band. *Nano Communication Networks, 3*(1), 74–81.
4. Yao, X.-W., Jornet, J. M. (2016). Tab-MAC: Assisted beamforming mac protocol for terahertz communication networks. *Nano Communication Networks, 9*, 36–42.
5. Han, C., Tong, W., & Yao, X. W. (2017). MA-ADM: A memory-assisted angular-division-multiplexing MAC protocol in terahertz communication networks. *Nano Communication Networks, 13*, 51–59.
6. Tong, W., & Han, C. (2017). MRA-MAC: A multi-radio assisted medium access control in terahertz communication networks. In *GLOBECOM 2017—2017 IEEE Global Communications Conference* (pp. 1–6).
7. Mohrehkesh, S., Weigle, M. C., & Das, S. K. (2015). DRIH-MAC: A distributed receiver-initiated harvesting-aware MAC for nanonetworks. *IEEE Transactions on Molecular, Biological and Multi-Scale Communications, 1*(1), 97–110.
8. Mohrehkesh, S., & Weigle, M. C. (2014). Optimizing energy consumption in terahertz band nanonetworks. *IEEE Journal on Selected Areas in Communications, 32*(12), 2432–2441 (2014)
9. Mohrehkesh, S., & Weigle, M. (2014). RIH-MAC: Receiver-initiated harvesting-aware MAC for nanonetworks. In *IEEE Transactions on Molecular, Biological and Multi-Scale Communications* (Vol. 1).

10. Xia, Q., Hossain, Z., Medley, M., J., & Jornet, J. M. (2015). A link-layer synchronization and medium access control protocol for terahertz-band communication networks. In *IEEE Transactions on Mobile Computing* (p. 1).
11. Xia, Q., & Jornet, J. M. (2019). Expedited neighbor discovery in directional terahertz communication networks enhanced by antenna side-lobe information. *IEEE Transactions on Vehicular Technology, 68*(8), 7804–7814.
12. Boulogeorgos, A.-A. A., & Alexiou, A. (2019). Performance evaluation of the initial access procedure in wireless THz systems. In *16th International Symposium on Wireless Communication Systems (ISWCS)* (pp. 422–426).
13. Kokkoniemi, J., Lehtomäki, J., & Juntti, M. (2018). Simplified molecular absorption loss model for 275–400 gigahertz frequency band. In *12th European Conference on Antennas and Propagation (EuCAP)* (London, UK).
14. Boulogeorgos, A.-A. A., Chatzidiamantis, N. D., & Karagiannidis, G. K. (2016). Spectrum sensing with multiple primary users over fading channels. *IEEE Communications Letters, 20*(7), 1457–1460
15. Boulogeorgos, A.-A. A., Papasotiriou, E. N., & Alexiou, A. (2019). Analytical performance assessment of THz wireless systems. *IEEE Access, 7*, 1–18.
16. Xia, Q., & Jornet, J. M. (2017). Cross-layer analysis of optimal relaying strategies for terahertz-band communication networks. In *2017 IEEE 13th International Conference on Wireless and Mobile Computing, Networking and Communications (WiMob)* (pp. 1–8).
17. Lee, S. J. (2018). Slotted CSMA/CA based energy efficient MAC protocol design in nanonetworks.
18. Piro, G., Grieco, L. A., Boggia, G., & Camarda, P. (2013). Nano-sim: simulating electromagnetic-based nanonetworks in the network simulator 3 (Vol. 2013, pp. 203–210).
19. Lin, J., & Weitnauer, M. A. (2014). Pulse-level beam-switching MAC with energy control in picocell terahertz networks. In *2014 IEEE Global Communications Conference*, (pp. 4460–4465).
20. Wang, P., Jornet, J. M., Malik, M. A., Akkari, N., & Akyildiz, I. F. (2013). Energy and spectrum-aware MAC protocol for perpetual wireless nanosensor networks in the terahertz band. *Ad Hoc Networks, 11*(8), 2541–2555
21. Afsharinejad, A., Davy, A., & Jennings, B. (2015). Dynamic channel allocation in electromagnetic nanonetworks for high resolution monitoring of plants. *Nano Communication Networks, 7*, 2–16.
22. Li, Z., Guan, L., Li, C., & Radwan, A. (2018). A secure intelligent spectrum control strategy for future THz mobile heterogeneous networks. *IEEE Communications Magazine, 56*(6), 116–123.
23. Cacciapuoti, A. S., Sankhe, K., Caleffi, M., & Chowdhury, K. R. (2018). Beyond 5G: Thz-based medium access protocol for mobile heterogeneous networks. *IEEE Communications Magazine, 56*(6), 110–115.
24. Cacciapuoti, A. S., Subramanian, R., Chowdhury, K. R., & Caleffi, M. (2017). Software-defined network controlled switching between millimeter wave and terahertz small cells. arXiv preprint arXiv:1702.02775.
25. You, L., Ren, Z., Chen, C., & Lv, Y.-H. (2017). An improved high throughput and low delay access protocol for terahertz wireless personal area networks.
26. Ren, Z., Cao, Y., Peng, S., & Lei, H. (2013). A MAC protocol for terahertz ultra-high data-rate wireless networks. *Applied Mechanics and Materials, 427–429*, 2864–2869.
27. Ren, Z., Cao, Y. N., Zhou, X., Zheng, Y., & Chen, Q. B. (2013). Novel MAC protocol for terahertz ultra-high data-rate wireless networks. *The Journal of China Universities of Posts and Telecommunications, 20*(6), 69–76.
28. Boulogeorgos, A.-A. A., Stratidakis, G., Papasotiriou, E., Lehtomaki, J., Kokkoniemi, J., Mushtaq, M. S., Hatami, M., Point, J.-C., & Alexiou, A. (2019). D4.2 THz-driven MAC layer design and caching overlay method, resreport, Terranova Consortium.

29. Xia, Q., & Jornet, J. M. (2017). Cross-layer analysis of optimal relaying strategies for terahertz-band communication networks. In *2017 IEEE 13th International Conference on Wireless and Mobile Computing, Networking and Communications (WiMob)* (pp. 1–8). IEEE.
30. Rong, Z., Leeson, M. S., & Higgins, M. D. (2017). Relay-assisted nanoscale communication in the THz band. *Micro & Nano Letters, 12*(6), 373–376.
31. Yu, H., Ng, B., & Seah, W. K. (2015). Forwarding schemes for em-based wireless nanosensor networks in the terahertz band. In *Proceedings of the Second Annual International Conference on Nanoscale Computing and Communication* (p. 17). ACM.
32. Pierobon, M., Jornet, J. M., Akkari, N., Almasri, S., & Akyildiz, I. F. (2014). A routing framework for energy harvesting wireless nanosensor networks in the terahertz band. *Wireless Networks, 20*(5), 1169–1183.
33. Liaskos, C., & Tsioliaridou, A. (2014). A promise of realizable, ultra-scalable communications at nano-scale: A multi-modal nano-machine architecture. *IEEE Transactions on Computers, 64*(5), 1282–1295.
34. Stelzner, M., Busse, F., & Ebers, S. (2018). In-body nanonetwork routing based on MANET and THz. in *Proceedings of the 5th ACM International Conference on Nanoscale Computing and Communication* (pp. 1–6).
35. Al-Turjman, F. (2017). A cognitive routing protocol for bio-inspired networking in the internet of nano-things (IoNT). In *Mobile Networks and Applications* (pp. 1–15).
36. Tsioliaridou, A., Liaskos, C., Ioannidis, S., & Pitsillides, A. (2015). CORONA: A coordinate and routing system for nanonetworks. In *Proceedings of the Second Annual International Conference on Nanoscale Computing and Communication* (p. 18). ACM.
37. Tsioliaridou, A., Liaskos, C., Dedu, E., & Ioannidis, S. (2017). Packet routing in 3d nanonetworks: A lightweight, linear-path scheme. *Nano Communication Networks, 12*, 63–71.
38. Lemic, F., et al., Survey on Terahertz Nanocommunication and Networking: A Top-Down Perspective, in *IEEE Journal on Selected Areas in Communications*, doi: 10.1109/JSAC.2021.3071837

Part X
Demonstrators and Experiments

Chapter 32
Real100G.RF

Pedro Rodríguez-Vázquez, Janusz Grzyb, and Ullrich R. Pfeiffer

Abstract Due to the wide practical use and its availability worldwide, wireless communication has been one of the main applications driving the development of microelectronic technology. This drive will persist in the future, where wireless networks with data rates of 100 Gb/s are already envisioned by the new IEEE 802.15.3d-2017 standard. Currently, the majority of wireless communication systems at near-THz frequencies rely on III–V semiconductors with very high f_{max}. However, their low integration level prevents them from being a feasible solution for mass production markets.

Contrary, silicon offers a higher integration level, including digital baseband (BB) processing. The main limitation of silicon was its poor performance at frequencies above 200 GHz. However, the latest advancements in process technology have opened new opportunities for silicon-based solutions at near-THz frequencies.

The main objective of the Real100G.RF project, founded by the special priority program of the DFG SPP1655, was to take advantage of this new technological breakthrough to leverage silicon-based economies of scale in wireless multi-gigabit communications above 200 GHz. Several RF front-ends were developed in this project, achieving data rates above 100 Gb/s and bringing the project to success. This chapter presents these RF front-ends and the wireless communication links established using them.

32.1 SISO Wireless Links at Near-THz Frequencies

Despite the improvement in the SiGe process technology, the design of such system remains very challenging. Circuit performance still deteriorates when the target frequencies are above 200 GHz. This leads not only to lower power efficiency and poorer link budgets, but it also introduces other impairments in the link [1], such as

P. Rodríguez-Vázquez · J. Grzyb · U. R. Pfeiffer (✉)
Bergische Universität Wuppertal, Wuppertal, Germany
e-mail: ullrich.pfeiffer@uni-wuppertal.de

T. Kürner et al. (eds.), *THz Communications*, Springer Series in Optical Sciences 234, https://doi.org/10.1007/978-3-030-73738-2_32

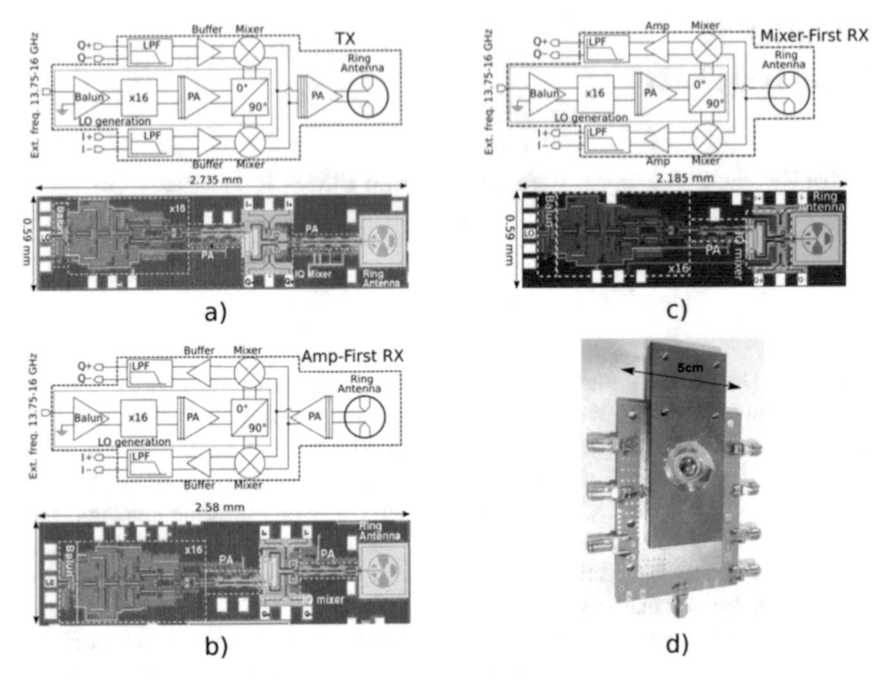

Fig. 32.1 Block diagrams and chip micrographs of (**a**) Tx, (**b**) amplifier-first Rx, and (**c**) mixer-first Rx. An example of a fully packaged RF front-end module is illustrated in (**d**). (From [3] © 2017 IEEE & [4] © 2018 IEEE; reproduced with permission)

sideband asymmetry. The sideband symmetry is a critical specification for achieving high data rates. Any asymmetry will cause the spectrum of the I channel to leak into the Q channel and vice-versa [2]. The symmetry condition is very hard to achieve, particularly for amplifier stages with a limited gain-BW product. This section presents two different single-input single-output (SISO) links established with the first generation of RF front-ends developed within the Real100G.RF project. In the first one, a typical amplifier first receiver is used, optimized for high conversion gain (CG) and low-noise figure (NF). In the second one, the receiver implements a mixer-first architecture [3], improving the sideband symmetry but worsening the CG and NF. The block diagrams of the RF modules, as well as the chip micrographs and a picture of one of the fully packaged RF modules, are displayed in Fig. 32.1.

The local oscillator (LO) generation path is common in all chips. It consists of four cascaded Gilbert-cell doublers followed by a three-stage power amplifier (PA) to provide enough signal power to drive the mixers. The quadrature is provided by a broadband hybrid coupler. In the Tx, the data streams at the I and Q channels are up-converted by a Gilbert-cell mixer, which is followed by a four-stage PA. In the amplifier-first Rx, the received signal is preamplified by a three-stage PA before it is down-converted by a Gilbert-cell mixer. All chips included a broadband on-chip ring antenna that radiates through the substrate into a hyper-hemispherical

9-mm high-resistivity silicon lens. Following an inexpensive chip-on-board (CoB) packaging scheme, the chip-on-lens assembly was placed in the recess of a Rogers 4350B PCB and wire-bonded. An eight-section step-impedance low-pass filter is implemented on the PCB to compensate for the inductance of the wire-bonds, offering a maximum baseband BW of 14 GHz. The antenna directivity of these modules increases monotonically from 25 to 27 dBi in the 200–280 GHz band.

At 230 GHz, the Tx shows a 3-dB BW of 28 GHz, a saturated output power of 8.5 dBm, and an IQ amplitude imbalance of 0.7 dB, while it consumes a power of 0.96 W. At the same carrier frequency, the amplifier-first Rx has a 3-dB RF BW of 23 GHz, a CG of 23 dB, an IQ imbalance below 0.5 dB, and a minimum estimated single-sideband (SSB) NF of 10 dB and dissipates a power of 1 W. The mixer-first Rx shows a 3-dB RF BW of 26 GHz, a CG of 8 dB with an IQ imbalance under 1 dB, and a minimum estimated SSB NF of 14 dB and consumes 450 mW. Contrary to the Tx and the amplifier-first Rx, the mixer-first Rx holds its performance for any carrier frequency ranging from 220 to 255 GHz [3].

For the wireless communication setup, the Tx and one of the Rxs were placed in a 1-m line-of-sight (LOS) setup. To generate the IQ data streams for the Tx, a set of two AWG70001A arbitrary waveform generators (AWG) were used, while the received signal was recorded using two DPO77002SX oscilloscopes and then analyzed in a vector signal analysis software (SignalVu). A root-raised cosine filter with a roll-off factor between 0.2 and 0.7 was applied to the BB signals to fit into the BW of the RF modules. In the link using the amplifier-first Rx, a maximum data rate of 90 Gb/s was achieved using 16-QAM modulation formats with an EVM of 15% [4, 5], while with a maximum modulation order of 64 QAM, the maximum achieved data rate was 81 Gb/s and an EVM of 8.1% [1]. For link using the mixer-first Rx, the maximum achieved data rate was 100 Gb/s with an EVM of 17% using a 16-QAM modulation [2], thanks to the mixer-first Rx RF improved BW. These results were achieved at a carrier frequency of 230 GHz, where the Tx shows the better sideband symmetry.

The maximum modulation order was found to be limited by the broadband phase noise (PN) floor. Here, the rms phase error in the LO path scales linearly with the modulation BW. The on-chip x16 LO up-conversion path was measured to scale the PN of the external driving PSG E8257D synthesizer, with a noise floor of -150 dBc/Hz, nearly ideal by a factor of 24.1 dB. The total rms phase error introduced by the up-converted double-sideband modulated signals with a BB BW of 14 GHz is 4°. This error prevents the system to operate with a modulation order higher than 64 QAM [1, 6].

32.2 MIMO Wireless Link at Near-THz Frequencies

Increases in the data rate are traditionally approached in three ways, broader bandwidth, higher modulation order, or higher number of channels. As was previously discussed, the gain-BW product for silicon circuits at this frequency band limits the first approach and the PN performance the second one. Therefore, the only remaining possibility is to increase the number of channels using a multiple-input multiple-output (MIMO) system. In this section, a MIMO link established using a pair of highly integrated single-chip dual-polarization (DP) Tx and Rx RF front-end modules is presented [7]. Both Tx and Rx chips accommodate two independent fundamentally operated direct-conversion IQ paths with separately tunable on-chip multiplier-based LO generation paths. A block diagram of the RF modules used for this link and the chip micrographs are depicted in Fig. 32.2.

The LO generation path is equivalent to the one employed for the single-polarization chips. The IQ up-conversion and down-conversion generation paths are similar to the previously presented Tx and mixer-first Rx, respectively. However, due to the limitations in the number of available lab equipment quantity to simultaneously drive four high-speed differential baseband signals on each link side, the Rx baseband outputs and Tx inputs needed to be operated as single-ended with a degraded balance in the output/input amplification stage resulting in a deteriorated signal-to-noise ratio (SNR) as well as an increased frequency dispersion in the operation bandwidth. Each of these modulated IQ data streams is then radiated as an orthogonal circular polarization, left-handed or right-handed circular polarization (LHCP and RHCP), with an isolation between them better than 18 dB for any relative orientation between the Tx and the Rx. This isolation allows for the link operation without the need for a high-speed depolarizer in the baseband.

The measurement setup is similar to the one presented in the previous session, adding two extra AWGs. This system allows for the transmission of two independent QPSK channels with an EVM of 32% at 230 GHz with an aggregated speed of 110 Gb/s and 80 Gb/s over 1-m and 2-m distance, respectively.

Table 32.1 compares all wireless links presented in this chapter.

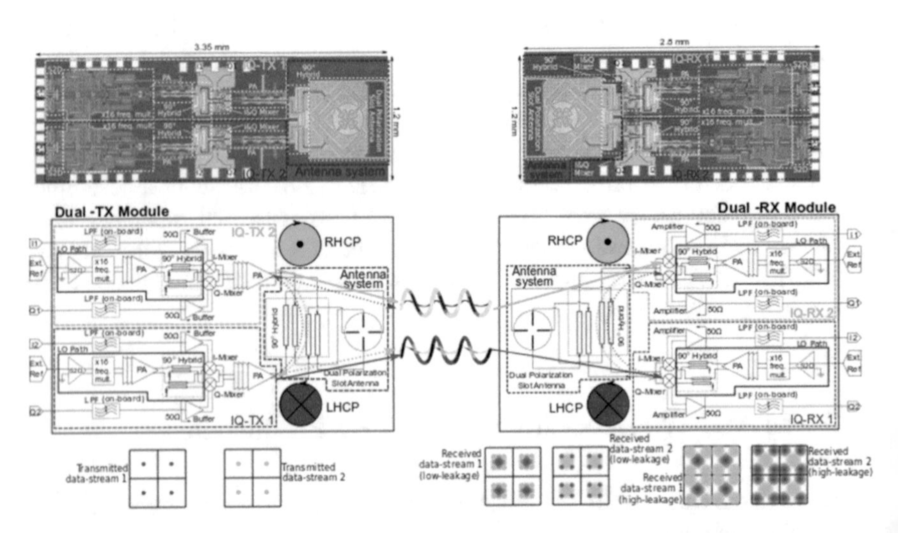

Fig. 32.2 Block diagram and chip micrographs of the DP Tx and the DP Rx. Distortions in the ideal circular polarization will lead into a leakage between the two data streams, increasing the EVM. (From Ref. [7])

Table 32.1 Summary of the near-THz wireless links developed by the BUW

Link	LO Freq (GHz)	# of IQ channels	Modulation	Distance (meters)	Data rate (Gb/s)	EVM_{rms} (%)	P_{DC} (W)
Tx + Amp-first Rx	225–255	1	16/64-QAM	1	90/81	15/8.5%	1.96
Tx + Mixer-first Rx	225–255	1	16 QAM	1	100	17%	1.41
DP Tx + DP Rx	225–255	2	QPSK	1/2	110/80	32%	2.85

References

1. Rodríguez-Vázquez, P., Grzyb, J., Heinemann, B., & Pfeiffer, U. R. (2019). *Optimization and performance limits of a 64-QAM wireless communication link at 220–260 GHz in a SiGe HBT technology*. IEEE Radio and Wireless Symposium (RWS), Orlando, FL, USA.
2. Rodríguez-Vázquez, P., Grzyb, J., Heinemann, B., & Pfeiffer, U. R. (2019). A 16-QAM 100-Gb/s 1-M wireless link with an EVM of 17% at 230 GHz in an SiGe technology. *IEEE Microwave and Wireless Components Letters*, 297–299.
3. Rodríguez-Vázquez, P., Grzyb, J., Sarmah, N., Heinemann, B., & Pfeiffer, U. R. (2017). *A 219–266 GHz fully-integrated direct-conversion IQ receiver module in a SiGe HBT technology*. 12th European Microwave Integrated Circuits Conference (EuMIC), Nuremberg, Germany.
4. Rodríguez-Vázquez, P., Grzyb, J., Sarmah, N., Heinemann, B., & Pfeiffer, U. R. (2018). *Towards 100 Gbps: A fully electronic 90 Gbps one meter wireless link at 230 GHz*. 48th European Microwave Conference (EuMC), Madrid, Spain.
5. Rodríguez-Vázquez, P., Grzyb, J., Heinemann, B., & Pfeiffer, U. R. (2018). *Performance evaluation of a 32-QAM 1-meter wireless link operating at 220–260 GHz with a data-rate of 90 Gbps*. Asia-Pacific Microwave Conference (APMC), Kyoto, Japan.
6. Grzyb, J., Rodríguez-Vázquez, P., Sarmah, N., Heinemann, B., & Pfeiffer, U. R. (2018). *Performance evaluation of a 220–260 GHz LO tunable BPSK/QPSK wireless link in SiGe HBT technology*. 48th European Microwave Conference (EuMC), Madrid, Spain.
7. Rodríguez-Vázquez, P., Grzyb, J., Heinemann, B., & Pfeiffer, U. R. (2020). A QPSK 110-Gb/s polarization-diversity MIMO wireless link with a 220–255 GHz tunable LO in a SiGe HBT technology. *IEEE Transactions on Microwave Theory and Techniques*, 3834–3851.

Chapter 33
TERAPAN: A 300 GHz Fixed Wireless Link Based on InGaAs Transmit-Receive MMICs

Iulia Dan

Abstract This chapter highlights a project which is based on a wireless communication point-to-point link operated in the low-terahertz range, at a center frequency of 300 GHz. The link is composed of all electronic components based on monolithic integrated circuits based on a metamorphic high-electron-mobility transistor technology. The project started in August 2013 and ended in December 2016, and it represents an important milestone in the integration of terahertz communication in industry-ready applications.

33.1 Motivation

The exponential increase in the wireless data demands new technological and innovative solutions to achieve the desired data speeds. [1] shows that the curve of maximum available data rates over time has the same exponential increase as Moore's law for semiconductors. The consortium of TERAPAN project consisting of three partners, Technische Universität Braunschweig (TUBS), Fraunhofer Institute of Applied Sciences (IAF), and the University of Stuttgart (USTUTT), has recognized early the potential of submillimeter wave-based wireless communication in future and beyond 5G networks. This project, which started in August 2013, represents one of the first efforts worldwide of integrating 300 GHz-based solid-state electronics into a communication link, in particular in a wireless personal area network (WPAN). In the meantime, the number of 300 GHz wireless links has increased significantly, many of these links showing a better performance in terms of data rates, transmission distance, transmission bandwidth, and signal processing capabilities. Nonetheless, this project represents an important milestone in the development of terahertz (THz) communication and its integration into industry-ready devices and components.

I. Dan (✉)
Qorvo Munich GmbH, Munich, Germany

Universität Stuttgart, Stuttgart, Germany

T. Kürner et al. (eds.), *THz Communications*, Springer Series in Optical Sciences 234, https://doi.org/10.1007/978-3-030-73738-2_33

33.2 Project and Demonstrator Description

The goal of the project was to show that a wireless link based on a transmit-receive system centered around 300 GHz can achieve data rates above 100 Gbit/s over a midrange indoor distance, between 1 and 10 m. An important and very ambitious aspect of this goal involved the demonstration of electrical beam steering using the proposed communication link.

Figure 33.1 shows both the setup of a wireless transmission and all the important devices and components that were developed in the frame of the project. At the core of the radio link are the transmitter and receiver, with an operation frequency centered around 300 GHz. Other crucial components of the link are the 300 GHz antennas. In comparison to radio links at lower frequencies deployed in commercial devices like phones and radios, the TERAPAN system uses a zero-IF also called direct conversion architecture. This means that the baseband signal is not up-converted to an intermediate frequency (IF), but directly connected to the mixer of the transmitter and receiver. This signal is generated using state-of-the-art analog-to-digital and digital-to-analog converters and has bandwidths of up to 25 GHz. The direct conversion approach allows for a large IF bandwidth and together with quadrature up- and down-conversion is also capable of transmitting bandwidth-efficient complex modulated data signals. The resulting RF transmitted signal is centered around 300 GHz and has bandwidths of up to 50 GHz.

The 300 GHz transmitter and receiver are based on millimeter-wave monolithic integrated circuits (MMICs) designed using a metamorphic high-electron-mobility transistor (mHEMT) technology. This technology is presented in detail in [2]. The MMICs are packaged in split-block waveguide modules, like the one presented in Fig. 33.2. V connectors and a liquid crystal polymer (LCP) substrate provide the MMICs with the in-phase (I) and quadrature (Q) baseband data signals. The LO signal at 100 GHz and the RF signal centered around 300 GHz use a rectangular waveguide interface to connect to the LO source, respectively, the antenna. Different

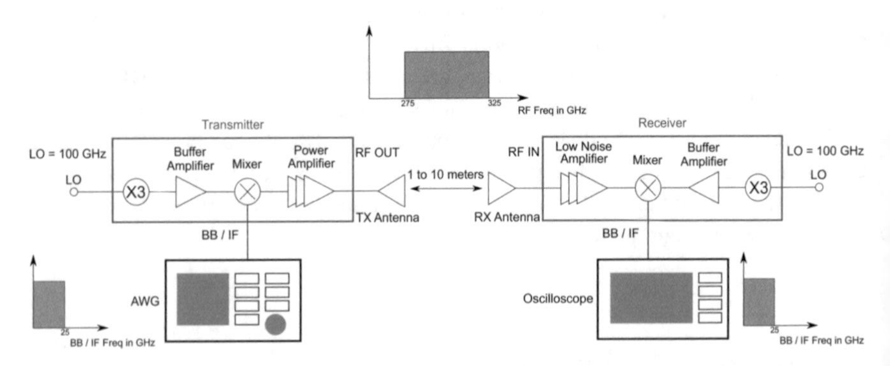

Fig. 33.1 Simplified schematic of TERAPAN wireless link setup using 300 GHz electronic transmitter and receiver

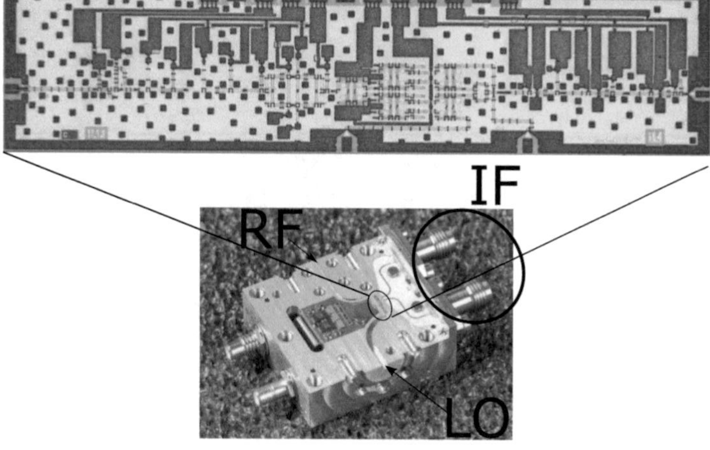

Fig. 33.2 Photograph of a split block waveguide module used in TERAPAN experiments. (Photo credits: Fraunhofer IAF)

generations of MMIC chip sets were designed in the frame of the project. Details and measurement results of these chipsets can be found in [3–5].

The project focused on two main demonstrations. The first one is similar to the generic case pictured in Fig. 33.1 and involves wireless data transmission from one transmitter to one receiver and is referred to as the single-input, single-output (SISO) demonstrator. The transmission is carried out over 1 m. With these configurations, data rates up to 64 Gbit/s can be transmitted and demodulated using off-line signal processing tools. Figure 33.3 shows the constellation diagram of the transmitted QPSK signal with a symbol rate of 32 Gbd, which corresponds to the data rate of 64 Gbit/s. Simple modulation formats like BPSK and QPSK are transmitted in this demonstration. [6–8] present in detail measurement results and deeper impairment analysis for this demonstration.

For the second main demonstration, a single-input, multiple-output (SIMO) concept was followed. For this purpose, the signal coming out of one transmitter is simultaneously received by a four-channel module. The 4 × 1 phased array antenna developed for this demonstration is presented in detail in [9] and in [10]. On the right-hand side of Fig. 33.4, a photograph of the four-channel receiver module and phased array antenna can be seen. Using this SIMO concept, electronic beam steering has been proven for the first time at 300 GHz in a live demonstration presented at the NGMN industrial exhibition in 2016. The achieved data rate was 12 Gbit/s and the transmission distance 60 cm.

The TERAPAN transmission links proved that 300 GHz indoor wireless communication is possible and can be integrated in applications like smart offices or data proximity showers.

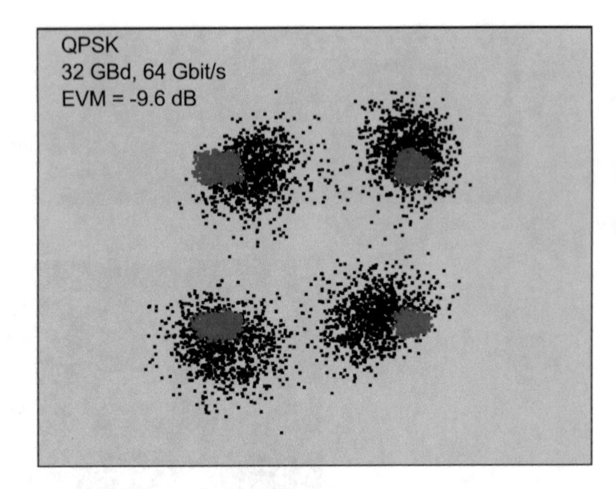

Fig. 33.3 One of the measurement results of the TERAPAN SISO demonstrator. In black are the measurement results of the 1 m transmission and in red the measurement of the reference signal transmitted directly from the signal generator to the oscilloscope

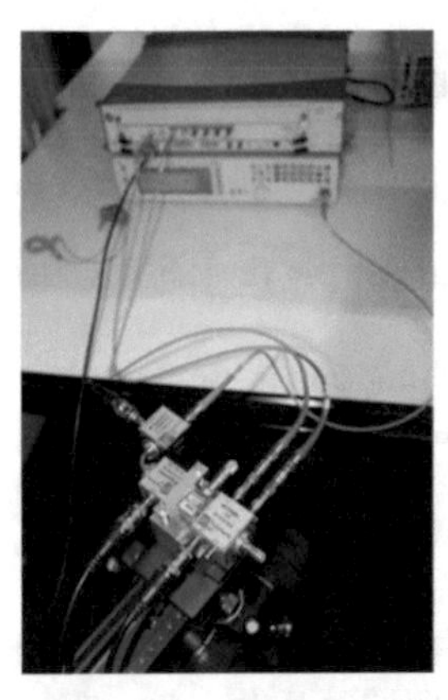

Fig. 33.4 Photographs of the modules used in the TERAPAN demonstrations. On the right-hand side the four-channel receiver module and on the left-hand side the one-channel transmitter module. (Photo credits: S. Rey)

References

1. Fettweis, G., & Alamouti, S. (2014). 5G: Personal mobile internet beyond what cellular did to telephony. *IEEE Communications Magazine, 52*(2), 140–152.
2. Leuther, A., Tessman, A., Massler, H., Losch, R., Schlechtweg, M., Mikulla, M., & Ambacher, O. (2008). 35nm metamorphic HEMT MMIC technology. In *Indium Phosphide and Related Materials*.
3. Antes, J., Lewark, U., Tessmann, A., Leuther, A., Wagner, S., & Kallfass, I. (2014). *Ultra-wideband single-balanced transmitter MMIC for future 300 GHz communication systems*. Proceedings of IEEE International Microwave Symposium (IMS2014), Tampa, FL, USA.
4. Kallfass, I., Harati, P., Dan, I., Antes, J., Boes, F., Rey, S., Merkle, T., Wagner, S., Tessmann, A., & Leuther, A. (2015). *MMIC chipset for 300 GHz indoor wireless communication*. IEEE international conference on Microwaves, Communications, Antennas and Electronic Systems (COMCAS 2015), Tel Aviv, Israel.

5. Dan, I., Schoch, B., Eren, G., Wagner, S., Leuther, A., & Kallfass, I. (2017). A 300 GHz MMIC-based quadrature receiver for wireless terahertz communications. In *IRMMW-THz*.
6. Dan, I., Rey, S., Merkle, T., Kürner, T., & Kallfass, I. (2017). *Impact of modulation type and baud rate on a 300GHz fixed wireless link*. 2017 IEEE Radio and Wireless Symposium (RWS), Phoenix.
7. Dan, I., Rosello, E., Harati, P., Dilek, S., Shiba, S., & Kallfass, I. (2017). *Measurement of complex transfer function of analog transmit-receive frontends for terahertz wireless communications*. 2017 47th European Microwave Conference (EuMC), Nuremberg, Germany.
8. Dan, I., Grotsch, C. M., Shiba, S., & Kallfass, I. (2017). *Investigation of local oscillator isolation in a 300 GHz wireless link*. 2017 IEEE international conference on Microwaves, Antennas, Communications and Electronic Systems (COMCAS), Tel Aviv, Israel.
9. Rey, S., Merkle, T., Tessmann, A., & Kürner, T. (2016). *A phased array antenna with horn elements for 300 GHz communications*. International Symposium on Antennas and Propagation (ISAP), Okinawa, Japan.
10. Rey, S., Ulm, D., Kleine-Ostmann, T., & Kuerner, T. (2017). *Performance evaluation of a first phased array operating at 300 GHz with horn elements*. 2017 11th European Conference on Antennas and Propagation (EUCAP), Paris, France.

Chapter 34
ThoR

Tetsuya Kawanishi and Shintaro Hisatake

Abstract ThoR (TeraHertz end-to-end wireless systems supporting ultra-high data Rate applications) is a joint EU-Japan project to provide technical solutions for the data networks beyond 5G based on 300 GHz RF wireless links. ThoR has a clear focus on the integration and operation of the first THz link in a real network environment. We propose to combine all-European and Japanese state-of-the-art photonic and electronic technologies, leveraging state-of-the-art gigabit modems and channel aggregation in the standardized 60 GHz and 70–80 GHz bands in order to demonstrate real-time THz wireless communications with a capacity × distance figure of merit larger than 40 Gbps × km.

Figure 34.1 shows ThoR's concept. ThoR's technical concept builds on the IEEE 802.15.3d standard [1] using state-of-the-art chip sets and modems operating in the standardized 60 and 70–80 GHz bands, which are aggregated on a bit-transparent high performance 300 GHz RF wireless link offering >100 Gbps real-time data rate capacity. The key enabling technologies are (1) high linearity, wideband, and high spectral purity photonic local oscillator (LO) generation, (2) medium power 300 GHz solid-state power amplifier, (3) high power 300 GHz traveling-wave tube amplifier, (4) multifunctional wideband and low noise 300 GHz solid-state receiver, (5) multifunctional and high linearity 300 GHz solid-state upconverter, and (6) channel aggregation at the 60 GHz and 70–80 GHz bands.

We are in the process of developing the key enabling technologies and, at the same time, validating the concept. In the proof-of-concept, data transmissions with various modulation formats are conducted using an AWG, before adopting modems operating in the standardized 60 and 70–80 GHz bands. The RF MMIC used in

T. Kawanishi (✉)
Waseda University, Shinjuku, Tokyo, Japan
e-mail: kawanishi@waseda.jp

S. Hisatake
Gifu University, Gifu, Japan
e-mail: hisatake@gifu-u.ac.jp

T. Kürner et al. (eds.), *THz Communications*, Springer Series in Optical Sciences 234, https://doi.org/10.1007/978-3-030-73738-2_34

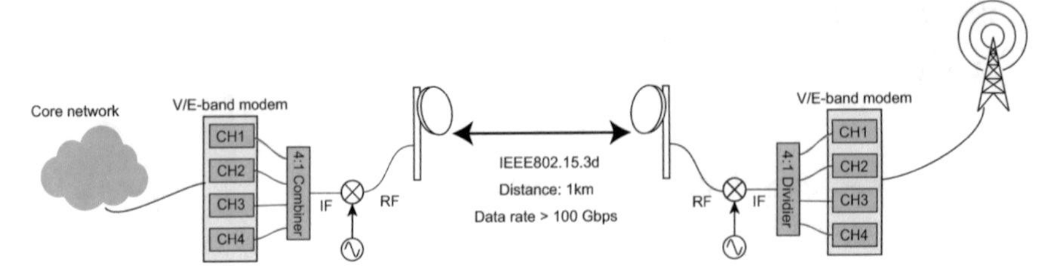

Fig. 34.1 ThoR concept

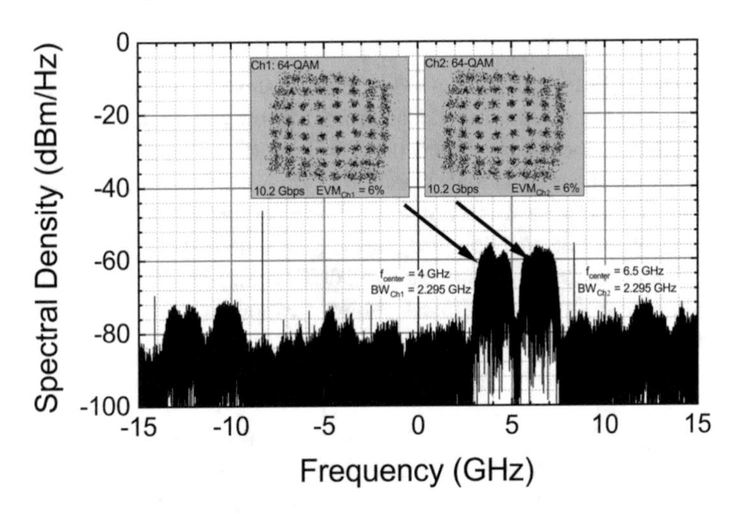

Fig. 34.2 Results of the channel aggression experiment

this proof-of-concept are fabricated in a 35 nm mHEMT InGaAs technology [2]. The highest data rate of 56 Gbps at 10 m transmission was achieved for the single-channel transmission with a 16-QAM [3]. The EVM was 12.2%. Figure 34.2 shows an example of the channel aggression result [4]. The transmission distance was 1 m. A channel data rate of 10.2 Gbps (total data rate of 20.4 Gbps) was achieved with the channel bandwidth of 2.295 GHz which is compatible to the new IEEE standard. The EVM for both channels was 6%. The current data rate is limited by the linearity of the system, which will be improved by a module currently under development. The successful transmission of aggregated channels shows the potential of ThoR concept.

References

1. IEEE Standard for High Data Rate Wireless Multi-Media Networks Amendment 2: 100 Gbps Wireless Switched Point-to-Point Physical Layer, IEEE-SA Standards Board Std.

2. Kallfass, I., Harati, P., Dan, I., Antes, J., Boes, F., Rey, S., Merkle, T., Wagner, S., Massler, H., Tessmann, A., & Leuther, A. (2015). MMIC chipset for 300 GHz indoor wireless communication. In *2015 IEEE International Conference on Microwaves, Communications, Antennas and Electronic Systems (COMCAS)* (pp. 1–4)
3. Dan, I., Ducournau, G., Hisatake, S., Szriftgiser, P., Braun, R. P., & Kallfass, I. (2019). A superheterodyne 300 GHz wireless link for ultra-fast terahertz communication systems. *International Journal of Microwave and Wireless Technologies, 12*(7), 578–587.
4. Dan, I., Ducournau, G., Hisatake, S., Szriftgiser, P., Braun, R. P., & Kallfass, I. (2019). A terahertz wireless communication link using a superheterodyne approach. *IEEE Transactions on Terahertz Science and Technology, 10*(1), 32–43.

Chapter 35
Terranova

Carlos Castro, Robert Elschner, Thomas Merkle, Francisco Rodrigues, José Machado, António Teixeira, and Colja Schubert

Abstract The available large and contiguous bandwidth at 300 GHz offers the opportunity to interconnect coherent THz wireless and fiber-optic networks using a slim analog baseband interface. The resulting hybrid fiber-optic/THz wireless links can be end-to-end (E2E) operated by fiber-optic real-time modems without the need of an additional DSP at the optical wireless analog baseband interface. This chapter summarizes the test beds and experiments that were conducted within the Terranova project to validate the key ideas of coherent THz wireless/fiber-optic links.

35.1 Beyond 5G Heterogeneous Network Architectures

The project Terranova envisions a heterogeneous, highly flexible fiber-optic/THz wireless network architecture to be a key enabling technology to implement beyond 5G networks at minimum cost and size, compatible to the already existing fiber-optic network infrastructure. THz wireless communications above 100 GHz are expected to bring wireless data rates on a par with fiber-optic systems by offering link throughputs in the range of hundreds of Gb/s or even Tb/s [1]. Terranova leverages on the digital-coherent fiber-optic transmission infrastructure with its highly advanced digital signal processing (DSP), enabling transmission at high symbol rates, high spectral efficiency, and elaborated impairment correction. The

C. Castro · R. Elschner · C. Schubert
Fraunhofer Heinrich Hertz Institute, Berlin, Germany
e-mail: robert.elschner@hhi.fraunhofer.de

T. Merkle (✉)
Fraunhofer Institute for Applied Solid State Physics, Freiburg, Germany
e-mail: thomas.merkle@iaf.fraunhofer.de

F. Rodrigues · A. Teixeira
PICadvanced S.A., Aveiro, Portugal

J. Machado
Altice Labs, Aveiro, Portugal

T. Kürner et al. (eds.), *THz Communications*, Springer Series in Optical Sciences 234, https://doi.org/10.1007/978-3-030-73738-2_35

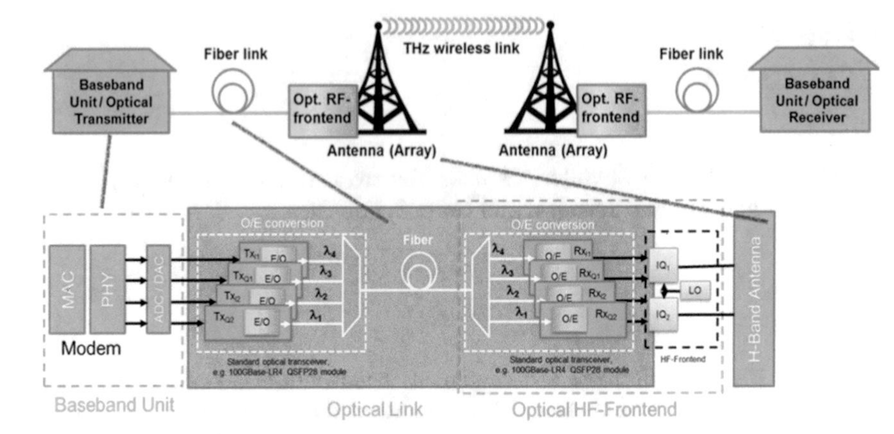

Fig. 35.1 The Terranova hybrid fiber-optic/THz wireless concept with E2E digital signal processing

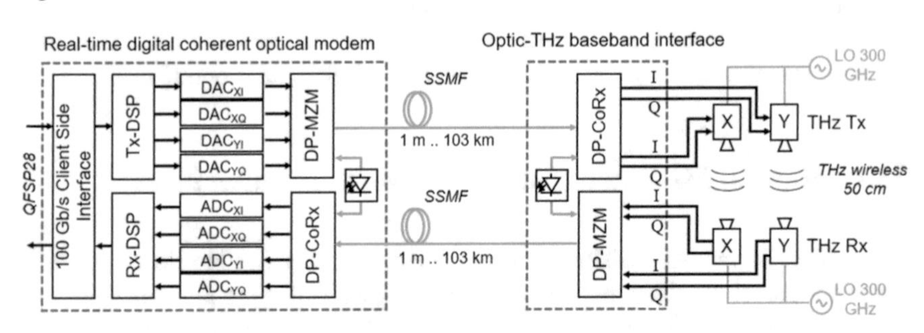

Fig. 35.2 Block diagram of the experimental setup for the investigation of real-time transmission over a fiber-optic/THz wireless/fiber-optic combined link with E2E digital signal processing

corresponding architecture is schematically depicted at link level in Figs. 35.1 and 35.2. While the idea is straightforward, it cannot be taken for granted that the E2E correction of the combined fiber-optic/THz wireless channel achieves the required stability and transparency under practically relevant conditions. Thus, Terranova was dedicated to develop real-time experiments, test beds, networks, and required hardware not only for the proof of concept in the lab [2–8] but also for the validation using different outdoor THz wireless links with up to 1 km distance [9].

35.2 Coherent THz Wireless Fiber Extenders

The concept of a hybrid fiber-optic/THz wireless link evolved step-by-step from the lab test bench [2, 3] to the implementation of an outdoor long-range link [9], as well as from a system using offline DSP to a real-time demonstrator [4, 5]. This includes also an experimental investigation of a 100G Ethernet transmission over

Fig. 35.3 Photograph of the implemented lab demonstrator corresponding to the schematic of Fig. 35.2

a THz wireless link using a real-time modem, showing a stable Ethernet frame throughput of up to 98 Gb/s with low frame loss rate and latency [6, 7]. With a similar system, the digital-coherent real-time transmission of a 100 Gb/s dual-polarized QPSK signal over two fiber-optic spans (up to 103 km total length), which were interconnected by a THz wireless fiber extender at 300 GHz carrier frequency, could be demonstrated for the first time, allowing to showcase an end-to-end video transmission over the hybrid fiber-optic/THz wireless link [6, 7]. Figure 35.3 shows a photograph of the employed test bed. A network-level demonstration with the THz wireless link operating in the 100G downlink of a commercial fiber-optic PON OLT is depicted in Fig. 35.4. Various solutions for the realization of the analog optical/wireless baseband interface were investigated, employing analog coherent optical as well as analog IM/DD optical transceiver front-ends [8]. While the use of analog coherent optical transceiver front-ends would require the least modifications of today's commercially available components, customized IM/DD optical transceiver front-ends might be an attractive approach in the future because of their simplicity [8]. The outdoor experiments showed also excellent agreement of the measured and modeled received THz power under varying weather conditions for dual-polarized QPSK transmission at 32 GBd over a THz wireless link section of 500 m [9], achieving a net data rate of 102 Gb/s.

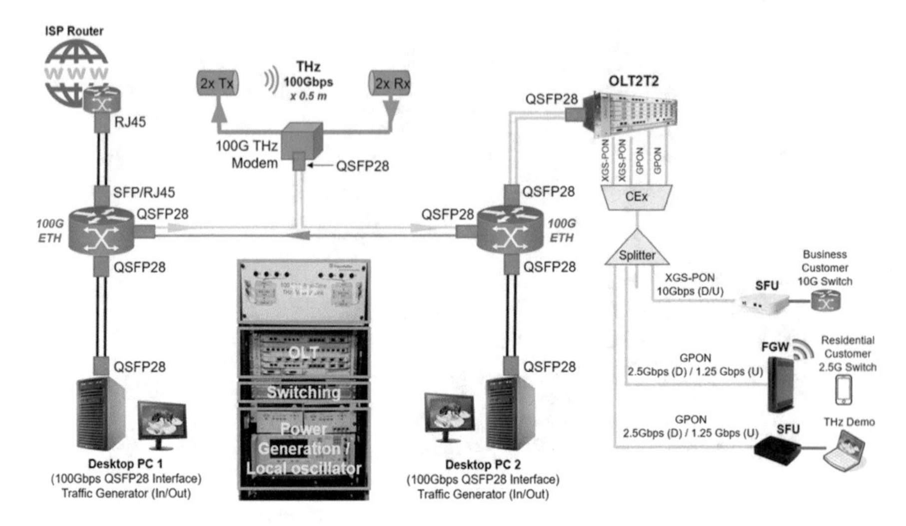

Fig. 35.4 Network-level demonstration with the high-speed THz wireless link operating in the 100G downlink of a commercial fiber-optic PON OLT

Acknowledgments This work was supported by the Fraunhofer Internal Programs under Grant No. MAVO 836 966 and by the EC Horizon 2020 Research and Innovation Program under Grant Agreement No. 761794.

References

1. Boulogerogos, A.-A., Alexiou, A., Merkle, T., Schubert, C., Elschner, R., et al. (2018). Terahertz technologies to deliver optical network quality of experience in wireless systems beyond 5G. *IEEE Communications Magazine, 56*(6), 144–151.
2. Castro, C., Elschner, R., Merkle, T., & Schubert, C. (2019). *100 Gbit/s terahertz-wireless real-time transmission using a broadband digital-coherent modem*. IEEE 5G World Forum, 30.09. – 02.10.2019, Dresden, Germany.
3. Nellen, S., Elschner, R., Sackey, I., Emmerich, R., Merkle, T., Globisch, B., de Felipe, D., & Schubert, C. (2019). *32 GBd 16QAM wireless transmission in the 300 GHz band using a PIN diode for THz upconversion*. Proceedings of OFC, paper: M4F.5, San Diego.
4. Castro, C., Elschner, R., & Schubert, C. (2019). *Analysis of joint impairment mitigation in a hybrid optic-THz transmission system*. 20th IEEE international workshop on Signal Processing Advances in Wireless Communications, Cannes.
5. Castro, C., Elschner, R., Merkle, T., & Schubert, C. (2019). *Experimental validation of coherent DSP for combined fibre-optical/terahertz wireless links*. 16th international symposium on Wireless Communication Systems (ISWCS), 27–30 August 2019, Oulu, Finland.
6. Castro, C., Elschner, R., Machado, J., Merkle, T., Schubert, C., & Freund, R. (2019). *Ethernet transmission over a 100 Gb/s real-time terahertz wireless link*. IEEE global communications conference, 9–13 December 2019, Waikoloa, HI, USA.
7. Castro, C., Elschner, R., Merkle, T., Schubert, C., & Freund, R. (2020). *100 Gb/s real-time transmission over a THz wireless fiber extender using a digital-coherent optical modem*. The optical networking and communication conference & exhibition (OFC), 8–12 March 2020, San Diego.

8. Rodrigues, F., Ferreira, R., Castro, C., Elschner, R., Merkle, T., Schubert, C., & Teixeira, A. (2020). *Hybrid fiber-optical/THz-wireless link transmission using low-cost IM/DD optics*. The Optical Networking and Communication Conference & Exhibition (OFC), 8–12 March 2020, San Diego.
9. Castro, C., Elschner, R., Merkle, T., Schubert, C., & Freund, R. (2020). Experimental demonstrations of high-capacity THz-wireless transmission systems for beyond 5G. *IEEE Communications Magazine, 58*(11), 41–47.

Chapter 36
ULTRAWAVE

Claudio Paoloni

Abstract This chapter describes the concept and achievements of the European Commission Horizon 2020 ULTRAWAVE "Ultra capacity wireless layer beyond 100 GHz based on millimeter wave Traveling Wave Tubes" aiming to produce the first D-band point-to-multipoint wireless system for backhaul of high-density cells.

The European Commission Horizon 2020 ULTRAWAVE "Ultra capacity wireless layer beyond 100 GHz based on millimeter wave Traveling Wave Tubes" project pushes the boundary of millimeter-wave and sub-THz technology to enable a novel Internet distribution paradigm by ultracapacity layers at roof level, to solve the high-density small cell backhaul [1]. ULTRAWAVE is the first backhaul architecture combining point-to-multipoint (PmP) distribution at D-band (141–148.5 GHz) connected by point-to-point (PtP) G-band transport links (275–305 GHz) (Fig. 36.1).

The ULTRAWAVE consortium is developing the full wireless system from C-band to G-band, including a transmission hub (schematic in Fig. 36.2) and terminals at D-band and a transmitter and receiver at G-band [3]. It is enabled by novel traveling wave tubes (TWTs) to provide about 10–12 watt saturated power at D-band and 1–3 watt saturated power at G-band [2] (see Chap. 27 for details on TWTs). The TWTs are in fabrication phase.

A transmission hub generates D-band sectors that can use up to 7.5 GHz of bandwidth to deliver up to 30 Gb/s at 64 QAM, over 250 MHz channels. A single sector, assuming 30-degree aperture angle and 600 m radius, could provide up to 300 Gb/s/km^2 area capacity. Terminals at D-band are arbitrarily allocated over the sector to flexibly receive the number of channels to satisfy the related traffic requirements. The D-band transmission hubs will be deployed to cover a wide area connected by a mesh network of G-band PtP links up to 600 m long with up 30 Gb/sec at QSPK.

C. Paoloni (✉)
Engineering Department, Lancaster University, Lancaster, UK
e-mail: c.paoloni@lancaster.ac.uk

T. Kürner et al. (eds.), *THz Communications*, Springer Series in Optical Sciences 234, https://doi.org/10.1007/978-3-030-73738-2_36

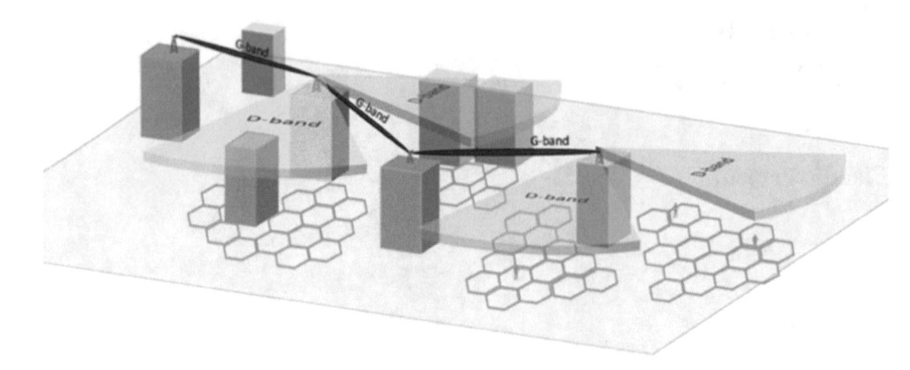

Fig. 36.1 ULTRAWAVE concept

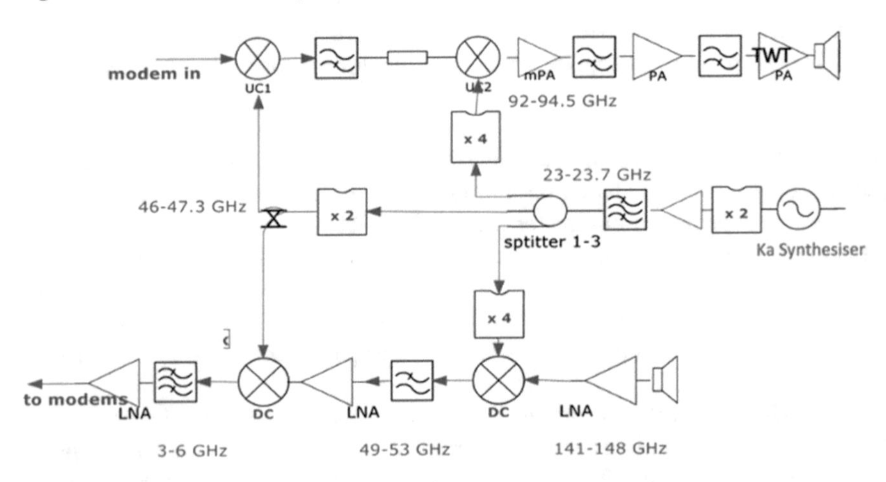

Fig. 36.2 Schematic of the D-band transmission hub [3]

The D-band MMIC chipset includes InP DHBT high-power (19 dBm) and medium-power amplifiers (better than 10 dBm), a GaAs low-noise amplifier (about 5 dB noise figure), and a GaAs 4x multiplier, upconverter, and downconverter FET mixers. The D-band LNA has 5 dB noise figure measured. The D-band MMIC chipset is used for both the transmission hub and the terminals with benefit for the final cost of the system. Local oscillator rejection filter, low-gain horn antennas for the transmission hub, and high-gain lens antennas for terminals were also designed, fabricated, and tested.

The transmission hub (Fig. 36.2) radio includes a transmitter and a receiver at D-band allocated on a single motherboard. It uses two levels of transposition between the transmitted signal at D-band and the modem at C-band (3–6 GHz) to ensure good image rejection capability and use a single synthesizer. One level is a conversion with local oscillator (LO) frequency at W-band (92–94.5 GHz) and a second with LO at Q-band (46–47.3 GHz). A Ka-band synthesizer at 11.6 GHz provides a high-stability LO signal multiplied by 4 and 8 by a chain of dedicated frequency multipliers. An indoor unit hosts the stack of modems.

The G-band transmitter is based on an optical signal generator with transmission power provided by the G-band TWT, driven by an InP DHBT power amplifier, connected to a high-gain antenna (39 dBi). The G-band receiver uses the same antenna, a GaAs LNA (8 dB noise figure), and a downconverter to D-band.

The project will be concluded by a field test of the D-band PmP wireless system in real environment at the University Polytechnic of Valencia, Spain.

The ULTRAWAVE system can be configured with maximum flexibility. The number of sectors and the number of channels per sector depend on the capacity given in the coverage area and on the capacity of the G-band links. ULTRAWAVE aims to a breakthrough for high-capacity backhaul as affordable and alternative solution to fiber deployments.

The ULTRAWAVE consortium includes Lancaster University (Coordinator Claudio Paoloni, UK), Ferdinand Braun Institute, HF Systems Engineering GmbH, Goethe University Frankfurt (Germany), OMMIC (France), Fibernova System, Universitat Politecnica de Valencia (Spain), and the University of Rome Tor Vergata (Italy). The project is funded the by European Union's Horizon 2020 under Grant Agreement No. 762119 (Budget: €2.9 M) for the period September 2017–May 2021.

References

1. ULTRAWAVE website [Online]. Available: http://ultrawave2020.eu
2. Basu, R., et al. (2018). Design of sub-THz traveling wave tubes for high data rate long range wireless links. *Semiconductor Science and Technology, 33*, 124009.
3. Hossain, M., et al. (2021, January). *D-band transmission hub for point to MultiPoint wireless distribution*. European microwave conference 2020.

Chapter 37
MILLILINK

Ingmar Kallfass

Abstract After the pioneering work in Japan on 120 GHz point-to-point links, the MILLILINK project, in the years 2010 to 2013, was among the first larger cooperative efforts targeting the application-oriented development of THz communication links. Based on a transmit and receive analog frontend implemented in metamorphic HEMT technology, it saw the transmission of 64 Gbit/s wireless data rate over a transmission distance of 850 m in a fixed wireless link operating at a center frequency of 240 GHz.

MILLILINK, acronym for "millimeter-wave wireless links in optical communication networks," received funding from the German Ministry of Research and Education in the years 2010 to 2013. After the pioneering work at NTT in Japan, which saw a point-to-point transmission of 10 Gbit/s over a distance of 5.8 km at a carrier frequency of 120 GHz [2, 10] using InP HEMT-based transmit and receive frontends [6], the MILLILINK project was among the first larger cooperative efforts targeting the application-oriented development of THz fixed wireless links. Based on initial transmission experiments at 220 GHz [3], a new chip set for an analog transmit and receive frontend operating at a center frequency of 240 GHz was developed [8]. The MMICs, integrating subharmonic quadrature up- and down-converter stages with RF post- and pre-amplifiers in the transmitter and receiver, respectively, were based on 50 nm and later also on 35 nm metamorphic HEMT technology developed by project partner Fraunhofer Institute of Applied Solid State Physics, with cutoff frequencies of over 500 GHz f_T and over 1 THz f_{max} [7]. WR-3 split-block metallic waveguide packaging technology with integrated supply voltage generation was adopted to make the MMIC performance available as user-friendly modules for analog frontend integration on the system level. Due to the integrated LNA stage, the receiver module provided a conversion of 3 dB without any IF amplification within a 3 dB gain frequency range from 228 to 252 GHz, while

I. Kallfass (✉)
University of Stuttgart, Stuttgart, Germany
e-mail: ingmar.kallfass@ilh.uni-stuttgart.de

T. Kürner et al. (eds.), *THz Communications*, Springer Series in Optical Sciences 234, https://doi.org/10.1007/978-3-030-73738-2_37

its minimum noise figure was measured to approximately 9 dB. The transmitter module could be operated with an output power of -3.6 dBm per IF channel in an IF bandwidth from 0 to 35 GHz [9]. A local oscillator signal with 7 dBm of power at 120 GHz, provided by a frequency-multiplier, was employed to drive the quadrature mixer stages.

The MILLILINK transmission experiments involved the generation of pseudo-random bit sequences using wideband arbitrary waveform generators and different complex modulation formats on the transmit side and the digitization, equalization, and synchronization using real-time sampling oscilloscopes with sufficient bandwidth on the receive side. A zero-IF approach was adopted in order to maximize the modulation bandwidth. The receive signal quality was analyzed by offline digital signal analysis in terms of the error vector magnitude as well as the actual bit error rate.

When operated in conjunction with WR3 horn antennas and collimating lenses, the 240 GHz link could be operated over a transmission distance of 40 m with data rates of up to 96 Gbit/s using 8PSK modulation [1]. The MILLINK receiver module was also used in conjunction with a photonic transmitter based on photonic mixing in a UTC photodiode, where a record data rate of 100 Gbit/s with 16QAM modulation could be demonstrated in an indoor 20 m link setup at the Karlsruhe Institute of Technology [5]. Finally, in combination with 25 cm diameter Cassegrain-type parabolic antennas provided by project partner Radiometer Physics GmbH, which provided a gain of roughly 55 dBi at the receiver and the transmitter, the link could be sustained with up to 64 Gbit/s data rate using QPSK modulation over a distance of 850 m between two rooftops at the University of Stuttgart (Fig. 37.1) [4].

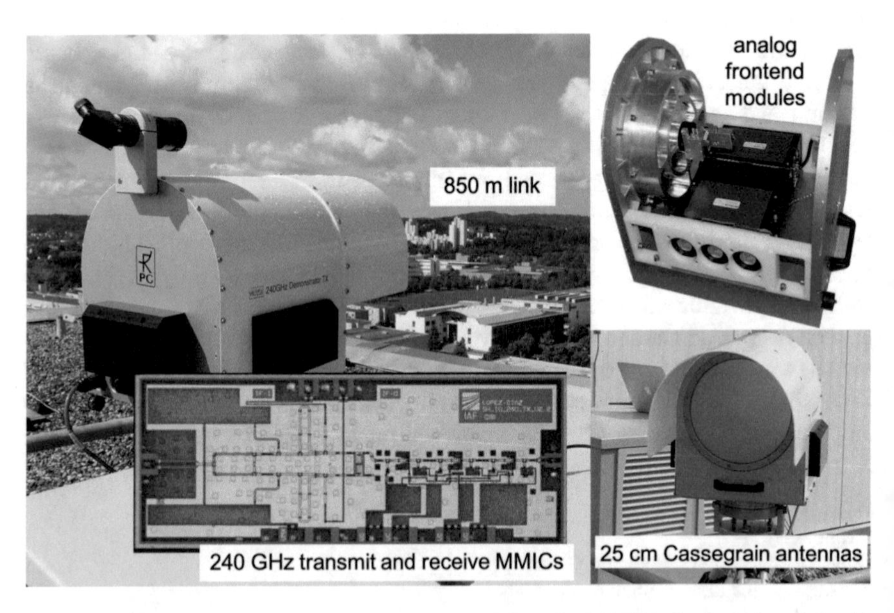

Fig. 37.1 The MILLILINK project saw the demonstration of a 240 GHz fixed wireless link with 64 Gbit/s data rate over a distance of 850 m. Adapted from [4]

References

1. Boes, F., Messinger, T., Antes, J., Meier, D., Tessmann, A., Inam, A., & Kallfass, I. (2014). Ultra-broadband MMIC-based wireless link at 240 GHz enabled by 64 GS/s DAC. In *Proc. 39th Int. Conf. on Infrared, Millimeter, and Terahertz Waves (IRMMW), Tuscon AZ* (pp. 1–4).
2. Hirata, A., Kosugi, T., Takahashi, H., Takeuchi, J., Murata, K., Kukutsu, N., Kado, Y., Okabe, S., Ikeda, T., Suginosita, F., Shogen, K., Nishikawa, H., Irino, A., Nakayama, T., & Sudo, N. (2010). 5.8-km 10-Gbps data transmission over a 120-GHz-band wireless link. In *2010 IEEE International Conference on Wireless Information Technology and Systems (ICWITS)* (pp. 1–4).
3. Kallfass, I., Antes, J., Schneider, T., Kurz, F., Lopez-Diaz, D., Diebold, S., Massler, H., Leuther, A., & Tessmann, A. (2011). All active MMIC based wireless communication at 220 GHz. *IEEE Transactions on Terahertz Science and Technology, 1*(2), 477–487.
4. Kallfass, I., Boes, F., Messinger, T., Antes, J., Inam, A., Lewark, U., Tessmann, A., & Henneberger, R. (2015). 64 Gbit/s transmission over 850 m fixed wireless link at 240 GHz carrier frequency. *Journal of Infrared Millimeter and Terahertz Waves, 36*(2), 221–233.
5. Koenig, S., Lopez-Diaz, D., Antes, J., Boes, F., Henneberger, R., Leuther, A., Tessmann, A., Schmogrow, R., Hillerkuss, D., Palmer, R., Zwick, T., Koos, C., Freude, W., Ambacher, O., Leuthold, J., & Kallfass, I. (2013). Wireless sub-THz communication system with high data rate. *Nature Photonics, 7*(12), 977–981.
6. Kosugi, T., Tokumitsu, M., Enoki, T., Muraguchi, M., Hirata, A., & Nagatsuma, T. (2004). 120-GHz Tx/Rx chipset for 10-Gbit/s wireless applications using 0.1/spl mu/m-gate InP HEMTs. *Compound Semiconductor Integrated Circuit Symposium, 2004. IEEE* (pp. 171–174).
7. Leuther, A., Tessmann, A., Dammann, M., Massler, H., Schlechtweg, M., & Ambacher, O. (2013). 35 nm mHEMT Technology for THz and ultra low noise applications. In *2013 International Conference on Indium Phosphide and Related Materials (IPRM)* (pp. 1–2).
8. Lopez-Diaz, D., Kallfass, I., Tessmann, A., Leuther, A., Wagner, S., Schlechtweg, M., & Ambacher, O. (2012). A subharmonic chipset for gigabit communication around 240 GHz. In *Proc. IEEE MTT-S Int. Microwave Symposium, Montreal.*
9. Lopez-Diaz, D., Tessmann, A., Leuther, A., Wagner, S., Schlechtweg, M., Ambacher, O., Kurz, F., Koenig, S., Antes, J., Boes, F., Henneberger, R., & Kallfass, I. (2013). A 240 GHz quadrature receiver and transmitter for data transmission up to 40 Gbit/s. in *Proc. European Microwave Week, Nuremberg* (pp. 1–4).
10. Yamaguchi, R., Hirata, A., Kosugi, T., Takahashi, H., Kukutsu, N., Nagatsuma, T. (2008). 10-Gbit/s MMIC wireless link exceeding 800 meters. In *2008 IEEE Radio and Wireless Symposium* (pp. 695–698).

Chapter 38
TERAPOD

Alan Davy

Abstract The TERAPOD project aims to investigate and demonstrate the feasibility of ultrahigh bandwidth wireless access networks operating in the Terahertz (THz) band. The proposed TERAPOD THz communication system will be developed, driven by end-user usage scenario requirements and will be demonstrated within a first adopter operational setting of a Data Centre. In this chapter, we define the full communications stack approach that is taken in TERAPOD, highlighting the specific challenges and aimed innovations that are targeted. We then provide an overview of the recent results that have emerged from the project. Finally, we will provide a short overview of the future research challenges in this area.

38.1 Introduction

The TERAPOD project aims to investigate and demonstrate the feasibility of ultrahigh bandwidth wireless access networks operating in the Terahertz (THz) band. The project demonstrates the TERAPOD THz communication system within a first adopter operational setting of a Data Centre and significantly progresses innovations across the full THz communications system stack. Data centres are an ideal target scenario for deployment of high bandwidth wireless technologies to augment the wired network due to the ridged physical network in place, the high bandwidth requirements and also networking issues such as spontaneous emergence of hotspots [1]. Traditionally, wireless technologies have not been considered within the Data Centre network due to insufficient capacity.

A. Davy (✉)
Waterford Institute of Technology, Waterford, Ireland
e-mail: adavy@tssg.org

T. Kürner et al. (eds.), *THz Communications*, Springer Series in Optical Sciences 234, https://doi.org/10.1007/978-3-030-73738-2_38

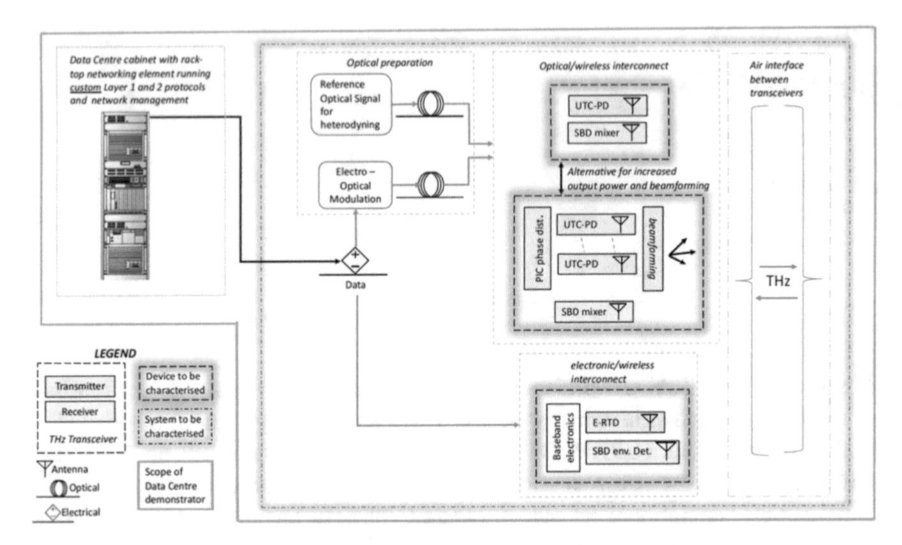

Fig. 38.1 Schematic representation of the TERAPOD technology application scenario

38.2 Objectives

The TERAPOD project focused on the development of several technologies, components, communications methods and architectures and protocols. In this sense, the project targets a proof-of-concept in a data centre deployment, which will be ultimately achieved using a small-scale test bed for the demonstration of at least 100 Gbps wireless communications between several stand-alone prototype nodes to be developed and manufactured within the project, including both optoelectronic and all-electronic interconnects at THz frequencies, in an operational environment.

Figure 38.1 provides a graphical description of the overall approach. The project focuses on two alternative technologies for THz sources, which are Resonant Tunnelling Diodes (RTD) and Uni-travelling Carrier Photodiodes (UTC-PD) and a sink device based on Schottky barrier diodes (SBD).

38.3 Technical Innovations

We present a number of technical advances that have been made by TERAPOD in order to progress towards a fully functional and integrated THz wireless system.

38.3.1 Demonstration of Reliable, Highly Efficient and High-Power THz RTD Sources

THz operation of RTD oscillators has been demonstrated by a few laboratories, with fundamental oscillations up to 1.92 THz recently being demonstrated [2, 3]. The main limitation of RTD oscillators has been their low output power (in the tens of microwatts range), but relatively high output power for some of these oscillators is now being reported, for example, up to 1 mW at 300 GHz with large micron-sized devices (by members of the present consortium) [4] and 0.6 mW at 620 GHz with a two-element array by a Japanese group [5].

As part of TERAPOD, we report on high-efficiency, high-power, and low-phase-noise resonant tunnelling diode (RTD) oscillators operating at around 30 GHz [6]. By employing a bias stabilization network, which does not draw any direct current (dc), the oscillators exhibit over a tenfold improvement in the dc-to-RF conversion efficiency (of up to 14.7%) compared to conventional designs (~0.9%). The oscillators provide a high maximum output power of around 2 dBm, and low phase noise of −100 and −113 dBc/Hz at 100 kHz and 1 MHz offset frequencies, respectively. The proposed approach will be invaluable for realizing very high efficiency, low phase noise, and high-power millimetre-wave (mm-wave) and terahertz (THz) RTD-based sources.

38.3.2 Demonstration of Power Combination of Multiple THz Sources

Although recent demonstrations have proven the capacity of THz communication systems for high throughput, most of the systems have not been tested within a real application environment and are still limited by the performances of the different components. In fact, the intrinsic high propagation loss at increasing carrier frequencies and the low power generated at these frequencies limits the THz wireless transmission distances to typically a few metres. To address this, TERAPOD will innovate upon the state of the art by demonstrating that the output of several UTC-PDs can be combined ultimately using an approach based on antenna arrays integrated with a photonic phase distribution system, which could also allow beam steering offering the possibility for electrical steering of receiver antenna and automatic search of the transmitter antenna. Early results show the feasibility of designing a photonic integrated circuit (PIC) phase distribution system for feeding a 1×4 UTC-PD array [7].

38.3.3 Demonstration of an Integrated THz Optical Wireless Bridge in a Data Centre Environment

In progressing THz wireless communication technology closer towards market uptake, near-market trials need to be carried out to determine the viability of such technology. TERAPOD carried out the world's first trial integration of a THz/optical wireless bridge within a data centre. The experiment consisted of developing a Top of Rack transmitter and Top of Rack receiver for a one-way THz/optical wireless bridge. The transmitter was a UTC-PD and SBD receiver. Further details of the experiment can be found here [8]. The results demonstrated that THz wireless links could be used to augment the wired network within a Data Centre to potentially alleviate congestion occurring due to the emergence of hotspots within the data centre, resulting in network bottlenecks.

References

1. Halperin, D., Kandula, S., Padhye, J., & Wetherall, D. (2011). Augment- ing data center networks with multi-gigabit wireless links. In *ACM SIGCOMM*.
2. Maekawa, T., et al. (2016). Oscillation up to 1.92 thz in resonant tunneling diode by reduced conduction loss. *Applied Physics Express, 9*(2).
3. Lee, J., et al. (2016). A 1.52 thz rtd triple-push oscillator with a w-level output power. *IEEE Transactions on Terahertz Science and Technology, 6*(2), 336–340.
4. Wang, J., et al. (2016). High performance resonant tunneling diode oscillators as terahertz sources. In *IEEE Microwave Conference (EuMC)* (pp. 341–344).
5. Suzuki, S., et al. (2013). High-power operation of terahertz oscillators with resonant tunneling diodes using impedance-matched antennas and array configuration. *IEEE Journal of Selected Topics in Quantum Electronics, 19*(1).
6. Al-Khalidi, A., et al. (2020). Resonant tunneling diode terahertz sources with up to 1 mW output power in the J-band. *IEEE Transactions on Terahertz Science and Technology, 10*(2), 150–157.
7. M. Garcia. *H2020 TERAPOD Results*. European Commission, 28 05 2018. [Online]. Available: https://ec.europa.eu/research/participants/documents/downloadPublic?documentIds=080166e5bb98ba59&appId=PPGMS. Accessed 30 Nov 2020.
8. S. Ahearne et al. (2019). Integrating THz wireless communication links in a data Centre network. In *IEEE 2nd 5G world forum*.

Chapter 39
iBROW

Joana S. Tavares, Scott Watson, Weikang Zhang, José M. L. Figueiredo, Horacio I. Cantu, Jue Wang, Abdullah Al-Khalidi, Anthony E. Kelly, Edward Wasige, Henrique M. Salgado, and Luis M. Pessoa

Abstract We present highlight results of resonant tunneling diode (RTD) oscillators developed in the iBROW project. Clear open eye diagrams were obtained using 300 GHz electrically modulated oscillators at 9 Gbit /s after wireless data transmission. One possible application for these oscillators consists of short-range high-frequency data transmission of audio and video signals. Thus, we present the first successful transmission of DVB-T audio/video signals through an optically modulated RTD.

39.1 Introduction

The demand for broadband content and services in both wired and wireless technologies has been growing at tremendous rates. We are living in an era in which broadband Internet with high bandwidth services and applications is in continuous evolution and demanding ever higher data rates [1]. To accommodate future data-rate requirements, operation at higher end of the spectrum is required, namely, above 60 GHz and up to 1 THz, since the frequency spectrum currently in use is not expected to be suitable to accommodate the predicted future data-rate requirements, in spite of the significant and continuous progress that has been achieved in spectrum efficiency techniques [2, 3].

J. S. Tavares · H. M. Salgado · L. M. Pessoa (✉)
INESC TEC and Faculty of Engineering, University of Porto, Porto, Portugal
e-mail: joana.s.tavares@inesctec.pt; luis.m.pessoa@inesctec.pt

S. Watson · W. Zhang · J. Wang · A. Al-khalidi · A. E. Kelly · E. Wasige
School of Engineering, University of Glasgow, Glasgow, UK

J. M. L. Figueiredo
CENTRA, Departamento de Física, Faculdade de Ciências, Universidade de Lisboa, Lisboa, Portugal

H. I. Cantu
CST-Global Ltd., Hamilton, UK

T. Kürner et al. (eds.), *THz Communications*, Springer Series in Optical Sciences 234, https://doi.org/10.1007/978-3-030-73738-2_39

To address this ever-growing demand, the iBROW project [4] aims at the development of energy-efficient and compact ultra-broadband short-range wireless communication transceiver technology, by exploiting resonant tunneling diode (RTD) oscillators that can be integrated into both ends of the wireless link, serving both source and detector functions. In this scenario, both downlink and uplink communication directions are considered, using all-electronic (e-RTD) and optoelectronic RTDs [5, 6]. The former is suitable for integration into cost-effective wireless portable devices, and the latter consists of monolithic integration between an RTD and a photodetector (RTD-PD) and hybrid integration with a laser diode (LD), which will be suitable for integration into mmwave/THz femtocell base stations connected to high-speed 40/100 Gbps fiber-optic networks.

An RTD device consists of a narrow band gap semiconductor material sandwiched between two thin wide band gap materials. These devices can be seamlessly integrated with optoelectronic components (photodetector and LD), since the same materials are used. RTDs have an intrinsic gain provided by its negative differential conductance region, which allows for a simple and energy-efficient implementation [7]. Another remarkable feature is the extremely high frequency operation, making them the fastest pure solid-state electronic device operating at room temperature with working frequencies exceeding 1 THz [8, 9].

Here, we present the highlight results of wireless transmission achieved using e-RTDs in Sect. 39.2 and the first successful transmission of digital video broadcasting—terrestrial (DVB-T) through an RTD-PD under optical modulation in Sect. 39.3.

39.2 e-RTDs for Wireless Data Transmissions

This work focused on the experimental evaluation of e-RTD oscillators aiming to achieve wireless data transmission using e-RTDs working in the W band and J-band, specifically at 84 GHz and 300 GHz, respectively. The laboratory testbed used to enable the demonstration of wireless data transmission based on electrically modulated RTDs is shown schematically in Fig. 39.1.

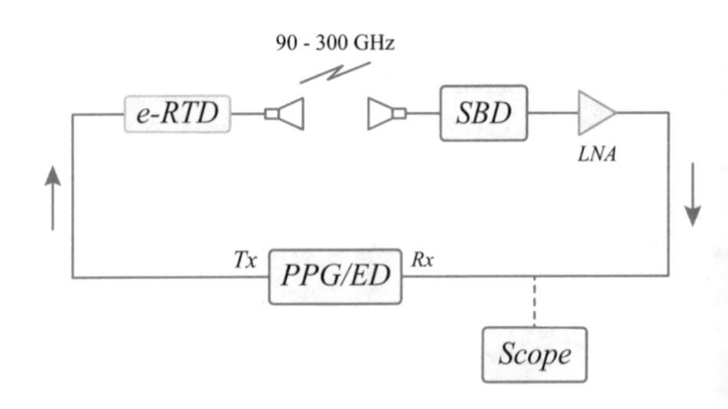

Fig. 39.1 Diagram of laboratory demonstration testbed using electrical link (e-RTD)

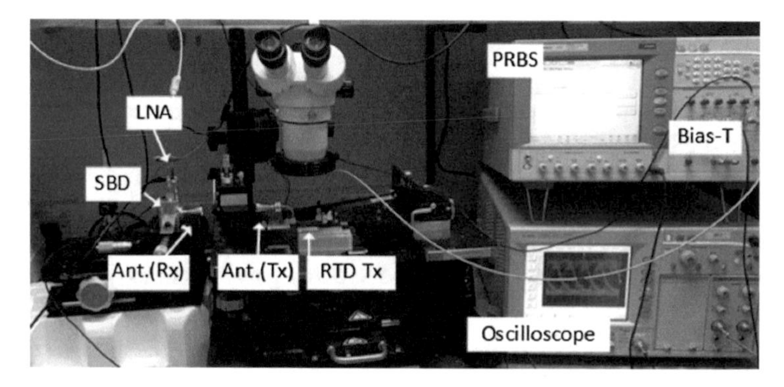

Fig. 39.2 Picture of the experimental setup used for the laboratory demonstration testbed using an e-RTD-based electrical link (at UGLA laboratory)

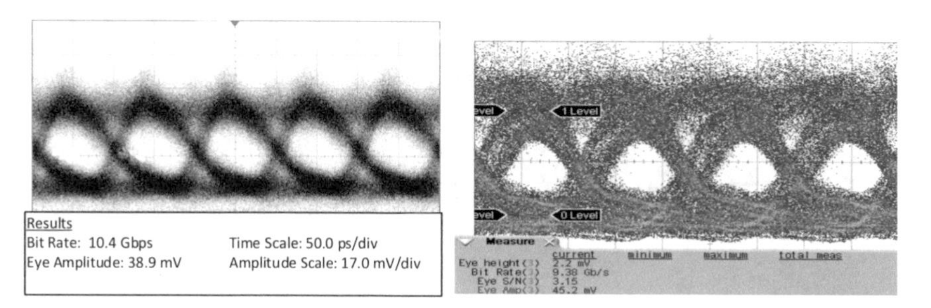

Fig. 39.3 Eye diagrams obtained with wireless transmission for an e-RTD oscillator at 84 GHz at 10 Gbit /s data rate (left) and at 300 GHz at 9 Gbit /s data rate (right)

A picture of the setup is shown in Fig. 39.2. The testbed can be described as follows: a pulse pattern generator and error detector (PPG/ED) is used to generate a pseudo-random binary sequence (PRBS) signal that drives directly the e-RTD, which is polarized within the negative differential resistance (NDR) region. Appropriate antennas are used to interface the RF signal produced by the RTD to the wireless domain. On the receiver side, a Schottky barrier diode (SBD)-based receiver is employed, connected to a similar antenna as used at the transmitter side, and followed by a low noise amplifier (LNA) that interfaces the received signal to the error detector module of the PPG/ED equipment, for bit-error-rate evaluation, or to an oscilloscope for eye diagram observation.

The modulation of the RTD with PRBS data within the NDR leads to an amplitude shift keying (ASK) modulation, which can be recovered using an envelope detector, in this case an SBD receiver, whose output voltage is proportional to the ASK modulated input signal power, therefore recovering the transmitted data stream. As shown in Fig. 39.3 clearly open eye diagrams were obtained using the 84 GHz RTD oscillator at 10 Gbit /s data rate and using the 300 GHz RTD oscillator at 9 Gbit /s data rate.

39.3 RTD-PD for Wireless Data Transmissions

The DVB-T transmission setup is represented in Fig. 39.4 and can be described as follows: a digital signal containing audio and video from the laptop HDMI port is converted by the modulator to a standard DVB-T signal, which is amplified and then converted to the optical domain by an LD before being injected into the RTD-PD optical window. The driver (or gain block) at the output of the modulator was used to amplify the signal, in order to achieve a suitable power level to ensure adequate LD modulation.

The transmitted signals are then down-converted by the spectrum analyzer and amplified before being demodulated by the television DVB-T tuner. A signal generator was used in this setup to perform injection locking [10]. The RTD-PD was biased with ~1 V of DC bias voltage and was oscillating at ~11 GHz.

To assess DVB-T transmission performance, we obtained the channel power in the spectrum analyzer and used the television to assess the strength and quality of the transmitted signals. Both audio and video are successfully presented by the television as shown in Fig. 39.5.

39.4 Conclusion

We have described the demonstration testbed for performance evaluation of e-RTDs and RTD-PDs for short-range high-frequency data transmission.

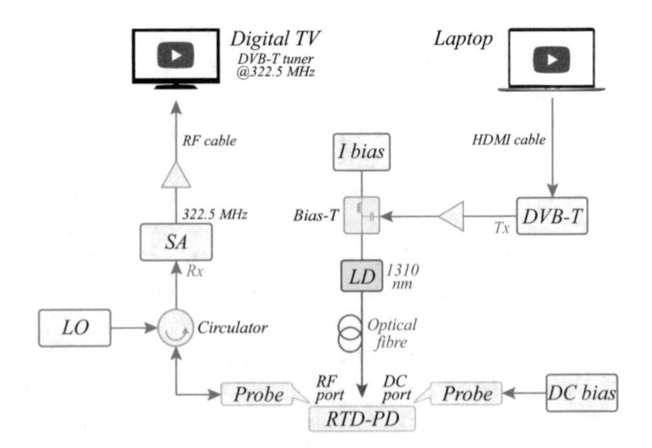

Fig. 39.4 Experimental setup diagram for DVB-T transmission using the RTD-PD with optical modulation. LO, local oscillator; SA, spectrum analyzer; LD, laser diode; Tx, transmitter; Rx, receiver

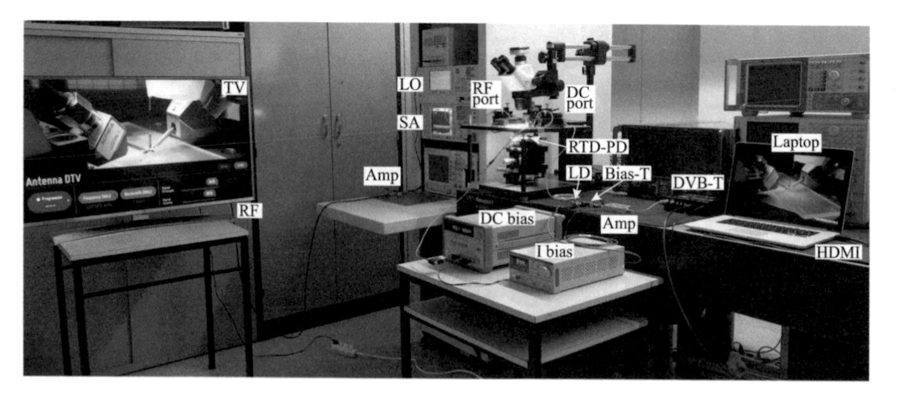

Fig. 39.5 Picture of the experimental setup used for the laboratory demonstration testbed using an RTD-PD-based optical fiber link for DVB-T transmission (at INESC TEC laboratory)

We presented laboratory demonstrations of wireless transmission based on an e-RTD at 300 GHz and an RTD-PD at around 10 GHz, with open eye diagrams shown at 10 Gbit/s and 50 Mbit /s, respectively.

We achieved a successful transmission of DVB-T audio/video signals through the RTD-PD under optical modulation, confirming the viability of using an RTD-PD as an optical to wireless interface, a novel worldwide achievement. While this was demonstrated using a 10 GHz RTD-PD oscillator, we expect that RTD-PDs fabricated to operate at much higher frequencies would provide similar results.

The results presented in this work allow to conclude that the RTD oscillator proposed in iBROW is a promising technology for the implementation of future high data-rate wireless communication systems, in short range applications, using the significant spectral availability in sub-THz frequencies.

References

1. Cisco visual networking index: Global mobile data traffic forecast update, 2016–2021 white paper. https://www.cisco.com/c/en/us/solutions/collateral/service-provider/visual-networking-index-vni/mobile-white-paper-c11-520862.html
2. Eisele, H. (2010). State of the art and future of electronic sources at terahertz frequencies. *Electronics Letters, 46*(26), 8–11.
3. Asada, M., & Suzuki, S. (2016). Room-temperature oscillation of resonant tunneling diodes close to 2 THz and their functions for various applications. *Journal of Infrared, Millimeter, and Terahertz Waves, 37*(12), 1185–1198.
4. iBROW: Innovative ultra-BROadband ubiquitous wireless communications through terahertz transceivers. http://ibrow-project.eu/
5. Tavares, J., Pessoa, L. M., Figueiredo, J., & Salgado, H. (2017). Analysis of resonant unnelling diode oscillators under optical modulation. In *2017 19th International Conference on Transparent Optical Networks (ICTON)* (pp. 1–4). IEEE.

6. Tavares, J., Pessoa, L., & Salgado, H. (2018). Experimental evaluation of resonant unnelling diode oscillators employing advanced modulation formats. In *2018 20th International Conference on Transparent Optical Networks (ICTON)* (pp. 1–4). IEEE.
7. Romeira, B., Pessoa, L. M., Salgado, H. M., Ironside, C. N., & Figueiredo, J. M. (2013). Photodetectors integrated with resonant tunneling diodes. *Sensors, 13*(7), 9464–9482.
8. Maekawa, T., Kanaya, H., Suzuki, S., & Asada, M. (2014). Frequency increase in terahertz oscillation of resonant tunnelling diode up to 1.55 THz by reduced slot-antenna length. *Electronics Letters, 50*(17), 1214–1216.
9. Maekawa, T., Kanaya, H., Suzuki, S., & Asada, M. (2016). Oscillation up to 1.92 THz in resonant tunneling diode by reduced conduction loss. *Applied Physics Express, 9*(2), 024101.
10. Kurokawa, K. (1973). Injection locking of microwave solid-state oscillators. *Proceedings of the IEEE, 61*(10), 1386–1410.

Chapter 40
120-GHz-Band Project

Akihiko Hirata

Abstract 120-GHz-band wireless link project in Japan is one of the early THz wireless communication researches. It achieved world-first 10-Gbps data transmission and uncompressed 8K transmission over radio wave links. This chapter presents the history and technologies of 120-GHz-band wireless link, and various field trials using the 120-GHz-band wireless link.

40.1 Introduction

Before 2000, it was impossible to transmit over 1-Gbps wireless data transmission. However, the demand for gigabit-class wireless link system had been increased with the spread of 10G Ethernet and 2K videos. In order to meet these 10-Gbps-class wireless link, 120-GHz-band wireless link that can transmit over 10-Gbps data had been developed in Japan (Fig. 40.1). The development of high-power photonic device and electronic devices enables the output power of over 10 dBm at >100 GHz and over 10-Gbps wireless data transmission over a distance of 1 km had been achieved. The 120-GHz-band wireless link employed uni-travelling carrier photodiode (UTC-PD) as photonic device, and InP high electron mobility transistor (HEMT) as electronic device. These specifications of the wireless link enable various kinds of field trials, such as uncompressed 8K video transmission or web conference over 10G Ethernet.

A. Hirata (✉)
Chiba Institute of Technology, Narashino, Japan
e-mail: hirata.akihiko@p.chibakoudai.jp

T. Kürner et al. (eds.), *THz Communications*, Springer Series in Optical Sciences 234, https://doi.org/10.1007/978-3-030-73738-2_40

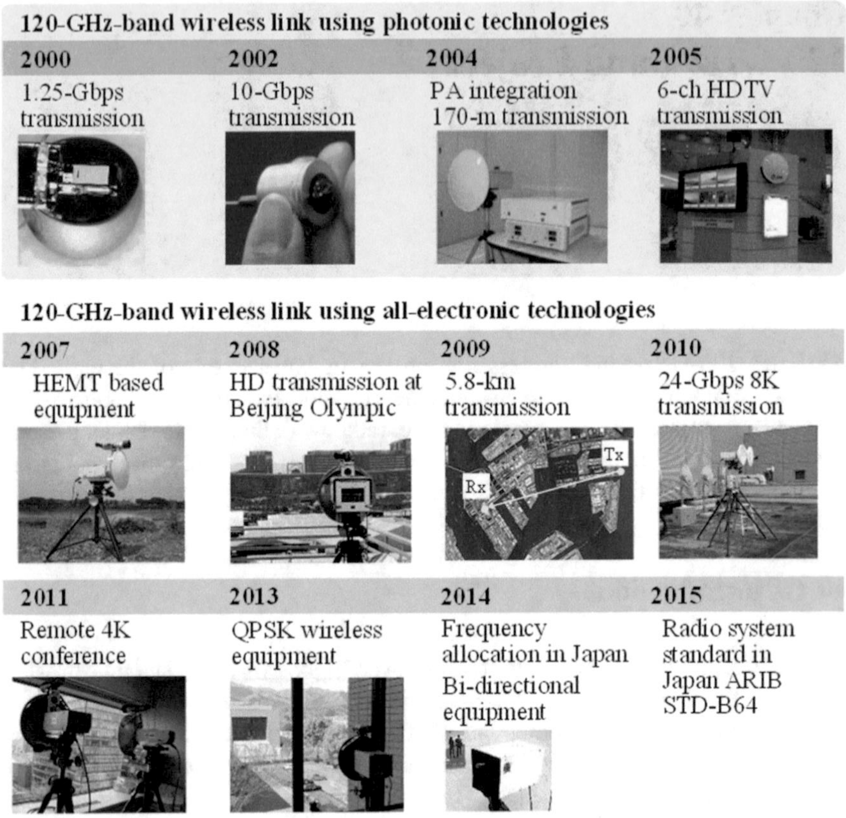

120-GHz-band wireless link using photonic technologies

2000	2002	2004	2005
1.25-Gbps transmission	10-Gbps transmission	PA integration 170-m transmission	6-ch HDTV transmission

120-GHz-band wireless link using all-electronic technologies

2007	2008	2009	2010
HEMT based equipment	HD transmission at Beijing Olympic	5.8-km transmission	24-Gbps 8K transmission

Tx
Rx

2011	2013	2014	2015
Remote 4K conference	QPSK wireless equipment	Frequency allocation in Japan Bi-directional equipment	Radio system standard in Japan ARIB STD-B64

Fig. 40.1 History of 120-GHz-band wireless link system

40.2 120-GHz-Band Wireless Link Using Photonic Technology

Research of 120-GHz-band wireless link started with indoor data transmission using photonics technologies. Photonic technologies enable us to generate MMW signals at >100 GHz with conventional photonic devices, and it is possible to modulate MMW signals at a rate of 10 Gbps by using optical modulators. Key devices for the 120-GHz-band wireless link are the UTC-PDs [1]. We used this small photonic emitter for the first 1-Gbps and 10-Gbps indoor wireless data transmission in 2000 [2] and 2002 [3], respectively. In these experiments, Si lens antennas were employed in the photonic emitter, and the output of the photonic emitter was collimated by Teflon lenses. In 2004, we obtained the first experimental radio station license at this band from the Ministry of Internal Affairs and Communications, Japan and conducted the first outdoor transmission experiments over a distance of 170 m [4]. In this system, the output of a single-mode laser is modulated at a frequency of

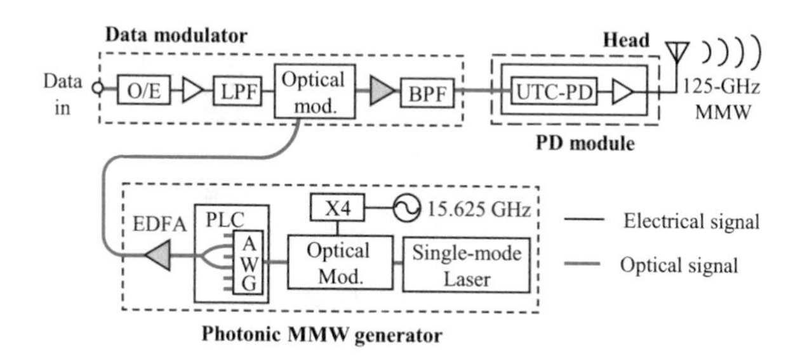

Fig. 40.2 Schematic diagram of transmitter using photonic technologies

62.5 GHz with an optical intensity modulator (Fig. 40.2). The modulated optical signals are fed into a planar lightwave circuit (PLC) that integrates an arrayed waveguide grating (AWG) and 3-dB coupler [5]. The PLC acts as an optical filter that outputs two modes whose frequency interval is 125 GHz. The optical MMW signal is modulated by data signals and input into the UTC-PD module. The head of the wireless equipment only acts as an O/E convertor, which makes the headlight and reduces its power consumption. In order to achieve outdoor data transmission experiments, 120-GHz-band power amplifier (PA) module and low-noise amplifier (LNA) that employed 0.1-μm-gate InP HEMT MMICs [6] were integrated in the transmitter and receiver, respectively. In 2005, a compact photonic wireless link was developed and the world's first transmission of six multiplexed 2K videos was achieved [7]. These 120-GHz-band wireless links used high-gain antennas, such as Gaussian optic lens antennas or Cassegrain antennas.

40.3 120-GHz-Band Wireless Link Using Electronic Technology

Since 2007, 120-GHz-band wireless signals were generated by using all-electronic technologies based on InP HEMT MMICs. The InP HEMT MMICs have a narrow operation bandwidth compared to photonic technologies; however, all-electronic systems have advantages of compactness and low cost, especially when the transceiver functions are implemented with MMICs. 0.1-μm-gate InP HEMT MMICs was employed to generate and detect 120-GHz-band wireless signals [6]. 125-GHz carrier signal was generated by multiplying 15.625-GHz CW signal, and it was modulated with data signals. Then, a PA module amplified the output up to 10 dBm (Fig. 40.3). In 2009, we developed a wireless equipment with an output power of 16 dBm by using composite-channel (CC) InP HEMT MMICs with improved breakdown voltage, and the error-free transmission of 10 Gbps and six multiplexed channels of uncompressed HD video signals over a distance of 5.8 km was achieved [8]. Table 40.1 shows the specifications of the photonic and electronic links. The head is lighter for the electronic system. However, the

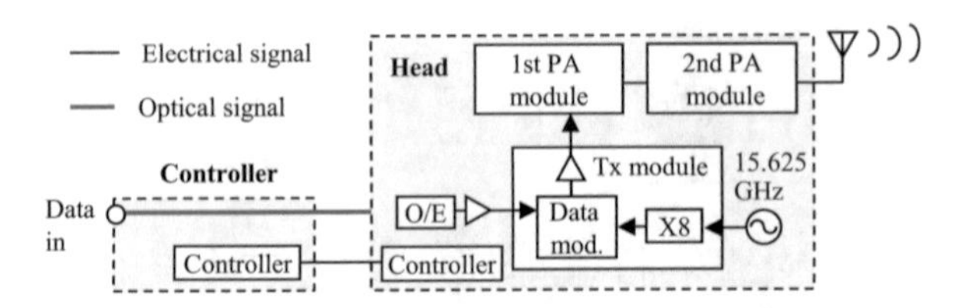

Fig. 40.3 Schematic diagram of transmitter using all-electronic technologies

Table 40.1 Comparison of specifications between UTC-PD system and InP HEMT MMIC system

	UTC-PD system	InP HEMT MMIC system
Center frequency	125 GHz	125 GHz
Occupied bandwidth	116.5–133.5 GHz	116.5–133.5 GHz
Output power	10 dBm	16 dBm
Modulation	ASK	ASK
Maximum data rate	11.096 Gbit/s	11.096 Gbit/s
Head size	W250 × D300 × H160 mm	W190 × D380 × H130 mm
Head weight	4.9 kg	7.3 kg
Controller size	W450 × D540 × H120 mm	W220 × D360 × H60 mm
Controller weight	20.1 kg	4.0 kg
Consumption power	<400 W	<100 W

electronic controller is much smaller than the photonic MMW generator, because in the photonic link millimeter-wave signals are generated in the head. Moreover, the electronic system consumes less power.

In order to improve spectral efficiency, 120-GHz-band wireless link that employs QPSK modulation scheme had been developed. 120-GHz-band wireless link employs a differentially coherent detection, which enables to construct a wireless link for real-time data transmission. 170-m-long data transmission at a rate of 11-Gbit/s and 22-Gbit/s back-to-back data transmission had been reported [9].

40.4 Field Trials of 120-GHz-Band Wireless Link

120-GHz-band wireless links have been used in various field trials. In 2008, 120-GHz-band wireless link was used for live-video transmission trials at the Beijing Olympics. The link transmitted 1.5 Gbps of uncompressed 2K video from the Beijing Media Center to the International Broadcast Center over a distance of 1 km. Transmitted data was used for 24 Fuji-TV live-relay programs in Japan [10]. In 2010, 120-GHz-band wireless link was connected to 10G Ethernet and had used 2K live-relay program in Sapporo Snow Festival in Japan.

A dual-green uncompressed 8K signal consists of 16-ch HD-SDI signals and its data rate is approximately 24 Gbit/s [11]. In order to double the data rate of 120-

GHz-band wireless link, the polarization multiplexing technologies were employed. One wireless link uses vertical polarization waves, and the other uses horizontal waves. The HD-SDI signal multiplexer/FEC encoder multiplexes 8-ch HD-SDI signals and transmitted the multiplexed approximately 11-Gbit/s serial data signal with forward error correction (FEC). Using the forward correction technologies, 8K signal transmission over a distance of 1.3 km had been achieved. In 2015, public viewing of uncompressed 8K signals using two pairs of 120-GHzband wireless links was conducted at Sapporo Snow festival [12].

The use of two links enables us to build a bidirectional wireless 10G Ethernet link. The 120-GHz-band wireless link was used for a remote academic symposium using 4K video. A conference hall in Kyoto was connected to the other hall in Tokyo with a 10G Ethernet network, and a part of the network employed the 120-GHz-band wireless link. The presentations held at each hall were shot by 4K cameras and transmitted to the other hall. Two pairs of 120-GHz-band wireless links were used to meet bidirectional communication over a 10G Ethernet.

References

1. Ishibashi, T., Furuta, T., Fushimi, H., Kodama, S., Ito, H., Nagatsuma, T., Shimizu, N., & Miyamoto, Y. (2000). InP/InGaAs uni-traveling-carrier photodiodes. *IEICE Transactions on Electronics, E83-C*(6), 938–949.
2. Hirata, A., Sahri, N., Ishii, H., Machida, K., Yagi, S., & Nagatsuma, T. Design and characterization of millimeter-wave antenna for integrated photonic transmitter. In *2000 Asia-Pacific microwave conference*, Sydney, Australia, 3–6 December 2000 (pp. 70–73).
3. Minotani, T., Hirata, A., & Nagatsuma, T. (2003). A broadband 120-GHz Schottky-diode receiver for 10-Gbit/s wireless links. *IEICE Transactions on Electronics, E86-C*(8), 1501–1505.
4. Hirata, A., Kosugi, T., Shibata, T., & Nagatsuma, T. (2004). High-directivity photonic emitter using photodiode module integrated with HEMT amplifier for 10-Gbit/s wireless link. *IEEE Transactions on Microwave Theory and Techniques, 52*(8), 1843–1850.
5. Hirata, A., Togo, H., Shimizu, N., Takahashi, H., Okamoto, K., & Nagatsuma, T. (July 2005). Low-phase noise photonic millimeter-wave generator using an AWG integrated with a 3-dB combiner. *IEICE Transactions on Electronics, E88-C*, 1458–1464.
6. Sugiyama, H., Kosugi, T., Yokoyama, H., Murata, K., Yamane, Y., Tokumitsu, M., & Enoki, T. (2008). High-performance InGaAs/InP composite-channel high electron mobility transistors grown by metal-organic vapor-phase epitaxy. *Japanese Journal of Applied Physics, 47*(4), 2828–2832.
7. Nagatsuma, T., Hirata, A., Kukutsu, N., & Kado, A. Y. (2007). Multiplexed transmission of uncompressed HDTV signals using 120-GHz-band millimeter-wave wireless link. In *International Topical Meeting on Microwave Photonics* (pp. 237–240).
8. Hirata, A., Kosugi, T., Takahashi, H., Takeuchi, J., Murata, K., Kukutsu, N., Kado, Y., Okabe, S., Ikeda, T., Suginosita, F., Shogen, K., Nishikawa, H., Irino, A., Nakayama, T., & Sudo, N. 5.8-km 10-Gbps data transmission over a 120-GHz-band wireless link. In *2010 IEEE International Conference on Wireless Information Technology and Systems (ICWITS)*, 207.1 2010.
9. Takahashi, H., Kosugi, T., Hirata, A., Takeuchi, J., Murata, K., & Kukutsu, N. (2013, December). 120-GHz-band fully integrated wireless link using QSPK for realtime 10-Gbit/s transmission. *IEEE Transactions on Microwave Theory and Techniques, 61*(12), 4745–4753.

10. Hirata, A., Takahashi, H., Kukutsu, N., Kado, Y., Ikegawa, H., Nishikawa, H., Nakayama, T., & Inada, T. (2009, March). Transmission trial of television broadcast materials using 120-GHz-band wireless link. *NTT Technical Review, 7*(3), 64–70.
11. Tsumochi, J., Okabe, S., Suginoshita, F., Takeuchi, J., Takahashi, H., & Hirata, A. Field experiments on super hi-vision signal transmission using 120-GHz-band FPU. *IEICE general conference 2014*, C-2-111, 2014.
12. https://www.nhk.or.jp/strl/results/annual2014/2014_chapter1.pdf

Chapter 41
300-GHz-Band InP IC Project

Hiroshi Hamada

Abstract In this chapter, two kinds of 300-GHz-band InP IC projects conducted in NTT Device Technology Labs are described. The first project is a 300-GHz kiosk download system. In this project, various building blocks for the system, such as a power amplifier, ASK transmitter, and hand-sized receiver, were developed by using the high-speed InP-HEMT technologies of NTT Corporation and Fujitsu Limited. Real-time 20-Gb/s wireless data transmission over a link distance of 1 m is successfully demonstrated by using these building blocks. The second project is an over-100 Gb/s 300-GHz transceiver for the beyond-5G era. In this project, a 300-GHz heterodyne transceiver is fabricated by using a broadband 300-GHz passive mixer and high linearity power amplifier. The transceiver shows a maximum data rate of 120 Gb/s using 16QAM with a link distance of 9.8 m. This is the highest data rate among electronics-based 300-GHz transceiver in 300-GHz-band as of 2020.

41.1 300-GHz 20-Gb/s Kiosk Downloader

One of the promising applications of MMW/THz is very high-speed wireless communication by using their wide bandwidth. The frequency range above 275 GHz was not allocated for wireless communication systems as of 2019 [1]. The atmospheric absorption is relatively low; less than 10 dB/km up to 320 GHz [2]. Therefore, there is a possibility of realizing high-speed wireless communication systems by using the 300-GHz band (275–320 GHz). In this chapter, the real-time 20-Gb/s 300-GHz kiosk download system experiment is introduced. The schematic and building blocks of the kiosk download system are shown in Fig. 41.1a. People can download large-sized data such as DVD data by placing their mobile terminal in front of the kiosk transmitter with high-speed 300-GHz wireless communication. The electronics to realize such a 300-GHz system are the key technology of this system.

H. Hamada (✉)
NTT, Kanagawa, Japan
e-mail: hiroshi.hamada.gs@hco.ntt.co.jp

T. Kürner et al. (eds.), *THz Communications*, Springer Series in Optical Sciences 234, https://doi.org/10.1007/978-3-030-73738-2_41

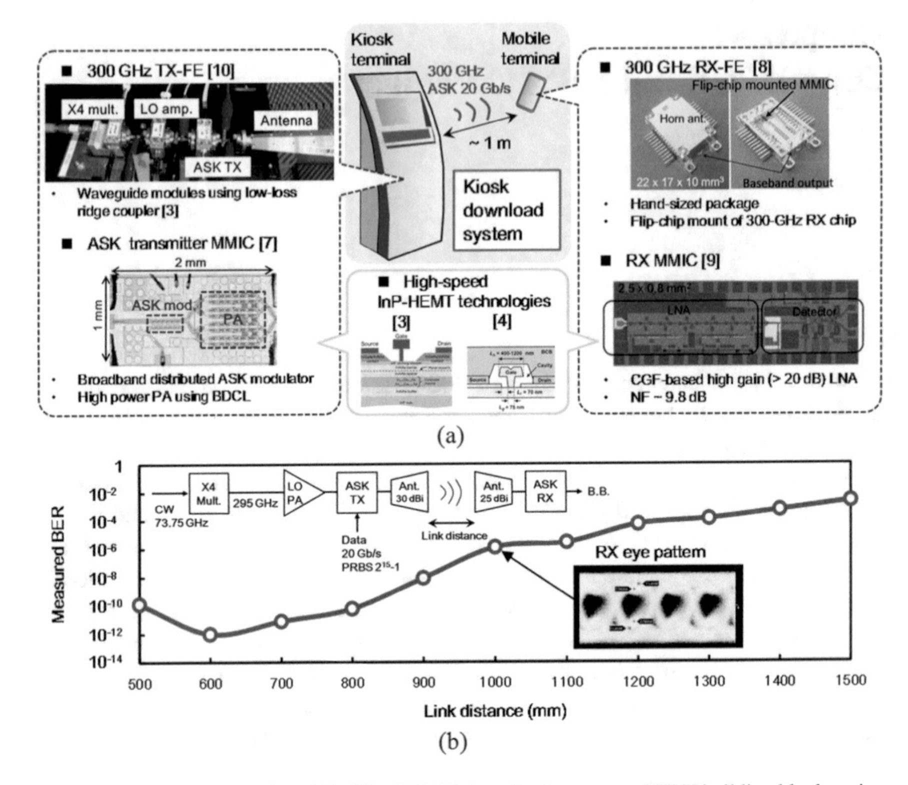

Fig. 41.1 (**a**) Schematic of the 300-GHz KIOSK download system and TRX building blocks using InP-HEMT technologies, and (**b**) BER measurement results against the link distance. (©2016 IEEE)

To implement 300-GHz monolithic microwave integrated circuits (MMICs), high-speed InP-based high electron mobility transistors (InP-HEMTs) [3, 4] are used for the transmitter (TX) and receiver (RX) front-end. High-speed and real-time communication is required for the kiosk download system. The amplitude-shift keying (ASK) transceiver (TRX) as shown in Fig. 41.1a was developed by NTT Corporation (TX), Fujitsu Limited (RX), and NICT (link budget) in the research and development program of Japan.

The transmitter front-end (TX-FE) is composed of individual waveguide-module-based building blocks of a X4 frequency multiplier, LO amplifier, and ASK transmitter (TX). The low-loss 300-GHz packaging is carried out by a ridge coupler [5]. These building blocks are fabricated by using the in-house InP process of NTT Corporation in which an 80-nm composite-channel InP-HEMT ($f_T = 300$ GHz, $f_{MAX} = 700$ GHz) [3] and 55-μm-thick thinned InP substrate with through-substrate vias (TSVs) [6] are supported. The X4 frequency multiplier generates a 295-GHz carrier signal from a 73.75 GHz continuous-wave signal. An ASK signal is generated by modulating this 295-GHz carrier signal by the ASK TX

MMIC shown in Fig. 41.1a [7]. An ASK modulator and power amplifier (PA) are integrated in the ASK TX MMIC. The ASK modulator is designed by using a travelling-wave passive switch that has low group delay dispersion ($< \pm 3.9$ ps) with moderate insertion loss (6 dB) to obtain wide modulation bandwidth [7]. The PA is designed to provide a large output power with the aid of an 8-way power combining network and loss reduction using the backside DC line (BDCL) technique [7]. Common-source FET (CSF) amplifiers are used as the unit amplifier of the PA. The measured gain and saturation output power of the PA TEG module are 14 dB and 9.5 dm at 290 GHz [7].

The receiver front-end (RX-FE) is designed to be as small as possible in anticipation of future integration into a hand-sized mobile terminal. Therefore, the flip-chip mounting of the 300-GHz RX MMIC and antenna-integrated small package [8] was developed as shown in Fig. 41.1a. The RX MMIC contains a 300-GHz low noise amplifier (LNA) and detector [9]. To attain high gain and wide bandwidth, a common-gate FET (CGF)-based amplifier was used in the LNA. The measured gain and noise figure (NF) of the LNA are over 20 dB and 9.5 dB, respectively.

To show the feasibility of the kiosk download system, a real-time 20-Gb/s wireless data transmission experiment is carried out using the TX-FE and RX-FE at the carrier frequency of 295 GHz [10, 11]. To generate 295 GHz, the X4 frequency multiplier based on NTT's InP-HEMT technology is used. A pulse-pattern generator (PPG) was used to generate 20-Gb/s pseudo-random binary sequence (PRBS) data (run length: $2^{15} - 1$). The antennas with a gain of 30 dBi and 25 dBi were attached to the TX and RX, respectively. The bit error rate (BER) was measured with several link distances, as shown in Fig. 41.1b. The BER is less than 10^{-6} up to a link distance of 1 m. By considering the forward error correction (FEC) implementation in the real situation, the link distance can be extended to more than 1.5 m with a BER of 2.4×10^{-3}. This is long enough for the kiosk download system.

41.2 300 GHz >100 Gb/s Wireless Transceiver for Beyond 5G

The demonstration of over-100 Gb/s wireless data transmission is described using the improved version of the building blocks of the kiosk downloader explained above. To support such a high data rate in the 300-GHz-band, a quadrature amplitude modulation (QAM) scheme is effective. There are two TRX architectures to generate a QAM signal in the 300-GHz-band: direct-conversion architecture and heterodyne architecture. The direct-conversion architecture has many advantages due to its simple constitution, such as low cost and small chip area. However, it is challenging to implement it in the 300-GHz-band because the high accuracy I-Q modulation/demodulation in the 300-GHz-band is generally difficult. On the other hand, the I-Q modulation/demodulation is carried out in a lower IF frequency band in the heterodyne architecture. Therefore, heterodyne architecture is often used in recently published 300-GHz-band TRXs [12–16] including the TRX described in this chapter [15, 16].

The photographs and schematic of the 300-GHz heterodyne TRX [15, 16] are shown in Fig. 41.2a. The generation and detection of the QAM signal (16QAM and 64QAM are used in this research) are carried out on the 20-GHz IF signal by the arbitrary waveform generator (AWG) and the digital storage oscilloscope (DSO). The frequency conversions between the IF signals and 300-GHz-band RF signals in both TX and RX are executed by the same 300-GHz InP-HEMT passive mixers [14, 16]. The fundamental mixer architecture is utilized in these mixers because of their higher conversion gain (CG) and lower LO harmonics than the harmonic mixer architecture. The 270 GHz, 5 dBm LO signals are used for both TX and RX and are generated by an X18 frequency multiplier and the in-house PA [7]. The photograph and measured CG are shown in Fig. 41.2b. The resistive topology is applied in the mixer; therefore, the commutative and broadband frequency-conversion characteristics are demonstrated. The measured up-conversion gain and 3-dB bandwidth are -15 dB and 31 GHz. For the PA in the TX and LNA in the RX, high-linearity InP-HEMT PAs [15, 16] with the BDCL technique [7] are used. In this research, the upper side band (290 GHz) is used as the RF signal. Therefore, in the TX, a high-pass filter (HPF) is inserted between the mixer and PA to cut out the lower side band image signal and also LO leakage of the mixer. In the RX, an IF amplifier with a gain of 23 dB is used after the mixer to provide a large enough IF signal to the DSO. A lensed horn antenna with a gain of 50 dBi is used for both the TX and RX.

The wireless data transmission experiment is conducted in an anechoic chamber (Fig. 41.2a). The link distance (antenna-to-antenna distance) is fixed to 9.8 m in the data transmission experiment. In this research, the signal equalization technique is applied to the received IF signal by the DSO. In the experiment, the signal to noise ratio (SNR) is used as the index of the communication quality, that is, the required SNR (SNR_{req}), which means the SNR value where the BER is equal to 10^{-3}, is used as the threshold value of the success/failure of the wireless communication. The value of the SNR_{req} of 16QAM and 64QAM are 16.5 dB and 22.5 dB, respectively. In the experiment, the variable attenuator (VATT) in the TX is used to virtually extend the link distance by decreasing the TX output power. The measured SNR versus equivalent link distance with several baud rates for the 16QAM and 64QAM are shown in Fig. 41.2c. For 16QAM, the data transmission was successfully demonstrated up to 30 Gbaud (120 Gb/s). The maximum link distances where the measured SNR is equal to the SNR_{req} of 16QAM for 30 Gbaud (120 Gb/s), 25 Gbaud (100 Gb/s), 20 Gbaud (80 Gb/s), and 15 Gbaud (60 Gb/s) are 10.5, 17.5, 29.5, and 42 m, respectively. The measurement results for 64QAM are also shown in Fig. 41.2c. The wireless data transmission was conducted up to 16.7 Gbaud (100.2 Gb/s). The maximum link distances of 16.7 Gbaud (100.2 Gb/s), 15 Gbaud (90 Gb/s), and 10 Gbaud (60 Gb/s) are 10.2, 13.7, 17.4 m, respectively.

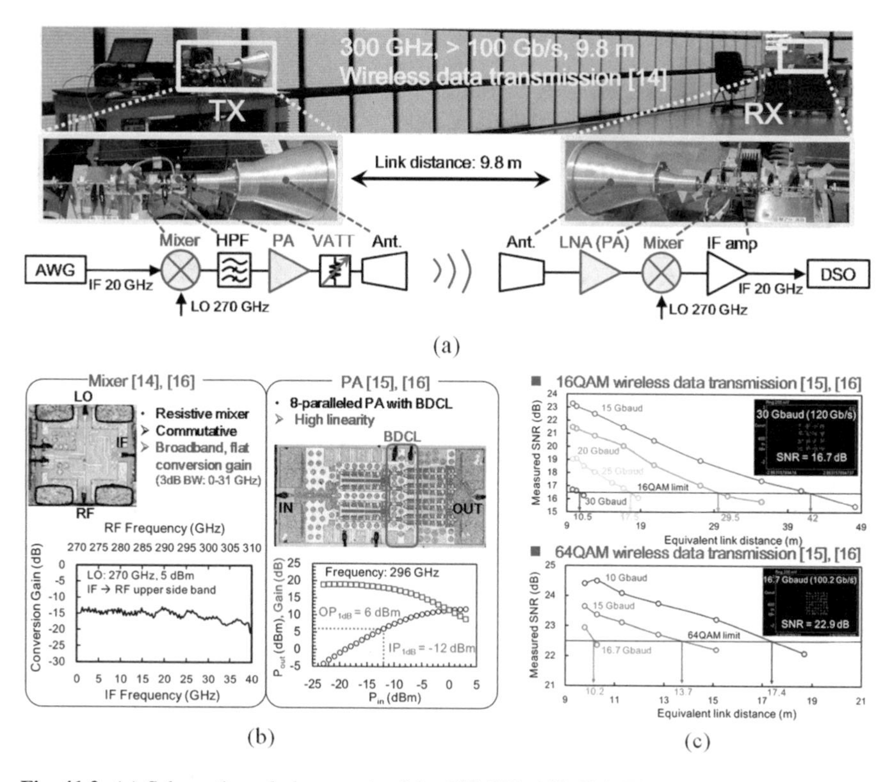

Fig. 41.2 (**a**) Schematic and photograph of the 300-GHz 120-Gb/s TRX with a link distance of 9.8 m, (**b**) the characteristics of the mixer and PA used in the TRX, and (**c**) the data transmission results of 16QAM and 64QAM. (©2018, 2019, 2020 IEEE)

Acknowledgments The author thanks T. Tsutsumi, H. Matsuzaki, H. Sugiyama, G. Itami, H.-J. Song, M. Yaita H. Nosaka (NTT Device Technology Labs), T. Fujimura, I. Abdo, A. Shirane, K. Okada (Tokyo Institute of Technology), Y. Kawano, T. Takahashi, Y. Nakasha (Fujitsu Limited), and A. Kasamatsu (NICT) for their efforts in developing the 300-GHz TRXs and our fruitful discussions. The author also thanks the continuous assistance of Y. Yago and R. Kamada. This work was supported in part by the research and development program for the expansion of radio resources promoted by the Ministry of Internal Affairs and Communications, Japan.

References

1. Kürner, T. THz communications – An option for wireless networks beyond 5G?. In *European Conference on Networks and Communications (EuCNC) 2018*. [Online].Available:https://www.eucnc.eu/2018/www.eucnc.eu/keynotes/index.html#1493814414886-8b2a7517-a544
2. Ulaby, F. T. (1973). Absorption in the 220 GHz atmospheric window. *IEEE Transactions on Antennas and Propagation, 21*, 266–269.
3. Sugiyama, H., Matsuzaki, H., Yokoyama, H., & Enoki, T. (2010). High-electron-mobility $In_{0.53}Ga_{0.47}As/In_{0.8}Ga_{0.2}As$ composite-channel modulation-doped structures grown by metal-

organic vapor-phase epitaxy. In *International conference on Indium Phosphide and Related Materials (IPRM)*.

4. Takahashi, T., Sato, M., Nakasha, Y., Hirose, T., & Hara, N. (2010). Improvement in noise figure of wide-gate-head InP-based HEMTs with cavity structure. In *International conference on Indium Phosphide and Related Materials (IPRM)*.
5. Kosugi, T., Hamada, H., Takahashi, H., Song, H.-J., Hirata, A., Matsuzaki, H., & Nosaka, H. (2014). 250–300 GHz waveguide module with ridge-coupler and InP-HEMT IC. In *IEEE Asia-Pacific Microwave Conference (APMC)*.
6. Tsutsumi, T., Hamada, H., Sano, K., Ida, M., & Matsuzaki, H. (2019). Feasibility Study of Wafer-Level Backside Process for InP-Based ICs. *IEEE Transactions on Electron Devices, 66*, 3771–3776.
7. Hamada, H., Kosugi, T., Song, H.-J., Yaita, M., El Moutaouakil, A., Matsuzaki, H., & Hirata, A. (2015). 300-GHz band 20-Gbps ASK transmitter module based on InP-HEMT MMICs. In *IEEE Compound Semiconductor Integrated Circuit Symposium (CSICS)*.
8. Kawano, Y., Matsumura, H., Shiba, S., Sato, M., Suzuki, T., Nakasha, Y., Takahashi, T., Makiyama, K., & Hara, N. (2014). Flip chip assembly for sub-millimeter wave amplifier MMIC on polyimide substrate. In *IEEE MTT-S International Microwave Symposium (IMS)*.
9. Kawano, Y., Matsumura, H., Shiba, S., Sato, M., Suzuki, T., Nakasha, Y., Takahashi, T., Makiyama, K., Iwai, T., & Hara, N. (2015). A 20 Gbit/s, 280 GHz wireless transmission in InP HEMT based receiver module using flip-chip assembly. In *European Microwave Conference (EuMC)*.
10. Song, H.-J., Kosugi, T., Hamada, H., Tajima, T., El Moutaouakil, A., Matsuzaki, H., Kawano, Y., Takahashi, T., Nakasha, Y., Hara, N., Fujii, K., Watanabe, I., Kasamatsu, A., & Yaita, M. (2016). Demonstration of 20-Gbps wireless data transmission at 300 GHz for KIOSK instant data downloading applications with InP MMICs. In *IEEE MTT-S International Microwave Symposium (IMS)*.
11. Hamada, H., Kosugi, T., Song, H.-J., Matsuzaki, H., El Moutaouakil, A., Sugiyama, H., Yaita, M., Tajima, T., Nosaka, H., Kagami, O., Kawano, Y., Takahashi, T., Nakasha, Y., Hara, N., Fujii, K., Watanabe, I., & Kasamatsu, A. (2016). 20-Gbit/s ASK wireless system in 300-GHz-band and front-ends with InP MMICs. In *URSI Asia-Pacific Radio Science Conference (URSI AP-RASC)*.
12. Hara, S., Katayama, K., Takano, K., Dong, R., Watanabe, I., Sekine, N., Kasamatsu, A., Yoshida, T., Amakawa, S., & Fujishima, M. (2017). A 32Gbit/s 16QAM CMOS receiver in 300GHz band. In *IEEE MTT-S International Microwave Symposium (IMS)*.
13. Dan, I., Ducournau, G., Hisatake, S., Szriftgiser, P., Braun, R.-P., & Kallfass, I. (2019, January). A Terahertz Wireless Communication Link Using a Superheterodyne Approach. *IEEE Transactions on Terahertz Science and Technology, 10*, 32–43.
14. Hamada, H., Fujimura, T., Abdo, I., Okada, K., Song, H.-J., Sugiyama, H., Matsuzaki, H., & Nosaka, H. (2018). 300-GHz, 100-Gb/s InP-HEMT Wireless Transceiver Using a 300-GHz Fundamental Mixer. In *IEEE MTT-S International Microwave Symposium (IMS)*.
15. Hamada, H., Tsutsumi, T., Itami, G., Sugiyama, H., Matsuzaki, H., Okada, K., & Nosaka, H. (2019, November). 300-GHz 120-Gb/s Wireless Transceiver with High-Output-Power and High-Gain Power Amplifier Based on 80-nm InP-HEMT Technology. In *IEEE BiCMOS and Compound Semiconductor Integrated Circuits and Technology Symposium (BCICTS)*.
16. Hamada, H., Tsutsumi, T., Matsuzaki, H., Fujimura, T., Abdo, I., Shirane, A., Okada, K., Itami, G., Song, H.-J., Sugiyama, H., & Nosaka, H. (2020). 300-GHz-band 120-Gb/s wireless front-end based on InP-HEMT PAs and mixers. *IEEE Journal of Solid-State Circuits, 55*, 2316–2335.

Chapter 42
300-GHz-Band Si-CMOS Project

Akifumi Kasamatsu

Abstract A radio transmission demonstration in the 300-GHz-band using transmitter and receiver employing silicon CMOS technology was carried out by a Japanese research group in a national project supported by the Ministry of Internal Affairs and Communications, Japan. 300-GHz-band front-end modules were fabricated using multilayer printed circuit boards with an RF front-end integrated circuit using Si-CMOS technology as a core device. Antennas and measuring instruments were connected to this module, and experimental transmission demonstrations were performed.

Multilayer printed circuit board (PCB) modules having a waveguide conversion structure that was connected to silicon CMOS transmitter or receiver have been developed [1, 2]. For the silicon CMOS transmitter and receiver, we used some of the novel technologies described in Chap. 21. Fig. 42.1 shows the multilayer structure of the glass epoxy PCB used in these radio communication modules. LO and IF signals are input and output via grounded coplanar waveguides (GCPW) fabricated on the same board. The GCPW are composed of the signal lines of the L1 metal layer with 50 μm width and the GND by the surrounding L1, L2, L3, and L4 metal layer. The signal input and output to the module were done via the V-band coaxial connector connected to the edge of the PCB.

In the wireless transmission experiments using the module and a standard horn antenna, we achieved a transmission rate of 20 Gb/s at a communication distance of 10 cm using 16QAM modulation. In the case of QPSK modulation, a communication speed of 2 Gb/s was realized at 75 cm. Fig. 42.2 shows results of constellation and the error vector magnitude (EVM).

Next, a Cassegrain antenna was developed as a high-gain antenna for extending the communication distance (Fig. 42.3) [3]. The diameter of a main reflection mirror was 60 mm, and a radome made with cycloolefin polymer (COP) was attached in

A. Kasamatsu (✉)
NICT, Tokyo, Japan
e-mail: kasa@nict.go.jp

T. Kürner et al. (eds.), *THz Communications*, Springer Series in Optical Sciences 234, https://doi.org/10.1007/978-3-030-73738-2_42

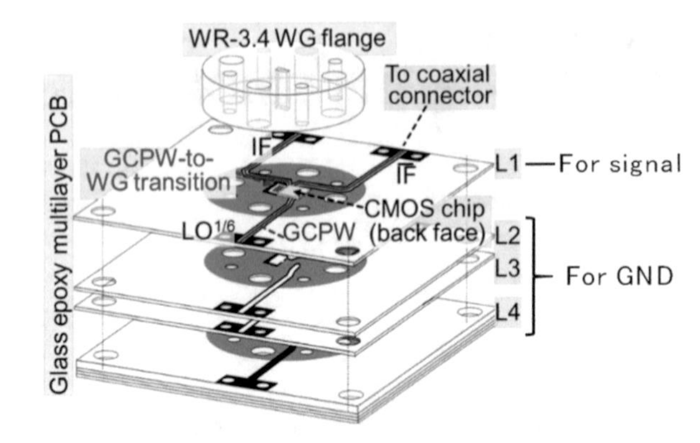

Fig. 42.1 Structure of multilayer printed circuit board

Fig. 42.2 Results of the transmission experiment using the module and a standard horn antenna

	QPSK	16-QAM
Constellation (Equalized)	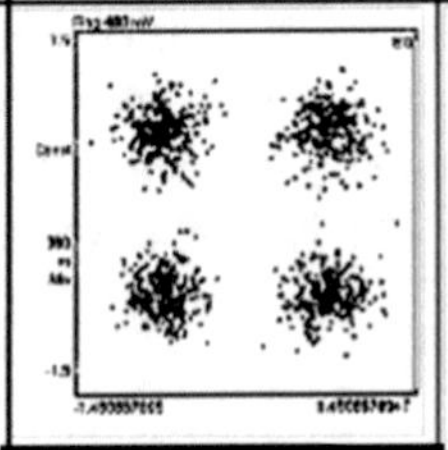	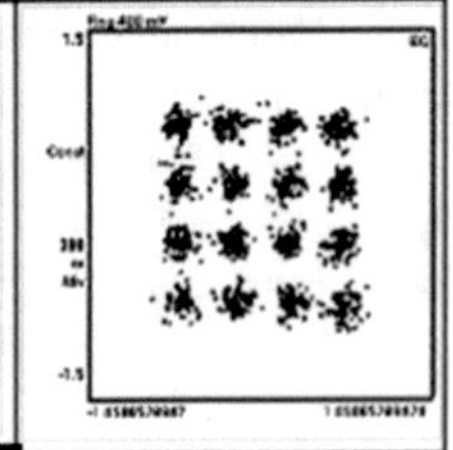
Distance	**75 cm**	10 cm
EVM	25.7%rms	12.2%rms
BER	5.1×10^{-5}	2.2×10^{-5}
Sym. rate	1 Gbaud	5 Gbaud
Data rate	2 Gbit/s	**20 Gbit/s**

front of the main mirror, and a sub mirror was fixed to center of the radome. A fabricated Cassegrain antenna achieved a gain of 39 dBi.

Moreover, to realize a small wireless transmitter/receiver equipment, a small-size power supplying control board was designed and integrated with the RF front-end module in a compact box whose volume is 10 cm × 10 cm × 20 cm (Fig. 42.4) [4]. A radio communication experiment of 1 m distance was carried out using the equipment. In the condition of 6.25 Gbaud (baseband bandwidth of 3.125 GHz) and 16QAM, 25-Gb/s radio communication with BER of less than 10^{-4} was realized. A multi-stream experiment was also performed with this system, and 100-Gb/s transmission can be achieved by assist of frequency and spatial multiplexing.

Finally, an experiment of real-time video transmission was performed using the radio communication system shown in Fig. 42.5. The experiment was carried out using an OFDM signal composed of eight component carriers with a bandwidth of 100 MHz. It was demonstrated that stable radio communication can be realized

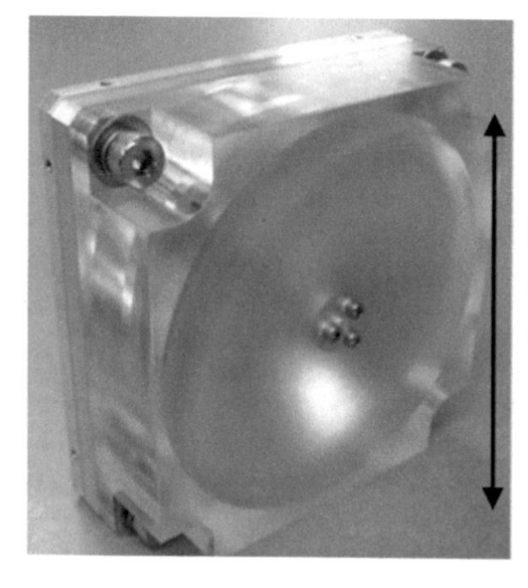

Fig. 42.3 Fabricated Cassegrain antenna

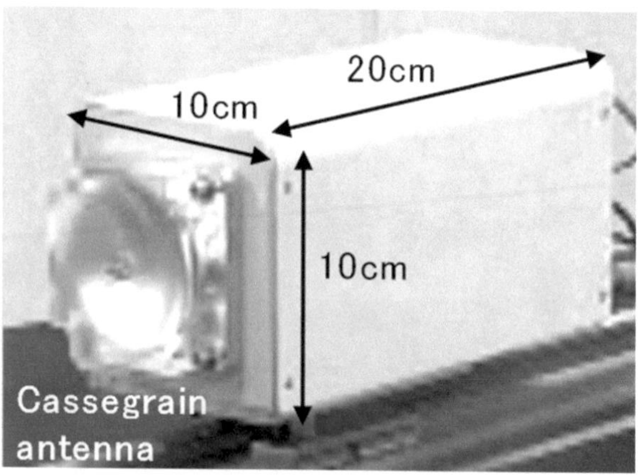

Fig. 42.4 A compact box of the 300-GHz-band transmitter and receiver

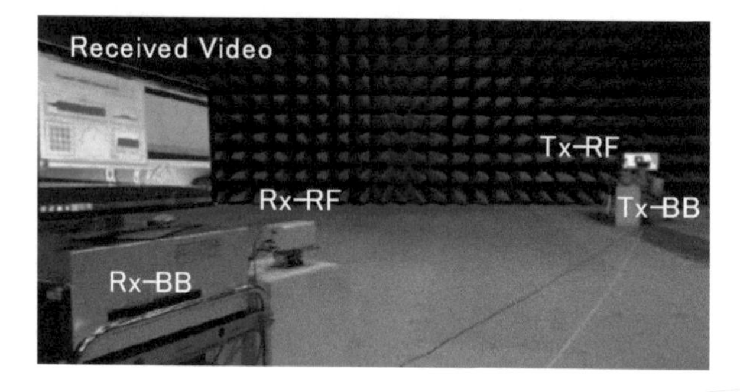

Fig. 42.5 Demonstration setup of 300-GHz-band real-time video transfer

under conditions of 16QAM and that real-time transmission of video can be performed even in the communication distance of 10m.

This work was done by collaboration with Panasonic, Hiroshima University, and National Institute of Information and Communications Technology, and was financially supported by the Ministry of Internal Affairs and Communications, Japan (JPJ000254).

References

1. Takano, K., Katayama, K., Hara, S., Dong, R., Mizuno, K., Takahashi, K., Kasamatsu, A., Yoshida, T., Amakawa, S., & Fujishima, M. (2018, January). 300-GHz CMOS transmitter module with built-in waveguide transition on a multilayered glass epoxy PCB. In *IEEE Radio and Wireless Symposium* (pp. 154–156).
2. Hara, S., Takano, K., Katayama, K., Dong, R., Mizuno, K., Takahashi, K., Watanabe, I., Sekine, N., Kasamatsu, A., Yoshida, T., Amakawa, S., & Fujishima, M. (2018, September). 300-GHz CMOS Receiver Module with WR-3.4 Waveguide Interface. In *Proceedings 2018 European Microwave Conference, Madrid* (pp. 396–399).
3. Sato, J., Morishita, Y., & Kashino, Y. (2018, September). A study of manufacturing accuracy of 300 GHz band cassegrain antenna. In *Proceedings Society Conference IEICE, C-2-46*.
4. Morishita, Y., Teraoka, T., Kashino, Y., Asano, H., Sakamoto, T., Shirakata, N., Takinami, K., & Takahashi, K. (2020, September). 300-GHz-Band selfheterodyne wireless system for real-time bideo transmission toward 6G. In *IEEE International Symposium on Radio-Frequency Integrated Technology*.

Chapter 43
Fully Electronic Generation and Detection of THz Picosecond Pulses and Their Applications

Babak Jamali, Sam Razavian, and Aydin Babakhani

Abstract Broadband generation and detection of THz electromagnetic waves enable a wide range of applications including high-speed wireless communication, material sensing, gas spectroscopy, hyper-spectral imaging, security imaging, and radars. In this chapter, we report our recent work on a laser-free fully electronic pulse source and detector technology that operates from 50 GHz up to 1.1 THz. The silicon-based source utilizes a novel technique of direct digital to impulse (D2I) radiation, in which a fast bipolar switch disconnects the DC current in a short period of time and converts it to a picosecond THz pulse that is radiated from an on-chip antenna. In the detector technology, which was utilized to demonstrate dual-comb spectroscopy, the picosecond pulses with tunable repetition rates are used as a reference frequency comb to down-convert the received millimeter-wave and THz signals to baseband frequencies.

43.1 Silicon-Based THz Pulse Radiation

Generating narrow pulses requires switches with ultra-short transients. Switching nonlinear devices have been used to produce ultra-short pulses in the direct digital to impulse scheme [1]. Moreover, PIN diodes (Fig. 43.1a), which are mostly used as RF/mm-wave switches, have been shown to be highly nonlinear in reverse recovery. In other words, when a PIN diode is switched off, a reverse current flows through the diode. After the charge is completely depleted from the space charge of the diode, the reverse current sharply falls to zero. Therefore, by driving this device with a large signal in the proper frequency range, PIN diodes can be pushed into a highly nonlinear region for pulse generation applications [2]. The schematic of a PIN diode-based pulse radiator is shown in Fig. 43.1b. A photograph of the chip, which was fabricated in a 130 nm SiGe BiCMOS process, is shown in Fig. 43.1c. In

B. Jamali · S. Razavian · A. Babakhani (✉)
University of California, Los Angeles, CA, USA
e-mail: aydinbabakhani@ucla.edu

T. Kürner et al. (eds.), *THz Communications*, Springer Series in Optical Sciences 234, https://doi.org/10.1007/978-3-030-73738-2_43

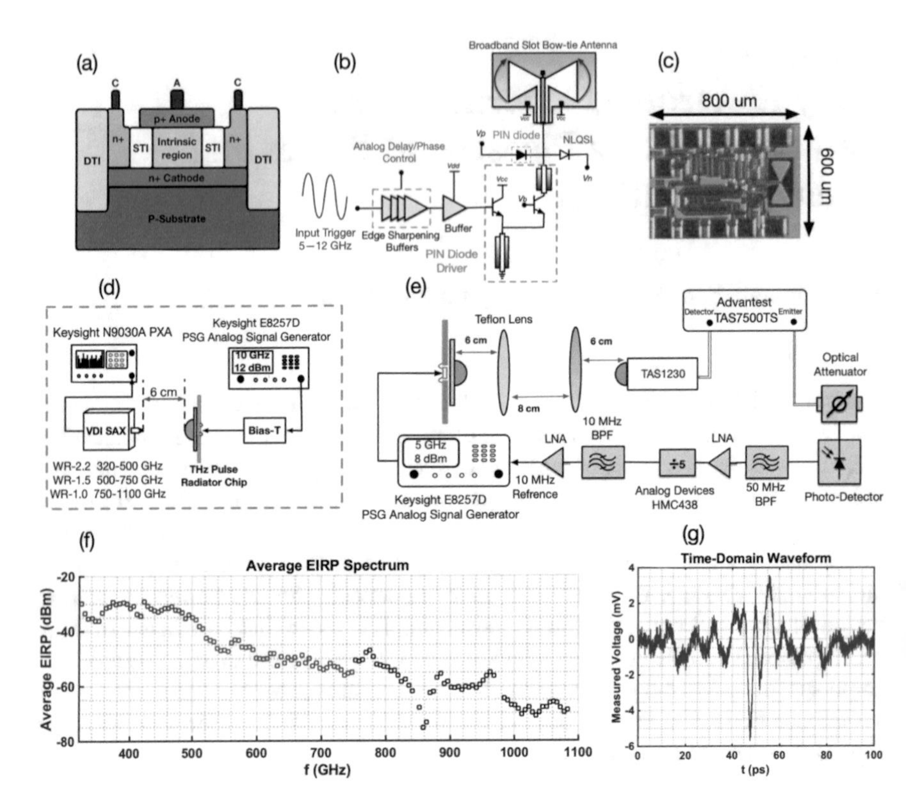

Fig. 43.1 A silicon-based pulse radiator: (**a**) The structure of a PIN diode in the SiGe BiCMOS process. (**b**) The schematic of the pulse radiator. (**c**) Micrograph of the implemented chip. (**d**) Frequency-domain measurement setup. (**e**) Time-domain measurement setup. (**f**) Measured EIRP of the radiator. (**g**) Measured time-domain pulses

order to characterize the chip in frequency domain and time domain, measurement setups in Fig. 43.1d, e have been utilized respectively. The measured time-domain waveform of the radiated pulses and its measured EIRP for frequencies up to 1.1 THz are shown in Fig. 43.1g, f.

43.2 Silicon-Based Frequency Comb Detection

Detecting picosecond pulses on a silicon-based platform demands a broadband solution that coherently detects millimeter-wave/THz signals over a bandwidth of larger than 100 GHz. To achieve this goal, a comb-based detection concept was introduced, which utilized a picosecond pulse train as the reference signal for down-converting the received signals [3]. Such picosecond pulses are generated with the D2I technique, which produces ultra-short pulses by fast switching of a nonlinear

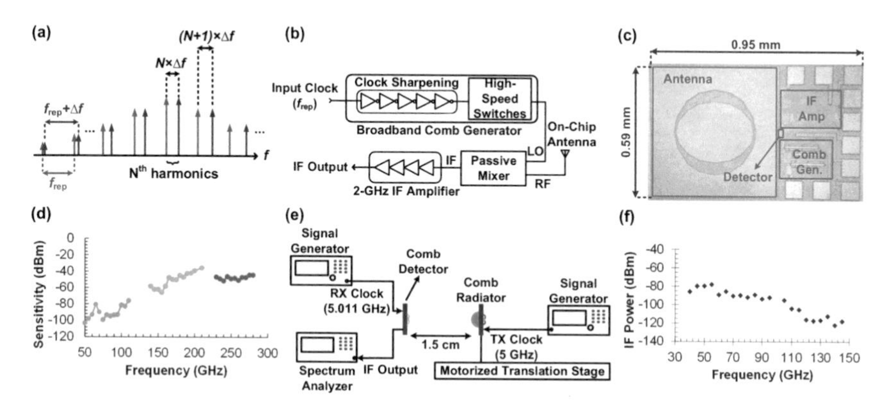

Fig. 43.2 A silicon CMOS frequency comb detector: (**a**) The operating concept. (**b**) The block diagram. (**c**) The implemented chip photograph. (**d**) Measured sensitivity of the detector. (**e**) The test setup to demonstrate dual-comb sensing. (**f**) The detected frequency comb from the dual-comb testing

device [1]. The frequency-domain representation of these pulses is a frequency comb, in which the comb spacing is equal to the tunable repetition rate of the pulse. As a result, by changing the comb spacing of the reference comb in the detector, the received signals can be detected over the comb bandwidth with a resolution that is only limited to the line-width of the comb tones. Such a comb-based coherent detector can also be used to perform dual-comb sensing and spectroscopy, which is illustrated in Fig. 43.2a. The block diagram and the chip photograph of the frequency comb detector are shown in Fig. 43.2b, c, in which a passive MOS-based mixer is used to down-convert the received signal by mixing it with the reference comb. The chip is fabricated in a 65 nm Si CMOS technology and consumes 34 mW dc power. The measured sensitivity of the detector is reported in Fig. 43.2d, and the dual-comb sensing setup and results are reported in Fig. 43.2e, f. These results indicate that a silicon-based frequency comb system is a viable, low-cost solution for broadband spectroscopy and sensing applications in the mm-wave/THz regime.

References

1. Assefzadeh, M. M., & Babakhani, A. (2017). Broadband oscillator-free THz pulse generation and radiation based on direct digital-to-impulse architecture. *IEEE Journal of Solid-State Circuits, 52*, 2905–2919.
2. Razavian, S., & Babakhani, A. (2019). A THz pulse radiator based on PIN diode reverse recovery. In *IEEE BiCMOS and Compound semiconductor Integrated Circuits and Technology Symposium (BCICTS), Nashville* (pp. 1–4). IEEE.
3. Jamali, B., & Babakhani, A. (2019). A fully integrated 50–280-GHz frequency comb detector for coherent broadband sensing. *IEEE Transactions on Terahertz Science and Technology, 9*, 613–623.

Chapter 44
RTD Transceiver Project

Masayuki Fujita, Julian Webber, and Tadao Nagatsuma

Abstract In this chapter, we review the history of terahertz (THz) communications using a resonant tunneling diode (RTD) as a transmitter and a receiver. Then, the key technologies to enhance the communication performance, including the circuit and the antenna design, the coherent detection, and the integration on a photonic-crystal platform are briefly described. As for state-of-the-art system using RTDs, the baseband radio-over-fiber THz communication and the direct THz communication with wireless and fiber links are presented.

44.1 Wireless Transmission

Data rate versus announcement date for RTD-based communications are categorized by transmitter (Tx), receiver (Rx), and Tx & Rx in Fig. 44.1. RTD Tx and Schottky barrier diode Rx with practical error-free (bit error rate (BER) $<1 \times 10^{-11}$) wireless transmission increased from 1.5 Gbit/s (300-GHz carrier-frequency) in 2011 [1] to 22 Gbit/s (490 GHz) by 2016 [12]. Meanwhile, RTD Tx-based forward error correction (FEC)-level performance (BER $<2 \times 10^{-3}$) were published in [4, 12, 14] with 3 (542), 34 (297) and 28 (490) Gbit/s (GHz), respectively. Error-free RTD Rx-based systems were reported in [2, 5–7, 10, 18] with 1.5 (300), 5.5 (300), 3.2 (290), 10 (317), 17 (297), and 27 (322) Gbit/s (GHz), respectively, while FEC-level performances were announced in [2, 5, 6, 15, 18, 22] with 2.5 (300), 6 (300), 3.8 (290), 1 (300), 32 (322), and 34 (350) Gbit/s (GHz), respectively.

RTD Tx & Rx-based error-free experiments were reported in [3, 8, 13, 17, 20, 21, 25, 26] with 1.5 (300), 2.5 (300), 9 (330), 1.5 (300), 1.5 (350), 11 (340), 30 (350), and 13 (345) Gbit/s (GHz), respectively. 4K-video transmissions were demonstrated in [13, 16, 17, 19, 21, 23–25]. Baseband radio-over-fiber (RoF) transmission followed by Tx & Rx RTD over 7.5 cm (350 GHz) with high-definition television transmission was reported in [20], and this result was increased to

M. Fujita · J. Webber · T. Nagatsuma (✉)
Osaka University, Toyonaka, Japan
e-mail: nagatuma@ee.es.osaka-u.ac.jp

T. Kürner et al. (eds.), *THz Communications*, Springer Series in Optical Sciences 234, https://doi.org/10.1007/978-3-030-73738-2_44

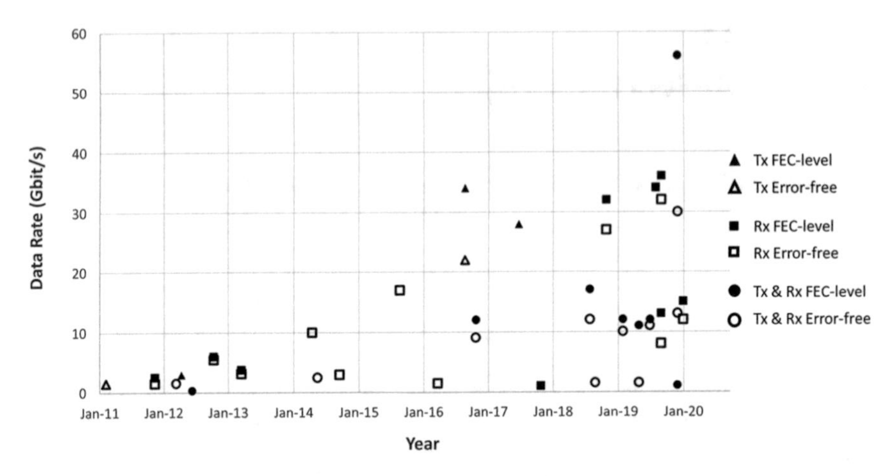

Fig. 44.1 Data rates versus publication date for RTD-based communications. Chart categorizes Tx, Rx, and Tx & Rx results with FEC-level as well as error-free performances

10 cm and 4K transmission (340 GHz) in [21]. All wireless experiments employed amplitude-shift keying modulation except [22] where quaternary phase-shift keying was demonstrated and a code-division-multiple access on-off-keying transmission was reported in [26]. The transmission distances, where reported, were under 8 cm except for [2, 13–15, 21] where 30 (300), 10 (286), 30 (490), 20 (283), and 10 (340) cm (GHz) distances, respectively, were achieved. RTD Tx & Rx-based FEC-level wireless transmission were announced in [13, 20, 21, 25] with 12 (286), 11 (350), 12 (340) and 56 (350) Gbit/s (GHz), respectively.

44.2 Photonic-Crystal-Based Transmission

Systems employing an RTD Rx integrated into a photonic-crystal structure were demonstrated with error-free transmission in [9, 11, 23, 24, 27] with 3 (330), 1.5 (332), 8 (330), 32 (350), and 12 (330) Gbit/s (GHz), respectively, while FEC-level performances were published in [23, 24, 27] for 13 (330), 36 (350), and 15 (330) Gbit/s (GHz), respectively. RTD Tx & Rx-based error-free transmissions were achieved in [16, 19] with 12 (340) and 10 (340) Gbit/s (GHz), respectively, while FEC-level results were reported in [16, 19] providing 17 (340) and 12 (340) Gbit/s (GHz), respectively.

44.3 Coherent Detection

The sensitivity of a RTD receiver is greatly enhanced by using coherent detection. Here, the Rx is locked to the Tx oscillation frequency. The frequency and phase of the local oscillator are aligned with those of the detected carrier signal. Conventionally, this is achieved by digital signal processor but is expensive to perform in terms of power consumption. The sensitivity of the Rx in [25] was enhanced by 40 dB at the maximum using coherent detection, and this permitted an error-free transmission of 30 Gbit/s.

44.4 Baseband Radio-over-Fiber with Tx-Rx RTD Demonstrator

A baseband signal was generated by an arbitrary waveform generator and amplitude modulated onto a laser carrier using an electro-optical modulator and amplified by an erbium-doped fiber amplifier before being transmitted over 1 km of fiber. The received optical signal was detected by a photodiode and then relayed over a 7.5 cm airgap by a Tx RTD. The Rx detected signal was pre-amplified and conditioned using a 11 Gbit/s limit amplifier [21]. The 3-dB baseband bandwidth of the RTD circuit is about 10.3 GHz. A silicon super-hemispherical lens of diameter 12 mm is affixed above the antenna as shown in Fig. 44.2a and improved the gain of the Tx to about 27 dBi. The RTD with baseband circuit and bow-tie antenna are shown in Fig. 44.2b. Error-free transmission at data rates up to 11 Gbit/s was achieved. The performance-limiting factors were considered to be the 3-dB bandwidths of the RTD baseband circuit and limit amplifier, as well as the signal-to-noise ration of the RTD Rx signal. This setup was then modified to demonstrate ultrahigh definition television transmission (3840 × 2160 pixels) at 30 frames per second which requires an error-free throughput of at least 6 Gbit/s.

44.5 Direct THz Communication with Wireless and Fiber Links

A THz fiber link of 1-m length together with RTD for wireless communication operating at 0.3 THz with photonic-crystal waveguide coupler was demonstrated in [23]. A tapered structure shown in Fig. 44.2c enhanced the coupling efficiency (~30%) between the photonic-crystal waveguide and the THz fiber allowing communications up to 10 Gbit/s [19]. Its envisaged uses can include remote sensing, security, and medical applications. The baseband circuit, RTD, and photonic crystal waveguide integration is shown in Fig. 44.2d.

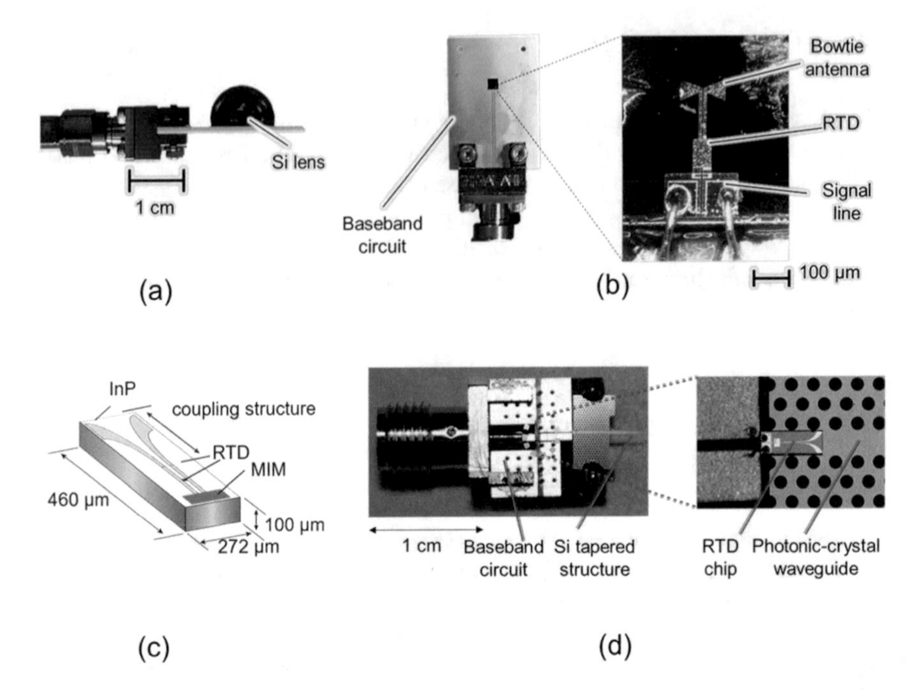

Fig. 44.2 (**a**) RTD with coaxial connector side view, (**b**) top view and photograph showing RTD with bow-tie antenna, and baseband matching circuit, (**c**) schematic of RTD chip to integrate photonic-crystal waveguide, and (**d**) photograph of RTD chip and photonic-crystal waveguide

44.6 Conclusion and Future Perspectives

This chapter has charted the research on RTDs and evolution of data rates achieved over the last decade. In 2011, a 2.5 Gbit/s system (FEC error level) with RTD Tx and Rx was reported, and performance had increased to 56 Gbit/s by 2019. Over the same period, error-free performance has increased from 1.5 Gbit/s to 32 Gbit/s. A description of experiments on baseband radio-over-fiber (RoF) and THz fiber integration were provided as two examples of systems employing an RTD.

Ongoing research is addressing the limited output power of RTD devices and investigating the embedding of RTD in the photonic-crystal platform as well as placement of multiple RTDs in a ring structure. The future potential will depend partly on the cost of deployment. There is the potential for RTD devices to be relatively low-cost devices in volume. Integration of an RTD and photodiode could further reduce cost, power consumption, and footprint in the Rx end of an RoF link.

References

1. Mukai, T., Kawamura, M., Takada, T., & Nagatsuma, T. (2011). 1.5-Gbit/s wireless transmission using resonant tunneling diodes at 300GHz. In *International conference on Optical Terahertz Science and Technology (OTST)*.
2. Shiode, T., Mukai, T., Kawamura, M., & Nagatsuma, T. (2011). Giga-bit wireless communication at 300 GHz using resonant tunneling diode detector. In *Asia-Pacific Microwave Conference (APMC)* (pp. 1122–1125).
3. Shiode, T., Kawamura, M., Mukai, T., & Nagatsuma, T. (2012). Resonant-tunneling diode transceiver for 300 GHz-band wireless link. In *Asia-Pacific Microwave Photonics Conference 2012 (APMP)*.
4. Ishigaki, K., Shiraishi, M., Suzuki, S., Asada, M., Nishiyama, N., & Arai, S. (2012). Direct intensity modulation and wireless data transmission characteristics of terahertz-oscillating resonant tunnelling diodes. *Electronics Letters, 48*, 582–583.
5. Kaku, A., Shiode, T., Mukai, T., Tsuruda, K., Fujita, M., & Nagatsuma, T. (2012). Characterization of resonant tunneling diodes as receivers for terahertz communications. In *International symposium on Frontiers in THz Technology (FTT)*.
6. Kaku, A., Shiode, T., Ishigaki, T., Mukai, T., Tsuruda, K., Fujita, M., & Nagatsuma, T. (2013). 3-Gbit/s error-free terahertz communication using resonant tunneling diode detectors integrated with MgO hyper-hemispherical lens. In *Asia-Pacific Microwave Photonics Conference (APMP)*.
7. Tsuruda, K., Kaku, A., Tsuji, D., Shiode, T., Ishigaki, T., Mukai, T., Fujita, M., & Nagatsuma, T. (2014). 10-Gbps error-free terahertz wireless communications using resonant tunneling diode receivers. In *7th Global Symposium on Millimeter-Waves (GSMM)*.
8. Nagatsuma, T., Fujita, M., Kaku, A., Tsuji, D., Nakai, S., Tsuruda, K., & Mukai, T. (2014). Terahertz wireless communications using resonant tunneling diodes as transmitters and receivers. In *3rd International Conference on Telecommunication and Remote Sensing (ICTRS)*.
9. Suminokura, A., Tsuruda, K., Mukai, T., Fujita, M., & Nagatsuma, T. (2014). Integration of resonant tunneling diode with terahertz photonic-crystal waveguide and its application to gigabit terahertz-wave communications. In *International. topical meeting on Microwave Photonics (MWP) and 9th Asia-Pacific Microwave Photonics Conference (APMP)*.
10. Nishio, K., Diebold, S., Nakai, S., Tsuruda, K., Mukai, T., Kim, J., Fujita, M., & Nagatsuma, T. (2015). Resonant tunneling diode receivers for 300-GHz-band wireless communications. In *2015 URSI-Japan Radio Science Meeting (URSI-JRSM)*.
11. Fujita, M., & Nagatsuma, T. (2016). Photonic crystal technology for terahertz system integration. *Proceedings of SPIE, 9856*, 98560P.
12. Oshima, N., Hashimoto, K., Suzuki, S., & Asada, M. (2016). Wireless data transmission of 34 Gbit/s at a 500-GHz range using resonant-tunnelling diode terahertz oscillator. *Electronics Letters, 52*, 1897–1898.
13. Diebold, S., Nishio, K., Nishida, Y., Kim, J., Tsuruda, K., Mukai, T., Fujita, M., & Nagatsuma, T. (2016). High-speed error-free wireless data transmission using a terahertz resonant tunneling diode transmitter and receiver. *Electronics Letters, 52*, 1999–2001.
14. Oshima, N., Hashimoto, K., Suzuki, S., & Asada, M. (2017). Terahertz wireless data transmission with frequency and polarization division multiplexing using resonant-tunneling-diode oscillators. *IEEE Transactions on THz Science Technology, 7*, 593–598.
15. Nishida, Y., Diebold, S., Tsuruda, K., Mukai, T., Kim, J., Fujita, M., & Nagatsuma, T. (2017). Injection-locked resonant tunneling diode receiver for terahertz communications. In *4th International symposium on Microwave/Terahertz Science and Applications (MTSA) & 8th International symposium on Terahertz Nanoscience (TeraNano 8)*.
16. Yu, X., Yamada, R., Kim, J., Fujita, M., & Nagatsuma, T. (2018). Integrated circuits using photonic-crystal slab waveguides and resonant tunneling diodes for terahertz communication. In *2018 Progress In Electromagnetics Research Symposium (PIERS)*.

17. Webber, J., Nishigami, N., Kim, J., Fujita, M., & Nagatsuma, T. (2018). Terahertz wireless communication using resonant tunneling diodes with radio-over-fiber technology. In *JSAP-OSA Joint Symposia 2018, (JSAP-OSA).*
18. Nishigami, N., Nishida, Y., Diebold, S., Kim, J., Fujita, M., & Nagatsuma, T. (2018). Resonant tunneling diode receiver for coherent terahertz wireless communication. In *Asia-Pacific Microwave Conference 2018 (APMC).*
19. Yu, X., Hosoda, Y., Miyamoto, T., Obata, K., Kim, J., Fujita, M., & Nagatsuma, T. (2019). Terahertz fibre transmission link using resonant tunnelling diodes integrated with photonic-crystal waveguides. *Electronics Letters, 55*, 398–400.
20. Webber, J., Nishigami, N., Yu, X., Kim, J., Fujita, M., & Nagatsuma, T. (2019). Terahertz wireless communication using resonant tunneling diodes and practical radio-over-fiber technology. In *2019 IEEE International Conference on Communication Workshops (ICC Workshops)* (pp. 1–5).
21. Webber, J., Nishigami, N., Kim, J., Fujita, M., & Nagatsuma, T. (2019). Terahertz wireless communications using resonant tunnelling diodes with radio-over-fibre. *Electronics Letters, 55*, 949–951.
22. Yamamoto, T., Nishigami, N., Kim, J., Fujita, M., & Nagatsuma, T. (2019). 34 Gbit/s QPSK wireless transmission using resonant tunneling diode receiver. In *13th Topical Workshop on Heterostructure Microelectronics (TWHM).*
23. Yu, X., Hosoda, Y., Miyamoto, T., Obata, K., Kim, J., Fujita, M., & Nagatsuma, T. (2019). Direct terahertz communications with wireless and fiber links. In *44th International Conference on Infrared, Millimeter Terahertz Waves (IRMMW-THz).*
24. Yu, X., Kim, J., Fujita, M., & Nagatsuma, T. (2019). Efficient mode converter to deep-subwavelength region with photonic-crystal waveguide platform for terahertz applications. *Optics Express, 27*, 28707–28721.
25. Nishida, Y., Nishigami, N., Diebold, S., Kim, J., Fujita, M., & Nagatsuma, T. (2019). Terahertz coherent receiver using a single resonant tunnelling diode. *Scientific Reports, 9*, 18125.
26. Webber, J., Nishigami, N., Kim, J., Fujita, M., & Nagatsuma, T. (2019). Terahertz wireless CDMA communication using resonant tunneling diodes. In *2019 IEEE Global Communication Conference (GLOBECOM Workshops)* (pp. 1–5).
27. Yu, X., Ohira, T., Kim, J., Fujita, M., & Nagatsuma, T. (2020). Waveguide-input resonant tunnelling diode mixer for THz communications. *Electronics Letters, 56*, 342–344.

Chapter 45
Photonics-Aided 300–500 GHz Wireless Communications Beyond 100 Gbps

Xianbin Yu, Lu Zhang, Xianmin Zhang, Shi Jia, Hao Hu, Toshio Morioka, Leif K. Oxenløwe, and Xiaodan Pang

Abstract We experimentally demonstrated a series of photonics-wireless transmissions in the 300–500 GHz band at the Technical University of Denmark (DTU) and the Zhejiang University (ZJU), with special efforts to drive the data rates beyond 100 Gbps and to extend the THz wireless reach by combining the techniques of state-of-the-art optoelectronic THz devices, spectrally efficient modulation formats and advanced digital signal processing.

45.1 Introduction

The proliferation of wireless connections and the emergence of bandwidth-consuming Internet services have significantly raised the demand of ultrahigh speed wireless communications. In this context, terahertz (THz, 0.3–10 THz)-based communication with extremely large bandwidth available is of great interest [1]. It has been well known that the photonic approach is superior to the electronic, particularly attributed to the development of state-of-the-art optoelectronic THz devices and the potential of seamless convergence to optical networks. In the recent

X. Yu (✉) · L. Zhang · X. Zhang
College of Information Science and Electronic Engineering, Zhejiang University, Hangzhou, China
e-mail: xyu@zju.edu.cn; zhanglu1993@zju.edu.cn; zhangxm@zju.edu.cn

S. Jia · H. Hu · T. Morioka · L. K. Oxenløwe
DTU Fotonik, Technical University of Denmark, Lyngby, Denmark
e-mail: shijai@fotonik.dtu.dk; huhao@fotonik.dtu.dk; tomo@fotonik.dtu.dk; lkox@fotonik.dtu.dk

X. Pang
Applied Physics Department, KTH Royal Institute of Technology, Stockholm, Sweden
e-mail: xiaodan@kth.se

T. Kürner et al. (eds.), *THz Communications*, Springer Series in Optical Sciences 234, https://doi.org/10.1007/978-3-030-73738-2_45

years, photonics-assisted THz communication has, therefore, progressed rapidly, and a series of photonics-wireless transmission experiments with beyond 100 Gbps data rates in the 300–500 GHz band have been successfully demonstrated at the Technical University of Denmark (DTU) and the Zhejiang University (ZJU).

In 2016, we implemented a high-speed 400 GHz photonic wireless communication system with an aggregated data rate of up to 60 Gbps, pushing the data rate envelope at frequencies above 300 GHz [2]. The system is flexible and scalable with line rates by employing Nyquist QPSK signals in a 12.5-GHz UD-WDM grid, which makes this scheme promising in bridging next-generation optical 100-GbE data rates into the wireless domain, for high data rate radio wireless applications. Subsequently, a photonic-wireless transmission of 160 Gbps QPSK was demonstrated by using a single THz emitter and 25 GHz-spaced 8 channels (20 Gbps per channel) in the 300–500 GHz band [3]. In 2017, the 16-QAM modulation was used to increase the spectral efficiency. By optimizing a spectral arrangement scheme and a tailored DSP routine, a single-transmitter/single-receiver THz link with a net data rate of up to 260 Gbps was successfully demonstrated based on six-channel 12.5 Gbaud 16-QAM signals [4]. Meanwhile, a single-channel THz wireless transmission at 106 Gbps net rate was achieved by improving the frequency efficiency without employing any spatial/frequency division multiplexing techniques [5].

As we know, THz power generated by photomixing is typically at tens of microwatt, which consequently limited the wireless propagation distance of high-speed THz communications. With regard to this issue, we recently placed a lot of efforts to extend the wireless reach. In 2018, a 350 GHz 25 Gbaud 16-QAM signal over a 2 m free space link was demonstrated, enabled by combining the techniques of cutting-edge narrowband uni-travelling carrier photodiode (UTC-PD), spectrally efficient modulation format and advanced digital signal processing algorithms [6]. In 2019, a 400 GHz wireless transmission of 131.21 Gbps net rate over 10.7-m wireless was demonstrated by employing 16-QAM-OFDM modulation and nonlinear DSP flow [7], and most recently, a 26.8 m wireless transmission of 120.97 Gbps was successfully achieved at the 350 GHz band without using any THz amplifiers, by combining the techniques of probabilistic shaping, Cassegrain antennas and advanced digital signal processing techniques [8].

45.2 Photonics-Aided THz Wireless Demonstrations at DTU and ZJU

Carrier frequency (GHz)	Line rate (single/multi channel)	Distance	Modulation Format	Best BER (offline)	THz power (dBm)	Year	References
400	60 Gbps (4-ch)	0.5 m	Nyquist QPSK	8.5e-4	−17	2016	[2]
300–500	160 Gbps (8-ch)	0.5 m	QPSK	4.7e-4	−21	2016	[3]
300–500	300 Gbps (6-ch)	0.5 m	16-QAM	1.8e-3	−21	2017	[4]
400	128 Gbps (single-ch)	0.5 m	16-QAM	2.5e-2	−12	2017	[5]
350	100 Gbps (single-ch)	2 m	16-QAM	3.5e-3	−12	2018	[6]
408	157.46 Gbps (single-ch)	10.7 m	16-QAM-OFDM	2.5e-2	−12	2019	[7]
350	120.97 Gbps (single)	26.8 m	16-QAM-OFDM	2.5e-2	−13	2019	[8]

45.2.1 260 Gbps Photonic-Wireless Transmission in the THz Band

In this work, the 16-QAM modulation format was used to increase the spectral efficiency, and an optimal frequency-band-allocation scheme was applied by grouping multichannel in pairs instead of equal spacing to reduce the number of guard bands. In addition, a tailored DSP routine with pre- and post-equalization was employed to accurately reconstruct the channel response. Three configurations were demonstrated with 2, 4 and 6 channels, modulated with 20, 16 and 12.5 Gbaud 16-QAM signals, respectively. In the 6-ch configuration, 3 channels at 391 GHz, 408 GHz and 442 GHz achieved BER performance below the 7%-OH hard-decision forward error correction (HD-FEC) threshold, and the other 3 channels below the 20%-OH hard-decision forward error correction (SD-FEC), resulting in a high post-FEC error-free net bit rate of 260 Gbps over 0.5 m (Fig. 45.1).

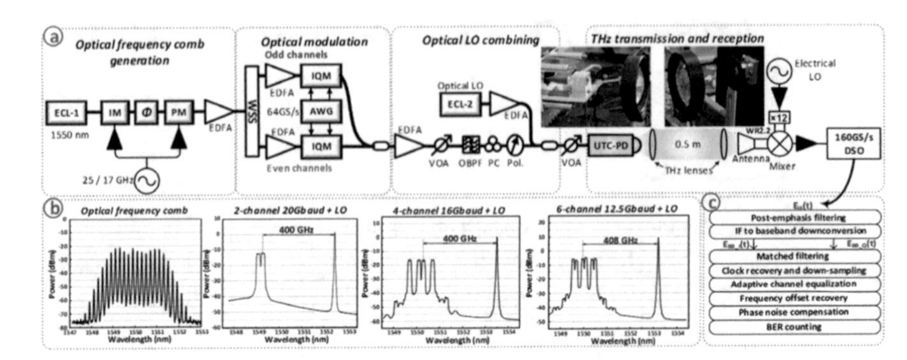

Fig. 45.1 (**a**) Experimental setup of the 300–500 GHz link. (**b**) Optical spectra of generated optical frequency comb, and multichannel signals with local oscillator (LO). (**c**) Structure of the DSP routine at the receiver

45.2.2 106 Gbps Single-Channel Bit Rate at 400 GHz

In this work, a single-channel THz photonic-wireless transmission system at the 400 GHz band was experimentally demonstrated based on a single pair of THz emitter and receiver without using any multiplexing techniques. The capacity of over 100 Gbps within a single channel was implemented by combining coherent photonic generation of a pure THz tone, 16-QAM modulation format, and a tailored DSP routine with pre- and post-equalization. By accurately reconstructing the frequency response, up to 32 Gbaud 16-QAM signals over 0.5 m wireless within a single broadband THz channel was successfully demonstrated in the experiment, resulting in a pre-FEC line rate of 128 Gbps and post-FEC error-free bit rate of up to 106 Gbps (Fig. 45.2).

45.2.3 26.8 M 350 GHz Wireless Transmission of Beyond 100 Gbps

In this work, 350 GHz photonic wireless transmission of 100.8 Gbps net rate over 26.8 m wireless was experimentally demonstrated without using any THz amplifiers. In the implementation, the probabilistic shaping orthogonal frequency division multiplexing techniques were used. In addition, a pair of high-gain Cassegrain antennas were employed to improve the link budget, and a Schottky mixer-based down-converter was used to improve the receiver sensitivity. After 26.8 m wireless transmission, the BER reaches the SD-FEC threshold of 2.7e-2 at 13 dBm incident optical power, resulting in a pre-FEC line rate of 120.97 Gbps and a net rate of 100.8 Gbps (Fig. 45.3).

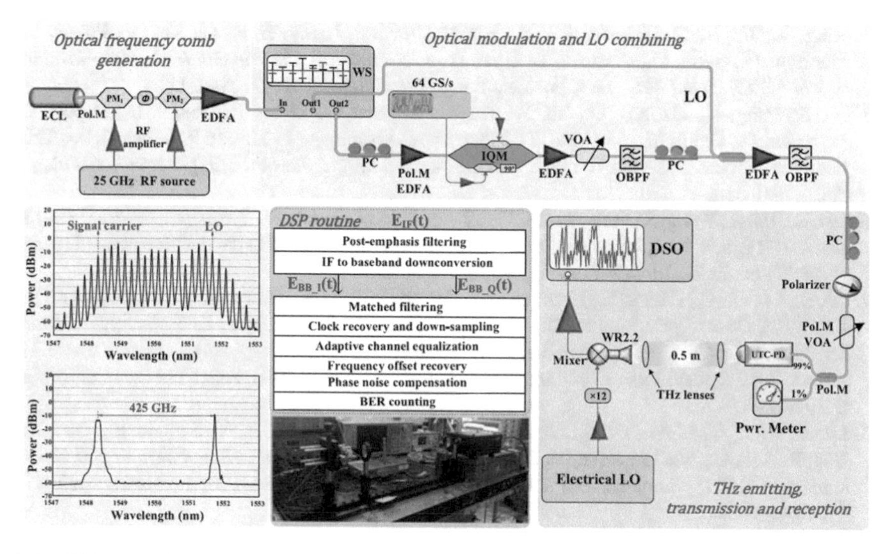

Fig. 45.2 Experimental setup of the single-channel photonic-wireless link in the 400 GHz band. Insets: optical spectra of generated optical frequency comb (up-left) and combined modulated signal and LO (down-left), and the DSP routine structure at the receiver (up-middle) and a picture of the actual setup (down-middle)

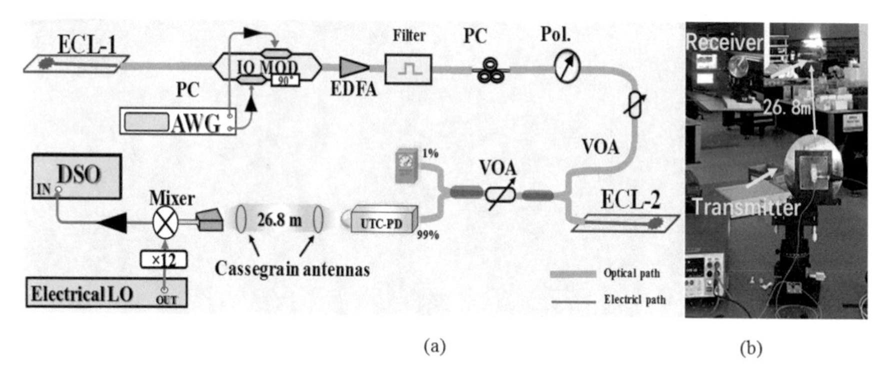

Fig. 45.3 (**a**) Experimental setup of THz wireless transmission system. (**b**) The picture of THz wireless link

References

1. Federici, J., & Moeller, L. (2010). Review of terahertz and subterahertz wireless communications. *Journal of Applied Physics, 107*(11).
2. Yu, X., Asif, R., Piels, M., Zibar, D., Galili, M., Morioka, T., Zhang, X., Jepsen, P. U., & Oxenløwe, L. K. (2016, November). 400 GHz wireless transmission of 60 Gbps Nyquist-QPSK signals using UTC-PD and heterodyne mixer. *IEEE Transactions on THz Science and Technology, 6*(6), 765–770.
3. Yu, X., Jia, S., Hu, H., Galili, M., Morioka, T., Jepsen, P. U., & Oxenløwe, L. K. (2016). 160 Gbit/s photonics wireless transmission in the 300-500 GHz band. *APL Photonics, 1*, 081301.

4. Pang, X., Jia, S., Ozolins, O., Yu, X., Hu, H., Marcon, L., Guan, P., Da Ros, F., Popov, S., Jacobsen, G., Galili, M., Morioka, T., Zibar, D., & Oxenløwe, L. K. *260 Gbit/s photonic-wireless link in the THz band*, IPC 2016, Postdeadline paper, October 2016, Hawaii, USA.
5. Jia, S., Pang, X., Ozolins, O., Yu, X., Hu, H., Marcon, L., Guan, P., Ros, F. D., Popov, S., Jacobsen, G., Galili, M., Morioka, T., Zibar, D., & Oxenlowe, L. K. (2018, January). 0.4 THz photonic-wireless link with 106 Gbit/s single channel bitrate. *Journal of Lightwave Technology, 36*(2), 610–616.
6. Liu, K., Jia, S., Wang, S., Pang, X., Li, W., Zheng, S., Chi, H., Jin, X., Zhang, X., & Yu, X. (2018, June). 100 Gbit/s THz photonic wireless transmission in the 350 GHz band with extended reach. *IEEE Photonics Technology Letters, 30*(11), 1064–1067.
7. Jia, S., Lo, M.-C., Zhang, L., Ozolins, O., Udalcovs, A., Kong, D., Pang, X., Yu, X., Xiao, S., Popov, S., Chen, J., Carpintero, G., Morioka, T., Hu, H., & Oxenløwe, L. K. *Integrated dual-DFB Laser for 408 GHz carrier generation enabling 131 Gbit/s wireless transmission over 10.7 meters*. 2019 Optical Fiber Communication Conference (OFC 2019), Paper Th1C.2, California, 2019 March 3–7.
8. Lu, Z., Wang, S., Li, W., Jia, S., Zhang, L., Qiao, M., Pang, X., Idrees, N., Saqlain, M., Gao, X., Cao, X., Lin, C., Wu, Q., & Yu, X. *26.8 m 350 GHz wireless transmission of beyond 100 Gbit/s supported by THz photonics*, ACP 2019, Postdeadline paper, Paper M4D.6, 2–5 November 2019, Chengdu, China.

Chapter 46
Ultra-Broadband Networking Systems Testbed at Northeastern University

Josep Miquel Jornet, Priyangshu Sen, and Viduneth Ariyarathna

Abstract Unleashing the potential of THz communications requires the development of nontraditional signal processing, communication, and networking solutions. The ultra-broadband networking systems testbed at Northeastern University is a programmable software-defined radio testbed spanning the 120–140 GHz, 200–240 GHz, and 1–1.05 THz bands, with baseband bandwidths ranging from 2 GHz to 32 GHz, which enables the validation of innovative solutions for THz networks.

46.1 Platform Specifications

The testbed, shown in Fig. 46.1, is integrated by two nodes (namely, a transmitter and a receiver) with RF front-ends at 120–140 GHz, 200–240 GHz, and 1–1.05 THz, making it the first THz communication testbed able to operate above 1 THz or *true* THz frequencies [1].

46.1.1 Terahertz Front-Ends

The front-ends consist of frequency-multiplying chains based on Schottky diode technology, utilized to up-convert independent local oscillators (LOs) generated by two Keysight PSG E8257 (up to 50 GHz). Sub-harmonic mixers based on the same technology are then used to modulate the THz carrier signals with the information signal in baseband (BB) or at an intermediate frequency (IF). The front-ends at 120–140 GHz and 1–1.05 THz are custom designed by Virginia Diodes Inc (VDI), and their output power is of 13 dBm (20 mW) and −15 dBm (30 μW), respectively. The front-end at 200–240 GHz has been jointly developed with the NASA Jet

J. M. Jornet (✉) · P. Sen · V. Ariyarathna
Ultrabroadband Nanonetworking Laboratory, Institute for the Wireless Internet of Things, Department of Electrical and Computer Engineering, Northeastern University, Boston, MA, USA
e-mail: jmjornet@northeastern.edu; sen.pr@northeastern.edu; viduneth@northeastern.edu

T. Kürner et al. (eds.), *THz Communications*, Springer Series in Optical Sciences 234, https://doi.org/10.1007/978-3-030-73738-2_46

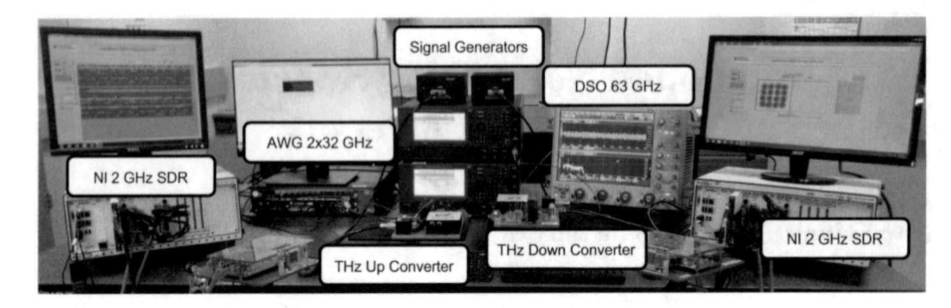

Fig. 46.1 Ultra-broadband networking systems testbed at Northeastern University

Propulsion Laboratory (JPL) and leverages NASA's patented multiplier technology based on on-chip power combining [2, 3]. The measured output power is a world-record-setting 23 dBm (200 mW). Directional horn antennas at the aforementioned frequencies with gains ranging from 25 to 55 dBi are available to the team.

46.1.2 Signal Processing Engines

The information-bearing signal at BB or IF can be generated at the transmitter and processed at the receiver following two different strategies, namely, an offline signal processing system based on an arbitrary waveform generator (AWG) and a digital storage oscilloscope (DSO), or a real-time signal processing platform based on state-of-the-art software-defined radios.

In the *offline engine* [4], a Keysight AWG M8196A with a baseband bandwidth of 32 GHz per channel (two channels) and a sampling rate of up to 92 Giga-Samples-per-second (GSas) with 8-bit resolution is utilized at the transmitter to generate the signals to be up-converted. Reciprocally, at the receiver, a Keysight DSO Z632A with a baseband bandwidth of 63 GHz and a sampling rate of up to 160 GSas is utilized to capture the received down-converted signals. To orchestrate the two, tailored framing, time synchronization, channel estimation with equalization, and single- and multi-carrier modulation techniques are implemented in software. MATLAB is utilized at both ends, which drastically simplifies the design of the different signal processing blocks and ensures a rapid transition from numerical analysis to experimental testing.

In the *real-time engine*, the National Instruments (NI) mmWave software-defined radio (SDR) platform with 2 GHz bandwidth is utilized as the starting point to design, implement, and test new physical and link layer solutions for THz communication networks. While the bandwidth of this platform is lower than that of the AWG/DSO setup, the NI platform allows the testing of new techniques in dynamically changing conditions. Among others, automatic gain control, automatic frequency offset control, phase noise compensation and modulation, and coding scheme selection for multi-GHz wide channels have been implemented at frequencies ranging from 120–140 GHz to 1–1.05 THz, both with single carrier and multi-carrier modulations.

46.2 Experimental Results

With the current platform, we have been able to obtain different results along three main categories:

- **Propagation and Channel Modeling:** We have conducted extensive ultra-broadband channel measurements at the different frequencies available in the testbed and over distances ranging from tens of centimeters at 1–1.05 THz [4] to beyond 1 km at 200–240 GHz [5]. This is possible thanks to the fact that synchronization is achieved over the air by processing of the IF signal, i.e., without any shared clock between the transmitter and the receiver. For example, in Fig. 46.2a, the channel frequency response for the 1–1.05 THz channel is shown for two different distances.
- **Physical Layer Design:** Tailored to the hardware response and channel behavior, we have implemented time, frequency, and phase synchronization algorithms; ultra-broadband channel estimation and equalization techniques; and single- and multi-carrier modulations (from BPSK to OFDM), at bit rates as high as 30 Gigabits-per-second (Gbps) and with bit error rates (BER) as low as 10^{-4}. The utilization of a software-defined engine enables the testing of nonconventional signal processing blocks able to compensate for the nonlinearities of frequency-multiplying chains. For example, in Fig. 46.2b, the constellation at 10 m for a 16 PSK transmission at 4 Gbps is shown. The measured BER was under 10^{-4}.
- **Link Layer Design:** Beyond channel modeling and physical layer design, enabled by the real-time signal processing engine, new networking solutions including medium access control (MAC) protocols [6] and neighbor discovery strategies [7] are being tested.

46.3 Conclusion

For many years, THz experimental research has been focused on characterizing the THz channel as well as testing and demonstrating new THz devices, usually implementing traditional communication techniques. Moving forward, nontraditional communication and networking solutions tailored to the capabilities of THz devices and the peculiarities of the THz channel that can be found in the related literature need to be experimentally tested and refined. The softwarized experimental platform presented in this chapter enables such research. It is our aim to not only utilize, maintain, and enhance the current platform but also to open the platform to the broader wireless communication research community.

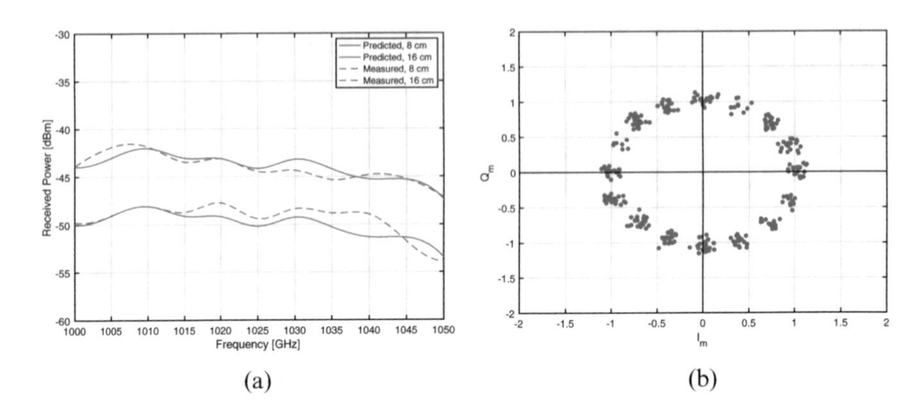

Fig. 46.2 Experimental results obtained with the ultra-broadband networking systems testbed at Northeastern University. (**a**) Channnel response at 1–1.05 THz.. (**b**) Constellation of a 16-PSK at 200–240 GHz over 10 m.

Acknowledgments This work was supported in part by the US Air Force Research Laboratory Grant FA8750-20-1-0200 and the US National Science Foundation Grant CNS-2011411.

References

1. Sen, P., Pados, D.A., Batalama, S.N., Einarsson, E., Bird, J.P., & Jornet, J. M. (2020). The TeraNova platform: An integrated testbed for ultra-broadband wireless communications at true Terahertz frequencies. *Computer Networks, 179*, 107370. https://doi.org/10.1016/j.comnet.2020.107370
2. Perez, J. V. S., Chattopadhyay, G., Lee, C., Schlecht, E. T., Jung-Kubiak, C. D., & Mehdi, I. (2015). On-chip power-combining for high-power Schottky diode based frequency multipliers. U.S. Patent 9,143,084.
3. Siles, J. V., Cooper, K. B., Lee, C., Lin, R. H., Chattopadhyay, G., & Mehdi, I. (2018). A new generation of room-temperature frequency-multiplied sources with up to 10× higher output power in the 160-GHz–1.6-THz range. *IEEE Transactions on Terahertz Science and Technology, 8*(6), 596–604
4. Sen, P., & Jornet, J. M. (2019). Experimental demonstration of ultra-broadband wireless communications at true terahertz frequencies. In *2019 IEEE 20th International Workshop on Signal Processing Advances in Wireless Communications (SPAWC)* (pp. 1–5). IEEE.
5. Jornet, J. M., Siles, J. V., Thawdar, N. (2019). Design, implementation and demonstration of a multi-Gbps link at 210–240 GHz beyond 1 km. *US AFRL Director Innovation Fund—Final Technical Report.*
6. Xia, Q., Hossain, Z., Medley, M., & Jornet, J. M. (2021). A link-layer synchronization and medium access control protocol for terahertz-band communication networks. *IEEE Transactions on Mobile Computing, 20*(1), 2–18.
7. Xia, Q., & Jornet, J. M. (2019). Expedited neighbor discovery in directional terahertz communication networks enhanced by antenna side-lobe information. *IEEE Transactions on Vehicular Technology, 68*(8), 7804–7814.

Chapter 47
Photonics-Based Projects at IEMN

Guillaume Ducournau, Jean-François Lampin, and Mohammed Zaknoune

Abstract The photonic technologies have enabled the generation of THz with the photomixing technique, which is essentially a down-conversion from optical domain to the THz. Working in pulse-mode or in continuous-wave, this approach relies on an efficient optical to THz converter. Such device can be a photoconductor or an ultra-fast photodiode. At 1.55 μm, the uni-travelling carrier photodiode (UTC-PD) has been proven to be efficient and scalable to reach the THz frequencies. We report here some examples of THz communication systems enabled by InGaAs/InP UTC-emitter, in the range 200–600 GHz. Among them, first-generation (2010–2015) passive THz hotspots based on bias-free UTC-PD at 200 GHz, broadband transmission at 400 GHz and first tests conducted in the 600 GHz band are described. The second generation (2016–2020) used a high-efficiency UTC-PD for 100 Gbit/s in the 300 GHz band leveraging on GaAs technology in the receiver part.

47.1 Passive THz Hot Spots Based on Bias-Free UTC-PD Coupled to Broadband Antenna

UTC-PD relies on InGaAs/InP heterostructure grown on InP substrate that can be engineered to enable the operation of the photodiode without any electrical bias. This is of particular importance for the seamless integration of future THz links fed by optical fibre networks. In that case, the THz hot spot can be only driven by the sole optical network. Moreover, transverse electromagnetic (TEM)-Horn antenna [1] enabled to reach frequencies between 200 and 600 GHz without any dispersion, which is very appealing for communication applications with high baud rates.

The use of this UTC-PD+antenna has been applied in [2] for passive 200 GHz emitter with real-time Gbit/s (BER $<10^{-9}$) over 2 m. In this system, the UTC-PD was fed by optical signals only. Leveraging the broadband TEM antenna, ultra-

G. Ducournau (✉) · J.-F. Lampin · M. Zaknoune
IEMN-CNRS, University of Lille, Lille, France
e-mail: guillaume.ducournau@univ-lille.fr

T. Kürner et al. (eds.), *THz Communications*, Springer Series in Optical Sciences 234, https://doi.org/10.1007/978-3-030-73738-2_47

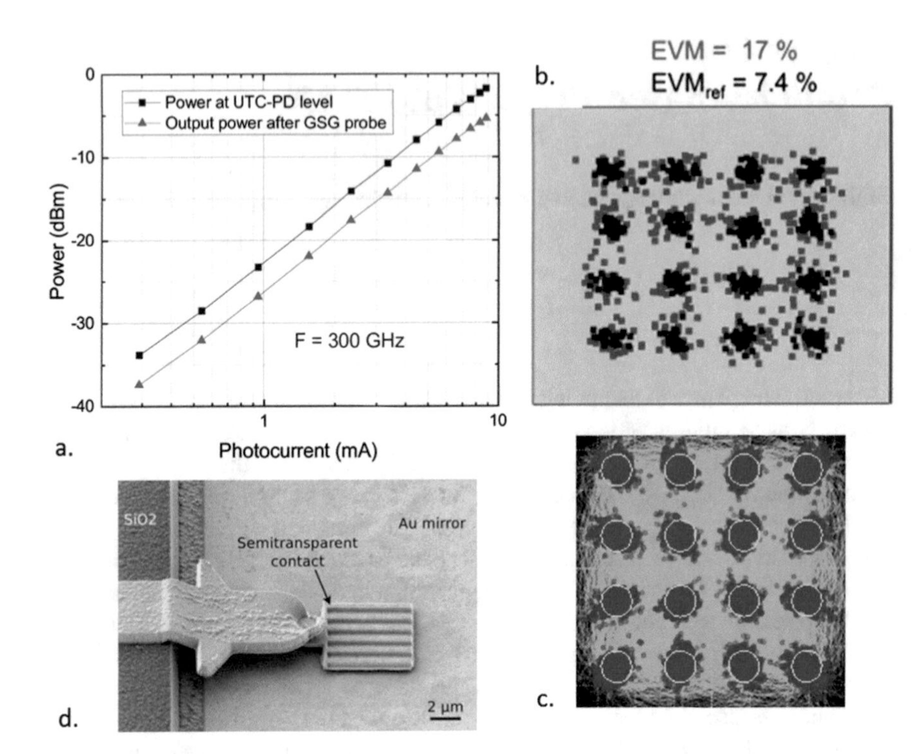

Fig. 47.1 (**a**) 300 GHz continuous-wave output power of the RCE UTC-PD [5, 8] (© 2020 IEEE, reproduced with permission). (**b**) 16QAM-100 Gbit/s constellation of back-to-back /15 m THz link. All details of the system can be found in [8] (© 2020 IEEE, reproduced with permission). (**c**) 16QAM-100 Gbit/s for a short range (50 cm) using a GaAs Schottky-based receiver (BER = 10^{-4}) [7] © 2018 IEEE, reproduced with permission. (**d**) SEM view of the RCE UTC-PD chip fabricated at the IEMN [5]. (© 2017 IEEE, reproduced with permission)

wideband amplitude modulations were also investigated at 400 GHz: in [3] and up to 46 Gbit/s signals were transmitted over 2 m using a mixer-based detection. Last, the same transmitter was used in [4] to reach the 600 GHz band. In that case, HDTV data (1.5 Gbit/s) were transmitted with real-time performances (no signal processing, no latency time in the THz system) thanks to the combination of waveguide-based sub-harmonic mixer combined to envelop detection at mixer output. The combination of a mixer+envelop detection enabled to reach a very good sensitivity (mixer-based detection) and the simplicity of envelop detection. The envelop detector is appealing to avoid the phase-locking between the THz transmitter and receiver. However, as residual IF carrier frequency at mixer output is not very high compared to the baud rate, custom design of the envelop detector had to be done to enable efficient detection of fast baud-rates modulated on IF.

47.2 100 Gbit/s Links Using High-Efficiency UTC-PDs

In the THz communications perspective, the power performance of any emitter is a key parameter, as well as the linearity of the whole transmission system. One advantage of the photonics approach is the good linearity of the photodiode when the device is operated below saturation region. As THz power increases with impinging 1.55 μm optical power, the optical to THz efficiency has to be optimized to limit the required optical pumping power to reach decent THz power at photomixer output.

In this perspective, resonant cavity enhanced (RCE) UTC-photodiodes were developed at the IEMN to improve the overall efficiency of the photomixer. This approach enabled to generate close to mW level at 300 GHz [5], with a high efficiency. Such photomixers, combining high output power with a good linearity response, were successfully used to generate up to 100 Gbit/s signals in the 300 GHz band, in the perspective of the IEEE 802.15.3d [6].

This was achieved using both waveguide-integrated sub-harmonic mixers [7] for short-range links. Recently, the combination with active receiver structures including 300 GHz LNAs enabled to reach 100 Gbit/s with a better system margin, and communication distances up to 15 m were demonstrated [8] in the framework of a transnational program between France and Germany (Fig. 47.1b).

References

1. Beck, A., et al. (2008, October 23). High-efficiency uni-travelling-carrier photomixer at 1.55 μm and spectroscopy application up to 1.4 THz. *Electronics Letters, 44*(22), 1320–1321.
2. Ducournau, G., et al. (2010, September 16). Optically power supplied Gbit/s wireless hotspot using 1.55 μm THz photomixer and heterodyne detection at 200 GHz. *Electronics Letters, 46*(19), 1349–1351.
3. Ducournau, G., et al. (2014, May). Ultrawide-bandwidth single-channel 0.4-THz wireless link combining broadband quasi-optic photomixer and coherent detection. *IEEE Transactions on Terahertz Science and Technology, 4*(3), 328–337.
4. Ducournau, G., et al. (2014, February 27). High-definition television transmission at 600 GHz combining THz photonics hotspot and high-sensitivity heterodyne receiver. *Electronics Letters, 50*(5), 413–415.
5. Latzel, P., et al. (2017, November). Generation of mW level in the 300-GHz band using resonant-cavity-enhanced Unitraveling carrier photodiodes. *IEEE Trans. on Terahertz Science and Technology, 7*(6), 800–807.
6. IEEE Standard for High Data Rate Wireless Multi-Media Networks–Amendment 2: 100 Gb/s Wireless Switched Point-to-Point Physical Layer. In IEEE Std 802.15.3d-2017 (Amendment to IEEE Std 802.15.3-2016 as amended by IEEE Std 802.15.3e-2017), vol., no., pp. 1–55, Oct. 18 2017.
7. Chinni, V. K. et al. Indoor 100 Gbit/s THz data link in the 300 GHz band using fast photodiodes. In 2018 25th International Conference on Telecommunications (ICT), St. Malo, 2018, pp. 288–290, https://doi.org/10.1109/ICT.2018.8464945.
8. Dan, I., et al. (2020, May). A 300-GHz wireless link employing a photonic transmitter and an active electronic receiver with a transmission bandwidth of 54 GHz. *IEEE Transactions on Terahertz Science and Technology, 10*(3), 271–281.

Chapter 48
Wireless THz Transmission Using a Kramers-Kronig Receiver

Wolfgang Freude, Christoph Füllner, Tobias Harter, Christian Koos, and Sebastian Randel

Abstract We discuss a wireless transmission experiment over 110 m using 16 QAM signalling for a net data rate of 115 Gbit/s. The carrier has a frequency of 0.3 THz. It is generated and modulated by photonic techniques. To allow Kramers-Kronig reception of *m*QAM data, a continuous-wave signal is transmitted at one edge of the signal spectrum. The receiver is a simple Schottky barrier diode. Its output current is digitally processed to recover both modulated quantities, amplitude and phase.

48.1 Introduction

High-speed wireless communication links must be technically simple and spectrally efficient. With carrier frequencies near 0.3 THz, the transmission of large data rates beyond 100 Gbit/s is possible. Atmospheric loss stays below 30 dB/km is even under adverse weather conditions. Distances of more than 100 m can be bridged, limited by the transmitter (Tx) power, by the receiver (Rx) sensitivity, and by the directivities of the antennae. Simple Schottky barrier diode (SBD) receivers reduce the hardware complexity. However, such an Rx detects amplitudes only, e.g., an *m*-level pulse amplitude modulation (*m*PAM, $m = 2, 3, \ldots, M$), and receives $r = \log_2 m$ bit per symbol. If more advanced modulation formats like *m*-state quadrature amplitude modulation (*m*QAM) are employed, a "local" oscillator (LO) has to be transmitted along with the data, or it has to be added at the receiver. This LO helps shifting the received THz spectrum to a more convenient frequency range, and it acts as a reference for the demodulation of the phase information. For a single-ended receiver, the LO has to be spectrally separated from the data by a guard band to avoid signal-signal beat interference (SSBI). The spectral efficiency is

W. Freude (✉) · C. Füllner · T. Harter · C. Koos · S. Randel
Institute of Photonics & Quantum Electronics (IPQ), Karlsruhe Institute of Technology (KIT), Karlsruhe, Germany
e-mail: w.freude@kit.edu

T. Kürner et al. (eds.), *THz Communications*, Springer Series in Optical Sciences 234, https://doi.org/10.1007/978-3-030-73738-2_48

therefore poor. A so-called Kramers-Kronig receiver as described in the next section overcomes the limitations imposed by reception with a single amplitude detector.

48.2 The Kramers-Kronig Receiver

The Kramers-Kronig (KK) receiver [1–4] relies on digital signal processing (DSP) and restores the transmitted phase information by evaluating the received intensity. A typical transmission setup is shown in Fig. 48.1. Two independent data signals delivered by an arbitrary waveform generator (AWG) drive an in-phase/quadrature (IQ) modulator that modulates an optical carrier (frequency $f_0 = 193.4865$ THz, vacuum wavelength $\lambda_0 = 1.55$ μm) independently by amplitude $\left|\underline{U}_{\mathrm{d}}(t)\right|$ and phase $\varphi_{\mathrm{d}}(t)$. The complex data $\underline{U}_{\mathrm{d}}(t) = \left|\underline{U}_{\mathrm{d}}(t)\right| \exp\left(\mathrm{j}\varphi_{\mathrm{d}}(t)\right)$ result in an mQAM formatted modulator output signal $u_{\mathrm{d}}(t) = \Re\left\{\underline{U}_{\mathrm{d}}(t) \exp\left(\mathrm{j}2\pi f_0\, t\right)\right\}$. This real signal is combined with two continuous-wave (CW) carriers at frequencies $f_1 = (193.4895 \ldots 193.5045)$ THz and $f_2 = 193.20$ THz. The carriers at f_0 and f_2 are provided by tunable laser sources with linewidths <100 kHz. The source f_1 has a linewidth < 0.1 kHz. Polarization controllers (not shown) align the polarization of the three sources. The signal spectrum near f_0 and the CW carrier f_1 (see Fig. 48.1 inset $\boxed{1}$) are down-converted to the THz range by photomixing with the CW carrier having a frequency f_2. The photomixer is a high-speed uni-travelling-carrier photodiode with a responsivity of 0.28 mA/mW operating in the H-band (0.220 ... 0.325) THz. Its total optical input power is (10 ... 14) dBm. The center frequency of the THz signal spectrum is $f_{\mathrm{THz,\,c}} = f_0 - f_2 = 0.2865$ THz. We transmit pulses with raised-cosine spectrum (roll-off factor $\beta = 0.1$). The bandwidth $B = (6, 12, 17, 23, 28, 34, 36)$ GHz is slightly larger than required by the symbol rate $R_{\mathrm{d}} = (5, 10, 15, 20, 25, 30, 33)$ GBd. At the upper edge of the THz signal spectrum, the CW carrier $f_{\mathrm{THz}} = f_1 - f_2 = f_{\mathrm{THz,\,c}} + B/2 \approx 0.3$ THz is positioned, which has a real amplitude U_0 and acts as a remotely supplied LO for KK reception. The LO frequency f_{THz} is offset from the THz center frequency $f_{\mathrm{THz,\,c}}$ by a spectral distance $B/2$. With respect to this LO, the resulting spectrum is a single-sideband spectrum (see Fig. 48.1 inset $\boxed{1}$). The link attenuation for a free-space distance of 110 m is 17dB. This propagation loss is compensated by a low-noise H-band THz amplifier with 25dB gain.

The amplified received THz voltage $u_{\mathrm{THz}}(t) = \Re\left\{\underline{U}(t) \exp\left(\mathrm{j}2\pi f_{\mathrm{THz}} t\right)\right\}$ (power P_{THz}) is rectified by the SBD having a responsivity of 2 V/mW. The SBD output bandwidth is 40 GHz. Its output current $i(t) = g\left(\left|\underline{U}(t)\right|\right)$ is recorded and subsequently evaluated by digital signal processing. We express the (instantaneous) inverse rectifier function $g^{-1}(i) = \sum_{n=0}^{N} a_n i^n$ by a power series in i for $0 \le i/\mathrm{mA} \le 1.2$. If the coefficients a_n are known after a calibration, the THz voltage amplitude $\left|\underline{U}_{\mathrm{R}}\right| = g^{-1}(i)$ can be reconstructed from the measured current i. For calibrating the SBD including possible saturation effects of the THz amplifier, we

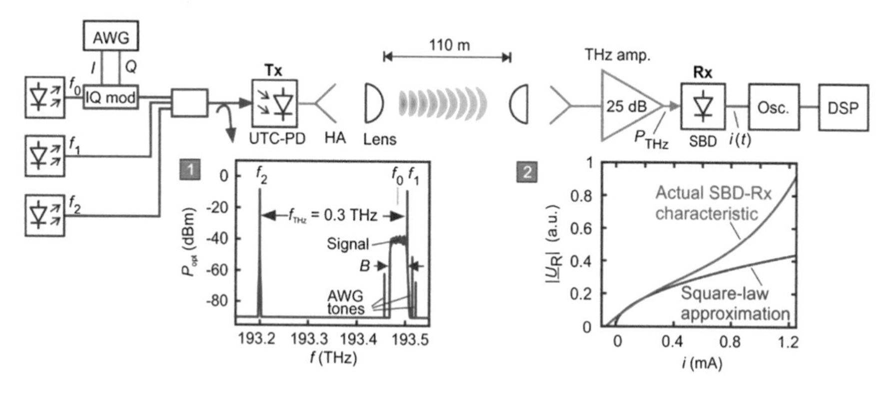

Fig. 48.1 *Wireless transmission and reception of mQAM data* generated with an IQ modulator, which is driven by an arbitrary waveform generator (AWG). Spectra near frequencies $f_0 = 193.4865$ THz, $f_1 = f_2 + f_{\mathrm{THz}}$ ($f_{\mathrm{THz}} \approx 0.3$ THz), and $f_2 = 193.20$ THz are photomixed in a uni-travelling-carrier photodiode (UTC-PD); see inset 1 (RBW = 180 MHz). The UTC transmitter (Tx) connects to a horn antenna (HA), and the emitted THz beam is focused with a polytetrafluoroethylene (PTFE) lens on the receiver (Rx) in a distance of 110 m. Here, a second PTFE lens collects the THz radiation and directs it to another horn antenna. After THz amplification, a Schottky barrier diode (SBD) mixes the THz signal spectrum with a remotely supplied THz local oscillator f_{THz} positioned at the upper edge of the spectrum. The SBD outputs a rectified current $i(t)$ which is recorded by a real-time oscilloscope with 80 GHz bandwidth and demodulated using digital processing (DSP). From a calibration step, the THz amplitude $|U_{\mathrm{R}}| \propto \sqrt{P_{\mathrm{THz}}}$ can be inferred from the SBD current i; see inset 2. (© 2020 SNPG [4])

apply a sequence of QPSK test signals with low symbol rate, vary the received THz amplitude $|\underline{U}|$, and measure the resulting current i. The carrier at f_1 remains switched off so that its amplitude is $U_0 = 0$. The coefficients a_n are determined by a least-squares fit (see Fig. 48.1 inset 2) ("actual characteristic"). Only in the low-current region $0 < i/\mathrm{mA} \leq 0.3$ the rectifier function g follows a square law so that $g^{-1}(i) = a\sqrt{i} = |U_{\mathrm{r}}| \approx |U_{\mathrm{R}}|$ ("square law", constant a); otherwise, $|U_{\mathrm{R}}| > |U_{\mathrm{r}}|$ holds. Because $a_0 \neq 0$, even a very small measured current results in a lower-limiting value of the estimated THz amplitude; this prevents large uncertainties in the phase retrieval. The next section discusses details.

48.3 Generalized Kramers-Kronig Processing

The received THz voltage $u_{\mathrm{THz}}(t)$ can be *interpreted* as the sum of the real parts of a single-sideband signal $\underline{U}_{\mathrm{s}}(t)\exp\left(\mathrm{j}2\pi f_{\mathrm{THz}}t\right)$ and of $U_0 \exp\left(\mathrm{j}2\pi f_{\mathrm{THz}}t\right)$. For a single-sideband spectrum, $\underline{U}_{\mathrm{s}}(t) = \underline{U}_{\mathrm{d}}(t)\exp\left(-\mathrm{j}\pi B\, t\right)$ must be an analytic signal where the imaginary part $\Im\left\{\underline{U}_{\mathrm{s}}(t)\right\} = \mathcal{H}\left\{\Re\left\{\underline{U}_{\mathrm{s}}(t)\right\}\right\}$ equals the Hilbert transform of the real part. The detection process delivers an output $\left|\underline{U}(t)\right| = g^{-1}\left(i(t)\right)$ with

$\underline{U}(t) = U_0 + \underline{U}_s(t) = |\underline{U}(t)| \exp(\mathrm{j}\Phi(t))$. We want to retrieve the phase of $\underline{U}(t)$ and eventually the amplitude and the phase of the data $\underline{U}_d(t)$.

We define an auxiliary quantity $\underline{s}(t) = \ln\left(|\underline{U}(t)/U_0| \exp(\mathrm{j}\Phi(t))\right)$, which in the form $\underline{s}(t) = \ln\left(|1 + U_s(t)/U_0|\right) + \mathrm{j}\Phi(t)$ is recognized as an analytic signal [5] as long as $\underline{U}(t)$ is minimum phase, i.e., if $|\underline{U}_s(t)|/U_0 < 1$. This is true with high probability if the carrier-to-signal power ratio CSPR $= 10\log_{10}\left(U_0^2/\langle |U_s(t)|^2 \rangle\right)$ is large enough. Because of the nonlinearity of the logarithm, the bandwidth of $\underline{s}(t)$ is larger than that of the data signal $\underline{U}_d(t)$. Real and imaginary parts of $\underline{s}(t)$ are related by the Hilbert transform expressed by a Cauchy principal value $(\mathcal{P})$ integral $\pi\ \Phi(t) = \mathcal{P}\int_{-\infty}^{+\infty} \ln\left(|\underline{U}(\tau)/U_0|\right)/(t-\tau)\,\mathrm{d}\tau$. Consequently, the phase $\Phi(t)$ of $\underline{U}(t)$ may be inferred from its magnitude $|\underline{U}(t)| = |U_0 + \underline{U}_d(t) \exp(-\mathrm{j}\pi B\ t)|$, which, however, must not be too small; otherwise, large errors could result due to the singularity of $\ln\left(|\underline{U}(t)/U_0|\right)$ at $|\underline{U}(t)| = 0$.

Now the complex data signal $\underline{U}_d(t)$ can be recovered from the measured amplitude $|\underline{U}(t)|_m$ and from the phase information that is retrieved by the Hilbert transform (subscript $\mathcal{H}$), $\underline{U}_d(t) = \left(|\underline{U}(t)|_m \exp(\mathrm{j}\Phi(t))_{\mathcal{H}} - U_0\right) \exp(+\mathrm{j}\pi B\ t)$.

We experiment [4] with the modulation formats QPSK, 16 QAM, and 32 QAM, and the results for 16 QAM are shown in Fig. 48.2. The green curves marked "w/o KK" result from single-ended heterodyne reception without a guard band. The blue curves marked "conv. KK" represent results from the conventional Kramers-Kronig (KK) processing [1, 2] as is common in optical communications, assuming an ideal power detection law $i(t) \propto |\underline{U}(t)|^2$. The red curves marked "gen. KK" illustrate the effect of an accurate description $i(t) = g\left(|\underline{U}(t)|\right)$ of the rectifying SBD in a generalized KK receiver [4].

Figure 48.2a shows [4] the influence of the carrier-to-signal power ratio on the bit error ratio (BER). The sum of carrier and average signal power is kept constant, so that the SBD input power is always $P_{\mathrm{THz}} = 370\ \mu$ W.

If the CSPR is too small, then the SSBI influence increases for heterodyne reception, and for KK processing is the probability of phase slips becomes larger; in both cases, the BER increases.If CSPR increases, then the signal power decreases, making the receiver noise more prominent and leading again to an increase in BER. The optimum is near CSPR = 7dB, and therefore this value is chosen in Fig. 48.2b, c.

For 16QAM signalling over a practically relevant distance of 110 m, the BER as a function of the symbol rate is shown [4] in Fig. 48.2b for an SBD input power of $P_{\mathrm{THz}} = 340\ \mu$ W. The horizontal broken lines mark the BER threshold and the overhead percentage for two different forward error corrections (FEC). We see that the BER increases in all cases, but generalized KK processing is always significantly better. The curves converge for large symbol rates, where receiver noise becomes

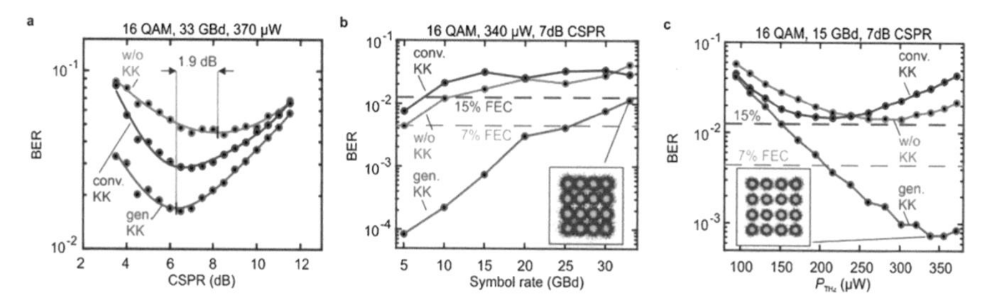

Fig. 48.2 *Experimental results.* Curves "w/o KK" indicate heterodyne reception, where SSBI increases the BER, while "conv. KK" stands for KK reception *assuming* $i \propto |U|^2$. The label "gen. KK" represents the generalized case, where the *measured* SBD characteristic is considered. (**a**) Bit error ratio (BER) as a function of the optical carrier-to-signal power ratio (CSPR) for a symbol rate 33 GBd at a constant SBD input power $P_{THz} = 370$ µ W. (**b**) BER vs. symbol rate, optimum CSPR = 7 dB at an SBD input THz power $P_{THz} = 340$ µ W after transmission over 110 m. For generalized KK reception, the BER stays below the 15% FEC limit, leading to a record (net) data rate 115 Gbit/s. Inset: constellation for a symbol rate 33 GBd. (**c**) BER vs. P_{THz} for a symbol rate 15 GBd and a CSPR = 7 dB after transmission over 110 m. (© 2020 SNPG [4])

more pronounced. At a symbol rate of 33GBd, the BER is still below the 15 % FEC limit, leading to a record-high net data rate of 115Gbit/s. When displaying [4] the BER as a function of the SBD input power P_{THz} for a symbol rate of 15GBd (Fig. 48.2c), the advantage of generalized KK processing becomes obvious.

References

1. Mecozzi, A., Antonelli, C., & Shtaif, M. (2016). Kramers–Kronig coherent receiver. *Optica, 3*, 1220–1227.
2. Füllner, C., Adib, M. M. H., Wolf, S., Kemal, J. N., Freude, W., Koos, C., & Randel, S. (2019). Complexity analysis of the Kramers–Kronig receiver. *Journal of Lightwave Technology, 37*, 4295–4307.
3. Mecozzi, A., Antonelli, C., & Shtaif, M. (2019). Kramers-Kronig receivers. *Advances in Optics and Photonics, 11*, 480–517.
4. Harter, T., Füllner, C., Kemal, J. N., Ummethala, S., Steinmann, J. L., Brosi, M., Hesler, J. L., Bründermann, E., Müller, A.-S., Freude, W., Randel, S., & Koos, C. (2020). Generalized Kramers–Kronig receiver for coherent terahertz communications. *Nature Photonics, 14*, 601–606.
5. See [1] (Appendix A): The Fourier transform of $s(t) = \ln\left(1 + x(t)\right) = \sum_{n=1}^{\infty} x^n(t)/n$ for $-1 < x(t) \leq +1$ is a multiply convolution of the spectra $\left\{\left[\breve{x}(f) * \breve{x}(f)\right] * \breve{x}(f)\right\} * \breve{x}(f) \cdots$, and if $x(t)$ is an analytic signal and its spectrum $\breve{x}(f)$ therefore causal, then $\breve{s}(f)$ is also causal, and $s(t)$ is an analytic signal, too.

Chapter 49
50 Gbps Demonstration with 300-GHz-Band Photonics-Based Link at ETRI

Kyung Hyun Park, Seung-Hyun Cho, Sang-Rok Moon, Eui Su Lee, Jun-Hwan Shin, Dong Woo Park, and Il-Min Lee

Abstract Based on photonics-based terahertz (THz) technologies, we are aiming to develop the key technologies for the next generation of telecommunication: THz communications. As the first step of such, we are currently conducting a 5-year project to develop the technologies for a photonics-based 100 Gigabits per seconds (Gbps) wireless THz link. Last year, as a preliminary result for the first year of the project, we have demonstrated a 50 Gbps signal transmission using our photonics-based THz emitter with silicon lens (Moon SR, et al., 50 Gb/s QPSK signal transmission using photonic-based THz signal emitter with silicon lens. Proceedings of MTSA 2019, Mo-POS-22, 2019; Lee ES, et al., J Lightwave Technol 36(2):274–283, 2017; Lee ES, et al., J Lightwave Technol 38(16):4237–4243, 2020). Recently, we are developing the entire core device modules for the THz short-range wireless transmissions including the GaAs mixer. In this chapter, we will briefly introduce the goal and perspectives of our research in ETRI and recent experimental achievements.

Recently, thanks to the rapid development in the performances of the key components, terahertz (THz) technologies in the fields of imaging, spectroscopy, and communications are about to escape from their laboratory-level demonstrations and bloom in the practical applications [1, 2]. With the launch of 5G (fifth-generation) commercial services and the natural growing interests to the B5G (beyond 5G) technology, photonics-based THz technologies gain intensive research since their advantageous features such as the capability in providing broadband and high speed of 1Tbps, transparency in connections with the fiber-optic networks, and competitiveness in their price and size [3]. In Electronics and Telecommunications

K. H. Park · E. S. Lee · J.-H. Shin · D. W. Park · I.-M. Lee (✉)
Terahertz Research Section, ETRI, Daejeon, South Korea
e-mail: ilminlee@etri.re.kr

S.-H. Cho · S.-R. Moon
Optical Networks Research Section, ETRI, Daejeon, South Korea

T. Kürner et al. (eds.), *THz Communications*, Springer Series in Optical Sciences 234, https://doi.org/10.1007/978-3-030-73738-2_49

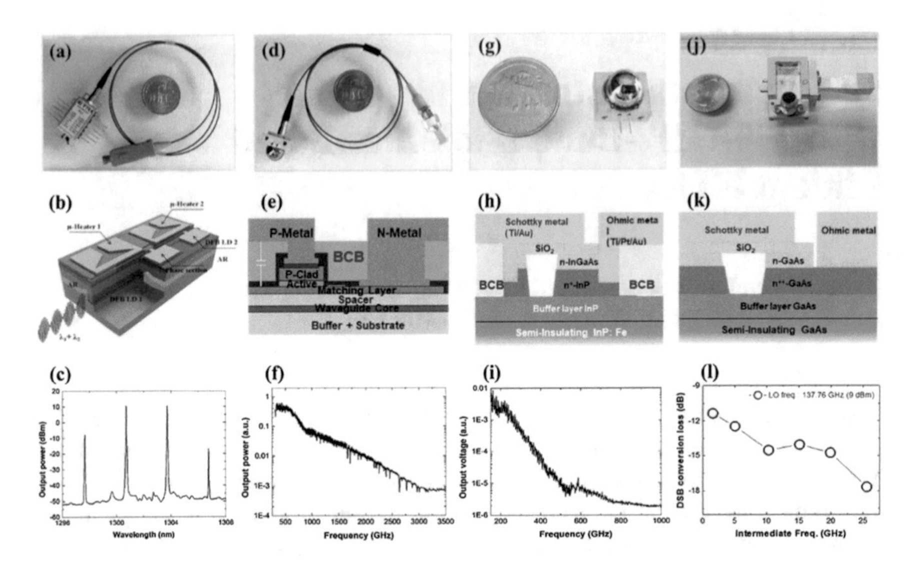

Fig. 49.1 Terahertz components of ETRI, (**a**) dual-mode laser (DML) module, (**b**) schematic of (DML) chip, (**c**) spectrum of DML, (**d**) UTC-PD hermetic module, (**e**) schematic of UTC-PD chip, (**f**) broadband characteristics of a UTC-PD, (**g**) Schottky barrier diode (SBD) module, (**h**) schematic of SBD chip, and (**i**) spectral response of an SBD, (**j**) WR3.4 subharmonic mixer module, (**k**) schematic of GaAs-based mixer chip, and (**l**) DSB (double-sideband) conversion loss of a mixer

Research Institute (ETRI), we are developing the key components for the THz communications: small but broadband-tunable dual-mode beating light sources, broadband antenna integrated UTC-PD (uni-traveling-carrier photodiode) as the broadband THz emitter, and broadband Schottky barrier diode (SBD) as detectors, for example. In addition to this, a GaAs-based subharmonic mixer (SHM) for phase extraction, which is essential for terahertz signal transmission, was developed and applied to the THz wireless link. We present some pictures of our module that are developed for the imaging and spectroscopic applications in Fig. 49.1.

For the applications of developed modules to the broadband THz wireless communications, we are developing 25 Gbps modulating beating source, GaAs SBD-based subharmonic mixer, comb laser for the coherent communications, and arrayed UTC-PD for the beam forming. We have a 5-year plan for developing the technologies for coherent THz communications including the short-distance M2M (machine to machine), M2H (machine to human), and kiosk communications and ultrashort-reach THz interconnections. Last year, as a preliminary result for the first year of the project, we have demonstrated a 50 Gbps signal transmission using our photonics-based THz emitter with silicon lens [1–3].

Experimental demonstration of photonics-based THz wireless link with 50 Gbps QPSK (quadrature phase shift keying) signaling in the 0.3 THz band: The experimental setup for THz wireless link with 50 Gbps QPSK signal based on photonics is shown in Fig. 49.2a. Two free-running lasers are employed to generate 0.3 THz-band beating signal. The laser diode (LD)-1 is operated with CW (continuous

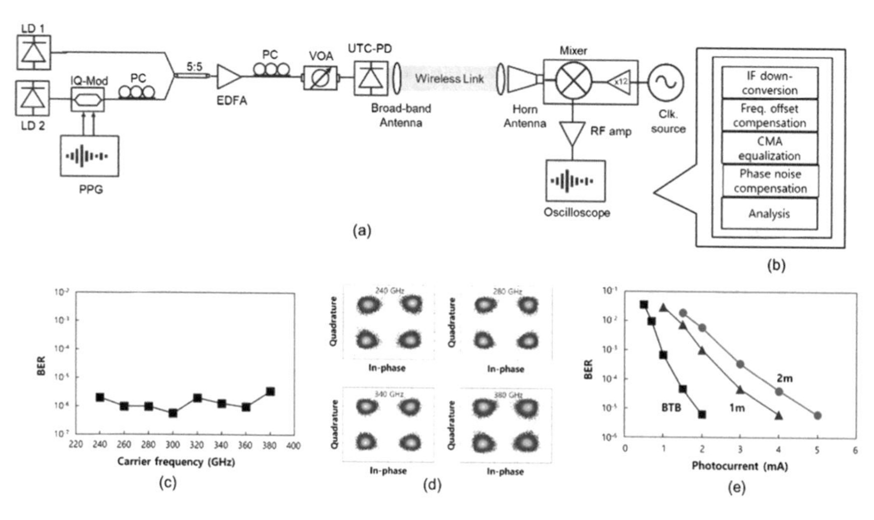

Fig. 49.2 Experimental setup and measured transmission performances of THz wireless link with 50Gbps QPSK signal based on photonics, (**a**) experimental setup, (**b**) detailed DSP configuration, (**c**) measured BER depending on the sub-THz frequency, (**d**) measured QPSK constellation at different center frequency, (**e**) measured BER for BTB configuration and over free-space transmission of 1 and 2 m at 300-GHz carrier frequency

wavelength) mode as an optical local oscillator (LO). The output of LD-2 is externally modulated with 50 Gbps QPSK signal by using optical IQ modulator. Two kinds of light are combined with 50:50 coupler and then amplified by erbium-doped fiber amplifier (EDFA). The wavelength difference between two laser diodes is adjusted to examine the frequency-dependent transmission characteristics. The output from EDFA is injected into the uni-traveling-carrier photodiode (UTC-PD) as a photo-mixer for generating THz wave. The THz wave is transmitted from the lens-type broadband antenna with UTC-PD to subharmonic mixer with horn antenna. Three lenses (2 for Tx, 1 for Rx) are used to compensate free-space path loss. The transmission distance of free space is approximately ~2 m. The received QPSK signal is frequency-downconverted by subharmonic mixer and then sampled with real-time oscilloscope. Finally, sampled signal is demodulated using offline digital signal processor (DSP), of which the detailed configuration is also shown in Fig. 49.2b. The measured BER as a function of THz-band frequency and constellations in back-to-back configuration are illustrated in Figs. 49.2c and 49.2d, respectively. There is no significant BER variations in measured THz frequency region which is determined by the specification of antenna with integrated waveguide feed (WR 3.4 and 2.8). We are able to obtain similar QPSK constellations at different THz frequency. As shown in these figures, 50 Gbps QPSK signal is successfully transmitted at 240 ~ 380-GHz band. We measure BER curves as a function of photocurrent with free-space transmission at 300-GHz carrier frequency. We observe slight BER degradations caused by free-space transmission of sub-THz wave, which is depicted in Fig. 49.2e. As a result, we demonstrate the

THz wireless link with 50 Gbps QPSK signal over a free-space distance of 2 m in the 0.3 THz band. In the near term, it should be necessary to improve the link budget in the THz band for successful commercialization of the THz wireless delivery system based on photonics. There would be two ways to increase the link budget. Firstly, it is essential to develop photo-mixer with higher-output power. Secondly, we need to have cost-effective THz amplifiers with larger bandwidth and gain for THz transceiver front-end. For long-term research goals, it is required to develop the technologies for THz beam forming/steering, photonic integrated circuits, and real-time DSP. We strongly hope that THz wireless link based on photonics for short-range wireless services such as indoor data showering and kiosk downloading will be realized in the near future.

References

1. Moon, S.-R., et al. (2019). *50 Gb/s QPSK signal transmission using photonic-based THz signal emitter with silicon lens*. Proceedings of MTSA 2019, Mo-POS-22.
2. Lee, E. S., et al. (2017). Semiconductor-based terahertz photonics for industrial applications. *Journal of Lightwave Technology, 36*(2), 274–283.
3. Lee, E. S., et al. (2020). High-speed and cost-effective reflective terahertz imaging system using a novel 2D beam scanner. *Journal of Lightwave Technology, 38*(16), 4237–4243.

Chapter 50
Brown University Test Bed

Daniel M. Mittleman

Abstract This chapter describes a THz link test bed that has been implemented at Brown University at frequencies from 100 GHz to 400 GHz, in particular to examine non-line-of-sight (NLOS) path conditions. This test bed has also been used the first studies of link security in the THz range.

50.1 THz Link Test Bed

In this chapter, we describe one of the recent experiments that have been performed at Brown University using a THz data link test bed [1]. The test bed can be used to generate a continuous wave (CW) signal at any one of four discrete frequencies: 100, 200, 300, and 400 GHz. The source is based on a frequency multiplier chain (Virginia Diodes), which up-converts a modulated baseband signal to the desired output frequency and radiates using a horn antenna, with vertical polarization. We can use a pulse pattern generator to produce a pseudorandom bit pattern for modulating this carrier wave at 1 Gb/s via on-off keying (OOK) modulation. Detection relies on a zero-bias Schottky diode, also coupled to a horn antenna. The Schottky signal is amplified to drive a bit error rate (BER) tester (Anritsu MP1764A) for real-time signal analysis. To improve the overall gain of the two antenna subsystems, we use transmissive (Teflon) lenses in front of the horn antennas ($f = 7.5$ cm). The half-power beam widths range from 10° to 13°, and the output power ranges from 9 dBm to 24 dBm, depending on carrier frequency.

We note that a directive antenna with a small HPBW can lead to the suppression of unintended multipath components, because multipath signals from outside of the maximum gain direction of the antenna are attenuated by the antenna radiation pattern. Even so, outdoor measurements can still reveal the effects of multipath interference if the range is large enough [1].

D. M. Mittleman (✉)
Brown University, Providence, RI, USA
e-mail: daniel_mittleman@brown.edu

T. Kürner et al. (eds.), *THz Communications*, Springer Series in Optical Sciences 234, https://doi.org/10.1007/978-3-030-73738-2_50

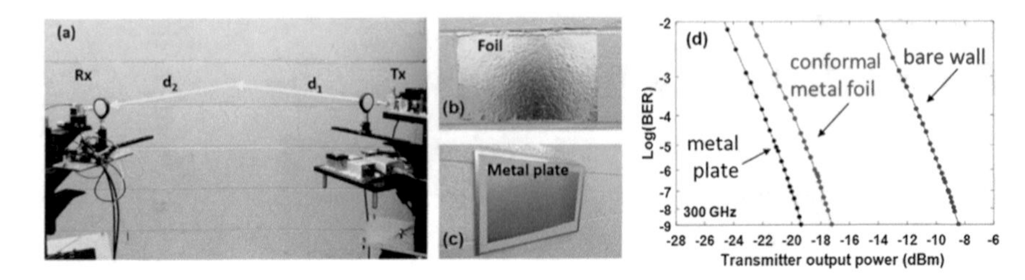

Fig. 50.1 (**a**) A photo of the experimental configuration for measurement of data transmission via a specular NLOS path, using a reflection from a conventional painted cinderblock wall, (**b**) a conformal metal foil which eliminates penetration of the THz wave into the material but approximately preserves the surface roughness, and (**c**) a smooth metal plate. (**d**) Measured bit error rate vs. transmitter power, for a 1-Gbps data stream at 300 GHz

50.2 Specular Non-line-of-Sight Links

One of the important questions in the implementation of THz wireless systems involves the possible use of non-line-of-sight (NLOS) links. In some circles, a conventional wisdom has persisted that the use of such links is impractical because of the prohibitively large power penalty due to absorption and scattering of the electromagnetic wave when it encounters a nonmetallic surface. To test this assumption, we have employed the test bed to test a data link which incorporates a specular reflection from an ordinary painted cinderblock wall. Figure 50.1a illustrates the Tx and Rx subassemblies, with plastic lenses, directing a beam toward a spot on the wall. The distance from the Tx/Rx antenna to the wall is both 1 m ($d_1 = d_2 = 1$ m), and the angle of incidence is 50°. We compare the results for three cases: The signal was reflected (a) by the bare painted cinderblock wall, (b) by a conformal metal foil attached on the wall, and (c) by a smooth metal plate. In the first case, we expect the signal to be degraded both by absorption in the cinderblock material and by diffuse random scattering from the rough surface [2, 3]. In the second case, the metal foil (which is assumed to act essentially as a perfect metal) eliminates absorption losses in the underlying cinderblock surface. However, the scattering losses remain since the thin foil conforms to the rough shape of the cinderblock surface. In the third case, both scattering and absorption losses are eliminated.

Our observations (e.g., Fig. 50.1d) indicate that, at all frequencies, the effect of scattering from the rough surface (i.e., the difference between the black and red curves) is significantly smaller than the effect of absorption (the difference between the red and blue curves). These data provide strong evidence that, contrary to most conventional wisdom, specular NLOS paths can realistically be used in indoor THz links even up to at least 400 GHz, with only moderate and manageable losses [1].

This test bed has been used for a variety of other measurements, including the first studies of link security in the THz range [4] and a demonstration of links with low BER using nonspecular NLOS scattering (i.e., diffusive scattering) [3].

References

1. Ma, J., Shrestha, R., Moeller, L., & Mittleman, D. M. (2018). Channel performance of indoor and outdoor terahertz wireless links. *APL Photon, 3*, 051601.
2. Piesiewicz, R., Jansen, C., Mittleman, D. M., Kleine-Ostmann, T., Koch, M., & Kürner, T. (2007). Scattering analysis for the modeling of THz communication systems. *IEEE Transactions on Antennas and Propagation, 55*, 3002–3009.
3. Ma, J., Shrestha, R., Zhang, W., Moeller, L., & Mittleman, D. M. (2019). Terahertz wireless links using diffuse scattering from rough surfaces. *IEEE Transactions on Terahertz Science and Technology, 9*, 463–470.
4. Ma, J., Shrestha, R., Adelberg, J., Yeh, C.-Y., Hossain, Z., Knightly, E., Jornet, J. M., & Mittleman, D. M. (2018). Security and eavesdropping in terahertz wireless links. *Nature, 563*, 89–93.

Chapter 51
Research at New Jersey Institute of Technology (NJIT)/Nokia Bell Labs

John Federici and Lothar Moeller

Abstract This chapter describes a laboratory test bed which emulates various weather conditions such as fog, dust, rain, and air turbulence. In many cases, weather conditions can be reproducibly generated in order to compare experimental measurements to theoretical predictions of THz channel performance. A colinear infrared channel enables direct comparison of THz and infrared channel performance.

51.1 Laboratory-Emulated Weather

Measurement and analysis of weather's impact on terahertz (THz) communication channels are challenging due to the required long observation times and the difficulty in obtaining reoccurrence of similar atmospheric conditions for comparison of independent measurements. For example, changing humidity levels slightly before and during falling snow present challenges in experimentally separating the effect of THz attenuation due to humidity from attenuation due to snow-aggregated airborne particulates [1]. Using this motivation, the NJIT/Bell Labs research group has opted to develop or adapt weather emulating chambers into a controlled laboratory environment [2–10]. In many cases, the weather conditions (e.g., rain drop size distribution) can be reproducibly generated in order to compare experimental measurements to theoretical predictions.

Weather chambers for fog, dust, rain, and air turbulence are developed. Fog is generated in a 62-cm-length chamber by dripping liquid nitrogen into a cup filled with hot water (temperature about ~80 °C) which is placed inside the fog chamber [7]. The fog density, controlled by the amounts of liquid nitrogen spilled into the cup, can be fluctuating over a large visibility range limited from a few centimeters

J. Federici (✉) · L. Moeller
NJIT, Newark, NJ, USA
e-mail: john.f.federici@njit.edu

T. Kürner et al. (eds.), *THz Communications*, Springer Series in Optical Sciences 234, https://doi.org/10.1007/978-3-030-73738-2_51

to several hundreds of meters. The fog droplets exhibit an average diameter of approximately 8 μm.

Bentonite powder, a mixture of clay formed from volcanic ash decomposition and largely composed of montmorillonite and beidellite, is used to produce dust particles with an average diameter of 8.6 μm. A known total mass of dust particles is injected at a high speed into a 30-cm diameter cylindrical dust chamber from its top plate [6]. Air flows into the chamber with constant volume speed through the holes from the bottom plate to keep the dust particles suspended in the air.

The rain chamber [4, 10] has a top plate machined with 3264 holes and 31-gauge needles epoxied into each hole. Pressurizing the chamber above the plate with air generates raindrops by forcing distilled water release through the needles. The raindrop size follows approximately a Gaussian distribution with an average diameter of 1.9 mm and a variance of 0.08 mm^2. The chamber rain rate is proportional to the pressure applied to the water reservoir and controllably varied from 0 to 500 mm/h by changing the air pressure in the chamber.

While the previously described chambers introduce various particulates into the air that attenuate and produce scintillations of the propagating beams, one can isolate the contribution of scintillations to signal degradation by only introducing air turbulence into the weather chamber. The turbulence chamber [5] consists of an enclosed chamber in which air is blown in through two holes in the top of the chamber using industrial heat guns. The air temperature and airspeed launched into the chamber are adjusted independently.

51.2 Channel Performance by Weather Impact

51.2.1 Co-propagating THz and Free-Space Infrared Link Test Bed

The communication test bed [5–8, 10] (Fig. 51.1) has a data rate of 2.5 Gbit/s at a carrier frequency of 625 GHz. A duobinary modulation technique, which enables signaling at high data rate with relatively compact spectrum, is utilized in the system to drive a THz source, which is based on a frequency multiplier chain with about 1 mW output power when operating in a CW mode. The THz beam propagates through a distance of ~3 m with a beam diameter of 20 mm and is then collected by a THz receiver horn similar to the transmitter antenna. The detected THz power from a Schottky diode is amplified and filtered by a quasi-Gaussian low-pass filter. The electrical power is split into a high-speed scope, a bit error rate tester (BERT), and a 2.5-Gbit/s NRZ clock recovery circuit that synchronizes the measurement equipment. At the amplifier output, RF powers of about −14 dBm for a BER around 10^{-9} are typical for 2^{13}–1 PRBS pseudorandom bit sequences.

The 1.5-micron-infrared (IR) transmitter is driven by the same 2.5 Gb/s data pattern as the THz source and outputs a beam with a diameter similar to that of the

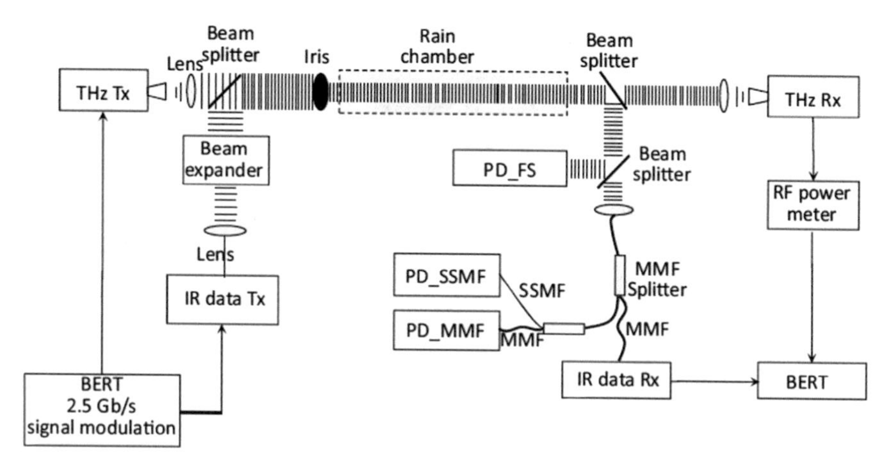

Fig. 51.1 Schematic diagram of the THz and IR wireless communication test bed [5–8, 10]

THz link. The co-propagating beams enable direct comparison of the degradations observed in two competing wireless solutions since they pass through the same weather conditions. The purpose of the multiple IR detectors with varying effective apertures is to detect the presence of scintillations in the IR channel. If scintillations were present, power fluctuations would increase as the effect detector aperture size is decreased.

51.2.2 Fog

While THz links suffer significantly more attenuation through humid air compared to free-space optical (FSO) IR links, the polar opposite is true with atmospheric fog. By co-propagating THz and IR wireless links through a fog chamber with a minimum visibility of a few centimeters [7], the attenuation of the IR light is so high (>40 dB) that it is essentially blocked by the fog, while the THz signal exhibits only a minor but measurable decrease (~0.6 dB attenuation) in power. Even "small" amounts of fog are sufficient to block the IR beam and prevent BER measurements of the IR signal, while the THz channel experiences only a relatively small increase in the BER.

51.2.3 Dust

As with fog, attenuation of THz waves due to dust is much smaller than the attenuation for FSO propagation [6]. This difference is a consequence of Mie scattering and the relatively small size of the airborne particles relative to the THz

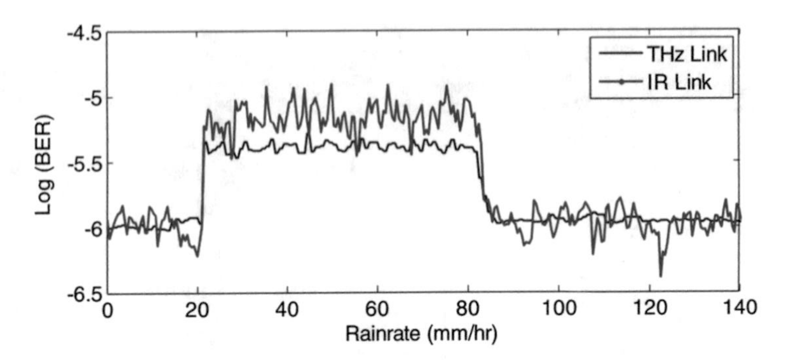

Fig. 51.2 Log (BER) of THz link (solid) and IR link (dash) in rain chamber as functions of time. Rain starts at a rate of 213 mm/h at ~20 s and stops at ~80 s [10]

wavelength. Attenuation of the IR beam is ~4.5 dB for dust particle densities of 4×10^9 m^{-3} compared to an attenuation of 0.045 dB in the THz range. The attenuation varies significantly with changing concentrations: larger concentration leads to higher attenuation and larger BER. Dust particles at these concentration levels have little impact on the THz signal compared to the IR signal across the limited propagation length.

51.2.4 Rain

Using the rain chamber, co-propagating THz and free-space optical links are sent through the same weather conditions [10]. At a rain rate of 213 mm/h, the THz link suffers a little bit higher attenuation (~0.1 dB) compared to the IR link, which is due to the slightly larger extinction cross-section of rain drops at THz frequencies. A typical evolution of the resulting BERs in the THz and IR links due to rain is shown in Fig. 51.2. After 80-second recording time, the rain stops, and the transmitted THz and IR signal powers approach their original level. For both channels, a strong correlation between their attenuation and their time-averaged (integration time 1 s) BER (Fig. 51.2) is observed. The slightly larger BER in the IR is attributed to slightly different receiver sensitivity curves in the THz and IR. The same amount of attenuation increases the BER of the THz channel less than the BER of the IR signal [10].

51.2.5 Air Turbulence and Scintillations

Scintillation effects are readily observable in free-space IR channels in the presence of fog, dust, and rain [6, 7, 10]. The key to detecting scintillations in the IR channel is to split the transmitted IR beam into three different detectors with three different effective aperture sizes. The effect of scintillations in the THz band is orders of magnitude smaller than in the IR band. In fog, dust, and rain, the reduction of THz link performance can be explained based purely on the additional attenuation in the beam path due to the particulates; scintillation effects in the THz band are negligible compared to the effects of attenuation [6, 7, 10]. By using the air turbulence chamber [5], the strong attenuation effects from particulates are removed enabling the observation of scintillation effects due to dominant real refractive index fluctuations in both the IR and THz channel. For the IR channel, a strong correlation between its attenuation and the recorded BER is clearly visible. The recorded BERs for the THz link verify that air turbulence has little impact on the THz compared to the IR channel. Scintillation effects in the experimental setup are mainly caused by temperature fluctuations which fit to a fundamental assumption made for outdoor scintillation models for the IR.

References

1. Renaud, D., & Federici, J. F. (2019). Terahertz attenuation in sleet and snow. *Journal of infrared, Millimeter, and Terahertz Waves, 40*(8), 868–877.
2. Federici, J., Moeller, L., & Su, K. (2013). Terahertz communication. In D. Saeedkia (Ed.), *Handbook of terahertz technology for imaging, sensing, and communications*. Cambridge: Woodhead Publishing.
3. Federici, J. F., Ma, J., & Moeller, L. (2016). Review of weather impact on outdoor terahertz wireless communication links. *Nano Communication Networks, 10*, 13–26. https://doi.org/10.1016/j.nancom.2016.07.006.
4. Ma, J., Vorrius, F., Lamb, L., Moeller, L., & Federici, J. F. (2015). Comparison of experimental and theoretical determined terahertz attenuation in controlled rain. *Journal of Infrared, Millimeter, and Terahertz Waves, 36*(12), 1195–1202. https://doi.org/10.1007/s10762-015-0200-6.
5. Ma, J., Moeller, L., & Federici, J. F. (2015). Experimental comparison of terahertz and infrared signaling in controlled atmospheric turbulence. *Journal of Infrared, Millimeter, and Terahertz Waves, 36*(2), 130–143. https://doi.org/10.1007/s10762-014-0121-9.
6. Su, K., Moeller, L., Barat, R. B., & Federici, J. F. (2012). Experimental comparison of terahertz and infrared data signal attenuation in dust clouds. *Journal of the Optical Society of America. A, 29*(11), 2360–2366.
7. Su, K., Moeller, L., Barat, R. B., & Federici, J. F. (2012). Experimental comparison of performance degradation from terahertz and infrared wireless links in fog. *Journal of the Optical Society of America. A, 29*, 179–184. https://doi.org/10.1364/JOSAA.29.000179.
8. Moeller, L., Federici, J., & Su, K. (2011). 2.5Gbit/s duobinary signalling with narrow bandwidth 0.625 terahertz source. *Electronics Letters, 47*(15), 856–858. https://doi.org/10.1049/el.2011.1451.
9. Federici, J., & Moeller, L. (2010). Review of terahertz and subterahertz wireless communications. *Journal of Applied Physics, 107*(11), 111101. https://doi.org/10.1063/1.3386413.
10. Ma, J., Vorrius, F., Lamb, L., Moeller, L., & Federici, J. F. (2015). Experimental comparison of terahertz and infrared signaling in laboratory-controlled rain. *Journal of Infrared, Millimeter, and Terahertz Waves, 36*(9), 856–865. https://doi.org/10.1007/s10762-015-0183-3.

Part XI
Standardisation and Regulation

Chapter 52
Standards for THz Communications

Thomas Kürner, Vitaly Petrov, and Iwao Hosako

Abstract In 2017, IEEE Std. 802.15.3d-2017 has been completed as the first wireless standard operating at the frequency 252 GHz to 321 GHz. This standard offers switched point-to-point connectivity with data rates of 100 Gbit/s and beyond. Application scenarios include short-range intra-device communication, kiosk downloading, wireless links in data centers, and wireless backhauling/fronthauling. This chapter describes the requirements for the target applications and usage scenarios and provides a summary of the specifics of the physical and medium access layers of IEEE Std. 802.15.3d-2017 including the results from the performance evaluation done during the standardization process.

52.1 Introduction

Up to now, standards for THz communication have been considered in IEEE 802 only. This may change in the near future with the development of 6G wireless systems, where THz communication is one of the candidates and is seen as a key enabler for applications demanding user data rates of several tens of Gbit/s [1, 2].

Already in 2008, IEEE 802.15 started activities towards a wireless standard operating at 300 GHz and beyond by chartering the terahertz interest group (IEEE 802.15 IG THz) with the goal to explore the feasibility of carrier frequencies at 300 GHz and beyond for wireless communications. These activities in conjunction with various hardware demonstrations available at that time showing the principle feasibility of wireless communications at 300 GHz lead to the creation of a study group

T. Kürner (✉)
Institute for Communications Technology, Technische Universität Braunschweig, Braunschweig, Germany
e-mail: t.kuerner@tu-bs.de

V. Petrov
Tampere University, Tampere, Finland

I. Hosako
NICT, Tokyo, Japan

T. Kürner et al. (eds.), *THz Communications*, Springer Series in Optical Sciences 234, https://doi.org/10.1007/978-3-030-73738-2_52

(IEEE 802.15 SG 100G) with the scope of determining the validity on 100 Gbit/s over beam switchable point-to-point links in July 2013 [3]. Although transmission technology for 300 GHz has been considered to be mature enough for developing a standard, the more complex tasks of device discovery, beam forming, and beam tracking were still in their infancy limiting THz communications to applications with known and fixed locations of the antennas at both ends of the link. Even with these limitations, a couple of interesting potential applications requiring ultrahigh data rates have been identified: wireless links in data centers, wireless intra-device communication, kiosk downloading, and wireless backhauling/fronthauling [4].

In March 2014, IEEE 802 formed the task group IEEE P802.15.3d to amend IEEE Std. 802.15.3 [5] with the definition a wireless switched point-to-point physical operating at a nominal PHY data rate of 100 Gbit/s with fallbacks to lower data rates as needed operating in the bands from 252 GHz to 325 GHz at ranges as short as a few centimeters and up to several 100 m [6]. The amendment was finalized in October 2017 and published as IEEE Std. 802.15.3d-2017 [7]. The development of this amendment significantly benefitted from the parallel development of IEEE Std. 802.15.3e-2017 [8] targeting high-rate close-proximity communication links at 60 GHz, which already introduced the concept of pairnet supporting point-to-point communications connecting only two devices with reducing the problem of interference and "fighting for access" [9]. Hence, IEEE Std. 802.15.3d-2017 inherited most of the MAC layer and the principal PHY-layer concepts from IEEE Std. 802.15.3e-2017.

52.2 Applications and Usage Scenarios

During the development of the standard, four (point-to-point) potential applications have been defined in an application requirement document [11] (see Fig. 52.1). These applications, usage scenarios, and requirements will be described in the following.

52.2.1 Kiosk Downloading

Kiosk downloading is defined as a sub-use case of close-proximity point-to-point applications and offers quick downloads of digital information to users' handheld devices. Examples of such information to be transferred are high-resolution videos from content providers. With kiosk downloading, the user's portable terminal and the network are connected via a kiosk terminal, and the transmission range is 50 mm or less. The kiosk terminals may be deployed in public areas such as train stations, airports, malls, convenience stores, rental video shops, libraries, and public telephone boxes. One such scenario is described in [10], where file downloading at toll gates in a train station shall enable an 859-MB file, which corresponds to a

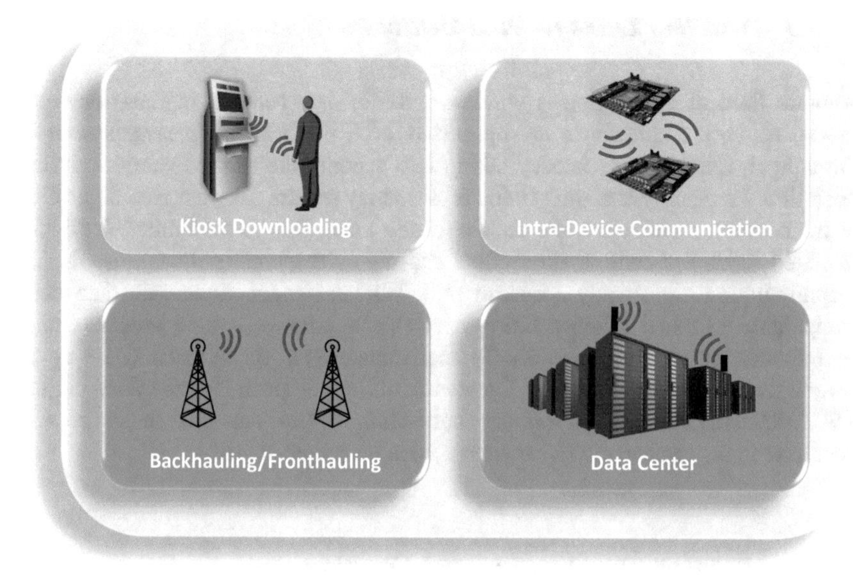

Fig. 52.1 Target applications for IEEE Std. 802.15.3d-2017

114-min HD video, to be downloaded within 248 ms when passing the toll gate. This download will require a data rate of 28 Gbit/s [11]. A first demonstration of this application at THz frequencies has been shown already in 2016 [12].

52.2.2 *Intra-device Communication*

During the generation of ultrahigh-definition content, for example, within a camera, data rates might easily exceed 100 Gbit/s [10], e.g., 8K resolution at 120 Hz and 36 bits per pixel requires 143 Gbit/s of gross data rate [13]. Wireless links inside of such devices have advantages over conventional copper-based technology by avoiding connectors and increasing flexibility when connecting the boards. The targeted transmission range is up to 10 cm in the air or through two layers of material reasonably transparent to terahertz wave (5 mm thickness) [10]. In 2019, a first demonstration of this application using a CMOS transceiver at 300 GHz has been published [14].

52.2.3 Wireless Links in Data Centers

Another field of application is wireless links in data centers. In those environments, network architectures are optimized w.r.t. performance measures such as throughput, reliability, or latency taking into account the traffic dynamics during operation. Frequently changing traffic profiles may require frequent reconfiguration of the network architecture and the associated links. Due to the limited flexibility of wired links, the configuration effort and the corresponding downtimes during reconfiguration are already a burden [15]. Wireless connections with data rates comparable to those of fiber links are viable alternative, where antennas may be mounted on top of the racks. Typical transmission distances may be up to several tens or meters. Within the framework of the European Horizon 2020 project TERAPOD, the feasibility of ultrahigh bandwidth wireless data links in data centers operating in the terahertz (THz) band has been investigated [16].

52.2.4 Wireless Backhauling and Fronthauling

Data traffic densities of several Tbit/s/km^2 are already predicted for 5G networks in the near future [17] requiring adequate solutions for backhauling and/or fronthauling with data rates of at least several tens of Gbit/s. Since fiber is not available at every base station, THz communication is a possible option for wireless backhauling/fronthauling. Transmission range is limited by weather conditions like rain, fog, clouds, and atmospheric attenuation. The typical transmission distances of 300 GHz may be up to several hundred meters. Already in 2015, the feasibility has been demonstrated at 240 GHz [18].

52.2.5 Impact of the Applications of MAC Requirements

The point-to-point nature of the considered applications and usage scenarios with static or quasi-static transmitters and receivers at known locations are properties all these applications and usage scenarios have in common. In combination with either the very short distance (intra-device communication and kiosk downloading) or the extremely narrow beams in case of deployments with larger distances between the communication nodes (wireless links in data centers and wireless backhauling/fronthauling), these properties have been a key enabler to inherit the MAC from IEEE Std. 802.15.3e-2017 [8] with a lower implementation complexity when compared to the IEEE 802.11 family standards [9]. These standards, in particular IEEE Std. 802.11ad-2014 [19] and IEEE P802.11ay [20], primarily target mobile point-to-multipoint wireless communications at 60 GHz with lower rates particularly. Although there is some overlap with applications and usage scenarios targeted by IEEE Std. 802.15.3d-2017, the latter not only has advantages in terms of reduced signaling (more applicable to inherently point-to-point connectivity) but also has the potential to achieve even higher data rates due to the much higher bandwidths available at 300 GHz.

52.3 MAC Layer

In contrast to the existing and prospective solutions for Wireless Local Area Networks (WLANs), e.g., IEEE Std. 802.11-2016 and IEEE P802.11ay, IEEE Std. 802.15.3d-2017 targets exclusively point-to-point connectivity between two nodes. Therefore, to facilitate the on-time development and the unification between the standards, many of the MAC layer features in IEEE Std. 802.15.3d-2017 are inherited from IEEE Std. 802.13.3e-2017 [8]. The described standard particularly utilizes the concept of a "pairnet," connecting two nodes:

- Pairnet coordinator (PRC)
- Pairnet device (PRDEV)

The high-level illustration for the MAC-layer signaling is presented in Fig. 52.2. There are two main stages of the communication process: (i) pairnet setup period (PSP) and (ii) pairnet-associated period (PAP). The PSP begins when PRC creates the pairnet and begins periodically sending beacon frames. When PRDEV appears in the PRC proximity and is willing to join the pairnet, it processes the received beacon to get the number and durations of the access slots. Then, PRDEV uses one of the defined access slots to transmit the *association request*. Once the *association request* is successfully received and processed by PRC, it stops sending periodic beacons and transmits the *association response*, thus ending the PSP.

Once the *association response* is successfully received and processed, the second communication period, PAP, starts. This period is dedicated to the bidirectional data exchange between the PRC and the PRDEV. During PAP, both nodes can transmit data frames and/or acknowledgments, confirming the successful reception of the previous data frame.

PAP may be terminated in two ways. First, when one of the nodes (either PRC or PRDEV) decides to terminate the connection, it sends the *disassociation request*. Second, the connection is terminated when PRC does not receive any data from PRDEV within a defined time-out (e.g., when PRDEV has left the PRC coverage area). In both cases, PRC finishes PAP and starts a new PSP period by sending a new beacon frame. In the event that PRDEV would like to keep the connection active but has no data to send, it can transmit the *probe request* that restarts the PRC time-out timer.

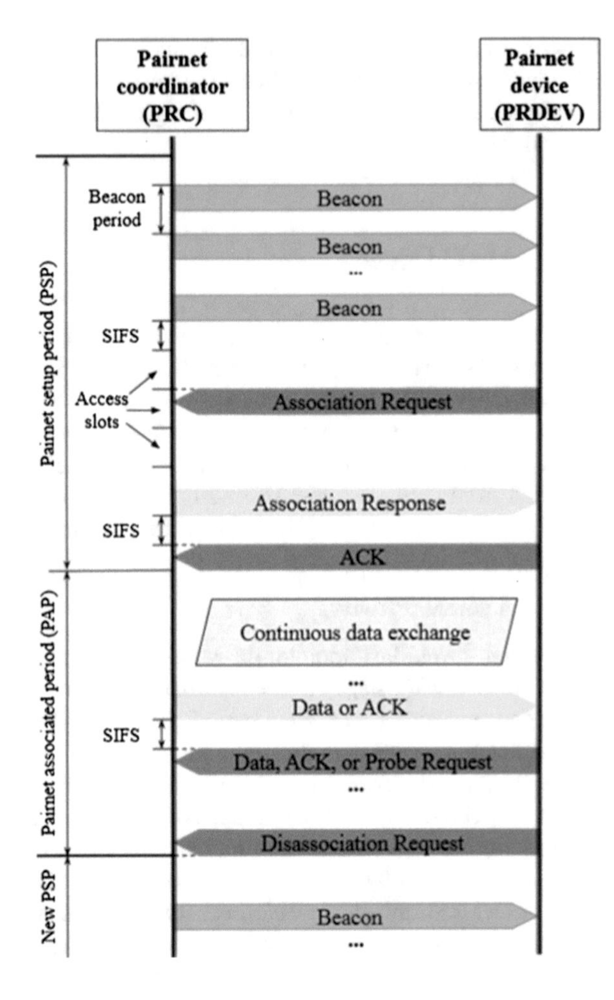

Fig. 52.2 MAC layer signaling utilized in IEEE Std. 802.15.3d-2017 [9] (© 2020 IEEE, reproduced with permission)

52.4 PHY Layer

52.4.1 Channels

The IEEE Std. 802.15.3d-2017 is designed to operate in the frequencies between 252.72 GHz and 321.84 GHz. Altogether, there are 69 channels available with the channel bandwidth ranging from 2.16 GHz up to 32 x 2.16 GHz = 69.12 GHz (see Fig. 52.3). The smallest bandwidth correspond to the one utilized in IEEE Std. 802.11-2016, while the larger ones target the data rates beyond the capabilities of the state-of-the-art solutions for 60 GHz frequencies. Channel number 41 around 290 GHz with the bandwidth of 2 x 2.16 GHz = 4.32 GHz is defined as a default channel for IEEE Std. 802.15.3d-2017.

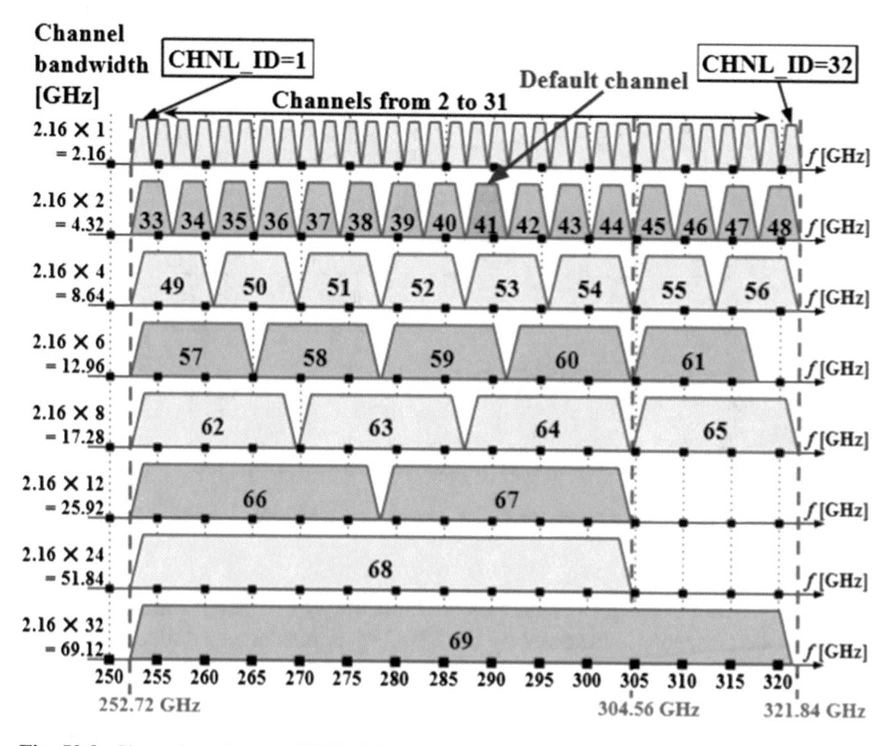

Fig. 52.3 Channels utilized in IEEE Std. 802.15.3d-2017 [9] (© 2020 IEEE, reproduced with permission)

52.4.2 Frame Structure

The IEEE Std. 802.15.3d-2017 PHY layer has two modes: (i) THz on-off keying mode (THz-OOK) and (ii) THz single-carrier mode (THz-SC). The later of these two modes, THz-SC, is designed to achieve the highest possible data rates, while the former one, THz-OOK, targets lower-complexity implementations, at cost of a reduced data rate.

The IEEE Std. 802.15.3d-2017 frame length ranges from 2048 bytes to ~2MB, not including the PHY-layer preamble or the base header. The format description is presented in Fig. 52.4. The first element is the physical preamble that facilitates synchronization, frame detection, and channel estimation. Two options for the PHY preamble are supported. The longer one is transmitted during PSP in order to increase the robustness of the frame detection and decoding, while the shorter one is used during PAP in order to reduce the overheads for the data exchange. During PSP, the overheads are less crucial as only seldom service frames (e.g., beacons) are transmitted.

The next item in the frame is the PHY-layer header that details the chosen bandwidth, the selected modulation and coding scheme (MCS), and other service

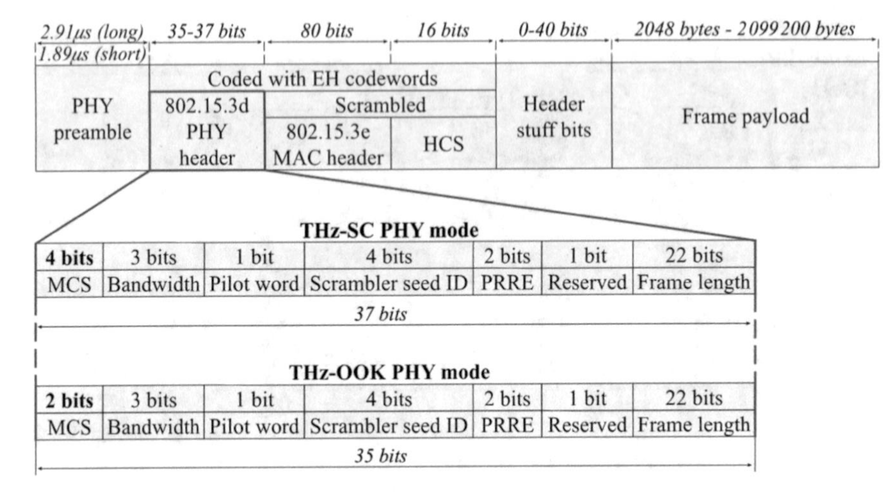

Fig. 52.4 Frame structure utilized in IEEE Std. 802.15.3d-2017 (based on [9])

parameters. The PHY-layer header also contains the 22-bit-long *frame length* field. The PHY-layer header is followed by the MAC-layer header as detailed in [5]. The header check sequence (HCS, particularly the CRC-16 cyclic redundancy check) is further inserted to protect both MAC- and PHY-layer headers. For the sake of robustness, the set of PHY-layer header, MAC-layer header, and HCS is encoded to concatenated code words of an Extended Hamming (EH) code. In order to get the appropriate size of the data block, *header stuff bits* may be added. The last field in the frame is the *payload* carrying the actual data.

52.4.3 Modulation and Coding

The modulation and coding schemes utilized for IEEE Std. 802.15.3d-2017 PHY-layer modes (THz-SC and THz-OOK) are illustrated in Fig. 52.5. The simpler THz-OOK supports one pulse-based low-complexity modulation scheme: on-off keying. With this scheme, the logical "one" in the data is represented as a pulse, while the logical "zero" is represented as silence. In contrast to the simpler THz-OOK, the more sophisticated THz-SC supports six modulations, including four-phase shift modulations (BPSK, QPSK, 8-PSK, and 8-APSK), as well as two quadrature amplitude modulations (16 QAM and 64 QAM). The first two modulations (BPSK and QPSK) are mandatory for the THz-SC, while the other four modulations are optional.

Forward error correction (FEC) for the THz-SC contains low-density parity-check (LDPC) codes: (i) 14/15 LDPC (1440,1344) and (ii) 11/15 LDPC (1440,1056). The first code is used to achieve the highest data rates, while the second one trades off some performance for reliability. For THz-OOK, only a

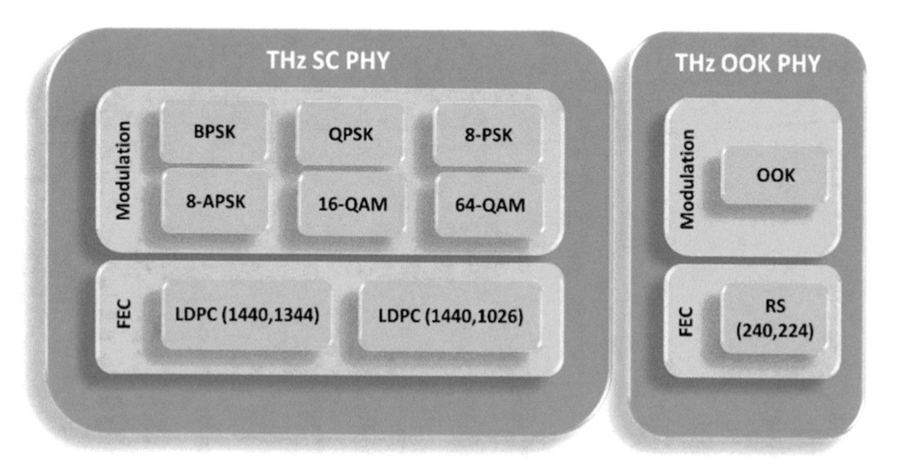

Fig. 52.5 PHY modes, modulation, and FEC schemes defined in IEEE Std. 802.15.3d-2017 [21]

single coding scheme – (240,224) Reed-Solomon (RS) code – is mandatory. The benefit of this scheme is its ability for the low-complexity decoding without the need for soft decision information. The described LDPC codes are optional for the THz-OOK.

52.5 Performance Evaluation

Here, a summary of the results from the performance evaluation done during the standardization process is provided [22].

52.5.1 Data Rate

We particularly focus on Fig. 52.6, illustrating the maximum PHY-layer data rates with the IEEE Std. 802.15.3d-2017 as a function of the THz PHY-layer mode, modulation scheme, and channel bandwidth. We assume a point-to-point connectivity with perfect beam alignment and high signal-to-noise ratio (SNR). Particularly, the SNR is high enough to maintain negligible error rate; hence, the communication link is assumed to be perfectly robust.

In the considered setup, the achievable data rates depend not on the channel conditions, but primarily on the limitations of the MCSs and signaling applied in IEEE Std. 802.15.3d-2017. For the THz-OOK, both RS and LDPC codes are applied (14/15 LDPC). For the THz-SC, only the high-rate 14/15 LDPC coding is used, as it provides the highest results in the considered setup.

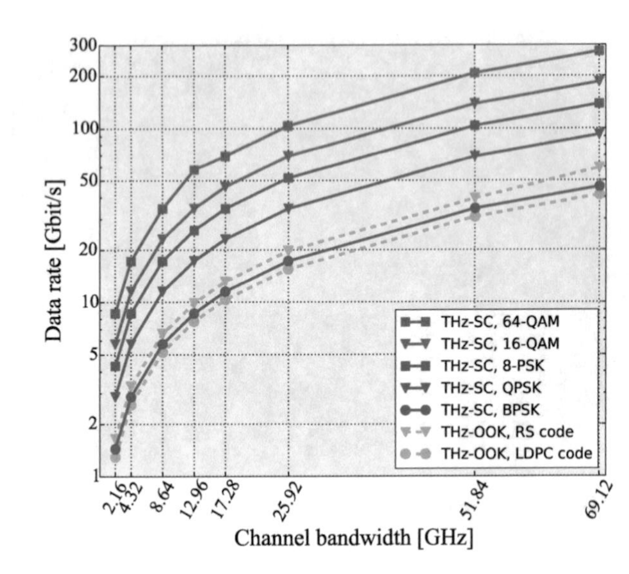

Fig. 52.6 Data rates with IEEE 802.15.3d-2017

As observed from Fig. 52.6, the achievable data rates with the THz-SC vary from 1.4 Gbit/s with 2.16 GHz bandwidth and BPSK up to 285 Gbit/s with 69.12 GHz bandwidth and 64 QAM. The lower-complexity THz-OOK features similar performance for the smallest 2.16 GHz bandwidth. At the same time, the data rate with largest 69.12 GHz bandwidth is limited to only 52.5 Gbit/s, which is approximately 18% of what can be achieved with THz-SC in the same channel.

52.5.2 *Communication Range*

Complementing the data rate study in Fig. 52.6, Table 52.1 presents the simulation results illustrating the communication range achievable with IEEE Std. 802.15.3d-2017 in different use cases [22]. **Bold** values in Table 52.1 present the conditions, where the achievable data rate is higher than 100 Gbit/s. As can be observed from this table, the communication range is inversely proportional to the selected channel bandwidth, decreasing, for example (backhaul), from 327 m for 2.16 GHz up to only 58 m for 69.12 GHz. At the same time, the observed values of the communication range correspond to the envisioned distances for the given use cases: particularly, only the cm-scale range is envisioned for the wireless intra-device connectivity with IEEE Std. 802.15.3d-2017.

Summarizing the chapter, IEEE Std. 802.15.3d-2017 can support the PHY-layer data rates considerably higher than 100 Gbit/s for the desired distances between the communicating nodes, thus notably exceeding the 5G requirements. The further development of the ideas, concepts, and solutions presented in the IEEE Std. 802.15.3d-2017 will contribute to the appearance of THz communications and also lay the foundation for the prospective 6G wireless systems.

Table 52.1 Communication range with IEEE Std. 802.15.3d-2017

Channel bandwidth [GHz]	Communication range [m]			
	Backhauling/Fronthauling	Data center	Kiosk downloading	Intra-device
2.16	327	58	3	0.16
4.32	231	41	2.12	0.12
8.64	163	29	1.5	0.08
12.96	133	24	1.22	0.07
17.28	116	21	1.06	0.06
25.92	94	17	**0.87**	**0.05**
51.84	**67**	**12**	**0.61**	**0.03**
69.12	**58**	**10**	**0.53**	**0.03**

References

1. Lattva-aho, M., & Lappänen, K.. Key drivers and research challenges for 6G ubiquitous wireless intelligence, http://urn.fi/urn:isbn:9789526223544
2. Boulogeorgos, A. A., et al. (2018). Terahertz technologies to deliver optical network quality of experience in wireless systems beyond 5G. *IEEE Communications Magazine, 56*(6), 144–151.
3. Kürner, T. (2013, November). *On the Scope of IEEE 802.15 SG 100G*. Doc. IEEE 802.15-13-0635-01-0thz, [online] https://mentor.ieee.org/802.15/dcn/13/15-13-0635-01-0thz-on-the-scope-of-ieee-802-15-sg-100g.pdf
4. Kürner, T., Hosako, I. (2017). IEEE 802.15.3d: Wireless Point-to-Point connections at 300 GHz. in M. Ulema et al., "Standards News," in *IEEE Communications Standards Magazine*, vol. 1(3): 13–19.
5. IEEE Standard for High Data Rate Wireless Multi-Media Networks. In *IEEE Std 802.15.3-2016 (Revision of IEEE Std 802.15.3-2003)*, vol., no., pp.1–510, July 25 201
6. [online] https://www.ieee802.org/PARs/2015_11/15-15-0682-01-003d-3d-par-change.pdf
7. IEEE Standard for High Data Rate Wireless Multi-Media Networks–Amendment 2: 100 Gb/s Wireless Switched Point-to-Point Physical Layer. In *IEEE Std 802.15.3d-2017 (Amendment to IEEE Std 802.15.3-2016 as amended by IEEE Std 802.15.3e-2017)*, vol., no., pp.1-55, Oct. 18 2017
8. IEEE Standard for High Data Rate Wireless Multi-Media Networks–Amendment 1: High-Rate Close Proximity Point-to-Point Communications. In *IEEE Std 802.15.3e-2017 (Amendment to IEEE Std 802.15.3-2016)* , vol., no., pp.1–178, June 7 2017
9. Petrov, V., Kürner, T., & Hosako, I. (2020). IEEE 802.15.3d: First Standardization Efforts for Sub-Terahertz Band Communications toward 6G. *IEEE Communications Magazine, 58*(11), 28–33. https://doi.org/10.1109/MCOM.001.2000273.
10. Kürner, T. et al (2015, May). *Applications Requirement Document (ARD)*. DCN: 15-14-0304-16-003d, IEEE 802.15 TG3d, https://mentor.ieee.org/802.15/documents
11. Kürner, T. (2018). THz Communications: Challenges and Applications beyond 100 Gbit/s," *2018 International Topical Meeting on Microwave Photonics (MWP)*, Toulouse, pp. 1–4
12. Song, H. et al. (2016). *Demonstration of 20-Gbps wireless data transmission at 300 GHz for KIOSK instant data downloading applications with InP MMICs*. 2016 IEEE MTT-S International Microwave Symposium (IMS), San Francisco, CA, pp. 1–4, doi: https://doi.org/10.1109/MWSYM.2016.7540141.
13. Shimamoto, H., Kitamura, K., Watabe, T., Ootake, H., et al. (2012). *120 Hz-frame-rate SUPER Hi-VISION Capture and Display Devices* (11 pages). Hollywood: SMPTE Annual Technical Conference & Exhibition.

14. Lee, S., et al. (2019). An 80-Gb/s 300-GHz-Band Single-Chip CMOS Transceiver. *IEEE Journal of Solid-State Circuits, 54*(12), 3577–3588.
15. Wu, K., Xiao, J., & Ni, L. M. (2012). Rethinking the architecture design of data center networks. *Frontiers of Computer Science, 6*(5), 596–603.
16. Davy, A. et al. (2017). Building an end user focused THz based ultra high bandwidth wireless access network: The TERAPOD approach. *2017 9th International Congress on Ultra Modern Telecommunications and Control Systems and Workshops (ICUMT)*, Munich, pp. 454–459.
17. NGMN 5G White Paper, https://www.ngmn.org/fileadmin/ngmn/content/downloads/Technical/2015/NGMN_5G_White_Paper_V1_0.pdf
18. Antes, J., et al. (Nov. 2015). Multi-Gigabit Millimeter-Wave Wireless Communication in Realistic Transmission Environments. *IEEE Transactions on Terahertz Science and Technology, 5*(6), 1078–1087.
19. IEEE Standard for Information technology–Telecommunications and information exchange between systems–Local and metropolitan area networks–Specific requirements-Part 11: Wireless LAN Medium Access Control (MAC) and Physical Layer (PHY) Specifications Amendment 3: Enhancements for Very High Throughput in the 60 GHz Band,“ in *IEEE Std 802.11ad-2012 (Amendment to IEEE Std 802.11-2012, as amended by IEEE Std 802.11ae-2012 and IEEE Std 802.11aa-2012)* , vol., no., pp.1-628, 28 Dec. 2012
20. IEEE P802.11ay - IEEE Draft Standard for Information Technology–Telecommunications and Information Exchange Between Systems Local and Metropolitan Area Networks–Specific Requirements Part 11: Wireless LAN Medium Access Control (MAC) and Physical Layer (PHY) Specifications–Amendment: Enhanced Throughput for Operation in License-Exempt Bands Above 45 GHz [online] https://standards.ieee.org/project/802_11ay.html
21. Kürner, T. (2019). *THz Communications – A Candidate for a 6G Radio?* The 22nd International Symposium on Wireless Personal Multimedia Communications (WPMC – 2019), Lisbon, Portugal, November 24–27, 2019, [online] https://doi.org/10.24355/dbbs.084-202008031406-0
22. [online] https://mentor.ieee.org/802.15/dcn/17/15-17-0039-04-003d-summary-of-results-from-tg3d-link-level-simulations.pdf

Chapter 53
Spectrum for THz Communications

Michael J. Marcus and Thomas Kürner

Abstract This chapter describes the framework for the use of spectrum by THz communications, which has to rely on the possibility to share the spectrum with passive service such as Earth Exploration-Satellite Service (EESS) and radio astronomy (RA). First, an introduction of the structure of the international and national policies is given taking into account spectrum beyond 50 GHz. The methods and procedures for sharing studies with passive services are described. This includes both the studies, which have been performed during the preparation of World Radiocommunication Conference (WRC) 2019 and novel concepts for future enhanced sharing studies. The chapter concludes with the details of the spectrum regulation for frequencies beyond 252 GHz based on the results of WRC 2019.

53.1 Introduction

The global availability for spectrum is a prerequisite for the future deployment of THz communication systems. The worldwide use of spectrum is regulated using a set of international and national spectrum policies, for which the International Telecommunications Union Radio Regulations (RR) [2] are the base document. The specific rules for the use of those frequencies are defined for bands under 275 GHz in specific allocations for various radio services and are influenced by a large degree by the terms of RR 5.340. For bands above 275 GHz there are presently no specific allocations but there are applicable provisions in RR 5.565 and RR 5.564A. These footnotes provide conditions for the various levels of protection of passive services such as Earth Exploration-Satellite Service (EESS [passive]) and radio astronomy service (RAS) from harmful interference by active

M. J. Marcus
Michael J. Marcus Spectrum Solutions, Washington, DC, USA

T. Kürner (✉)
Institute for Communications Technology, Technische Universität Braunschweig, Braunschweig, Germany
e-mail: t.kuerner@tu-bs.de

T. Kürner et al. (eds.), *THz Communications*, Springer Series in Optical Sciences 234, https://doi.org/10.1007/978-3-030-73738-2_53

services, such as THz communications operated in the same bands. In the bands covered by RR5.340, at present "all emissions are prohibited". EESS perform spaceborne measurements of the molecular absorption lines in the THz band for meteorology, climatology, and atmospheric chemistry. EESS antennas are generally on low Earth orbit (LEO) satellites and either point tangentially ("limb sensing") on the satellite orbit or downward toward the Earth's surface. Relevant frequencies are multiple bands between 100 and 3000 GHz, some of which have several tens of GHz bandwidth, while others are only a few GHz [3–5]. The RAS rely on the reception of electromagnetic (EM) waves over the entire THz range from 275 to 3000 GHz to study EM radiation from molecular gas clouds, remote stars, or galaxies. In contrast to the EESS, RAS activities are primarily ground-based, and the antennas are facing upward at positive elevation angles. Thus, while RAS antennas do not get interference in their high-gain main beam from terrestrial sources, they can get interference from satellite sources or through antenna sidelobes. At THz frequencies, RAS systems lose sensitivity at low altitudes, and in humid environments due to atmospheric absorption of the signals, they seek to receive. Thus, most, but not all, RAS sites, with THz capability, are sited in high-arid locations, e.g., Northern Chile, for maximum performance. THz RAS sites are generally rare in urbanized areas.

A closer look at these bands reveals that almost all bands above 275 GHz are in potential use by either EESS or RAS. Hence, any use of spectrum beyond 275 GHz for THz communications with bandwidths of up to several tens of GHz as in [6] is possible only if the spectrum is shared with at least one of the abovementioned passive services [7]. Intensive sharing studies have been performed as part of the preparation of the World Radiocommunication Conference (WRC) 2019 resulting in the identification of frequency bands for use by fixed service (FS) and land mobile service (LMS).

This chapter describes the framework for the use of spectrum by THz communication and is organized as follows: Section 57.2 introduces the structure of the international and national spectrum policies. Methods and procedures for sharing studies with passive services at 275–450 GHz are described in Section 57.3. The final Section 57.4 contains the details of the spectrum regulation based on the results of WRC 2019.

53.2 Structure of International and National Spectrum Policies

The beginning of spectrum regulation at the national and international levels was greatly influenced by the sinking of the steamship *Titanic* in 1912. While ITU traces its origins back to 1865, its early decades focused on European telegraph issues. The potential role of early radiotechnology in enhancing maritime safety was obvious in the missed opportunities that contributed to the large death toll in the *Titanic* disaster. From the earliest days of spectrum, regulation frequency scarcity was an issue. In those days, the maximum practically usable frequency was rather low, and modulations were not as efficient as today's. The frequencies in early use had long propagation ranges, and little or no directionality was possible in antennas of the era. Thus, early regulations paid attention to what types of radio use were permitted, what technologies were deemed efficient enough, and what steps were needed to prevent harmful interference between users. This is a tradition of *prescriptive* regulation which was developed where new technologies would have to run the regulatory gauntlet to come into general use.

While international and national spectrum regulations have undergone significant deregulation in the past century, remnants of the early regulatory regime linger along with many earlier frequency allocations which decide what applications are allowed in each band.

The ITU Radio Regulations are treaty obligations of the 193 countries that are members unless they specifically exempt themselves from a provision by taking a "reservation" on it. Article 4.4 of the Radio Regulations (RR 4.4) provides an exemption from following the prescribed use of each band if the nonstandard usage does not actually cause harmful interference to a properly authorized radio system of another country compliant with ITU RR or if the other affected countries agree with the nonstandard usage. However, many ITU members never use terms of RR 4.4 provision and question this interpretation although the language is quite clear.

In general, a given band of spectrum has more than one radio service allocations, some of which may be "primary" and some of which may be "secondary." In general, a secondary allocation's use may not cause interference to a primary allocation's use although footnotes to the allocation table can modify the relationship of different uses in each band. Administrations (in ITU usage, "administration" means the national spectrum regulator of a member country) may decide to implement all the ITU RR allocations for a given band or may decide to use only some of them within its jurisdiction.

While the allocation of purely passive bands in the mmWave area goes back to the 1979 ITU World Administrative Radio Conference where "all emissions" were prohibited in 51.4–54.25 GHz, 58.2–59 GHz, 64–65 GHz, 86–92 GHz, 105–116 GHz, and 216–231 GHz, the current basic prohibitions in RR footnote 5.340 were adopted at the 1997 World Radio Conference although they have since been updated a little. The current bands in 50–275 GHz that are subject to an "all emissions are prohibited restriction" are given in Table 53.1.

At WRC-2000, many of the bands above 100 GHz were added to RR.5.340, and Resolution 731 was adopted calling for studies of the feasibility of sharing in 71–275 GHz between the passive services and various types of active services. The resolution called for studies in both the passive bands listed in RR5.340 and other passive bands that have co-primary passive allocations with intersatellite communications. To date, ITU-R has not completed studies on the feasibility of sharing passive spectrum in 71–275 GHz as requested in Res. 731 although the Conference Preparatory Meeting for WRC-2000 concluded:

Table 53.1 List of prohibited bands between 50 GHz and 252 GHz

Prohibited bands above 50 GHz in RR 5.340 (GHz)
50.2–50.4
52.6–54.25
86–92
100–102
109.5–111.8
114.25–116
148.5–151.5
164–167
182–185
190–191.8
200–209
226–231.5
250–252

Table 53.2 Impact of RR footnote 5.340, exclusively passive allocations in different spectrum regions

Band	Frequency (GHz)	Number of passive blocks	Fraction of band passive
UHF	0.3–3	2	1%
SHF	3–30	3	2%
EHF	30–300	15	15%

ITU-R studies indicate that sharing between EESS (Passive) and the fixed service is generally feasible in bands of high atmospheric absorption, Sharing in many bands may require constraints on the fixed service in order to protect spaceborne sensors.[8]

There are no formal allocations at present above 275 GHz, so there are no bands there with such total restrictions now. RR footnote 5.565 "identifies bands for use by administrations for passive service applications" but also states "The use of the range 275-1000 GHz by the passive services does not preclude use of this range by active services. Administrations wishing to make frequencies in the 275-1000 GHz range available for active service applications are urged to take all practicable steps to protect these passive services from harmful interference until the date when the Table of Frequency Allocations is established in the above-mentioned 275-1000 GHz frequency range." As is discussed in Section 57.4, RR footnote 5.564A also "identifies" bands in 275–450 GHz for land mobile and fixed service applications.

The impact of the passive bands in EHF, 30–300 GHz, is vastly greater than their impact in lower-spectrum regions as shown in Table 53.2.

While there are passive bands in UHF and SHF, their impact is negligible in terms of how they fragment the spectrum to limit maximum achievable bandwidth and in terms of the fraction of spectrum they occupy. However, the impact in EHF is much larger in both respects. This is shown in Table 53.3.

It can be seen that the largest contiguous band for *any* transmitters in 100–275 GHz region is the 32.5 GHz in 116–148.5 GHz. However, even that spectrum has

Table 53.3 Nonpassive spectrum between passive bands

Lower edge of passive band (GHz)	Upper edge of passive band (GHz)	Nonpassive bandwidth above passive band (GHz)
100	102	7.5
109.5	111.8	2.45
114.25	116	32.5
148.5	151.5	12.5
164	167	15
182	185	5
190	191.8	8.2
200	209	17
226	231.5	18.5
250	252	

Table 53.4 Bands in the 100–275 GHz with existing ITU allocations for fixed and mobile use

Band (GHz)	Bandwidth (GHz
102–109.5	7.5
111.8–114.25	2.45
122.25–123	0.75
130–134	4.0
141–148.5	7.5
151.5–164	12.5
167–174.8	7.8
191.8–200	8.2
209–226	17.0
232–235	3.0
238–241	3.0
252–275	23.0

allocation problems for terrestrial transmitters as some of the allocation blocks within it do not at present contain *fixed* or *mobile* allocations, so their use in countries where the administration is reluctant to use the provisions of RR 4.4 is problematical.

Table 53.4 shows the bands with existing ITU *fixed* and *mobile* allocations in 100–275 GHz. At present, all bands above 100 GHz which have ITU fixed allocations also have ITU mobile allocations. Administrations may implement fixed and/or mobile services in these bands immediately as long as they implement a framework that protects other co-primary allocations in each band which exists in most of them. It is interesting to note that the largest contiguous bandwidth in any of these bands is 23 GHz, and that is only at the upper band. Below 200 GHz, the largest contiguous bandwidth now with a fixed or mobile allocation is 12.5 GHz.

53.3 Sharing Studies with Passive Service in 275–450 GHz

RR footnote 5.565 of the Radio Regulations defines the protection of passive services in 275–1000 GHz, while footnote 5.340 defines it at lower bands. Although the pre-WRC 2019 regulatory provisions already allow the operation of active services (which include THz communications) in the frequency range 275–1000 GHz, each national radio administration has to decide what are the "practical steps" defined in [2] in order to protect the identified EESS and RAS bands. In a worst-case scenario, this could lead to different decisions in different countries. On the other hand, the use of ultra-large bandwidths of several tens of GHz for THz communications is only possible when spectrum can be shared between THz communications and the passive service. A first investigation on potential interference scenarios between THz communications and passive services has been performed by Priebe et. al [3] (see also Figure 53.1), concluding that with multiple interference stations in the field of reception of the satellite, an interference margin may become necessary without mitigation countermeasures like highly directive antennas. In order to enable the simultaneous use of spectrum by both THz communications and passive services, detailed interference studies between active transmission systems and passive services must be provided. From the THz communication systems' point of view, the main aim is to identify and demonstrate means for interference-free coexistence, so that the entire THz range can potentially be used for data transmission without limitations [3].

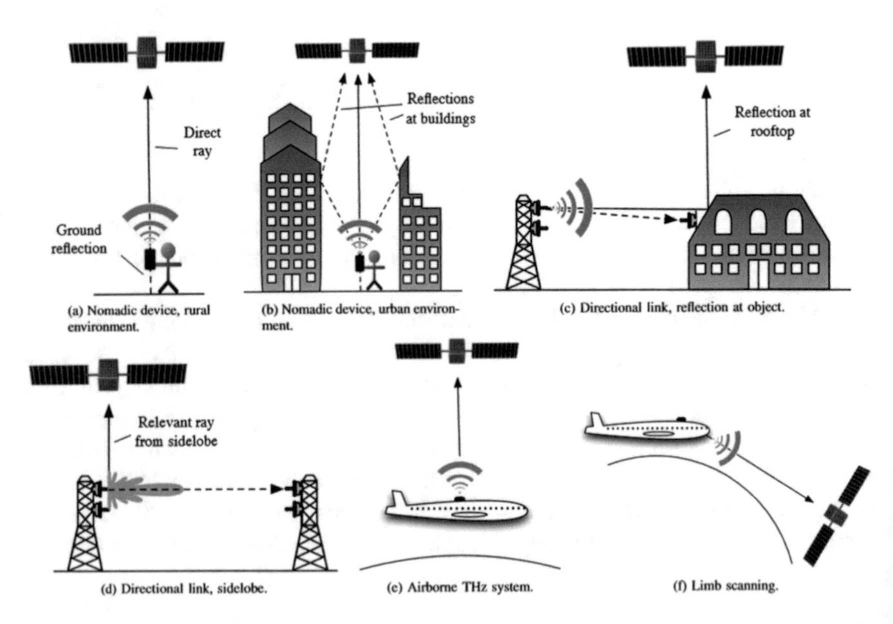

Fig. 53.1 Possible interference scenarios. (**a**) Nomadic device, rural environment. (**b**) Nomadic device, urban environment. (**c**) Directional link, reflection at object. (**d**) Directional link, sidelobe. (**e**) Airborne THz system. (**f**) Limb scanning [3]. (© 2012 IEEE, reproduced with permission)

In the following, two approaches for the corresponding sharing studies are briefly described. In Subsection 57.3.2, the general framework and approach for the sharing studies carried out during the preparation of WRC 2019 are described, and exemplary results for a sharing study between FS and EESS performed by the EU-Japan Horizon 2020 project ThoR [9] are provided. Section 57.3.2 provides an alternative approach for future sharing studies.

53.3.1 Sharing Studies in Preparation of WRC 2019

The World Radiocommunication Conference (WRC) 2015 had invited ITU-R to perform "Studies towards an identification for use by administrations for land-mobile and fixed services applications operating in the frequency range 275–450 GHz" [11], with the goal to define more detailed rules at the World Radiocommunication Conference (WRC) 2019 under its AI (agenda item) 1.15. The focus of these sharing studies has been on EESS [9, 12, 13], since earlier studies taking into account the operational characteristics of the passive services [3, 14] came to the conclusion that potential interference to EESS is more critical compared to RAS, where the ground-based large antennas are typically located in very remote area, pointing up the sky. This has been confirmed in [15], where the main conclusion is that sharing between radio astronomy and active services in the range 275–3,000 GHz is possible if atmospheric characteristics as a function of height above sea level, as well as transmitter antenna directivity, are taken into account. Especially for FS separation, distances and/or avoidance angles between RAS stations and FS stations should be considered depending on the deployment environment of FS stations [16]. The sharing studies between EEES on one side and FS and LMS on the other side have been based on the technical characteristics of these systems [13, 17, 18].

In the following, the method and the results of a sharing study for FS an EEES carried out within the Horizon 2020-EU-Japan project ThoR are briefly sketched [9, 10]. The study is based on the technical characteristics described in [13, 17] and uses a Monte Carlo simulation approach: A number of links are randomly deployed for several sets of defined densities of FS links in an area of the size of the field of view (FOV) for the EEES (see Fig. 53.2).

The position of the satellite has been calculated based on its nadir angle and its altitude further assuming an azimuth of 0° relative to the middle of the FOV. The path loss between the middle of the FOV and satellite is calculated based on ITU-R recommendation ITU-R P.525 [19] and ITU-R P.676 [20] taking into account the slant path length and the angle of elevation.

The investigations revealed that the conical-type EEES ICI system is the most critical one wrt interference. Simulation results for these systems are presented in Fig. 53.3 for various ink densities and elevation angles of the FS links. It is shown that in the frequency bands 296 to 306 GHz, 313 to 319 GHz, and 333 to 354 for some of the link densities and elevation angles, the allowed limit of interference is

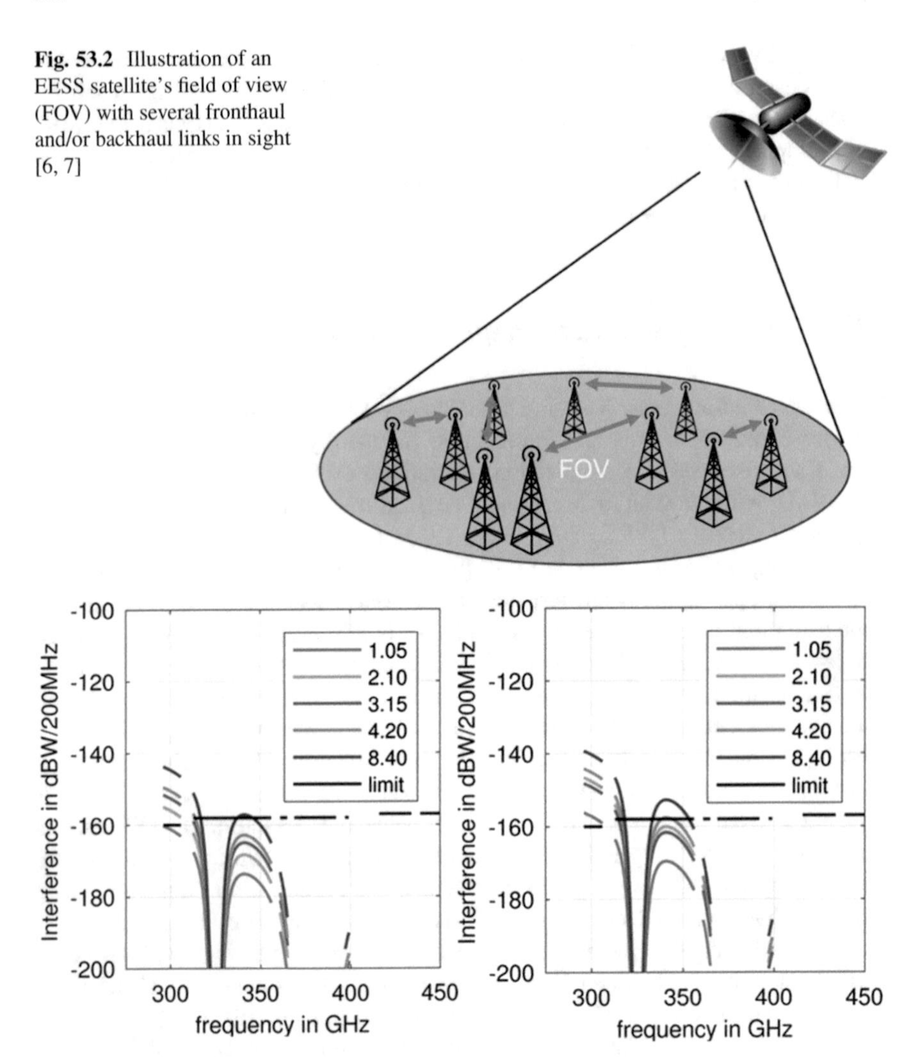

Fig. 53.2 Illustration of an EESS satellite's field of view (FOV) with several fronthaul and/or backhaul links in sight [6, 7]

Fig. 53.3 Maximum simulated interference power densities to an ICI-type system for elevation angles between -20° and +20° (left) and -65° to +65° (right) for systems for several link densities in links per square kilometer (color) and the maximum interference level to the EESS [9, 10]

exceeded. The conclusions from this study are that in those two frequency bands – in contrast to all other frequency bands investigated – unconditional sharing between FS and EEES is not possible.

A similar conclusion taking into account also sharing between LMS and EEES has been drawn in most of the studies documented in [18], where in the bands 275–296 GHz, 306–313 GHz, 320–330 GHz, and 356–450 GHz only, no specific conditions to protect EESS are necessary, for systems operating within the parameters given in the referenced ITU-R reports [13, 16, 17]. Specific conditions such as power limits, shielding requirements, elevation angle restrictions, etc. have not been investigated in [9, 18]. This is something left open for future studies potentially enabling sharing with EESS in the remaining frequency bands between 275 and 450 GHz.

WRC-19 updated the terms of Resolution 731 [20] which requests ITU-R studies of spectrum sharing above 71 GHz. The terms of this request to ITU-R include the following provisions:

- Continue studies to determine if and under what conditions sharing is possible between active and passive services in the frequency bands above 71 GHz, such as, but not limited to, 100–102 GHz, 116–122.25 GHz, 148.5–151.5 GHz, 174.8–191.8 GHz, 226–231.5 GHz, and 235–238 GHz.
- Conduct studies to determine the specific conditions to be applied to the land mobile and fixed service applications to ensure the protection of EESS (passive) applications in the frequency bands 296–306 GHz, 313–318 GHz, and 333–356 GHz.
- Develop recommendations specifying sharing criteria for those frequency bands where sharing is feasible.

Thus, these deliberations in ITU-R WP1A will be the major ongoing international forum to review these protection issues.

Regulatory actions by the FCC in the USA and Ofcom in the UK show the introduction of new techniques to limit unintended emissions from reaching passive satellites. The FCC decision in March 2019 [22] authorized unlicensed use of 116–123 GHz, 174.8–182 GHz, 185–190 GHz, and 244–246 GHz. While none of these bands are subject to the restrictions of 5.340, there are allocation issues. 116–122.25 GHz, 174.8–182 GHz, and 185–190 GHz have no terrestrial allocations and co-primary allocations for EESS and Space Research (passive) as well as intersatellite. While the later two services have no real interference risk, the EESS does. 116–123 GHz is also adjacent to the 114.25–116 band with an RR 5.340 restriction. 244–246 GHz is part of a larger band with only Radio Astronomy and Radiolocation (Radar) allocations along with a secondary amateur radio allocation. The FCC decision limits the interference potential by allowing only terrestrial use and restricting the transmission EIRP to 40 dBm unless a high-gain antenna is used. It then allows fixed point-to-point transmitters whose "the average power of any emission shall not exceed 82 dBm and shall be reduced by 2 dB for every dB that the antenna gain is less than 51 dBi." The effect of these provision is to allow an increase of EIRP from 43 dBm to 82 dBm if it is achieved by using high-gain antennas (>51 dBi) which focus the power on the destination antenna and thus reduce incidental satellite illumination.

The Ofcom (UK) decision of October 2020 [23] takes a similar approach with some interesting differences. It will license the terrestrial use of three bands in the 100–200 GHz range, 116–122 GHz, 174.8–182 GHz, and 185–190 GHz – using three of the four FCC authorized bands. Ofcom authorizes 55 dBm EIRP in these bands for indoor use. For outdoor use, the same 55 dBm limit applies, but there are also a main beam maximum elevation angle limit of 20° and a maximum antenna relative gain limit for sidelobes relative to the main beam gain. Ofcom requires power reductions for bandwidth less than 100 MHz but states that such narrow bandwidths are unlikely in practice in these EHF bands as these would serve little practical function. Thus, Ofcom has chosen to regulate more directly than FCC the incidental illumination of the sky which is the main threat to EESS. It is puzzling why the net maximum EIRP for FCC is 82 dBm and for Ofcom is 55 dBm.

As was shown in Table 53.1c, there are bands in 100–275 GHz with both ITU mobile and fixed co-primary allocations that could be used by administration immediately for licensed uses and possibly for unlicensed uses. Most of these have co-primary allocations in other services where there is some expectation of protection. In some cases, e.g., 174.5–174.8 GHz, the co-primary service is just intersatellite links where there is little concern of interference, but in others, e.g., *209–217* GHz, radio astronomy is co-primary which will require interservice coordination in a very small number of locations in populated areas.

Pending progress in studies on WRC Resolution 731 and changes to ITU Radio Regulations, which is unlikely before 2027 at the earliest, the existing allocations and RR 5.340 will place serious limitations on available contiguous bandwidth below 275 GHz.

53.3.2 Alternative Approach for Future Studies

New studies in ITU-R on the sharing questions in Res. 731 will likely address alternative approaches to sharing. The WRC-19 studies used to consider sharing above 275 GHz had an implicit assumption that terrestrial fixed service antennas would be scaled from the parabolic dish antennas commonly used in much lower bands and the link density requirements would be estimated from link densities at lower bands. This meant that sharing compatibility would be judged on whether existing technology and usage could be extrapolated to a much higher band. But Res. 731 is clearer than WRC AI 1.15 was on the goal at hand – it is to determine whether any sharing is possible in passive bands above 71 GHz not just whether existing usage can be moved up a few octaves. Thus, it is likely that Res. 731 studies will not presume the deployments and antenna technologies of lower bands but will study what combination of parameters for *fixed* service such as data rate, communication distance, antenna sidelobe levels, link density, and maximum elevation angles are compatible with interference-free sharing with EESS. While all finite size antennas must have sidelobes, in terrestrial *fixed*/EESS sharing scenarios, the main threat for interference is from high-elevation angle sidelobes from the *fixed* transmitter. The AI 1.15 studies assumed antennas with sidelobe levels in the -20 to -25 dBi range typical of production parabolic antennas in lower bands. This is not a physical limit and is exceeded in both radio astronomy antennas and military radar antennas where sidelobe suppression is vital.

Dynamic approaches might also be considered. Two basic possibilities exist. A grid network use in THz bands would be more robust due in the fast of weather-related propagation impairments where information could be rerouted over a different path in the grid. If the EESS satellite orbit was known, the choice of routing path could also be based on an estimate of the illumination of the satellite from different link paths.

Finally, a variant of MIMO antennas might be considered (see also chapter 18). MIMO antennas evolved from early technology for jam-resistant antennas that used multiple elements with phase and amplitude weights to null out interference from an undesired direction. MIMO antennas also use multiple elements, but the amplitudes and phases are computed to maximize received signal. But it is possible to change the control algorithms of such an antenna to achieve two goals: minimizing the EIRP in the direction where the satellite is on its path across the sky and optimizing received power at the receiver. They will require more antenna elements than a single purpose MIMO antenna but if that enables the use of broader bands of spectrum it may be practical.

53.4 Radio Regulations After WRC 2019

When considering the situation for THz communications in the RR, possible future communication use in the 100–275 GHz bands should be considered along with the frequency bands between 275 and 450 GHz as well as the frequency bands 252–275 GHz. These bands are used by IEEE Std. 802.15.3d-2017 and have already an allocation for FS and LMS on a co-primary basis [2]. In its final acts of WRC 2019, [19] has identified the bands 275 to 296 GHz, 306 to 313 GHz, 318 to 333 GHz, and 356 to 450 GHz. Together with the already allocated spectrum between 252 and 275 GHz totally, 160 GHz of spectrum with contiguous bandwidths of 44, 7, 15, and 94 GHz are available for THz communications providing a sound basis for the deployment of THz communication systems. All details and the bands and the conditions for their use by THz communications are provided in Table 53.5. Due to its atmospheric conditions, the band 252 to 296 GHz is favorable for fixed outdoor links with its requirements to several hundred meter link distances, whereas the other three bands can be also used, for example, for short-range indoor applications, e.g., wireless links in data centers [6].

Table 53.5 Overview of identified spectrum for THz communications above 252 GHz [5]

Frequency in GHz	Status in the radio regulations [2]
252–275	Allocation for land mobile and fixed service on a co-primary basis
275–296	Identification for use for the implementation of land mobile and fixed service according to FN 5.564A. No specific conditions are necessary to protect Earth Exploration-Satellite Service (passive) applications
306–313	
318–333	
356–450	
296–306	May only be used by fixed and land mobile service applications when specific conditions to ensure the protection of Earth Exploration-Satellite Service (passive) applications are determined in accordance with Resolution 731 (Rev.WRC-19) [20]
313–318	
318–356	

References

1. ITU, CPM Report on technical, operational and regulatory/procedural matters to be considered by the 2000 World Radiocommunication Conference Section 4.1.2.2.1 [Online]: http://search.itu.int/history/HistoryDigitalCollectionDocLibrary/4.126.51.en.101.pdf at p. 225
2. World Radiocommunication Conference, "Radio Regulations, Edition of 2019," [Online].: https://www.itu.int/pub/R-REG-RR.
3. Priebe, S., Britz, D. M., Jacob, M., Sarkozy, S., Leong, K. M. K., Logan, J. E., Gorospe, B., & Kürner, T. (2012). Interference Investigations of Active Communications and Passive Earth Exploration Services in the THz Frequency Range. *IEEE Transactions on Terahertz Science and Technology, 2*(5), 525–537.
4. Panel on Frequency Allocations and Spectrum Protection for Scientific Users, National Academy of Sciences (USA), Handbook of Frequency Allocations and Spectrum Protection for Scientific Uses, 2015. [Online]: https://www.nap.edu/catalog/21774/handbook-of-frequency-allocations-and-spectrum-protection-for-scientific-uses
5. Panel on Frequency Allocations and Spectrum Protection for Scientific Users, National Academy of Sciences (USA), Spectrum Management for Science in the 21st Century, 2010. [online]: https://www.nap.edu/catalog/12800/spectrum-management-for-science-in-the-21st-century
6. IEEE Standard for High Data Rate Wireless Multi-Media Networks–Amendment 2: 100 Gb/s Wireless Switched Point-to-Point Physical Layer," in IEEE Std 802.15.3d-2017 (Amendment to IEEE Std 802.15.3-2016 as amended by IEEE Std 802.15.3e-2017) , pp.1-55, Oct. 18, 2017.
7. Kürner, T., & Hirata, A. (2020). *On the impact of the results of WRC 2019 on THz communications*. 2020 Third International Workshop on Mobile Terahertz Systems (IWMTS), Essen, Germany, pp. 1–3. https://doi.org/10.1109/IWMTS49292.2020.9166206.
8. ITU, CPM Report on technical, operational and regulatory/procedural matters to be considered by the 2000 World Radiocommunication Conference Section 4.1.2.2.1 [Online]: http://search.itu.int/history/HistoryDigitalCollectionDocLibrary/4.126.51.en.101.pdf at p. 225
9. Rey, S., Initial results on sharing studies, Thor Deliverable D5.1; [online]: https://thorproject.eu/results/
10. Kürner, T. (2019, June 18–21), Regulatory Aspects of THz Communications and Related Activities Towards WRC 2019. *Proceedings of European Conference on Networks and Communications*, 18-21 Jun-2019; Valencia, Spain; electronic (2 pages), [online] https://doi.org/10.24355/dbbs.084-201908211524-0

11. RESOLUTION 767(WRC-15): Studies towards an identification for use by administrations for land-mobile and fixed services applications operating in the frequency range 275-450 GHz [Online]: https://www.itu.int/dms_pub/itu-r/oth/0c/0a/R0C0A00000C0016PDFE.pdf
12. ITU-R, Recommendation ITU-R RS.2017: Performance and interference criteria for satellite passive remote sensing. [Online]: https://www.itu.int/rec/R-REC-RS.2017/en.
13. ITU-R Report RS.2431-0: Technical and operational characteristics of EESS (passive) systems in the frequency range 275-450 GHz; [Online]: https://www.itu.int/dms_pub/itu-r/opb/rep/R-REP-RS.2431-2019-PDF-E.pdf
14. Clegg, A., Sharing between radio astronomy and active services at THz frequencies, [online] https://mentor.ieee.org/802.15/dcn/10/15-10-0829-00-0thz-sharing-between-active-and-passive-services-at-thz-frequencies.ppt
15. ITU-R Report RA.2189-1 (09/2018): ITU-R Sharing between the radio astronomy service and active services in the frequency range 275-3 000 GHz [Online]: https://www.itu.int/dms_pub/itu-r/opb/rep/R-REP-RA.2189-1-2018-PDF-E.pdf
16. ITU-R Report Report SM.2450; Sharing and compatibility studies between land-mobile, fixed and passive services in the frequency range 275-450 GHz [Online]: https://www.itu.int/dms_pub/itu-r/opb/rep/R-REP-SM.2450-2019-PDF-E.pdf
17. ITU-R, Report ITU-R F.2416: Technical and operational characteristics and applications of the point-to-point fixed service applications operating in the frequency band 275-450 GHz. [Online]: https://www.itu.int/dms_pub/itu-r/opb/rep/R-REP-F.2416-2018-PDF-E.pdf
18. ITU-R, Report ITU-R F.2417: Technical and operational characteristics of land-mobile service applications in the frequency range 275-450 GHz. [Online]: https://www.itu.int/dms_pub/itu-r/opb/rep/R-REP-M.2417-2017-PDF-E.pdf
19. ITU-R, Recommendation ITU-R P.525: Calculation of free-space attenuation. [Online]: https://www.itu.int/rec/R-REC-P.525/en
20. ITU-R, Recommendation ITU-R P.676: Attenuation by atmospheric gases. [Online: https://www.itu.int/rec/R-REC-P.676/en
21. World Radiocommunication Conference 2019 (WRC-19) Final Acts; [online]: https://www.itu.int/dms_pub/itu-r/opb/act/R-ACT-WRC.14-2019-PDF-E.pdf
22. Federal Communications Commission (USA), First Report and Order, Docket 18-21, March 15, 2019 [online]: https://docs.fcc.gov/public/attachments/FCC-19-19A1_Rcd.pdf
23. Ofcom(UK), STATEMENT: Supporting innovation in the 100-200 GHz range-Increasing access to Extremely High Frequency (EHF) spectrum, October 2020 [online]: https://www.ofcom.org.uk/__data/assets/pdf_file/0024/203829/100-ghz-statement.pdf

Chapter 54
Outlook on Standardization and Regulation

Thomas Kürner

Abstract This chapter provides a brief outlook on future tasks and activities in standardization and regulation. This includes a discussion on the activities of the IEEE 802.15 Standing Committee THz, perspectives for standardization in 3GPP and ETSI, and potential regulatory activities toward the WRC 2023 and WRC 2027.

A first wireless standard for 300 GHz has been published [1], and 137 GHz of spectrum beyond 275 GHz has been identified at the World Radiocommunication Conference (WRC) 2019 [2], as described in more details in earlier sections of this book (see Chaps. 52 and 53). Even so, at the time this book is written, standardization and regulatory activities for THz communications are still at the very early stages.

At IEEE 802.15, the Standing Committee on THz is monitoring the development of THz communication in order to initiate projects for the development of further amendments of IEEE Std. 802.15.3d-2017 or even develop completely new standards. Amendments to include all the newly identified spectrum at WRC 2019 up to a carrier frequency of 450 GHz will be of interest; the high-frequency portion of this range is not covered by the current standard. New projects could also consider the progress made in forward error correction as described in Chap. 30 of this book. Of specific interest for new projects will be standards covering applications with at least one end of the link being mobile, as described in Chap. 1 of this book. This will specifically require methods for [3]:

- *Initial access and device discovery*, where the devices can efficiently find each other with pencil beams
- *Beam tracking and beam forming*, where a prerequisite is the design of electronically steerable THz antennas

T. Kürner (✉)
Institute for Communications Technology, Technische Universität Braunschweig, Braunschweig, Germany
e-mail: t.kuerner@tu-bs.de

T. Kürner et al. (eds.), *THz Communications*, Springer Series in Optical Sciences 234, https://doi.org/10.1007/978-3-030-73738-2_54

– *Efficient and low-complexity methods for interference mitigation and handling multiple access*

These concerns define current important research questions, which need to be resolved before any standardization activity can begin. These methods are already subject to ongoing research (see, for example, [4, 5] and Chap. 31 of this book). Once these methods are mature enough to initiate a standard, THz frequencies might be also of interest for an advanced WLAN (Wireless Local Area Network) standard in IEEE 802.11.

Based on the landscape of ongoing projects on beyond 5G (B5G) as presented in Part X of this book and some discussion surrounding 6G, where THz communication is a candidate for at least one of its components [6], THz communications may also be part of the definition of B5G and 6G at 3GPP (3rd Generation Partnership Project) [7, 8]. The European Telecommunications Standards Institute (ETSI) is already considering frequencies well beyond 100 GHz for fixed point-to-point applications in the W-band (92–114.5 GHz) [9]. Further ideas for the use of THz frequencies have already emerged by combining communication, localization, and sensing [10] and may significantly enlarge the scope of potential applications.

Based on the outcome of WRC 2019, regulatory activities with respect to THz communications are twofold [11]:

– In the frequency range 275 to 450 GHz, 38 GHz of spectrum may only be used by THz communications when specific conditions to ensure the protection of Earth Exploration-Satellite Service (passive) applications have been determined, in accordance with Resolution 731 (Rev.WRC-19). With this resolution, ITU-R is invited to conduct studies to determine the specific conditions to be applied to the land mobile and fixed service applications to ensure the protection of Earth Exploration-Satellite Service (passive) applications in the frequency bands 296–306 GHz, 313–318 GHz, and 333–356 GHz. This might be potentially discussed at WRC 2023 (see also Chap. 53 of this book).
– A potential agenda item at WRC 2027 relating to the identification of spectrum for radio location applications in the range 275–700 GHz were already identified at WRC 2019. This will require sharing studies with THz communications as the incumbent application.

At the time the book is written, the development of THz communications is at the dawn of intense standardization and regulation activities. This may keep the THz community busy for at least a decade.

References

1. IEEE Standard for High Data Rate Wireless Multi-Media Networks–Amendment 2: 100 Gb/s Wireless Switched Point-to-Point Physical Layer. In IEEE Std 802.15.3d-2017 (Amendment to IEEE Std 802.15.3-2016 as amended by IEEE Std 802.15.3e-2017), pp.1-55, Oct. 18, 2017.

2. World Radiocommunication Conference 2019 (WRC-19) Final Acts; [online]: https://www.itu.int/dms_pub/itu-r/opb/act/R-ACT-WRC.14-2019-PDF-E.pdf
3. Petrov, V., Kürner, T., & Hosako, I. (2020). IEEE 802.15.3d: First Standardization Efforts for Sub-Terahertz Band Communications toward 6G. *IEEE Communications Magazine, 58*(11), 28–33. https://doi.org/10.1109/MCOM.001.2000273.
4. Doeker, T., Reddy, P., Negi, P. S., Rajwade, A., Kürner, T. (2021, March 22–26). Angle of arrival and angle of departure estimation using compressed sensing for Terahertz communication. *Proceedings of 15th European Conference on Antennas and Propagation*, 5 pages
5. Xia, Q., & Jornet, J. M. (2019). Expedited neighbor discovery in directional Terahertz communication networks enhanced by antenna side-lobe information. *IEEE Transactions on Vehicular Technology, 68*(8), 7804–7814.
6. Lattva-aho, M., & Lappänen, K. Key drivers and research challenges for 6G ubiquitous wireless intelligence, http://urn.fi/urn:isbn:9789526223544
7. Huq, K. M. S., Busari, S. A., Rodriguez, J., Frascolla, V., Bazzi, W., & Sicker, D. C. (2019). Terahertz-enabled wireless system for beyond-5G Ultra-Fast networks: A brief survey. *IEEE Network, 33*(4), 89–95. https://doi.org/10.1109/MNET.2019.1800430.
8. Polese, M., Jornet, J. M., Melodia, T., & Zorzi, M. (2020). Toward end-to-end, full-stack 6G Terahertz networks. *IEEE Communications Magazine, 58*(11), 48–54. https://doi.org/10.1109/MCOM.001.2000224.
9. ETSI GR mWT 018 V1.1.1 (2019-08) Analysis of Spectrum, License Schemes and Network Scenarios in the W-Band, https://www.etsi.org/deliver/etsi_gr/mWT/001_099/018/01.01.01_60/gr_mWT018v010101p.pdf
10. Sarieddeen, H., Saeed, N., Al-Naffouri, T. Y., & Alouini, M. (2020). Next generation terahertz communications: a rendezvous of sensing, imaging, and localization. *IEEE Communications Magazine, 58*(5), 69–75. https://doi.org/10.1109/MCOM.001.1900698.
11. Kürner, T., & Hirata, A. (2020). *On the Impact of the Results of WRC 2019 on THz Communications*. 2020 Third International Workshop on Mobile Terahertz Systems (IWMTS), Essen, Germany, pp. 1–3, https://doi.org/10.1109/IWMTS49292.2020.9166206.

Printed in the United States
by Baker & Taylor Publisher Services